Studies in Surface Science and Catalysis 123

CATALYSIS: AN INTEGRATED APPROACH

Second, revised and enlarged edition

Studies in Surface Science and Catalysis 123

Advisory Editors: B. Delmon and J.T. Yates

Vol. 123

CATALYSIS: AN INTEGRATED APPROACH

Second, revised and enlarged edition

Editors

R.A. van Santen
Schuit Institute of Catalysis, Eindhoven University of Technology, Eindhoven, The Netherlands

P.W.N.M. van Leeuwen
Department of Inorganic Chemistry, University of Amsterdam, Amsterdam, The Netherlands

J.A. Moulijn
Section for Process Catalysis, Delft University of Technology, Delft, The Netherlands

B.A. Averill
E.C. Slater Institute, University of Amsterdam, Amsterdam, The Netherlands

ELSEVIER
Amsterdam — Lausanne — New York — Oxford — Shannon — Singapore — Tokyo

ELSEVIER SCIENCE B.V.
Sara Burgerhartstraat 25
P.O. Box 211, 1000 AE Amsterdam, The Netherlands

First edition 1993
Second, revised and enlarged edition 1999
Second impression 2000

Library of Congress Cataloging-in-Publication Data

Catalysis : an integrated approach / editors, R.A. van Santen.--2nd, rev. and enl. ed.
p. cm. -- (Studies in surface science and catalysis ; vol.123)
Includes bibliographical references and index.
ISBN 0-444-82963-6 (alk. paper)
1. Catalysis. I. Santen, R. A. van (Rutger A.)

QD505 .C3813 1999
541.3'95 21--dc21

99-044946

ISBN: 0 444 82963 6

Transferred to digital printing 2005

Printed and bound by Antony Rowe Ltd, Eastbourne

Preface

This book has undergone a thorough revision. The first reason was that, in our opinion, biocatalysis could no longer be left out of the integrated approach that is our aim with this book. Several chapters have been extended by incorporation of biocatalytic details. One new chapter has been written especially for this book (Chapter 7, Biocatalysis), whereas extensive biocatalytic sections have been added to other chapters (e.g. Chapter 1.3, Biocatalysis, and Chapter 4.5 & 4.6, Bonding and Elementary Steps in Catalysis).

The other major reason for a revision was the development of catalysis science in general, which necessitated rewriting sections or complete chapters. There is virtually no chapter left untouched by this revision operation. Special chapters on catalyst characterization (Chapter 12) and on adsorption methods (Chapter 13) highlight these new insights. Adding biochemistry to this bunch of flowers has added a new scent, or , to change the metaphor, taught us to communicate in a new language, which we all enjoyed learning. And also, pasting together all these bits and ends proved to be more time consuming than was originally envisaged.

The NIOK Integrated book is one of the most concrete signs of a research school which, since it first came into being, has grown into one of Europe's leading institutes. We repeat our wish that it may continue to be of service to teachers, students and other scholars in catalysis.

Rutger A. van Santen
Piet W.N.M. van Leeuwen
Jacob A. Moulijn
Bruce A. Averill

Preface to the first edition

Catalysis is a fascinating science for those who are actively involved in it, but perhaps even more so for those who know only a little about it. Chemical reactions, for which one would expect half-life times as long as centuries, can be accomplished in minutes to hours with the magic power of a mysterious black box containing a catalyst. Most of our liquid fuels and some 80% of our chemical products are manufactured with the aid of catalytic conversions. In the near future most of our automotive exhaust gases will be cleaned by a catalyst. The ever-increasing yield of fuel from a barrel of crude oil, the continuing increase of the quality of our polymeric materials, the steadily decreasing production cost of bulk and fine chemicals, etc., can be ascribed, to a great extent, to improvements in catalytic systems. Nevertheless, to the layman catalysis does not play a glamorous role, or at least it didn't until the catalytic converter in automobiles entered his life. Even to the modern chemist who is not familiar with it, catalysis has the aura of an empirical discipline, containing some of the ingredients of a black art. Indeed, entirely new catalysts and processes have been discovered mostly by accident, but the majority of the enormous improvements in catalyst performance which we have seen have been brought about by progress in the underlying sciences.

To a large extent catalysis is concerned with the transformations of organic molecules. Nevertheless, the subject is not the realm of organic chemists but rather that of catalysis specialists who are interested in the (often simple) conversions of basic organic molecules while using highly complex catalyst systems. Apart from a large body of acid–base catalysis and catalysis by 'organic' catalysts, heterogeneous and homogeneous (metal) catalysts have been developed in isolation from organic chemistry. There are several factors which have a bearing on this: (i) the synthesis of a catalytic material requires a knowledge of inorganic and organometallic chemistry, surface science, and spectroscopic techniques; (ii) the industrial importance of catalysis ensured that a great deal of the empirical development of the catalysts was not readily available to the scientific community, which concentrated its efforts on more fundamental aspects of the catalyst; (iii) industrial catalysis deals with simple organic molecules, the synthesis of which, quite justifiably, did not attract the interest of modern organic chemists; and (iv) the differing requirements imposed by the industrial manufacture and small-scale laboratory synthesis of chemicals made the gap even larger. In this book we will concentrate on industrially relevant reactions which are catalyzed by heterogeneous and homogeneous catalysts. This is a somewhat arbitrary choice, but some choice is necessary. Unfortunately, this leaves large areas of industrial biocatalysis and catalysis science uncovered. In keeping with this industrial focus, the topic of catalytic reaction engineering has been included. In industrial applications reaction engineering forms an integral part of catalysis; the constraints of the catalyst material are determined by chemical and reactor engineering aspects, and the related problems have to be solved in close harmony.

Clearly, catalysis is a multidisciplinary activity and this is reflected in the present book. We have chosen a novel combination of basic disciplines which we hope will be of value to our readers. Homogeneous catalysis by metal complexes is treated jointly with heterogeneous catalysis using metallic and non-metallic solids. In both areas the high degree of sophistication of spectroscopic techniques and theoretical modelling has led to an enormous increase in our understanding at the molecular level. This holds for the kinetics of the reactions and the reactivities of the catalysts, as well as for the syntheses of the catalytic materials. In homogeneous catalysis, the time span between the discovery of a new catalytic process and the delineation of the reaction mechanism on a molecular level has usually been relatively short. Hence, while homogeneous catalysis with organometallics is a young branch of catalysis science, its rapid development can be ascribed to a successful molecular approach in the early stages of its development. Heterogeneous catalysis has to a great extent been developed within the context of catalytic reaction engineering. The more recent advances in surface science have contributed to an understanding of the catalytic materials and processes on a molecular level. It therefore seems timely to deal with the organometallic chemistry of heterogeneous and homogeneous catalysis in a concerted manner. The analogies between the bonding of discrete complexes on surfaces and in solution has been noted many times. Many studies have been published on the analogies and differences between bonding in cluster compounds and metal surfaces. There are, however, also fundamental differences between elementary events on metal surfaces and elementary steps on isolated metal complexes. The study of the basic principles is refreshing and inspiring to both worlds.

Outline

The *introductory section* (Chapter 1) presents a brief survey of the history of industrial heterogeneous and homogeneous catalysis. The survey shows that the availability of the feedstocks has had a decisive influence on the catalysts that have been developed. In some instances the chemical 'leads' for new catalysts have lain dormant in the literature for a number years before they were developed into industrial processes. Subsequently (Chapter 2) a very limited selection of current industrial catalytic processes is described. These include the industrial context of the features to be discussed later in the sections on fundamental and applied catalysis. A broad spectrum of important catalytic applications is presented. Each summary contains the basic chemistry, some engineering aspects, feedstock sources and product utilisation. In Chapter 3 the kinetic principles are treated, since in a way they form the basis for catalysis.

The section on *fundamental catalysis* starts off (Chapter 4) with a description of the bonding in complexes and to surfaces. The elementary steps on complexes and surfaces are described as they are relevant to catalysis. The chapter on heterogeneous catalysis (Chapter 5) deals with the mechanistic aspects of three groups of important reactions: syn-gas conversions, hydrogenation, and oxidation. These three reactions give us an opportunity to present the main principles of metal and metal oxide catalysis. Likewise, in the chapter on homogeneous catalysis (Chapter 6) we concentrate on only three reactions, representing examples from three areas: hydroformylation, polymerization, and asymmetric catalysis. Identification by *in-situ* techniques has been included, since it was felt to be more appropriate to outline this here than to devote separate chapters to the preparation and characterization of organometallic catalysts (*vide infra* for heterogeneous catalysts).

Many constraints on the industrial use of a catalyst have a macroscopic origin. In *applied catalysis* (Chapter 7) we show how catalytic reaction engineering deals with such macroscopic considerations. The transport and kinetic phenomena in both model reactors and industrial reactors are outlined.

The section on *catalyst preparation* (Chapters 8 and 9) is concerned with the preparation of catalyst supports, zeolites, and supported catalysts, with an emphasis on general principles and mechanistic aspects. For the supported catalysts the relation between the preparative method and the surface chemistry of the support is highlighted. The molecular approach is maintained throughout.

The first chapter (Chapter 10) in the section on *catalyst characterization* summarizes the most common spectroscopic techniques used for the characterization of heterogeneous catalysts, such as XPS, Auger, EXAFS, etc. Temperature programmed techniques, which have found widespread application in heterogeneous catalysis both in catalyst characterization and the simulation of pretreatment procedures, are discussed in Chapter 11. A discussion of texture measurements, theory and application, concludes the section on the characterization of solid catalysts (Chapter 12).

The final chapter (Chapter 13) gives an outline of *current trends* in catalysis. Two points of view are adopted: the first one focuses on developments in process engineering. Most often these have their origin in demands by society for better processes. The second point of view draws attention to the autonomous developments in catalysis, which is becoming one of the frontier sciences of physics and chemistry.

Final remarks

The main theme of the book is the molecular approach to industrial catalysis. The integrated way in which the subject matter has been treated involves many disciplines; as a consequence, the writing of the book has been entrusted to a number of authors. The editors envisaged a careful planning and harmonization of the contents of the chapters. The latter is important, because a rigorous selection of subject matter is necessary in order to keep the size of the book within the limits of a practicable textbook.

The editors have enjoyed working together on this project. We have learned a great deal about our mutual fields while discussing which essential elements should be included and what reactions should serve as instructive examples. The order of the editors' names does not in any way reflect the outcome of these discussions, nor the magnitude of our efforts! We are very grateful to our authors, who have contributed a number of very good chapters. Some of them were well in time with their responses, while others enjoyed the privilege of having written the most up to date contributions!

When we started putting together the national course on catalysis, on which this book is based, we were a loosely organised foundation for the promotion of catalysis. In the meantime we have organised ourselves into the Netherlands Institute for Catalysis Research, NIOK. We express our wish that this book may contribute in shaping the new Dutch School of Catalysis by providing a broad, general training for our Ph.D. students in the essential elements of catalysis.

Jacob A. Moulijn
Piet W.N.M. van Leeuwen
Rutger A. van Santen

Contents

Chapter 5
HETEROGENEOUS CATALYSIS
B.K. Hodnett, F.J.J.G. Janssen, J.W. Niemantsverdriet, V. Ponec, R.A. van Santen and J.A.R. van Veen

Chapter 6
HOMOGENEOUS CATALYSIS WITH TRANSITION METAL COMPLEXES
G. van Koten and P.W.N.M. van Leeuwen

Chapter 10
PREPARATION OF SUPPORTED CATALYSTS
J.W. Geus and J.A.R. van Veen

CATALYST CHARACTERIZATION

Chapter 11
CATALYST CHARACTERIZATION WITH SPECTROSCOPIC TECHNIQUES
J.W. Niemantsverdriet

Chapter 12
CATALYST CHARACTERIZATION AND MIMICKING PRETREATMENT PROCEDURES BY TEMPERATURE- PROGRAMMED TECHNIQUES
F. Kapteijn, J.A. Moulijn and A. Tarfaoui

Chapter 13
ADSORPTION METHODS FOR THE ASSESSMENT OF THE SPECIFIC SURFACE AREA AND THE PORE SIZE DISTRIBUTION OF HETEROGENEOUS CATALYSTS
J.A. Lercher

List of contributors

B.A. Averill
E.C. Slater Institute, University of Amsterdam, Plantage Muidergracht 12, 1018 TV Amsterdam, The Netherlands

K.P. de Jong
Institute for Inorganic Chemistry, Utrecht University, Sorbonnelaan 16, 3584 CA Utrecht, The Netherlands

E.B.M. Doesburg
(Retired from Netherlands Energy Research Foundation ECN), Balderikstraat 43, 7581 VS Losser, The Netherlands

J.W. Geus
(Retired from Utrecht University), Gezichtslaan 100, 3723 GJ Bilthoven, The Netherlands

B.K. Hodnett
Department of Chemical and Environmental Sciences, University of Limerick, Limerick, Ireland

F.J.J.G. Janssen
Kema Nederland BV, Utrechtseweg 310, 6812 AR Arnhem, The Netherlands

F. Kapteijn
Section for Process Catalysis, Delft University of Technology, Julianalaan 136, 2628 BL Delft, The Netherlands

A.P.G. Kieboom
Gist-brocades BV, Postbus 1, 2600 MA Delft, The Netherlands

N.W.M. Laane
Laboratory for Biochemistry, Wageningen Agricultural University, Postbus 8128, 6700 ET Wageningen, The Netherlands

J.A. Lercher
Institute for Chemical Technology, Technische Universität München, Lichtenbergstraße 4, D-85748 Garching, Germany

G.B. Marin
Labo voor Petrochemische Techniek, Universiteit Gent, Krijgslaan 281, B-9000 Gent, Belgium

J.A. Moulijn
Section for Process Catalysis, Delft University of Technology, Julianalaan 136, 2628 BL Delft, The Netherlands

J.W. Niemantsverdriet
Schuit Institute of Catalysis, Eindhoven University of Technology, Postbus 513, 5600 MB Eindhoven, The Netherlands

V. Ponec
(Retired from Leiden University), Sternstraat 5, 2352 EH Leiderdorp, The Netherlands

I.M.C.M. Rietjens
Laboratory for Biochemistry, Wageningen Agricultural University, Postbus 8128, 6700 ET Wageningen, The Netherlands

R.A. Sheldon
Section for Organic Chemistry and Biocatalysis, Delft University of Technology, Julianalaan 136, 2628 BL Delft, The Netherlands

A.J.J. Straathof
Section for Biokinetics, Delft University of Technology, Julianalaan 136, 2628 BL Delft, The Netherlands

A. Tarfaoui
10 Rue de Brésil, 5ième étage, 1002 Tunis Belvedère, Tunesia

J. Tramper
Department for Process Technology, Wageningen Agricultural University, Postbus 8129, 6700 EV Wageningen, The Netherlands

J.H.C. van Hooff
(Retired from Eindhoven University of Technology), Joachim Patinirstraat 2, 5613 LZ Eindhoven, The Netherlands

G. van Koten
Section for Organic Synthesis, Utrecht University, Padualaan 8, 3584 CH Utrecht, The Netherlands

P.W.N.M. van Leeuwen
Department of Inorganic Chemistry, University of Amsterdam, Nieuwe Achtergracht 166, 1018 WV Amsterdam, The Netherlands

R.A. van Santen
Schuit Institute of Catalysis, Eindhoven University of Technology, Postbus 513, 5600 MB Eindhoven, The Netherlands

J.A.R. van Veen
Shell Research and Technology Centre Amsterdam, P.O. Box 38000, 1030 BN Amsterdam, The Netherlands

R. Wever
E.C. Slater Institute, University of Amsterdam, Plantage Muidergracht 12, 1018 TV Amsterdam, The Netherlands

Introductory section

Chapter 1

History of catalysis

1.1 INTRODUCTION

The name 'catalysis' was coined by Berzelius in 1836. He concluded that besides 'affinity' a new force is operative, the 'Catalytic Force'. Reaction occurred by catalytic contact. The word 'catalysis' stems from the Greek: it has the sense of 'down' and 'loosen'. At that time affinity was known as a chemical driving force, but no understanding existed, on a molecular level, of reaction rates. Catalysis—as a tool for carrying out reactions—had already been exploited much earlier. It has been applied for thousands of years in processes such as fermentation. An interesting example is the production of sulphuric acid. In the Middle Ages this was synthesised in small quantities in glass equipment by burning sulphur with nitric acid in humid air. In 1746 lead was used as a construction material for the reaction chambers and larger production volumes became possible. In 1793 Clement and Desormes showed that the quantity of nitre could be much reduced by admitting additional air to the reactor. They were aware of the fact that the nitrous vapours are only intermediates and that the oxidising agent is air. Their conclusion was correct, and nitre is what we would nowadays call a catalyst.

At the end of the eighteenth and the beginning of the nineteenth century the influence of metals and oxides on the decomposition of several substances was studied by many scientists. It was noticed that contact with different substances gives very different products. An example is the decomposition of alcohol: in the presence of copper or iron, carbon and an inflammable gas are produced. In the presence of pumice stone decomposition into ethene and water was observed. In other words, selectivity was demonstrated. Many other important milestones can be mentioned: we limit ourselves to a few. Thenard investigated the dissociation of ammonia in contact with metals. In 1813 he found that the dissociation occurs over various metals, provided they are hot. Later he systematically studied the dissociation of hydrogen peroxide. He concluded that some of the solids studied changed and others did not. Humphry Davy performed systematic investigations with the objective of developing a miner's safety lamp. He discovered that the oxidation of coal gas is catalysed by platinum. He found that palladium was also active, whereas copper, silver and gold did not show any catalytic activity. Davy found that the platinum wires he used had to be hot and, as a consequence, he concluded that the action of platinum was to heat the reacting mixture. The fact that only platinum and palladium were active was explained on the basis of their low heat capacity and low thermal conductivity. Later it was found that finely dispersed platinum is active even at room temperature and Davy's explanation was thus not correct. Platinum catalyses the combustion of the components of coal gas, in particular CO and CH_4, which in the absence of a catalyst can lead to an explosion. Copper and silver are also active in the oxidation of CO, but they are not sufficiently active in CH_4 oxidation. Of course, depending on the construction, a wire can function as a flame arrester in which radical terminations take place and, as a consequence, explosions are avoided.

In 1834 Faraday proposed that the reactants have to adsorb simultaneously at the surface, but he did not really explain the catalytic action. Of course, neither did Berzelius give an explanation, but he nicely generalised many results in a simple description. Later, Ostwald gave the definition that a catalyst does not influence the thermodynamic equilibrium of reactants and products but affects the rates of the chemical reactions. The conclusions of Berzelius and Faraday proved to be correct.

Industrial catalysis, based on man-made catalysts, developed its own history which is briefly summarised in Section 1.2. The roots of biocatalysis and biocatalytic processes are completely different, notwithstanding the early recognition that fermentation processes are in essence of a catalytic nature. The relationship between biocatalysis and organic chemistry is close. It is really quite recently that biocatalysis and man-made catalysis have become part of the same context. This is mainly the result of increasing interest in catalytic applications to fine chemicals.

An historical review of biocatalysis is given in Section 1.3.

1.2 INDUSTRIAL CATALYSIS

Industrial catalysis is an old practice. Catalysts have always been used in the production of wine and beer. Among the first industrial catalytic processes are a few inorganic oxidation processes, viz. the Deacon process (oxidation of HCl into Cl_2) and the production of sulphuric acid. These processes were developed before a scientific basis of chemical reactivity was established. Only after the formulation of the theory of chemical equilibria by van 't Hoff did a framework for catalyst development become available. This had a major impact on the development of a process for the synthesis of ammonia at the beginning of the twentieth century, allowing a systematic, scientifically based search for a good catalyst to be performed. It also initiated the development of chemical process engineering as we know it today.

1.2.1 *Sulphuric Acid*

The production of sulphuric acid was commercialised in the mid-18th century. In the so-called lead chamber process the oxidation of SO_2 into SO_3 was catalysed by NO. The acid produced is not very concentrated. The raw material used was elemental sulphur from Sicily. Later, pyrite was used because of its lower price. One of the consequences was a much higher impurity level in the feed to the reactor. As early as 1831 a process was patented in which SO_2 was oxidised in the presence of finely divided platinum. The commercial application, however, was strongly delayed due to technical difficulties, the major one being catalyst poisoning.

In the First World War the explosives industry demanded highly concentrated sulphuric acid, which required a different technology. A logical way to go was to use the heterogeneous metal catalysts which had just been developed at that time. Initially, platinum supported on silica or asbestos was used as a catalyst. Subsequently, platinum was substituted by V_2O_5-based catalysts because of the high price of the precious metal and the fact that it is easily poisoned by the pollutants present in the raw material, such as arsenic. The raw material has also changed. Nowadays, it is once again mainly elemental sulphur, which, to a considerable degree, is a product of the hydrotreatment of oil.

1.2.2 *Ammonia Synthesis*

Initially, the source of ammonia was coke oven gas and Chile saltpetre. In Germany, in particular, it was recognised as early as the turn of the 20th century that insufficient ammonia was available for agricultural needs. Moreover, the use of ammonia for the manufacture of explosives increased dramatically due to the beginning of the First World War. Extensive efforts were made by teams in many countries, but particularly in Germany, to synthesise NH_3 directly from N_2. Non-catalysed routes were discovered and were commercialised, but they were very inefficient. The breakthrough was the development of a catalytic process.

In 1905 Haber reported a successful experiment in which he succeeded in producing NH_3 catalytically. However, under the conditions he used (1293 K) he

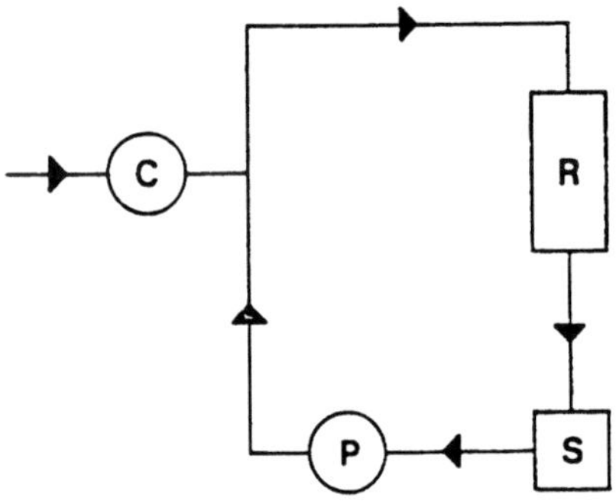

Fig. 1.1. Process scheme for the production of ammonia according to Haber.

found only minor amounts of NH_3. He extrapolated his value to lower temperatures (at 1 bar) and concluded that a temperature of 520 K was the maximum temperature for a commercial process. This was the first application of chemical thermodynamics to catalysis, and precise thermodynamic data were not then known. At that time Haber regarded the development of a commercial process for ammonia synthesis as hopeless and he stopped his work. Meanwhile, Nernst had also investigated the ammonia synthesis reaction and concluded that the thermodynamic data Haber used were not correct. He arrived at different values and this led Haber to continue his work at higher pressures. Haber tried many catalysts and found that a particular sample of osmium was the most active one. This osmium was a very fine amorphous powder. He approached BASF and they decided to start a large program in which Bosch also became involved.

The process development studies were carried out in a systematic way. A good catalyst had to be formulated; the reactor was to be scaled up; and an integrated process had to be designed, including the production of sufficiently pure synthesis gas. Haber envisaged the process scheme given in Fig. 1.1.

Systematic studies were carried out in order to discover a suitable catalyst. Iron catalysts were especially tried, because it was known that iron catalyses the decomposition of ammonia, which is the reverse of the reaction being studied. It was discovered that iron alone was only slightly active but its activity could be improved (promoted) or worsened (poisoned) by additives. In their studies over 10 000 catalysts were prepared and over 4000 were tested.

When the laboratory studies gave promising results, scale-up studies commenced. The major problem was the construction of a reactor which was able to withstand the reaction conditions. Because of the high pressure, high-strength carbon steel was used; however, this steel is corroded by H_2 under the severe reaction conditions used, and consequently loses its strength. Bosch designed a reactor which contained an outer wall of high carbon steel, lined internally with soft, low carbon steel (Fig. 1.2). The catalyst was also heated internally, so that the external wall was continuously kept at a low temperature.

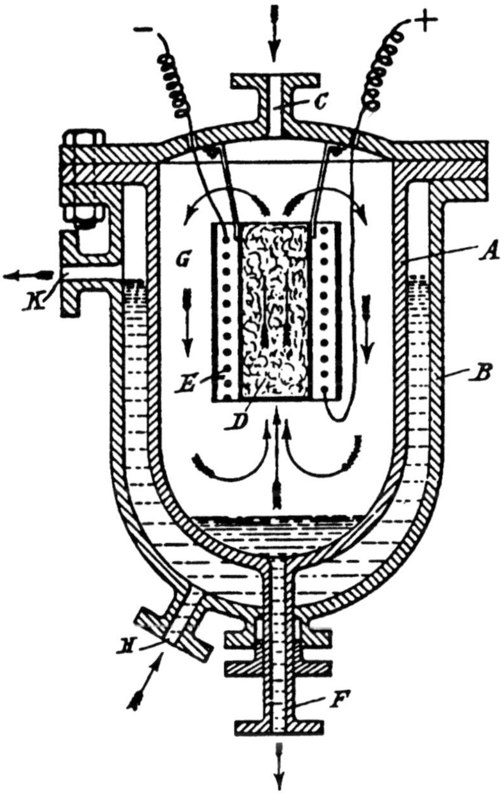

Fig. 1.2. Reactor for the synthesis of ammonia (BASF 1910).

Very pure, electrically generated H_2 was available for the pilot studies, but for a commercial plant, this was of course not practical. The solution was found in coal gasification:

$$C + H_2O \rightarrow CO + H_2$$

followed by the water–gas shift reaction:

$$CO + H_2O \rightarrow CO_2 + H_2$$

Haber received the Nobel prize in 1919 for his work on ammonia synthesis. Bosch was awarded the 1931 Nobel prize for his achievements in high-pressure technology.

Up to the 1950s, the technology for ammonia synthesis did not change very much. Fundamental changes became possible in the '60s. These changes were of both a chemical and a technological nature.

Technically, the most attractive hydrocarbons for the production of H_2 are methane and, to a lesser extent, oil ('CH_2'):

$$CH_4 + H_2O \rightarrow CO + 3\,H_2$$

$$'CH_2' + H_2O \rightarrow CO + 2H_2$$

These reactions are carried out catalytically using nickel catalysts. However, a side reaction is carbon deposition. Since the rate of this side reaction increases with pressure, the production of synthesis gas at high pressure was not feasible. The addition of promoters, which catalyse the carbon steam reaction, suppresses carbon deposition, thus allowing higher pressures. This made a revolution possible in the ammonia synthesis industry. At the same time progress in the development of centrifugal compressors enabled large, single-train plants to be constructed. These developments have led to much more efficient ammonia plants. Figure 1.3 illustrates the process development since the beginning of commercial ammonia production by nitrogen fixation.

Still further process improvements may be possible, but potential improvements should be analysed by considering the whole process integrally.

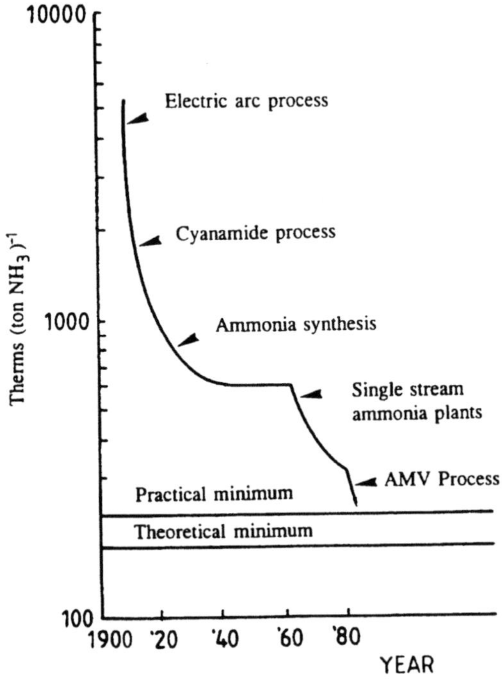

Fig. 1.3. Efficiency of nitrogen fixation [1].

1.2.3 Coal, Oil, Natural Gas

Industrial catalysis has always been closely connected with changes in society. If we confine ourselves to the last 100 years, it is appropriate to highlight the fuel market. In this period a transition is observed from coal to oil and, more recently, to natural gas.

1.2.3.1 Coal

Essentially, in all coal utilisation processes the first step is a heating step and a variety of pyrolysis products are formed, depending on the technology. This is illustrated in Fig. 1.4 [2].

A major application of coal was the production of coke. This process is a mild pyrolysis process and, besides coke, large quantities of coke oven gas are also produced containing many of the base materials for the chemical industry, e.g., ammonia, acetylenes and aromatics.

In Germany, before and during the Second World War, a large amount of ethylene was produced from acetylene derived from coke oven gas. Coke ovens also provided essentially all the aromatics in Germany. Coal was also gasified,

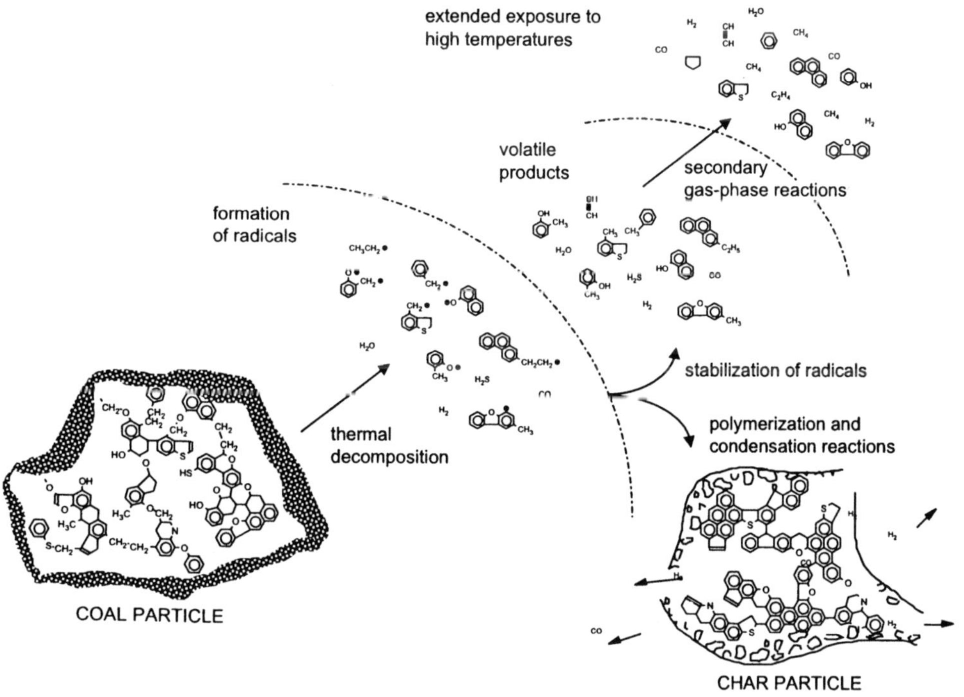

Fig. 1.4. Reactions occurring during the pyrolysis of coal [2].

giving synthesis gas, a mixture of CO and H_2. Products such as ammonia and methanol were catalytically produced from this synthesis gas. Fischer–Tropsch technology was also developed to convert synthesis gas to motor fuels.

Coal was also directly hydrogenated into liquids which could be used as feedstocks in the chemical industry. The first catalytic hydrogenation of coal was reported by Berthelot in 1869, using HI as a catalyst. The success in developing catalytic high pressure processes for ammonia and methanol also stimulated process development in catalytic coal hydrogenation. Catalyst deactivation appeared to be a major problem. The solution was found in carrying out the process in two steps: first a liquid phase hydrogenation was carried out in a slurry of coal particles, oil produced from coal and highly dispersed catalyst. Many catalysts were tried, e.g. iron oxide and sulphide. In the second stage the oils from the first stage were hydrogenated in a fixed-bed reactor. As early as 1924, it was known that sulphides of Mo, W, Co and Fe were suitable hydrogenation catalysts which were not poisoned by the heteroatoms (S, N, O) in the feedstock. Later, in the development of hydrotreating processes for the oil refinery industry, the same catalysts were applied, but now in the treatment of oils.

The extensive activity in Germany in the 1940s is well illustrated by Table 1.1.

TABLE 1.1

German hydrogenation plants 1943/44

Start-up date	Location	Raw material	Pressure (bar)		Product capacity 1943/44 (t per year)
			Liquid phase	Vapour phase	
1927	Leuna	Brown coal Brown coal tar	200	200	650 000
1936	Boehlen	Brown coal tar	300	300	250 000
1936	Magdeburg	Brown coal tar	300	300	220 000
1936	Scholven	Bituminous coal	300	300	230 000
1937	Welheim	Coal tar pitch	700	700	130 000
1939	Gelsenberg	Bituminous coal	700	300	400 000
1939	Zeitz	Brown coal tar	300	300	280 000
1940	Luetzkendorf	Tar oils	500	500	50 000
1940	Poelitz	Bituminous coal oils	700	300	700 000
1941	Wesseling	Brown coal	700	300	250,000
1942	Bruex	Brown coal tar	300	300	600 000
1943	Blechhammer	Bituminous coal and tar	700	300	420 000
	12 Plants				about 4 million

1.2.3.2 Oil and Natural Gas

In the U.S. oil and natural gas were playing a significant role, together with coal, as early as the 1920s. It was obvious that, compared with coal, oil was a superior chemical feedstock since the hydrocarbons are present in a gaseous or liquid form. The problem was that crude oil is very unreactive; this applies particularly to the paraffins which are main constituents of crude oil. The situation changed when thermal and, later, catalytic cracking processes were developed.

Distillation of crude oil does not result in large amounts of gasoline of a satisfactory octane number. Cracking of larger molecules, isomerisation and aromatisation have to be carried out in order to produce sufficient amounts of high quality gasoline.

As early as 1915, $AlCl_3$ was used in a batch process. It is not surprising that there was a severe problem in the disposal of the sludges consisting of spent catalyst dissolved in hydrocarbons. At the same time activated clays were found to exhibit catalytic activity. A major breakthrough was the work of Houdry—a mechanical engineer and an automobile racing driver. He concluded that the limitations on the engine performance were not of a mechanical nature; the constraints were due to the performance of the gasoline. He developed a fixed bed process in which activated clay was the catalyst. Activated clays had been used before to produce kerosene for lamp oil. The cracking unit consisted of parallel reactors which were operated cyclically. A scheme of an early design is given in Fig. 1.5.

Three catalytic converters are present in order to enable the process to operate continuously. Under typical conditions a reactor was on stream for 10 min. Subsequently, it was purged for 10 min and regenerated for 10 min. During regeneration the coke deposited on the catalyst was removed by burning, so continuous operation was possible. Heat was supplied by a molten salt system. In a next generation an adiabatic process was implemented, thus eliminating the molten salt heat transfer system. The heat produced during the regeneration was stored in the bed and provided the heat needed for heating the feedstock to the reaction temperature and providing the heat of reaction. Most of the aviation gasoline used in the Battle of Britain was produced in Houdry units. The next generation of catalytic crackers was developed immediately after the commercialisation of the Houdry units. These were based on the principle that the catalyst particles move between a cracking zone and a regeneration zone. Conveyor type reactors have been used and commercialised. A particularly successful approach, developed mainly by the Standard Oil Company of New Jersey, has been the Fluid Catalytic Cracking (FCC) technology. This technology was introduced at the beginning of the '40s. The reactor and the regenerator were fluidized beds and silica-alumina catalysts were used. The FCC process has been continuously improved both with respect to the catalyst and to the design of the

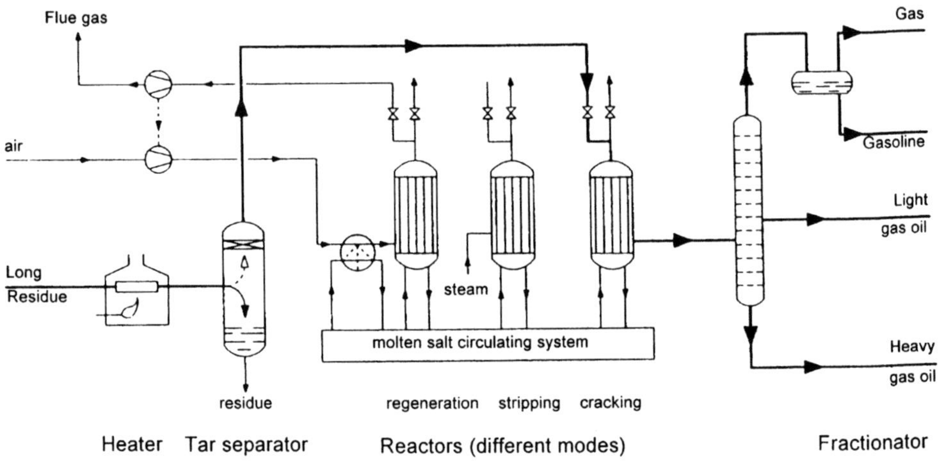

Fig. 1.5. Houdry catalytic cracking process.

reactor and the process. The acid-treated clays were gradually replaced by synthetic silica-alumina catalysts. In the early '60s these were replaced by zeolite-based catalysts. The fluid cracking reactor was replaced by a riser reactor after the mechanical strength and the activity of the catalyst particles had been sufficiently improved.

Catalytic cracking was not the only technology being developed for gasoline needs: iso-Octane was produced as early as 1930 by oligomerisation of lower olefins followed by hydrogenation. Ipatieff discovered paraffin alkylation in 1935, in which isobutane was alkylated with butenes or propene, the catalysts being sulphuric acid and hydrofluoric acid. This is an example of homogeneous catalysis. Alkylation was the basis of large volumes of aviation gasoline in World War II. While sufficient amounts of alkenes were available, there was a shortage of isobutane. Isomerisation processes were commercially applied to convert *n*-butane into isobutane with a catalyst consisting of a mixture of $AlCl_3$ and HCl.

It was not only gasoline production that was fundamental in the late '30s; synthetic rubber was just as important. In 1934 BASF and Bayer had developed Buna-S rubber, which was a copolymer containing styrene. Processes were also developed in the U.S. to produce styrene.

1.2.4 Catalytic Reforming

During the Second World War methylcyclohexane was transformed into toluene over molybdena/alumina catalysts. An attempt was made to use the same process for the 'catalytic reforming' of naphtha in order to increase the octane number by carrying out isomerisation, cyclisation and dehydrogenation of paraffins into aromatics. The stability of the catalyst was unsatisfactory because a large amount of coke was deposited. The breakthrough came with a catalyst developed at Universal Oil Products (about 1950). This catalyst was Pt supported on an acidic support. The process was called 'Platforming'. This catalyst could be used over a period of 6–12 months and could be regenerated by careful oxidation. It was observed that small amounts of sulphur did not poison the catalyst but even enhanced the reaction rate. Although a scientific success, further catalyst development was needed because the catalyst contained 3% Pt and this was considered far too expensive. Catalysts with satisfactory activity were developed with Pt concentrations between 0.2 and 0.7%. A major improvement was the application of bimetallic catalysts. The first commercially successful one, developed by Chevron, contained Re besides Pt, and was much more stable. The process based on this catalyst is called 'Rheniforming'. Catalytic reforming has had a great impact: in 1950 about 2% of the U.S. gasoline pool was from catalytic reforming; by the late 1950s this figure had risen to 30%.

1.2.5 *Hydrorefining*

Hydrorefining is directly related to the technology developed in coal liquefaction. In this process heteroatoms, in particular S and N, are removed by catalytic hydrogenation. This is done in order to abate pollution and to prevent extensive catalyst deactivation in further catalytic processing. The most common catalysts are $CoS–MoS_2/Al_2O_3$ and $NiS–MoS_2/Al_2O_3$. This technology has seen widespread application only since the late 1950s. A recent development is the processing of residual oils. Catalyst poisoning is more severe, here, in particular because these feedstocks contain large amounts of metals which deposit on the catalysts. The most abundantly occurring metals are Ni and V. A typical feature of these feedstocks is the presence of large hydrocarbon molecules. The answer has been found in the production of catalysts supported on wide-pore carriers.

1.2.6 *Acetaldehyde*

Acetaldehyde is an intermediate in acetic acid and vinyl acetate production. Since 1916 it has been produced from the addition of water to acetylene, a reaction catalysed by divalent mercury in sulphuric acid (20%)/water. Acetylene was made from coal. In Germany in particular, a lot of research was carried out on the use of acetylene as a chemical feedstock.

$$HC{\equiv}CH + H_2O \rightarrow CH_3CHO$$

Coal was also the feedstock for synthesis gas (*vide infra*). Many contributions to acetylene chemistry are due to Reppe. His work on new homogeneous metal (mainly nickel) catalysts for acetylene conversion, carried out in the period from 1928 to 1945, was not published until 1948. Under the influence of nickel iodide catalysts, acetylene, water and CO were found to give acrylic acid. A process based on this chemistry was commercialised in 1955.

When the oil era started, ethene became a much cheaper C_2 feedstock than acetylene and a search was started for new catalytic conversion routes for ethene. In their attempts to epoxidize ethene over a palladium catalyst, Smidt and coworkers at Wacker-chemie discovered that ethene and water produced acetaldehyde. They rediscovered a stoichiometric reaction that had been known since 1894 (Phillips):

$$H_2C{=}CH_2 + PdCl_2 + H_2O \rightarrow CH_3CHO + Pd + 2\,HCl$$

Smidt combined this reaction with a redox system, which led to a catalytic process ('Wacker process'):

$$Pd + 2\,CuCl_2 \rightarrow PdCl_2 + 2\,CuCl$$

$$2\,CuCl + 2\,HCl + 1/2\,O_2 \rightarrow 2\,CuCl_2 + H_2O$$

The overall reaction reads:

$$H_2C{=}CH_2 + 1/2\,O_2 \rightarrow CH_3CHO$$

1.2.7 Butanol

Since the twenties and thirties butanol together with acetone was produced by fermentation of carbohydrates (corn). In the sixties the process was replaced by the hydroformylation of propene. In the OXO process alkenes react with synthesis gas in the presence of a homogeneous catalyst to give a mixture of branched and linear aldehydes:

$$RHC{=}CH_2 + CO + H_2 \rightarrow a\,RCH_2CH_2CHO + (1-a)\,RCH(CHO)CH_3$$

In formal terms, the addition of formaldehyde across the double bond has taken place. The reaction was discovered by Roelen while working on the conversion of synthesis gas to hydrocarbons. The process was first patented in 1938 and a commercial plant was briefly operated at the end of the war after which it was dismantled. In 1948 a commercial plant was put on stream by Exxon in the U.S.

The primary aldehyde product is reduced to the desired butanol, or it is subjected to a base-catalysed aldol condensation and then hydrogenated to give 2-ethylhexanol. The phthalic ester of the latter is used as a plasticiser in PVC. The first process was based on a $Co_2(CO)_8$ catalyst, a precursor of $HCo(CO)_4$. The pressure is high, ca. 200–300 bar, in order to maintain the catalyst's stability. In the '60s Shell developed a process using phosphine ligands which allowed the use of lower pressures. The catalyst is less active but it directly produces alcohols with a somewhat higher linearity.

A breakthrough occurred in the mid-seventies when Union Carbide and Celanese introduced Rh/phosphine catalysts in commercial processes. This catalyst is based on the work by Wilkinson's group; he received the Nobel prize for his work in 1973. Rhodium-based catalysts are much more active than cobalt catalysts and, under certain conditions, at least for 1-alkenes, they are also more selective. The processes for the hydroformylation of higher alkenes (detergent alcohols) still rely on cobalt catalysis. A new development is the use of water-soluble complexes obtained through sulphonation of the ligands (Ruhrchemie). The new hydroformylation plants for the production of butyl-aldehyde from propene are all based on rhodium catalysts.

1.2.8 Acetic Acid

Originally, acetic acid was produced by fermentation; this is still the major process for the production of vinegar. Modern production is by acetaldehyde oxidation, liquid phase hydrocarbon oxidation and preferentially by methanol carbonylation. The latter process is to be preferred because of the low raw material and energy costs. As early as 1913 BASF described the carbonylation of methanol at high temperature and pressure:

$$CH_3OH + CO \rightarrow CH_3COOH$$

Due to the extreme conditions, commercialisation was not successful. Corrosion was very important; only gold and graphite appeared to be resistant enough. Even gold-lined autoclaves were used. In the early '30s the process was abandoned for economic reasons.

Carbonylation with homogeneous nickel catalysts was very extensively studied by Reppe in the thirties and forties. This work was not published until 1953. Based on these results, an acetic acid manufacturing process was commercialised in 1955.

In 1941 Reppe demonstrated the potential of many metal carbonyls in several reactions including hydroformylation. This work resulted in 1960 in a process based on CoI_2 operating at 700 bar and 250°C. Corrosion problems were overcome by applying Hastelloy C.

In 1968 Monsanto reported a chemically related process based on rhodium iodide complexes. Due to its high reaction rates, high selectivity and different kinetics, the process differs substantially from the cobalt process. Commercialisation was achieved in 1970. Operating conditions are remarkably mild: 30 bar, 180°C.

1.2.9 Polymerisation

The first polymerisations were free radical reactions. In 1933 researchers at ICI discovered that ethene polymerises into a branched structure that is now known as low density polyethene (LDPE). In the mid-'50s a series of patents were issued for new processes in which solid catalysts were used to produce polyethene at relatively low pressures. The first was granted to scientists at Standard Oil (Indiana) who applied nickel oxide on activated carbon and molybdenum oxide on alumina. Their research did not lead to commercial processes. In the late '40s Hogan and Banks of Phillips were assigned to study the di- and trimerisation of lower olefins. The objective was to produce high octane motor fuels. When they tried a chromium salt as promoter of a certain catalyst (Cr was a known reforming catalyst) they found solid white material which appeared to be polyethene. It is worth mentioning that the same scientists discovered the metathesis of alkenes! The Phillips catalyst has been improved and is used in many commercial plants. About a year later, Ziegler's group patented a catalyst based on $TiCl_4$ and aluminium alkyls.

Initially, the activity of these catalysts was poor and, as a result, a high concentration of catalytic material was present in the product. An expensive catalyst removal step was necessary. Catalysts were later improved and a process became feasible in which the polymer formed precipitates and catalyst removal is no longer needed.

In 1956 Natta discovered that propene can be polymerised with a titanium catalyst and that a highly regular polymer is formed. Ziegler and Natta shared the Nobel prize in 1963.

1.2.10 Metathesis

A novel catalytic reaction was discovered independently by scientists of Phillips Petroleum Company, du Pont and Standard Oil in Indiana. They found that alkenes can react according to the following stoichiometry:

$$R_1HC{=}CHR_2 + R_1HC{=}CHR_2 \rightleftharpoons R_1HC{=}CHR_1 + R_2HC{=}CHR_2$$

The mechanism has been subject to much debate. The research in this field has had a large impact on transition metal catalysis. It is also interesting that in this field the usually large gap between heterogeneous and homogeneous catalysis does not exist. There was never much doubt that the reaction mechanisms were not the same in both cases.

The first large-scale application was the Phillips Triolefin Process (1966) in which propene was converted into ethene and 2-butene. Due to market changes the reverse process, in which propene is produced, became more attractive later. This process has been in operation since 1985. Another process is the Shell Higher Olefin Process (SHOP) in which ethene is oligomerised and the products are metathesised into detergent range olefins. The same company developed a process in speciality chemicals in which alpha-, omega-dienes are formed from cyclic alkenes.

Metathesis is not limited to acyclic alkenes. Cyclic alkenes also undergo metathesis with the formation of polymers. Homogeneous catalysts are used in most cases. New polymers are produced in this way by Hüls, CdF Chimie and Hercules.

1.2.11 Motor Vehicle Emission Control

The development of catalytic reactors for cleaning the exhaust gases of motor vehicles has been stimulated enormously by Californian legislation. In 1959 and 1960 laws on motor vehicle emission standards were enacted which would become operative when at least two different devices were developed which could measure up to these standards. This gave a large impetus for the research. However, in the mid-'60s, the car industry announced that engine modifications were successful and that the standards could be met without catalytic converters. This delayed further research in the development of catalytic converters. In the late '60s a further tightening of the standards was announced and catalytic research increased considerably. The Clean Air Act of 1970 set standards that went well beyond existing technology and it was clear that catalytic mufflers would be essential.

The initial objective was to reduce the emission of CO and hydrocarbons (HC). The mixture in the engine was fuel-rich and under those conditions the concentrations of CO and hydrocarbons are relatively high, whereas the NO_x concentration is low. The exhaust gases were oxidised in a converter to which air was supplied. Somewhat later the NO_x standards tightened and also the amount of NO had to be reduced. So both oxidation and reduction reactions were required. In the beginning, a so-called dual-bed system with separate reduction and oxidation catalysts was envisaged. The engine would be operated under fuel-rich conditions in the first bed; reduction would take place. After air injection into the second bed, CO and the hydrocarbons would be oxidised. The next generation of catalysts could catalyse all reactions simultaneously, provided the fuel–air ratio in the gas mixture was stoichiometric. A typical example of the influence of the dimensionless air–fuel ratio (λ) is given in Fig. 1.6.

At $\lambda < 1$ the activity for NO reduction is high, but not for the oxidation of CO and hydrocarbons. At $\lambda > 1$ the reverse is the case. A special control system had to

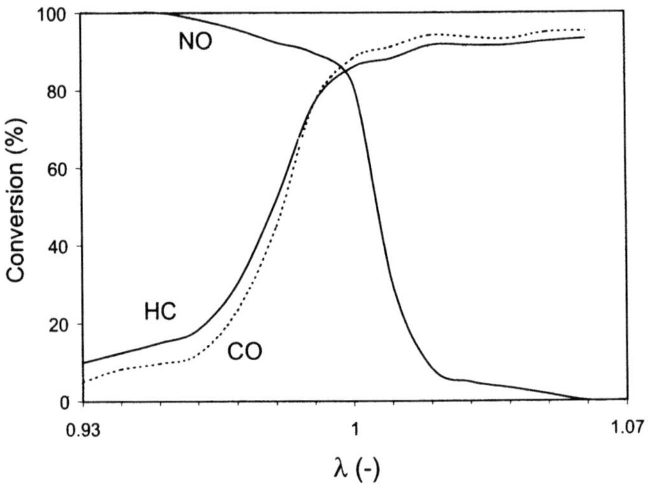

Fig. 1.6. Conversion of CO, NO and HC as a function of λ, the dimensionless air–fuel ratio.

be developed to guarantee the desired exhaust gas composition. The control system appears to be more critical than the catalyst itself.

The catalysts in use contain Pt and Rh as major constituents. A converter typically contains 1–2 g Pt and 0.2–0.3 g Rh. The design of the reactor has received a lot of attention. Beads were also applied initially, but nowadays monolithic honeycombs are used almost exclusively.

The introduction of catalytic converters has had a tremendous impact on the composition of gasoline. The catalysts used became poisoned by small amounts of impurities; in particular the lead compounds present in high octane gasoline were detrimental. Processes which produce high octane number compounds were therefore stimulated. First, cracking and reforming increased in importance. More recently, the aromatics content is also expected to have to decrease and alternative processes are in use or under way, e.g. MTBE (methyl tertiary-butyl ether) is already produced on a large scale (see Section 2.12).

1.3 BIOCATALYSIS

1.3.1 Fermentation

In a broad sense, fermentation can be defined by all chemical transformations brought about by living microorganisms, starting with carbohydrates and some N, P, S and other nutritients as feedstock. World-wide use of carbohydrates for fermentation is about 30 million tons/year, of which 20 million is sucrose, the remainder glucose, lactose and starch.

Several thousands of different catalysts (enzymes) are available in nature to perform these transformations, which has been classified by six major groups of reactions together with the corresponding enzyme type:

Oxidation and reduction	Oxidoreductases
Transfer of functional groups	Transferases
Hydrolysis	Hydrolases
Addition and elimination	Lyases
Isomerisation	Isomerases
Carbon bond formation	Ligases

A cell of a living microorganism can be considered as a small factory consisting of different reactors containing different catalysts (enzymes). In this way, the cell can perform multi-step conversions without the need for recovery or purification of intermediate products. Through this so-called metabolic pathway, i.e. the consecutive conversion of the starting material into the end product, one fermentation can replace a multi-step chemical and/or catalytic procedure. In addition, as the enzymes are chiral catalysts, only one of the two optical isomers is produced.

In fact, fermentation is both the oldest and newest type of catalytic processing. In the very old days—back to several thousand years BC—beer, bread and wine-making were typical fermentation processes. In the present, modern recombinant DNA technology allows us to reconstruct the cell factories, i.e. the metabolic pathways, and so to develop novel multi-step conversions by fermentation that are not available in nature.

1.3.2 Microbial transformation

Apart from fermenting a carbohydrate into the desired end product, it is also possible to use microorganisms to convert other substrates, often in order to accomplish a highly region- and/or enantio-selective conversion that cannot be simply achieved by other chemical or catalytic means. In particular, microbial conversions are applied in the case of desired oxidation/reduction or C-bonding reactions by oxidoreductases and ligases, as these enzymes often require co-factors like NAD(P)H and ATP, the regeneration of which is taken care of by the living cell.

Examples are found for the selective hydroxylation, dehydrogenation and oxidative removal of side chains of steroids, and for the selective oxidation of sorbitol into L-sorbose in the vitamin C synthesis.

In fact, one might say that here only part of the available cell factory is used for the conversion desired and, in most cases, the overall conversion is restricted to one step.

1.3.3 *Enzymatic transformation*

Instead of using living cells, one may use the enzyme activity of killed microorganisms or use enzymes isolated from them. The use of such (crude or pure) enzyme preparations is largely restricted to enzymes that do not require cofactors such as hydrolases, transferases, lyases and isomerases.

The use of (isolated) enzymes for conversions has the advantage that fairly normal processing conditions known for catalytic conversions can be applied, without the need for careful conditions to keep microorganisms alive. This allows a broader range of temperature, pH and solvent mixtures (up to working in almost aqueous-free environments) and the possible immobilisation of the biocatalyst for continuous processing or reuse of the biocatalyst. Furthermore, fermentative production of the biocatalyst followed by recovery and use of the biocatalyst for the actual conversion process prevents part of the biomass production (an undesired side reaction during fermentation processes) and allows conversions to take place under much more concentrated conditions (leading to higher yield per reactor volume).

Examples are found for the glucose to fructose isomerisation process using immobilised glucose isomerase in a continuous column reactor by passing through a highly concentrated glucose syrup at 70°C and for the hydrolysis of Penicillin-G into 6-aminopinillicinium acid with immobilised acylase in a stirred tank fed batch reactor with easy separation and reuse of the, rather expensive, acylase catalyst.

Finally, enzyme isolation from microorganisms allows the application of biocatalysts in many other fields and products such as laundry detergents (with the washing machine as reactor), food (cheese, bread), drinks (juice, beer, wine), and feed (enzymes as digestion aid). The food and feed industries in particular replace more and more additives by enzymes, i.e. eatable and digestible catalysts, that result in comparable or even better functional products.

1.4 SUMMARY

The discussion of the history of catalysis in this chapter has been limited to the major events, so many interesting developments could not be mentioned. Moreover, the selection of topics has been subject to a personal bias. A more comprehensive overview of the development of industrial catalytic processes is given in Table 1.2. A similar summary of biocatalytic processes is given in Table 1.3. The list is impressive and it is also clear that catalysis remains a very fascinating area.

Many reasons underlie the development of the science and technology of catalysis. One of the driving forces that should not be overlooked is the availability of the raw materials. Originally, biomass was predominant. Later coal became the most important industrial feedstock. Coke oven gas components played the role of base chemicals. Subsequently, oil took over the place of coal and the technological scene changed profoundly. More recently, natural gas resources appear to have become much more important than had been expected. Therefore it is not surprising that a lot of research in catalysis is aimed at processes based on natural gas. It is also clear that biomass is experiencing a revival. Figure 1.7 illustrates this point: it shows that for a number of important chemicals, the development of processes is linked to the availability of different raw materials. In recent years environmental considerations have been the major driving force for novel processes, often catalytic processes.

TABLE 1.2

Historical summary of the development of industrial processes

Year	Process	Catalyst
1750	H_2SO_4 lead chamber process*	NO/NO_2
1870	SO_2 oxidation	Pt
1880	Deacon process (Cl_2 from HCl)	$ZnCl_2/CuCl_2$
1885	Claus process (H_2S and SO_2 to S)	Bauxite
1900	Fat hydrogenation	Ni
	Methane from syngas	Ni
1910	Coal liquefaction	Fe
	Upgrading coal liquids	WS_2
	Ammonia synthesis (Haber–Bosch)	Fe/K
	NH_3 oxidation to nitric acid	Pt
1920	Methanol synthesis (high pressure process)	Zn, Cr oxide
	Fischer–Tropsch synthesis	Promoted Fe,Co
	SO_2 oxidation	V_2O_5
	Acetaldehyde from acetylene*	Hg^{2+}/H_2SO_4
1930	Catalytic cracking (fixed bed, Houdry)	Clays
	Ethene epoxidation	Ag
	Polyvinyl chloride*	Peroxide
	Polyethene (low density, ICI)*	Peroxide
	Oxidation of benzene to maleic anhydride	V
	Alkylation*	HF/H_2SO_4
1940	Hydroformylation, alkene to aldehyde*	Co
	Catalytic reforming (gasoline)	Pt
	Cyclohexane oxidation (nylon 66 production)*	Co
	Benzene hydrogenation to cyclohexane	Ni, Pt

Continued

TABLE 1.2 (*Continuation*)

Year	Process	Catalyst
	Synthetic rubber, SBR*	Li, peroxide
	BNR*	Peroxide
	Butylrubber*	Al
1950	Polyethylene (high density), Ziegler–Natta	Ti
	Phillips	Cr
	Polypropene, Ziegler–Natta	Ti
	Polybutadiene, Ziegler–Natta	Ti
	Hydrodesulphiding (HDS)	Co, Mo sulphides
	Naphthalene oxidation to phthalic anhydride	V, Mo oxides
	Ethene oxidation to acetaldehyde*	Pd, Cu
	p-Xylene oxidation to terephthalic acid*	Co, Mn
	Ethene oligomerization*	Co
	Hydrotreating of naphtha	Co-Mo/Al_2O_3
1960	Butene oxidation to maleic anhydride	V, P oxides
	Acrylonitrile via ammoxidation of propene (Sohio)	Bi, Mo oxides
	Propene oxidation to acrolein/acrylic acid	Bi, Mo oxides
	Xylene hydroisomerization	Pt
	Propene metathesis	W, Mo, Re
	Adiponitrile via butadiene hydrocyanation*	Ni
	Improved reforming catalysts	Pt, Re/Al_2O_3
	Improved cracking catalysts	Zeolites
	Acetic acid from MeOH (carbonylation)*	Co
	Vinyl chloride via ethene oxychlorination	Cu chloride
	Ethene oxidation to vinyl acetate	Pd/Cu
	o-Xylene oxidation to phthalic anhydride	V, Ti oxides
	Propene oxidation to propene oxide*	Mo
	Hydrocracking	Ni-W/Al_2O_3
	HT water–gas shift process	$Fe_2O_3/Cr_2O_3/MgO$
	LT water–gas shift process	$CuO/ZnO/Al_2O_3$
1970	Methanol synthesis (low pressure, ICI)	Cu-Zn-Al oxide
	Acetic acid from MeOH (carbonylation, low pressure process, Monsanto)*	Rh
	Improved process for xylene isomerization	Zeolite

Continued

TABLE 1.2 (*Continuation*)

Year	Process	Catalyst
	α-Alkenes via ethene oligomerization*/isomerization/metathesis (SHOP)	Ni, Mo
	Improved hydroformylation*	Rh
	Auto exhaust gas catalysts	Pt/Rh
	L-DOPA (Monsanto)*	Rh
	Cyclooctenamer (metathesis)*	W
	Hydroisomerization	Pt/zeolite
	Selective reduction of NO (with NH_3)	V_2O_5/TiO_2
1980	Gasoline from methanol process (Mobil)	Zeolite
	Vinyl acetate from ethene and acetic acid	Pd
	Methylacetate (carbonylation)*	Rh
	Methylacrylate via t-butanol oxidation	Mo oxides
	Improved coal liquefaction	Co, Mo sulphides
	Diesel fuel from syngas	Co
1990	Polyketone (from CO and ethene)*	Pd

The data refer to activities of a pilot plant scale at least.
*Homogeneous catalyst system.

TABLE 1.3
Historical summary of the development of industrial biocatalytic processes

Year	Process	Micro-organism/Enzyme
–6000	Beer brewing by malting procedure	Malt enzymes
–3000	Wine making by fermentative conversion of grape juice sugars	Yeast a.o.
–2000	Alcohol(ic drinks) by fermentation various carbohydrate sources	Yeast a.o.
	Bread and other leavenend bakery products by carbon dioxide and alcohol	Baker's yeast
	Vinegar (acetic acid) by aerobic fermentation of ethanol	Acetobacter
–800	Cheese making (coagulation) by casein hydrolysis with calf stomach extract	Calf rennet (chymosin)
1850	Yoghurt formation from milk by lactose to lactic acid conversion	Lactobacillus

Continued

TABLE 1.3 (*Continuation*)

Year	Process	Micro-organism/Enzyme
1870	Bread baking by using pure baker's yeast	Baker's yeast
1880	L-Lactic acid formation by fermentation of glucose or lactose	Lactobacillus
1900	Beer production by using pure brewer's yeast	Brewer's yeast
	Foodstuff and flavouring production from soy protein	Fungal amylase
1910	Acetone-butanol-ethanol (3:6:1) mixture by anaerobic fermentation of glucose	*Clostridium acetobutylicum*
1920	Citric acid fermentation	*Aspergillus*
	Textile desizing (removal of starch after weaving)	Bacterial amylase
1930	D-Gluconic acid by oxidation of glucose	*Acetobacter suboxydans, Asp. Niger*
	Ephedrine intermediates through acyloin condensation	Yeast
	Fruit juice clarification by hydrolysis of pectins, hemicellulose and starch	Pectinase a.o.
	Sorbitol to L-sorbose oxidation (in vitamin C production)	*Gluconobacter suboxydans*
	Vitamin B2 (riboflavin) fermentation	Ashbya, Eremothecium
1940	Invert sugar by hydrolysis of sucrose	Invertase
	Penicillin (beta-lactam antibiotic) fermentation	Penicillinium
1950	Bacitracin (antibiotic) fermentation	Streptomyces
	Cephalosporin (beta-lactam antibiotic) fermentation	Cephalosporium
	D-Ribose (vitamin B2 precursor) by glucose oxidation	*Bacillus subtilus*
	L-Amino acids (miscellaneous) through fermentation	Brevi-, Corynebacterium species
	L-Glutamate fermentation	Corynebacterium
	L-Leucine fermentation	Brevibacterium
	L-Tartaric acid by glucose oxidation	*A. suboxydans*
	L-Tryptophan fermentation	Brevibacterium
	Maltodextrin syrup production through endo-starch hydrolysis (gelatinization)	Alpha amylase
	Mycin (erythro-, nio-, strepto-) fermentation	Streptomyces
	Partial hydrolysis of proteins (food and beverages processing)	Proteases
	Steroid (pharma) oxydation and reduction reactions	*Rhizopus, Aspergillus, Penicillinium*
	Tetracycline (antibiotic) fermentation	Streptomyces

Continued

TABLE 1.3 (*Continuation*)

Year	Process	Micro-organism/Enzyme
1960	Glucose syrup production by maltodextrin hydrolysis (saccharification)	Amyloglucosidase
	Itaconic acid by glucose fermentation	*Aspergillus terreus*
	Laundry detergent presoaking method (Biotex)	Alkaline protease
	Vitamin B12 fermentation	*Pseudomonas denitrificans*
	Xanthan gum by aerobic glucose fermentation	*Xathomonas campestris*
1970	6-Aminopenicillanic acid by hydrolysis of penicillin	Bacterial/Fungal acylase
	Cheese making (coagulation) by casein hydrolysis using pure microbial rennet	Chymosin
	Cheese making (emulgation by fat hydrolysis)	Lipase
	Cyclodextrins (cyclic hexa-, hepta-, octa-glucoses) from amylose (starch)	*Bacillus circulans, macerans*
	Dextrane (1,6-polyglucose) from sucrose by glucose transfer reaction	*Leuconostoc mesentroides*
	Glucose to fructose isomerization (HFCS production)	Glucose isomerase
	Lactose hydrolysis in milk (milk products for lactose intolerant people)	Lactase
	L-Aspartic acid by ammonia addition to fumaric acid	Aspartase (*E. coli*)
	L-Lysine fermentation	Brevibacterium
	L-Phenylalanine fermentation	Brevibacterium
	L-Phenylalanine from racemic N-acyl derivative	Aminoacylase
	L-Threonine fermentation	*Serratia narcescens*
	Maltose syrup production through exo-starch hydrolysis	Beta amylase
	Protein hydrolysis during laundry wash	Alkaline protease
	Starch glue (dirt binding) hydrolysis during laundry wash	Alpha amylase
1980	Amylopectin into linear amylose fragments by hydrolysis	Pullunase
	Beta-(R)-hydroxylation of carboxylic acids (pharma precursors)	*Candida rugosa*
	D-p-Hydroxyphenylglycine by hydrolysis of the racemic hydantoin	D-specific hydantoinase
	Insulin fermentation	Bacillus
	L-Alanine by decarboxylation of L-aspartic acid	Aspartate beta-decarboxylase
	L-Amino acids by enzymatic N-deacylation of racemates	L-specific aminoacylase
	L-DOPA from ammonia, pyruvate and catechol	Bacterial enzymes

Continued

TABLE 1.3 (*Continuation*)

Year	Process	Micro-organism/Enzyme
	L-Lysine by combined hydrolysis and racemization of aminocaprolactam	Aminoacylase, Racemase
	L-Malic acid by water addition to fumaric acid	Fumarase (Brevibacterium)
	L-Tartaric acid by cis-epoxysuccinic hydrolysis	Epoxide hydrolase
	L-Tryptophan from ammonia, pyruvate and indole	Bacterial enzymes
	L-Tyrosine by fermentation	Brevibacterium
	L-Tyrosine from ammonia, pyruvate and phenol	Bacterial enzymes
	Testosterone formation by reduction 17-keto into 17-beta-hydroxy group	Baker's yeast
1990	Acrylamide by hydration of acrylonitrile	Nitrilase
	Animal fat and vegetable oil hydrolysis during laundry wash	Alkaline lipase
	Aspartame (sweetener) by regio- and enantioselective dipeptide coupling	Thermolysin (protease)
	Cellulose hydrolysis (softening) during laundry wash	Alkaline cellulase
	D- and L-Amino acids (miscellaneous) by amino acid amide hydrolysis	L-specific amidase
	Enantioselective ketopantoate reduction (in vitamin B5 synthesis)	Agrobacterium
	Human insulin from porcine insulin by alanine-threonine interconversion	Trypsin
	L-Serine by enantioselective condensation of glycine and formaldehyde	*Klebsiella aerogenes*
	Neosugar (fructo-olisaccharide) from sucrose	Fructosyl transferase
	Nicotinic acid 6-hydroxylation (insecticide precursor)	Achromobacter
	Palatinose (isomultulose) by 'isomerization' of sucrose	Glucosyl transferase
	Phytate hydrolysis into inositol and phassphate (feed)	Phytase
	Poly-beta-(R)-hydroxybutyrate and –valerate by fermentation	Alcaligenes bacteria
	Vegetable oil extraction ;by aqueous break-down of cell walls	Cell-wall degrading enzymes
	(2R,3S)-glycidate ester (Diltiazem precursor) by enantioselective ester hydrolysis	Lipase
	(R)-Carnitine formation by a dehydrogenase-blocked mutant strain	Agrobacterium

Continued

TABLE 1.3 (*Continuation*)

Year	Process	Micro-organism/Enzyme
	(R)-glycidol by enantioselective hydrolysis of racemic glycidyl butyrate	Lipase
	(R)-Solketal by enantioselective oxidation of racemic solketal	Oxidase
	(S)-2-chloropropionic acid by enantioselective dehalogenation racemate	(R)-specific dehalogenase
2000	Bleach boosting (lignin removal) paper pulp through hemicellulose hydrolysis	Xylanase
	Fructose by inulin (polyfructose) hydrolysis	Inulinase
	Partial hydrolysis of feed components (to enhance digestability by animals)	Cell-walll degrading enzymes
	Upgrading natural fats and oils by transesterification	Lipase
	(S)-Naproxen and (S)-ibuprofen by enantioselective ester hydrolysis	Esterase (bacillus)

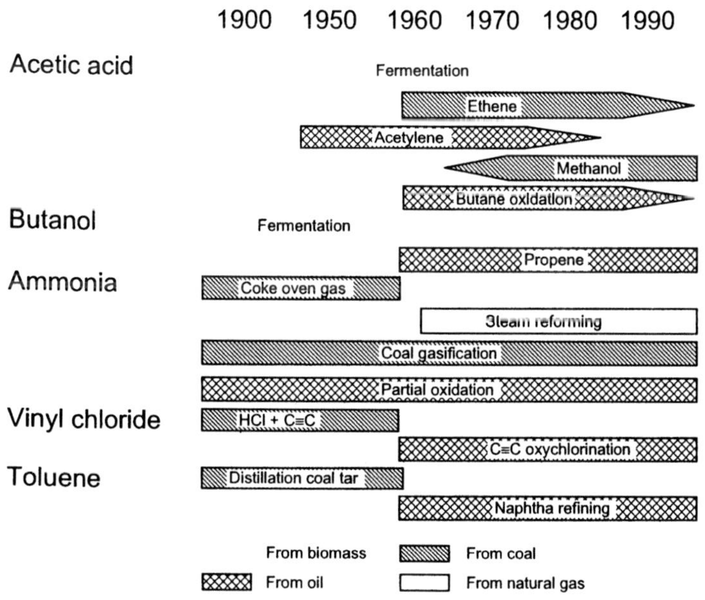

Fig. 1.7. Change in raw materials for some important chemicals.

REFERENCES

1 S.A. Topham, The History of the Catalytic Synthesis of Ammonia, in J.R. Anderson and M. Boudart (Editors), *Catalysis, Science and Technology*, Springer Verlag, Berlin, 1985, Vol. 7, pp. 1–50.
2 P.J.J. Tromp and J.A. Moulijn, Slow and Rapid Pyrolysis of Coal, in Y. Jürüm (Editor), *New Trends in Coal Sciences*, Kluwer Acad. Publ., Dordrecht, 1988, pp. 305–309.

GENERAL LITERATURE

H.S. Bell, *American Petroleum Refining*, 3rd edition, Constable and Company Ltd., Lancaster, PA, USA, 1945.

E.E. Donath, History of Catalysis in Coal Liquefaction, in J.R. Anderson and M. Boudart (Editors), *Catalysis, Science and Technology*, Springer Verlag, Berlin, 1982, Vol. 3, pp. 1–38.

H. Heinemann, A Brief Hisory of Industrial Catalysis, in J.R. Anderson and M. Boudart (Editors), *Catalysis, Science and Technology*, Springer Verlag, Berlin, 1981, Vol. 1, pp. 1–42.

A.J.B. Robertson, The Early History of Catalysis, *Platinum Metals Review*, 19(2) (1975) 64–69.

J. Trofast, J.J. Berzelius and the Concept of Catalysis, in *Perspectives in Catalysis: In commemoration of Jöns Jacob Berzelius*, CWK Gleerup, pp. 9–17.

Kirk-Othmer, *Encyclopedia of Chemical Technology*, 4th edn. John Wiley & Sons, New York, 1991–1998.

Chapter 2

Catalytic processes in industry

2.1 INTRODUCTION

In this chapter a dozen industrial catalytic processes will be described with the aim of giving the reader an impression of the use of catalysis. We have chosen to do this before going into the scientific aspects of catalysis, catalyst preparation and characterization, and catalytic reaction engineering. In this way we set the scene with the industrial applications, so that the reader may keep it in mind when studying the various aspects of catalysis. The choice of the processes is arbitrary, of course. We have tried to represent as many elements as possible within this limited number of processes. From the viewpoint of applications we have looked for catalytic processes from the oil industry, petrochemical industry, exhaust gas technology, fine chemical industry, polymer preparation, etc. A broad variation in chemical reactions has also been pursued, ranging from cracking to denoxing and asymmetric catalysis. Within this limited number of examples we have tried to introduce as many reactor types as possible. Separation of catalyst and product is often a point of interest. The descriptions include a brief survey of the chemistry, the catalyst, the process and the reactor. Occasional reference is made to alternative processes, and the use of the product is briefly indicated.

2.2 CATALYTIC PROCESSES IN THE OIL REFINERY

Oil refineries are complex plants. Besides physical processes such as distillation and extraction, a large number of different chemical conversion processes are applied. Catalysis plays a crucial role in these processes. The most important ones are:

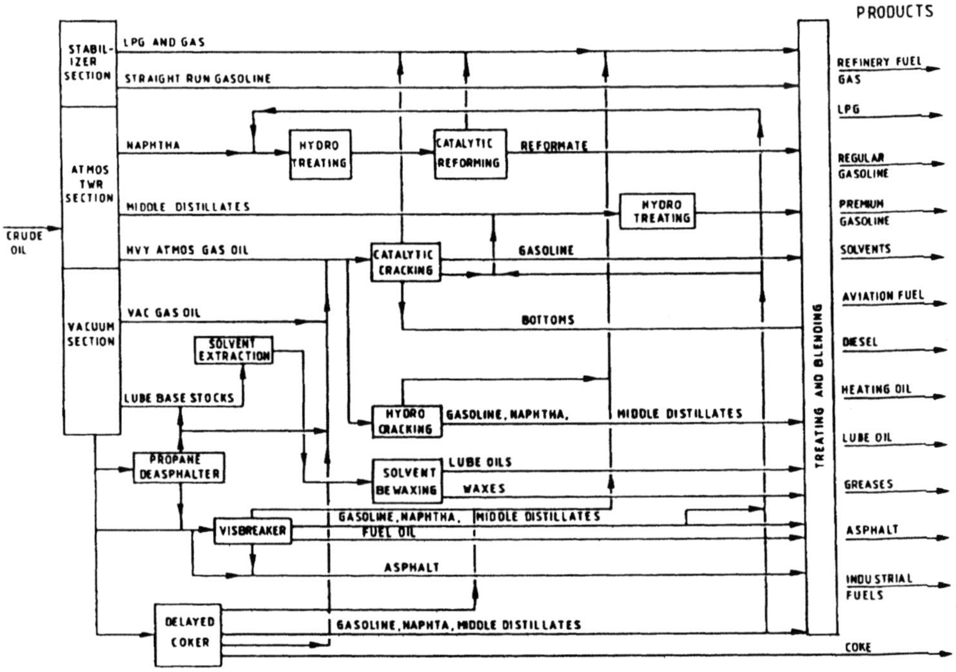

Fig. 2.1. Advanced refinery.

- catalytic reforming;
- hydrotreatment;
- fluid catalytic cracking;
- alkylation.

The role of these processes in an oil refinery can be seen in the simplified scheme in Fig. 2.1. The crude oil is first separated by distillation into a number of fractions. The naphtha fraction has the right boiling point for gasoline, but does not have an acceptable performance. The 'octane number' is too low and has to be increased. This is achieved by 'catalytic reforming'. The catalysts for this process contain acid sites and noble metals, which are poisoned by nitrogen-containing bases and sulphur-containing compounds. That is why a hydrotreatment step precedes the reforming reactors. A catalytic hydrogenation is carried out in this hydrotreatment step, removing nitrogen and sulphur as NH_3 and H_2S. These gaseous compounds can easily be recovered by absorption in a special liquid.

Cracking reactions are carried out in order to reduce the molecular size and to produce more valuable transport fuel fractions (gasoline and diesel). Fluid catalytic cracking is acid catalyzed (zeolites) and a complex network of carbenium ion reactions occur leading to size reduction and isomerization (see Chapter 4, Section 4.4). Hydrogenation also takes place in hydrocracking, as well as cracking.

2.2.1 *Catalytic Reforming*

The purpose of catalytic reforming is to boost the octane number by converting molecules with a low octane number into molecules with a high octane number (see Table 2.1).

TABLE 2.1

Overview of octane number of different hydrocarbons

Hydrocarbon	Octane number	Boiling point (K)
n-pentane	62	309
2-methylbutane	90	301
cyclopentane	85	322
n-hexane	26	342
2-methylpentane	73	333
2,2-dimethylbutane	93	323
1-hexene	63	337
2-hexene	81	341
benzene	>100	353
cyclohexane	77	354
n-octane	0	399
2-methylheptane	13	391
2,2,4-trimethylpentane	100	372
1-octene	35	395
2-octene	56	398
3-octene	68	396
xylenes	>100	–
ethylbenzene	98	410
1,2-dimethylcyclohexane	79	403
ethylcyclohexane	41	403
methyl tert.-butyl-ether (MTBE)	118	328
ethyl tert.-butyl-ether (ETBE)	118	345
tert.-amyl methyl-ether (TAME)	115	359

The capacity of catalytic reforming is very large. In the western world alone it amounts to over 1 million t/d. The major reactions occurring are:

Isomerization

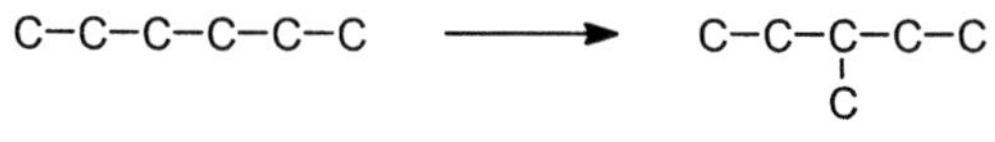

ΔH (298 K) = ± – 12 kJ/mol

Aromatization

ΔH (298 K) = 206 kJ/mol

Isomerization and aromatization

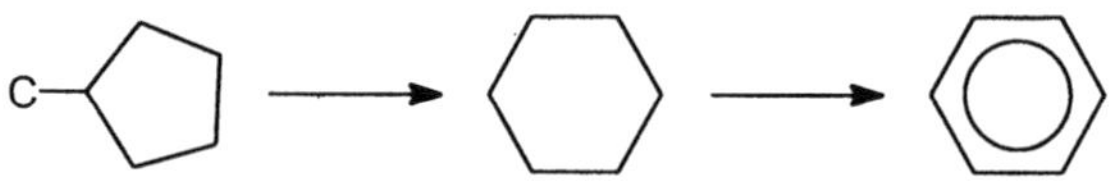

ΔH (298 K) = 189.6 kJ/mol

Clearly, favourable reaction conditions are a high temperature and a low pressure. However, a low pressure is not possible because large carbon deposits are formed. Carbon deposition is lowered by increasing the hydrogen partial pressure. Typical practical conditions are 800 K and 30 bar. Under these conditions, catalyst activity is typically retained over a period of 3 to 6 months. Regeneration is performed by removing the carbon deposits by careful burning in diluted air.

In practice catalytic reforming is usually carried out in fixed bed reactors (see Fig. 2.2). Because the reactions are endothermic, heat has to be introduced. A conventional scheme is based upon a series of three or four adiabatic fixed bed

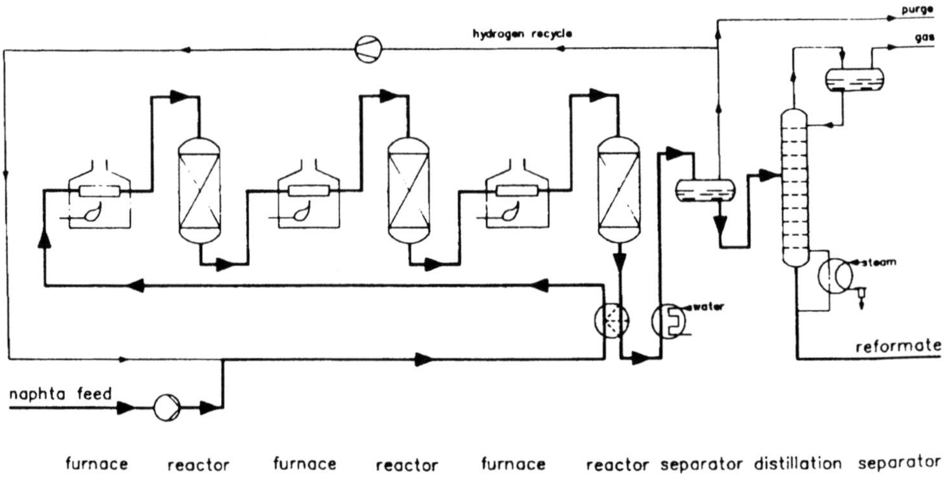

Fig. 2.2. Catalytic reforming.

reactors with interstage heating. Part of the hydrogen produced is recycled to maintain high hydrogen partial pressures. The product mixture from the last reactor is cooled and separated into a gas phase and a liquid phase. The latter is purified by distillation.

The catalyst is called 'bifunctional'; both the carrier and the metallic particles dispersed over the carrier exhibit different catalytic functions. The carrier contains chlorine ions and, as a consequence, it has acid properties and exhibits isomerization and cyclization activities. The metal particles consist of alloys of, for example, Pt/Re which exhibit hydrogenation/dehydrogenation activity.

2.2.2 *Catalytic Cracking*

The purpose of catalytic cracking is to increase the fraction of gasoline. Originally cracking was performed thermally, but cracking in the presence of a catalyst predominates nowadays. Catalytic cracking is one of the largest applications of catalysis: worldwide 500 million ton/y. Catalytic cracking was the first large-scale application of fluid beds. This explains the name 'fluid catalytic cracking' (FCC) which is still used, although the modern catalytic cracking reactor is an entrained flow reactor.

The mechanism

In catalytic cracking many reactions take place simultaneously. Cracking occurs by C–C bond cleavage of paraffins, dealkylation etc. Isomerization and even condensation reactions take place. These reactions occur via positively charged hydrocarbon ions (carbocations). The nature of the carbocations is the subject of debate. For the cracking of paraffinic hydrocarbons it is usually assumed that carbenium ions are the crucial intermediates, which decompose via beta fission into olefins and (smaller) carbenium ions (see Chapter 4, Section 4.4). A typical reaction mechanism for catalytic cracking (and hydrocracking) under the relatively mild conditions used in FCC is shown overleaf.

The catalyst causes a 'classical' carbenium ion to be formed by acid catalyzed activation reactions. The classical carbenium ion is transformed into the key intermediate which can be described as a protonated cyclopropane structure. After some rearrangements cracking occurs. The formation of branched paraffins is very fortunate since branched paraffins have high octane numbers and the isobutane produced can be used in alkylation. The preferred products are those of which the formation proceeds via tertiary carbenium ions. Carbenium ions can also be generated by intermolecular hydride transfer reactions between alkane and carbenium ions that are not able to form tertiary carbenium ions (see Chapter 4, Section 4.4). Under more severe conditions lower paraffins can also be cracked.

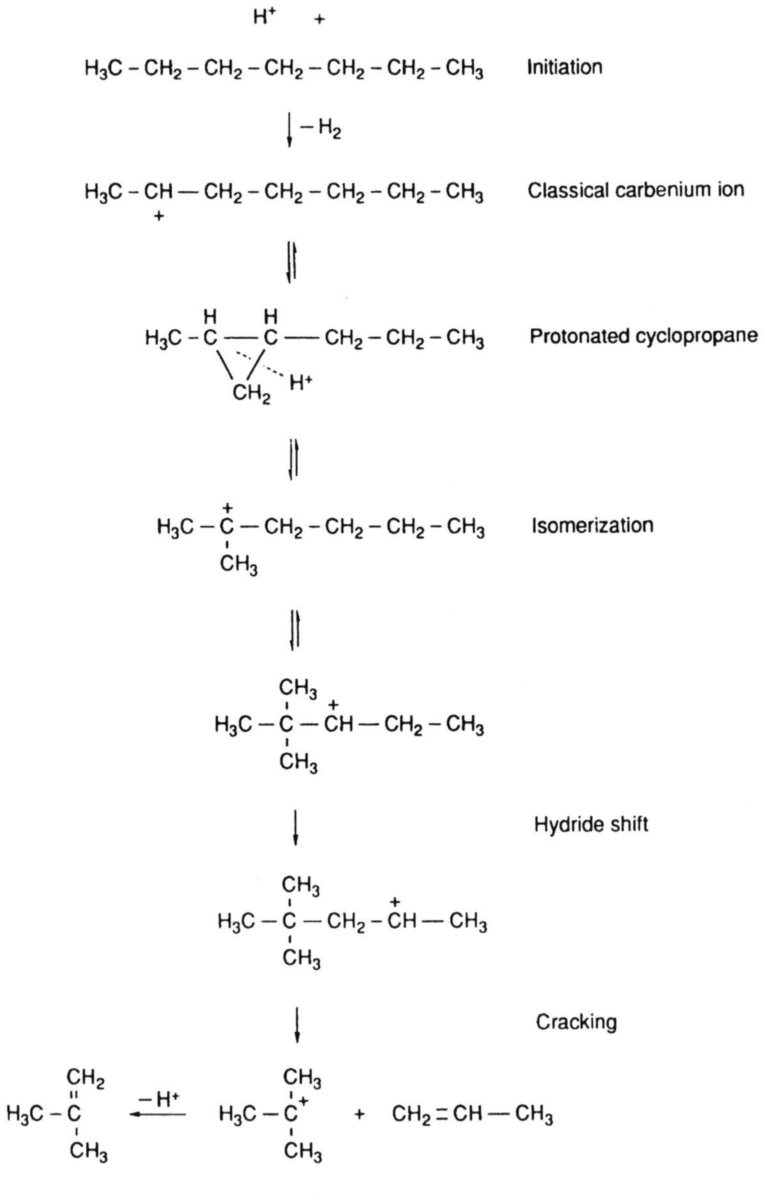

Process description

Extensive coke deposition takes place during catalytic cracking, resulting in loss of activity. Typically, the catalyst loses 90% of its activity within one second. An elegant solution has been found for this problem. The clue to this solution is a combination of a reactor in which cracking takes place with a reactor used for regeneration of the catalyst by burning the deposited coke. In this set-up coke is

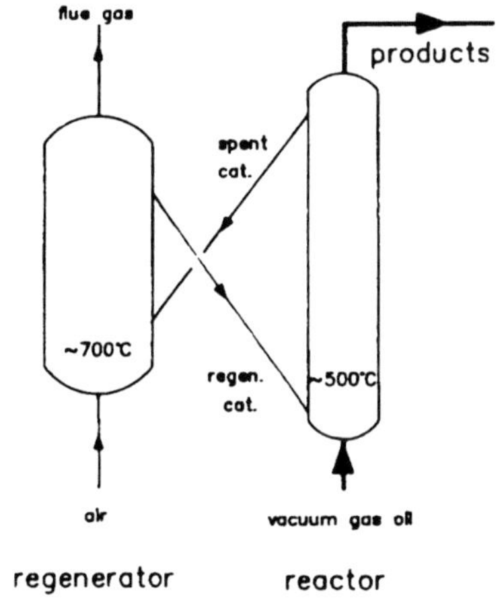

Fig. 2.3. The principle of catalytic cracking.

often referred to as a 'mixed blessing' because, although it poisons the catalyst, it also enables the production of the heat required for the endothermic cracking reactions.

The principle is as follows (see Fig. 2.3). The feed is vaporized and entrains the catalyst particles (size ca. 70 μm). The residence time in the riser is about one second. Such risers are very high: typically 30 m (reactor diameter 1 m). The catalyst particles are separated by a cyclone and transferred to the regenerator, in which the coke is burned. After regeneration the catalyst particles are returned to the riser. So a catalyst particle is actually a batch reactor. It enters the riser at a high activity and temperature. While it is travelling through the riser, it acts as a catalyst, its temperature drops due to the endothermic nature of the cracking reactions, coke is deposited and its activity decreases strongly. Subsequently, it is transferred to the regenerator and coke burning starts. It recovers its activity and the temperature is increased by the exothermic combustion.

A diagram of a cracking unit is shown in Fig. 2.4.

Catalyst

The catalyst particles consist of a silica–alumina matrix containing zeolite crystals. Zeolites contain pores of molecular dimensions, and as a consequence they have to be very small to give the reactant molecules the opportunity to reach the active sites. Typically, the zeolite crystals have a diameter of 1–5 μm. Such a fine powder cannot be handled in the reactors and therefore the particles used are composed of a matrix containing the zeolite crystals. The matrix itself has large pores and also exhibits activity. Special additives are added, among other things to catalyze the CO oxidation reaction since when large amounts of CO are present, a separate burner is needed to recover the energy present in the gas.

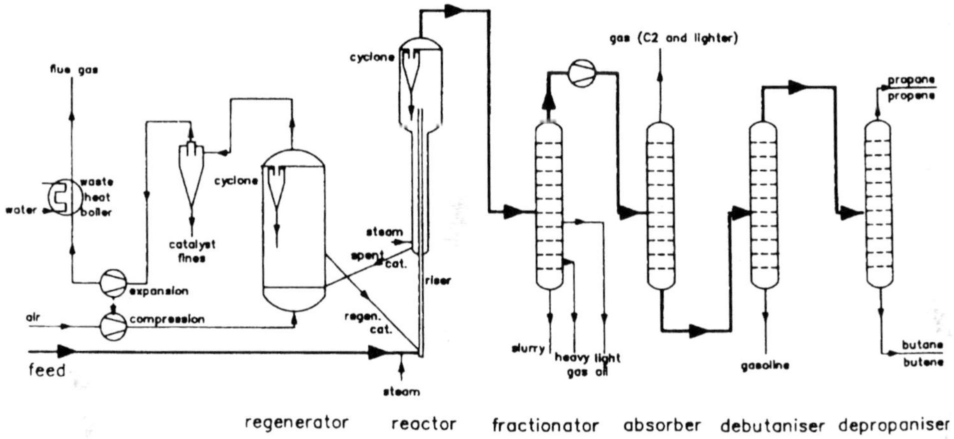

Fig. 2.4. Scheme of a catalytic cracking unit.

Environmental aspects

A catalytic cracking plant emits dust, nitrogen oxides and sulphur oxides. The first stream is derived from attrition of the catalyst particles in the regenerator, which is a fluidized bed. Nitrogen oxides are formed in the regeneration and the sulphur oxides arise from the sulphur compounds present in the feed. Part of the sulphur is released in the riser, and part in the regenerator. From an environmental point of view these two modes have quite different impacts. In the riser S is released in the form of H_2S and is recovered downstream in the gas cleaning sections. In the regenerator the gas phase is oxidized and S is released as sulphur oxide in a highly diluted gas stream containing as major component the large amount of N_2 from the air used for burning the coke. In contrast to H_2S from the regenerator gas effluent, the recovery of the sulphur oxides from the diluted regenerator gas is difficult. Recently attention has been given to the question of what to do with the spent catalyst.

2.2.3 *Hydrotreatment*

Hydrotreatment belongs to the class of conversions involving reactions with hydrogen. In this section 'hydrotreatment' is limited to hydrogenation and hydrogenolysis reactions in which removal of heteroatoms (in particular S,N) and some hydrogenation of double bonds and aromatic rings take place. In these reactions—in contrast to hydrocracking, in which size reduction is the objective—the molecular size is not drastically altered. Hydrotreatment as defined here is sometimes called hydropurification.

The major objectives are:
- protection of downstream catalysts;
- improvement of gasoline (odour, colour, stability, corrosion);
- protection of the environment.

Typical reactions that occur are:

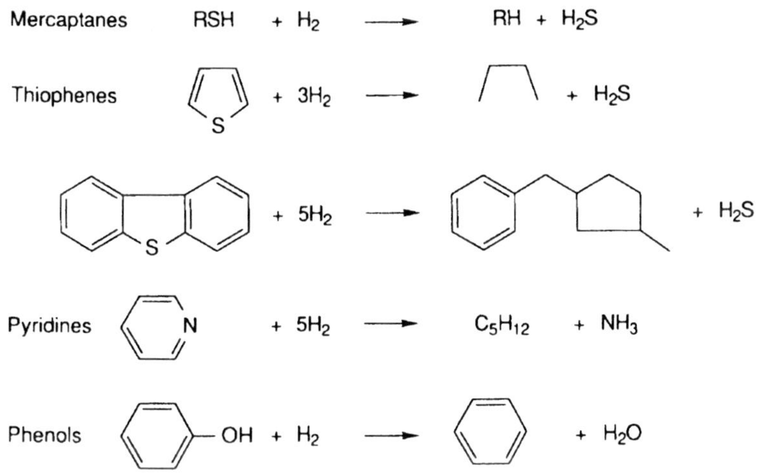

In naphtha reforming, hydrotreatment is always applied to protect the platinum-containing catalyst against sulphur poisoning. The specification of sulphur content for the feed for the reforming unit is less than 1 ppm.

The hydrogen used in naphtha hydrotreatment is a by-product of the catalytic reforming. When hydrotreatment of heavy residue is performed, separate H_2 production units are often required.

Unstable byproducts are formed in naphtha cracking, and these are hydrotreated in order to increase their stability. One of the important reactions occurring is the saturation of diolefins.

Process description

In simple naphtha hydrotreatment the feed is vaporized and led through a fixed bed reactor. When heavier feed stocks are treated, vaporization is not possible and trickle phase operation will be more suitable. In this mode the liquid and gas flow concurrently downwards. Most of the hydrotreatment of heavy residues is performed in trickle bed reactors. The major concern is to guarantee a complete wetting of the catalyst particles, because dry regions will cause extensive coking. When more extensive hydrogenation occurs (often from the hydrogenation of aromatic rings) interstage cooling is often applied.

Typical process conditions are:

Temperature:	350–420°C
Pressure:	40–100 bar
LHSV:	1–6 h^{-1}

The severity of the processing conditions depends on the feed; for light petroleum fractions it will be milder than for heavy residues. Moreover, it is common practice to compensate for deactivation of the catalyst by increasing the temperature of the reactor. A simplified flow scheme involving a trickle flow reactor is shown in Fig. 2.5.

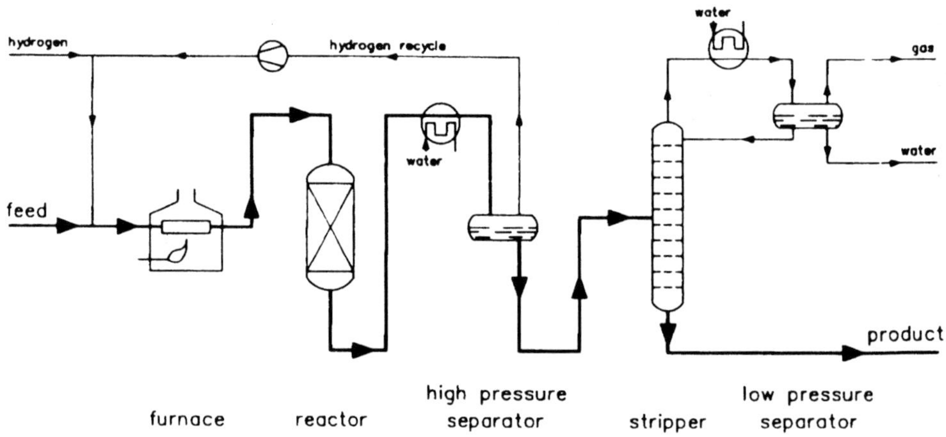

Fig. 2.5. Simplified scheme of hydrotreatment involving a trickle flow reactor.

Catalysts

Catalysts are mixed metal sulphides on a carrier. The major examples are supported mixed phases of CoS and MoS_2, of NiS and MoS_2 and NiS and WS_2. There is as yet no full agreement on the structure. For the first catalyst the best description is small MoS_2 clusters with CoS units at the edges. Typical dimensions of these clusters are 1 nm and a stacking between 1 and 3.

Environment

From the point of view of clean technology hydrotreatment is gaining increasing interest. A recent example is diesel fuel. Sulphur-containing compounds in diesel fuel cause serious difficulties in catalytic cleaning of the exhaust gases. Undoubtedly, in the near future, novel processes will be needed for the deep desulphurization of diesel.

2.3 TOTAL ISOMERIZATION PROCESS OF PARAFFINS

Overall Reaction

Hydroisomerization of C_5 and C_6 paraffins (250°C, Pt–H–mordenite as the catalyst) combined with separation of normal and branched paraffins (over zeolite CaA).

Feedstock

Straight run C_5–C_6 paraffins ('light ends') from refinery distillation.

Product Use

As gasoline component with improved octane number, cf. Table 2.1.

Scale

Over 25 plants are operating the total isomerization process (TIP) and this number is growing continuously. Many millions of tons of isomerized paraffins are produced annually by this process.

Process Description

The TIP process is a combination of Shell's Hysomer process and Union Carbide's ISOSIV process. Both processes are zeolite based.

In the Hysomer process (Fig. 2.7) the catalyst is a Pt(O)-loaded zeolite of the H–mordenite type. The temperature applied has a strong bearing on the composition of the equilibrium mixture obtained, see Fig. 2.6 for the hexane isomers: the lower the temperature applicable, the more dimethylbutanes are present. The temperature dependence is understood and it will be clear that a low temperature is favourable for the purpose of the process.

Hydroisomerization is thought to start with dehydrogenation of paraffins, giving carbenium ions and alkenes. Secondary carbenium ions can rearrange to tertiary cations, probably via protonated cyclopropane derivatives. Transfer of hydride towards the tertiary carbenium ions provides the isomerized branched paraffins (see Chapter 4, Section 4.4). A side reaction is C–C scission of carbenium ions leading to cracked products. Hydrogenolysis on the metal function can also contribute to this undesired reaction. The larger the alkane, the more hydrocracking takes place, which may be the reason that the TIP process so far has focused on the C_5 and C_6 fractions.

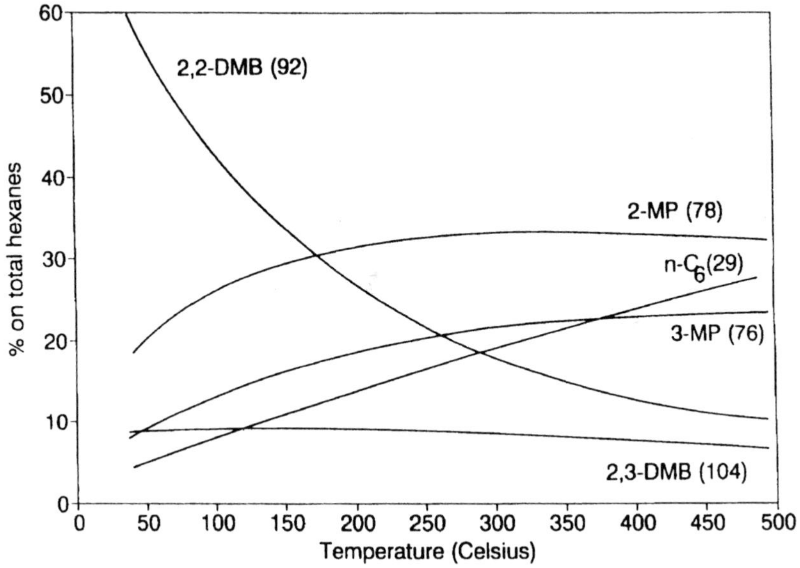

Fig. 2.6. Thermodynamic equilibria for hexane hydroisomerization (RON).

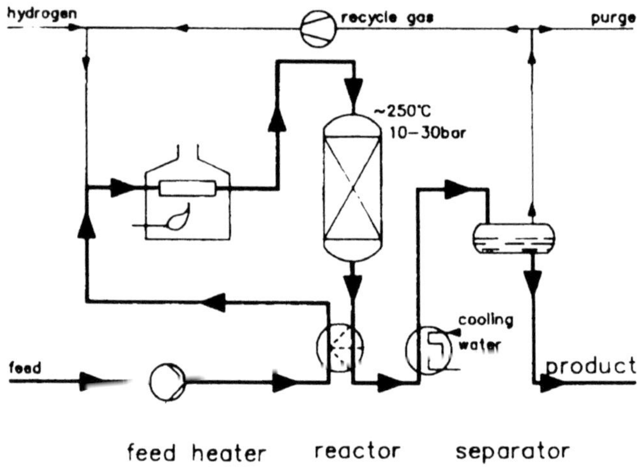

Fig. 2.7. Diagram of Shell Hysomer process.

It should be noted that H-mordenite itself is also able to isomerize alkanes such as pentane. However, the presence of platinum greatly improves its selectivity and stability, as shown in Fig. 2.8. Apparently the hydrogenation function prevents oligomerization etc. leading to coke formation.

In the TIP process the Hysomer process is combined with the ISOSIV process which separates normal paraffins from branched ones by selectively adsorbing the normal fraction into zeolite CaA (pressure swing adsorption). After de-

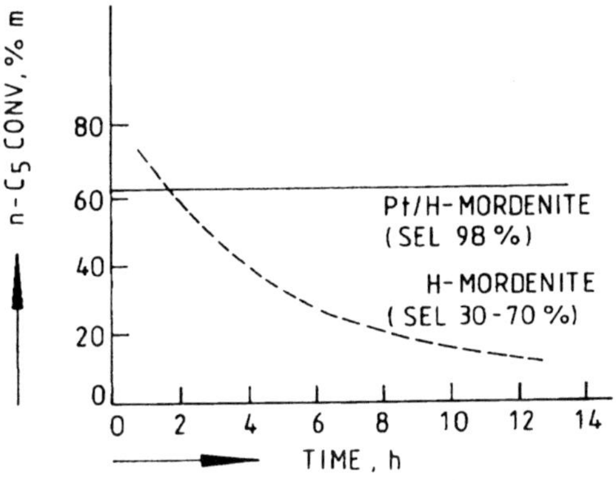

Fig. 2.8. Influence of metal function in pentane hydroisomerization.

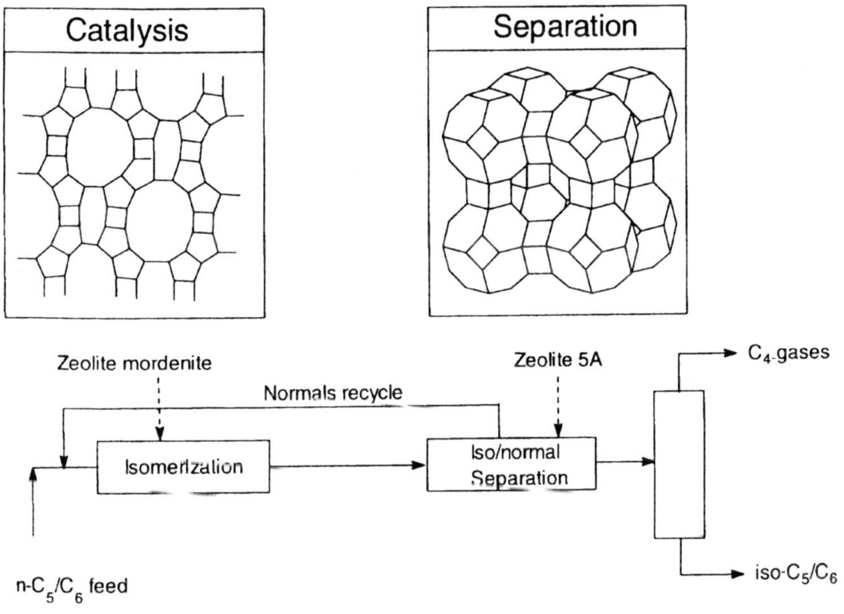

Fig. 2.9. Paraffin total isomerization process (TIP).

sorption (by applying vacuum) the normal paraffins are recycled. A schematic view of the TIP process, including a structural picture of the two zeolites involved, is given in the Fig. 2.9.

The pore openings of zeolite A just allow normal alkanes to enter the pore system of the crystal. It might be of interest to have a zeolite available which adsorbs the normals and the mono methyl-branched paraffins and leaves the high octane dimethyl branched compounds unadsorbed.

Catalyst and Adsorbent

Mordenite is a versatile zeolite, the Si/Al ratio of which can easily be changed from 5 to higher values. Its parallel channels having a somewhat elliptical cross section (0.65×0.7 nm) allow all C_5 and C_6 isomers to enter and to leave. Platinum is introduced by ion exchange with $Pt(NH_3)_4^{2+}$. Competitive exchange with NH_4^+ (excess) is recommended in order to obtain a high Pt dispersion. A careful calcination follows in which local high temperatures in the zeolite crystal have to be prevented. Finally the Pt catalyst is reduced with hydrogen.

The selective adsorbent CaA is prepared in the sodium form (the detergent zeolite NaA) and subsequently partially exchanged with Ca(II).

2.4 ISOTACTIC POLYPROPYLENE

Overall Reaction

$$\text{Propene} \xrightarrow[\text{titanium catalyst}]{} \text{Polypropylene}$$

Isotactic polypropylene is a stereoregular polymer with all methyl groups pointing in the same direction when the backbone is stretched. In syndiotactic polypropylene the methyl groups alternate along the chain. The atactic polymer lacks regularity. All three forms retain a perfect head-to-tail structure.

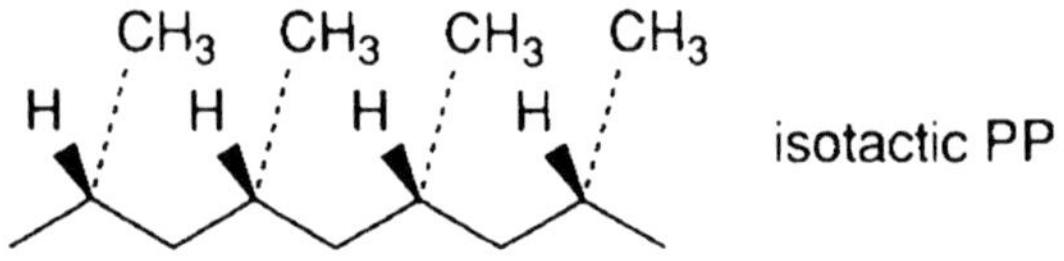

Fig. 2.10. Isotactic PP.

Feedstock

Propene from oil refinery crackers.

Product Use

Polypropylene is a plastic which finds many end uses with a variety of product property requirements. In general, isotactic polypropylene has a high tensile strength, a low density, a good heat and abrasion resistance, and it has a nice looking optical gloss. The melting point is 162°C, the density is 0.9 g/cm^3. The chemical properties, determined in the manufacture, include stereoregularity, molecular weight and molecular weight distribution. The process gives a powder that is first extruded into pellets (nibs). During the processing many organic and inorganic chemicals may be added, such as antioxidants, UV stabilizers, fillers, reinforcing fibres, dyes, etc. Processing may involve injection moulding, film extrusion, and pipe extrusion. The processing parameters in the second transformation—temperature, cooling rate, etc.—determine the mechanical properties of the consumer product. The product has a crystallinity of 60%. The crystalline spherulites may have different morphological forms which strongly depend on the processing conditions.

Scale

Twenty million tons (1996) of polypropylene are produced annually. In 1973 this was only 3 million tons.

Process Description

The performance of the catalyst is of crucial importance to the design of the plant. Figure 2.11 shows the scheme of a high-yield liquid process for which the catalyst was pioneered by Montedison and Mitsui. The catalyst is a titanium tetrachloride supported on magnesium dichloride which, in the presence of various modifying donor compounds, is reduced by triethylaluminium. This leads to a highly active catalyst, a product with a high stereoregularity, and a high bulk density. The propene feed has to be thoroughly purified. Traces of contaminants such as water, oxygen, carbon monoxide and sulphur compounds would destroy the catalyst. The components are fed to the reactor (60–70°C, 10 bar), a CSTR (see Section 8.3.3). The reaction is highly exothermic (2500 kJ/kg, from gaseous C_3H_6). Cooling coils are used in slurry processes and the feed may be additionally cooled. In the second step the unreacted propene is removed by

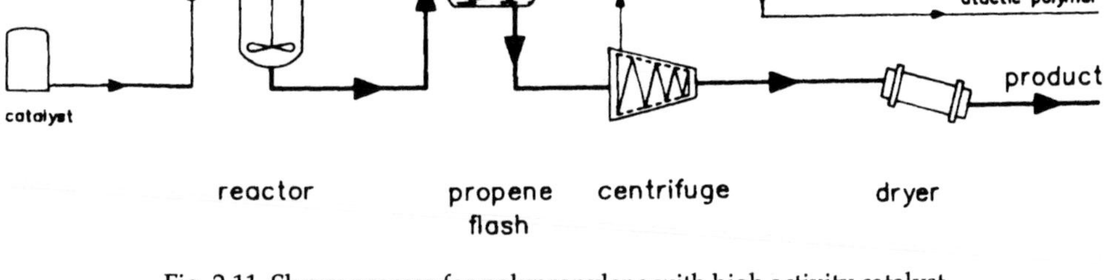

Fig. 2.11. Slurry process for polypropylene with high activity catalyst.

flashing and recycled. The solid isotactic product is separated by centrifugation. Depending on the catalyst, the liquid from the centrifugation stage may contain dissolved atactic polypropylene which is collected by removal of the solvent with steam. The isotactic product from the centrifugation is dried and the solvent is recycled after thorough drying. In this process the catalyst concentrations are low and no catalyst removal ('de-ashing') is needed. The fluffy powder is very sensitive to oxidation and a stabilization package is added to the product before it is exposed to air. The dried powder is pelletized in an extruder.

Alternatives

The similar, older slurry process uses a less active catalyst. The monomer is dissolved in isooctane, the titanium catalyst and aluminium cocatalyst are added and this mixture is fed to the reactor which is maintained at 70°C. The inorganic corrosive (Cl) residues are removed in a washing step with alcohols. The atactic material is removed by extraction. A third process employs propene as the liquid in combination with a high activity catalyst. The Himont 'Spheripol' process, which uses spherical catalyst particles, gives spherical polymer beads of millimetre size that need no extrusion for certain purposes. A more recent development is the gas-phase polymerization using an agitated bed. All processes are continuous processes, where the product is continuously removed from the reactor. Over the years we have seen a reduction of the number of process steps. The process costs are very low nowadays, propene feed costs amounting to more than 60% of the total cost.

The Catalyst

The catalyst was discovered by Ziegler who mixed $TiCl_4$ and aluminium alkyls and thus polymerized ethene. Natta recognized the peculiar structural properties that a substituted alkene might give to the polymer. Commercial production of both polymers with the new catalysts started very soon after its discovery in 1954. Titanium is reduced to the trivalent state and it is generally accepted that propene insertion into the titanium alkyl bond leads to the polymer. The stereoregularity is determined by the geometry of the site of the catalytic centre. This is known as the Cossee–Arlman mechanism. The solid $TiCl_3$ catalysts contain different sites, some of which give atactic and some of which give isotactic polymer. The major improvement in the second generation catalysts was in the preparation of a high surface area $TiCl_3$, free of cocrystallized $AlCl_3$. The modern high-yield catalysts consist of very small crystallites of $MgCl_2$ with supported $TiCl_4$ and a Lewis base, e.g. dimethyl phthalate. They are reduced by $AlEt_3$.

The productivity of the catalyst has been the topic of continuous research. Table 2.2 gives a review of the development of the catalyst in the last four decades.

TABLE 2.2

Development of PP catalyst activity

PP catalysts	Year	Yield (kg/g)	Isotacticity (%)
$TiCl_3$, 1st generation	1954	4	92
$TiCl_3$, 2nd	1971	16	96
$MgCl_2$, $TiCl_4$, 3rd	1975	325	92
$MgCl_2$, $TiCl_4$, 3rd	1981	1300	96
$MgCl_2$, $TiCl_4$, 3rd	1998	5000	98

In addition to the molecular properties, the macroscopic growth of the polymer particles is important; they should be sufficiently large and dense, otherwise there is no means of collecting and processing the product. What is more, the capacity of the plant also increases with the bulk density of the product. The theoretical bulk density for ideal uniform spheres is 540 kg/m^3. Values obtained in modern plants are of the order of 500 kg/m^3. The morphology is also determined by the solid catalyst preparation. The morphology of the macroscopic catalyst particles is 'reproduced' by the growing polymer particles.

2.5 CATALYSTS FOR AUTOMOTIVE POLLUTION CONTROL

Automobiles contribute significantly to air pollution, especially in urban areas. Typical concentrations in the exhaust gases of a gasoline engine are given in Fig. 2.12.

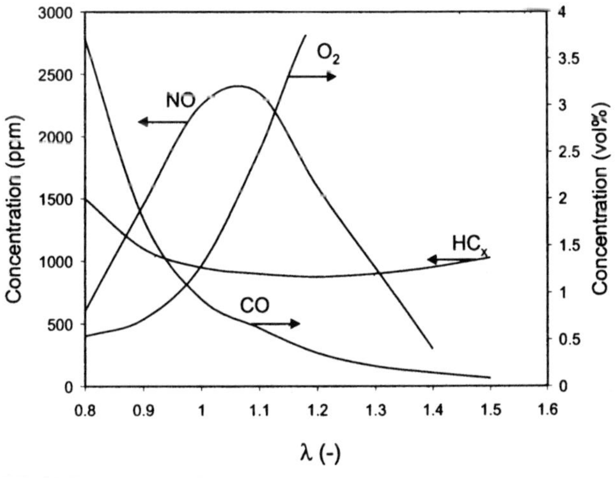

Fig. 2.12. The concentration of CO, NO, HC, and O_2 emitted by a gasoline engine as a function of λ (the dimensionless air-to-fuel ratio).

Here λ is the dimensionless air to fuel ratio ($\lambda = 1$ corresponds to stoichiometric combustion). For low values, the amount of air is substoichiometric and a lot of CO and hydrocarbons are produced. At higher values of λ ($\lambda > 1$), combustion is more complete and much less unburned hydrocarbons and CO are emitted. However, the temperature in the cylinder increases and, as a consequence, more NO_x is emitted. At still higher values of λ, because of the large excess of air, the temperature drops and less NO_x is formed. Operation under a large excess of air is referred to as 'lean' burning. With respect to the amounts of CO and NO formed, lean burning is very attractive. However, lean burning also has disadvantages: emission of unburned hydrocarbons increases and ignition gives problems. The latter problem can be solved by applying a higher compression ratio. Recently, an attempt has been made to develop and manufacture engines working at high λ values. If lean-burning engines can be used, only oxidation of hydrocarbons would be needed in the catalytic converter.

Catalysis has been remarkably successful in reducing emissions. Table 2.3 shows typical emissions for Otto (gasoline) engines with and without a 'three-way catalyst'. Data for diesel engines are also included.

Clearly, catalysis essentially solves the problem of the air pollution. The data illustrate the good performance of diesel engines as far as the emissions of NO_x and CO are concerned. Until recently, the NO_x emission was below the legislation limits in the developed countries. More recent limits are stricter, however. Also, it is clear that for diesel engines the emission of particulates (mainly soot) is very large compared to that of Otto engines. As diesel engines are considerably more efficient than Otto engines, intensive research is being carried out in order to eliminate the environmental problems of diesel engines. Both primary measures (engine modifications) and secondary measures (catalyzed NO_x reduction and soot combustion) are adopted. Diesel emissions are also specifically associated with the presence of SO_2. SO_2 emissions are low in gasoline engines because gasoline is thoroughly desulphurized. For diesel fuel this is not yet the case, but it is expected that this will change in the near future.

TABLE 2.3

Emission of particulates, hydrocarbons, NO_x, and CO for gasoline engines, gasoline engines with a catalyst and diesel engines

Engine	Particulates (mg/mi)	Hydrocarbons (g/mi)	NO_x (g/mi)	CO (g/mi)
Otto engines without catalyst	15–30	2.2–3.8	2.6–8.0	13–33
Otto engines with three-way catalyst	5–10	0.1–0.3	0.2–0.6	1–4
Diesel engines	200–650	0.1–0.8	1.0–2.0	1–3

mi = mile, measured for 18 engines in a standard driving cycle (FTP); data from Volkswagen, 1988.

Catalysts

For three-way catalysts in converters for Otto engines Pt/Rh supported on La-stabilized alumina are used. Promoters are present, e.g., CeO_2. The alumina is usually situated as a 'washcoat' layer in a monolith (typically with a cell density of 400 cells/in^2) consisting of cordierite. In practice, the catalysts appear to be very stable. In fact, the durability of the electronic control system (for the regulation of the λ value in the converter) is more critical than the catalyst itself.

Recently, in academic circles and in industry, a renewed interest can be observed in the development of other metal catalysts, e.g., Cu, Ag, and Ce-based catalysts. If lean-burning systems were to become available, oxidation would be the only reaction to be catalyzed and catalyst composition will certainly be changed accordingly. Catalytic converters are still in the development stage for diesel engines.

2.6 ETHENE OXIDE

Overall Reaction

$$\text{Ethene} + \tfrac{1}{2}O_2 \longrightarrow \text{ethene oxide}$$

Side reaction:

$$\text{Ethene} + 3\,O_2 \xrightarrow{\text{silver catalyst}} 2\,CO_2 + 2\,H_2O$$

$$(250°C,\ 15\ \text{bar}).$$

Feedstock

Ethene from oil refinery crackers.

Product Use

The main use of ethene oxide is as ethylene glycol, half of which is used as anti-freeze and the other half is used in the production of polyesters. Other products are glycol ethers, polyurethanes and poly(ethylene glycols). We find these in many consumer products such as fibres, foils, bottles, solvents, plasticizers, adhesives, detergents, brake fluids, etc.

Scale

Ten million tons of ethene oxide are produced annually.

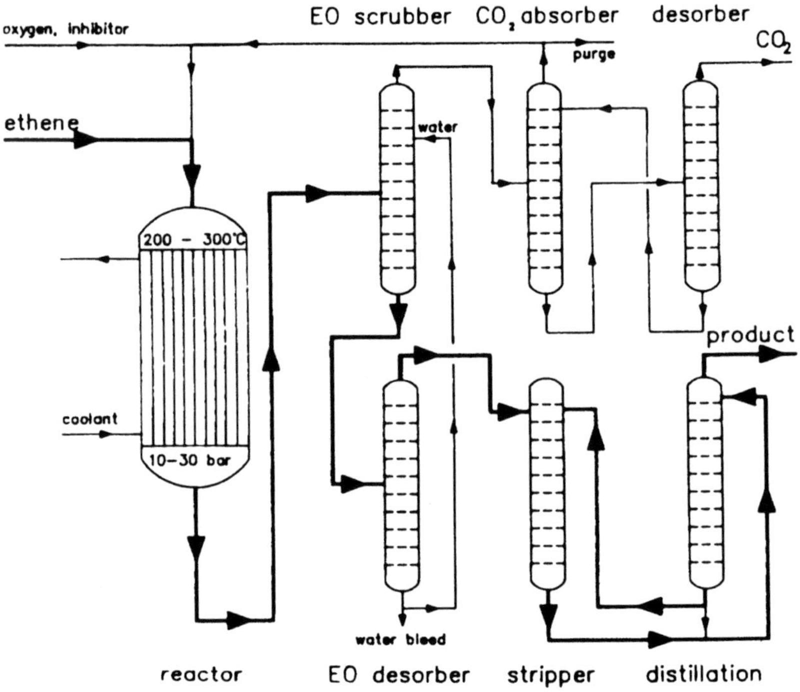

Fig. 2.13. Oxidation of ethene with oxygen (heat exchangers omitted).

Process Description

Here we will describe the process using pure oxygen. The catalyst used is a silver catalyst supported (±10%) on a highly pure porous alumina with a low surface area (~1 m^2). The side reaction is highly exothermic (1300 kJ/mol of ethene under the conditions given, and with a selectivity of 80% most of the heat is produced by the side reaction) and the reactor therefore consists of a huge bundle of pipes of a few centimetres in diameter and several meters long to obtain good heat transfer. The tubes are cooled by water or a high-boiling hydrocarbon. The heat of the heat transfer liquid is used to heat the reactants and for steam generation. The mass conversion of ethene is of the order of 10%. The concentration of ethene oxide in the gas that leaves the reactor is only 1 or 2% which is necessary for suppression of complete combustion. The inlet gas is made up from the recycle gas and, in addition to ethene (±25%) and oxygen (±8%), contains carbon dioxide, argon and methane. After the gas is cooled it is sent to a scrubber where ethene oxide is removed with water. It is then led through a basic solution to remove the carbon dioxide. The ethene oxide (b.p. 10.8°C at 100 kPa) is removed from the water. Likewise the carbon dioxide is removed in a desorber.

Alternatives

The alternative process uses air instead of oxygen. The operation of this system is totally different since now a large amount of inert nitrogen would end up in the recycle. Therefore a large amount of the gas is vented, which simultaneously circumvents the removal of carbon dioxide. The conversion per pass, however, must be much higher, and the amount of ethene in the off-gas is still too high for direct venting into the atmosphere and hence a secondary reactor is needed.

Another process, the chlorohydrin process, is now obsolete. On a weight basis this process produced more calcium chloride than ethene oxide!

The Catalyst

The silver catalyst was discovered by Lefort in 1931. Today's catalyst is much more active and selective. It contains silver particles with a diameter of 0.1–1 μm. Alkali metal salts, cesium for instance, may be added as promoters. The feed also contains organic chlorides such as vinyl chloride, which suppress the side reaction. The selectivity in commercial plants has never exceeded 86%, according to published records. This has long been regarded as a magic number, since it was argued that only half of the dioxygen molecule could be utilized for epoxidation, and that the other oxygen atom should be removed from the catalyst by 'burning' ethene. Laboratory experiments have now proved that this is not necessarily the case and that, in principle, higher selectivities may be expected (see Section 5.4.2).

Propene oxide cannot be made directly from propene and oxygen with this catalyst since the total combustion predominates. Instead, it is produced using hydroperoxides as the oxidant, see the next section (Section 2.7).

2.7 STYRENE AND PROPYLENE OXIDE (SMPO PROCESS)

Overall Reactions

The process involves four steps (see Fig. 2.14):

(a) alkylation of benzene with ethene using H-ZSM-5 (Mobil-Badger Process, gas phase, 420°C, 15 bar);

(b) autoxidation of ethylbenzene to the corresponding hydroperoxide (EBHP, 130°C, liquid phase, 30 mbar);

(c) liquid phase catalytic epoxidation of propylene with EBHP, affording propylene oxide and 1-phenylethanol (110°C, liquid phase, Ti on silica, or homogeneous Ti or Mo catalyst);

(d) dehydration of 1-phenylethanol over a titania catalyst at 180–280°C to give styrene.

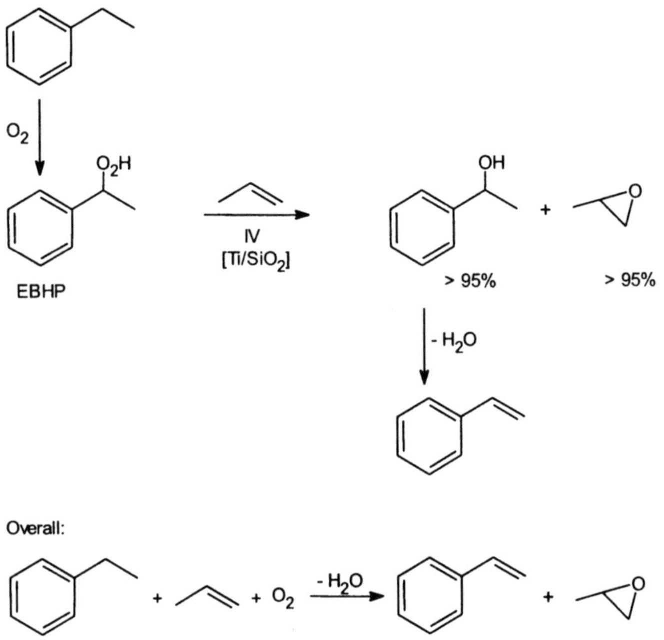

Fig. 2.14. The SMPO process.

Feedstock

Ethene, benzene, oxygen and propene.

Product Use

Styrene is the monomer for polystyrene, styrene–butadiene rubber and ABS (copolymer of acrylonitrile, butadiene, and styrene). Propylene oxide is mainly used for the manufacture of propylene glycol and polyurethanes.

Scale

About 14 million tons of styrene and 3.5 million tons of propylene oxide are produced annually, worldwide. Most of the styrene is still produced by dehydrogenation of ethylbenzene and it is estimated that ca. 1 million tons of styrene and 500,000 tons of propylene oxide are produced by the SMPO process.

Process Description

A fundamental problem in the alkylation of benzene is that the product is more

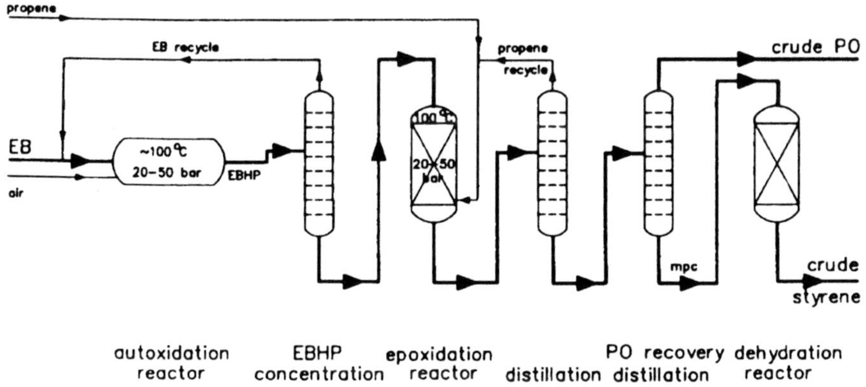

Fig. 2.15. Simplified flow diagram of the SMPO process.

reactive than benzene, thus giving di- or tri-alkylated products. The latter have to be recycled. The reaction is operated at conversion levels of 40% with ethene: benzene ratios of 0.35 to 0.55. The Mobil-Badger Process gives low conversions to diethyl benzenes compared to older processes employing Lewis acids.

Two variations of step (c) are used, the only essential difference being the catalyst employed in the epoxidation step. In the Arco (Atlantic Richfield) process a homogeneous molybdenum catalyst is used. The Shell process employs a heterogeneous titanium/silica catalyst.

The stoichiometry of the reaction is such that ca. 2 tons of styrene are produced per ton of propene oxide. The SMPO route to styrene is 'cleaner' and much less energy intensive than the conventional dehydrogenation route. The amount of styrene that can be produced is limited by the amount of propylene oxide that can be sold. The flow diagram for the process is shown in Fig. 2.15.

The Catalyst

The catalyst for the alkylation is ZSM-5 in the H^+ form. This zeolite is prepared using the tetrapropylammonium ion as the template. After the synthesis the template is removed by calcination. Several other zeolites, H-Y, H-MCM-22, and H-Beta are industrially applied for the ethylation and isopropylation of benzene. Also a highly de-aluminated H-mordenite has been reported to be an excellent alkylation catalyst.

Alternatives

An alternative to the use of the ethylbenzene hydroperoxide is tertiary-butyl hydroperoxide obtained from isobutane and dioxygen. The product alcohol is converted with methanol to MTBE (methyl tertiary butylether).

2.8 HIGHER OLEFINS

Overall Reaction

$$\text{Ethene} \xrightarrow{\text{nickel complex}} \alpha\text{-olefins}$$

$$\text{lowest + highest fraction} \xrightarrow{\text{isomerisation}} \text{internal olefins (low and high fraction)}$$

$$\xrightarrow{\text{metathesis}} \text{full range internal olefins}$$

Feedstock

Ethene from oil refinery crackers.

Product Use

The desired α-alkenes in the C_{10}–C_{20} range are used as feedstock in the production of detergent alcohols and in the synthesis of lubricants, the C_6-C_{10} alkenes are used as plasticizer alcohols in phthalates for PVC, and as comonomers in low-density polyethylene, and a variety of other products. The internal by-products are mainly used for the production of alcohols via a cobalt-catalyzed hydroformylation.

Scale

The typical size of the Shell process plant described here is 250,000 tons per year. The total production of higher olefins includes several alternative routes and is estimated to be 2 million tons annually. A large part of the alkenes is produced for captive use.

Process Description

Many transition metal hydrides will polymerize ethene to polymeric material or, alternatively, dimerize it to butene. Fine-tuning of these catalysts to one that will give a mixture of, say, predominantly C_{10}–C_{20} oligomers is not at all trivial. Nickel complexes have been extensively studied by Wilke and his coworkers for their activity as alkene oligomerization catalysts. In the late sixties Keim and coworkers at Shell discovered a homogeneous nickel catalyst which selectively oligomerizes ethene to higher homologues. A simplified flow diagram is shown in Fig. 2.16. The catalyst is prepared in a prereactor from nickel salts with boron hydrides as the reducing agent under a pressure of ethene, and a ligand is added. Polar solvents, such as alcohols (1,4-butanediol), are used for the dissolution of the catalyst. The catalyst solution and ethene are led to the reactor, a stirred

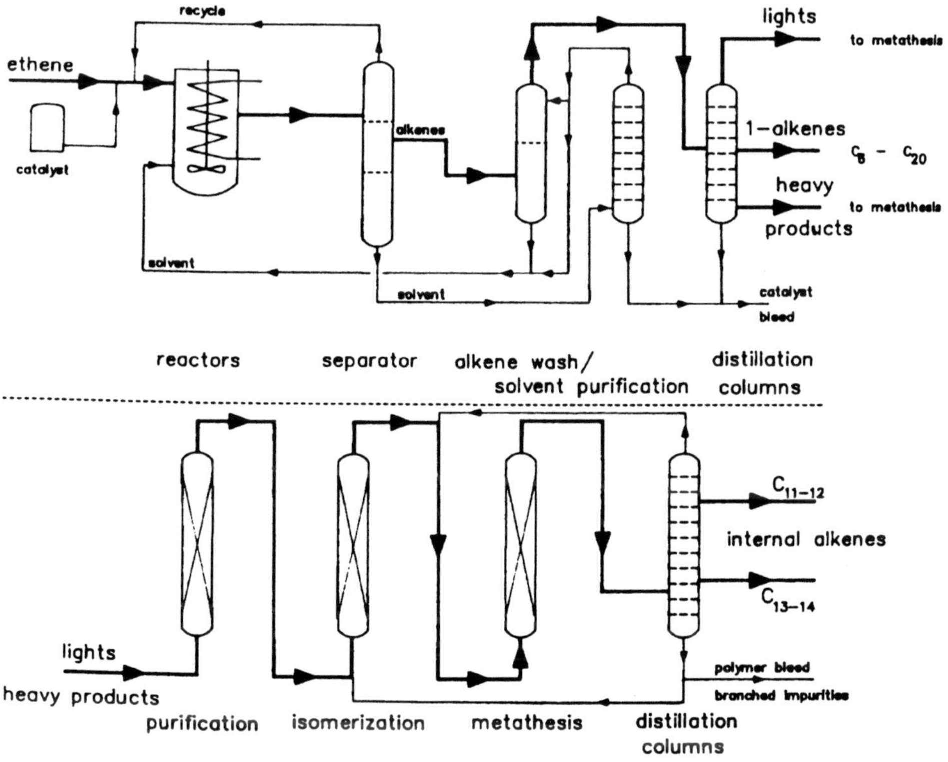

Fig. 2.16. Shell higher olefins process.

autoclave, which is maintained at 80–120°C and 100 bar of ethene. The product alkenes are insoluble in the alcohol and phase separation takes place. After settling, the alcohol layer goes to a regeneration unit. The alkene layer is washed and ethene is recycled to the reactor. The products are distilled and the desired fractions are collected. The lower alkenes, the composition of which depends on the demand, and the heavy alkenes must be 'disproportionated' to give the full range of alkenes. Direct metathesis of the higher and lower 1-alkenes is not the way to go, since metathesis of e.g. C_{22} and C_4 1-alkenes would give ethene and an internal C_{24}, which is no improvement. The light and heavy alkenes are sent to an isomerization reactor after having passed a purification bed, a simple absorbent to remove alcohol and ligand impurities. The isomerized mixture is then passed over a commercial molybdenum metathesis catalyst (CoMo), also in a fixed bed reactor, to give a broad mixture of internal alkenes. After distillation the C_{11-13} fraction (±15%) is used as a feedstock for alcohol production via cobalt catalysts (hydroformylation and reduction). The product is now a mixture of internal, linear alkenes. The bottom and the top ends of this mixture are again recycled, and eventually all material leaves the isomerization/metathesis plant as internal C_{11-13} alkenes. This 'by-product' constitutes a large part of the production, and only half of the production or less amounts to the more valuable 1-alkenes. In spite of the high selectivity of the oligomerization reaction, there is a build-up of the branched alkenes in the isomerization–metathesis recycle, which may be as high as 50%. A bleed stream of the heavies controls the amount of branched material and the amount of (continuously growing) polymers.

Alternatives

Two aluminium-based processes are in use. The one-step process is operated at high temperature and with aluminium alkyls as the catalyst. This process resembles the nickel process discussed above; as a matter of fact Ziegler's group was working with this aluminium catalyst when they accidentally discovered the influence of nickel which, under these drastic conditions, gave the unwanted butenes! The second aluminium-based process is a two-step process that employs aluminium stoichiometrically in each pass; the alkyls are grown at a lower temperature, decomposed in the next reactor at higher temperature, and the aluminium complexes are recycled. This procedure leads to 'peaking' in the C_{10}–C_{16} range.

Higher alkenes can be obtained from thermal cracking of wax, and although a thermodynamic mixture of internal alkenes might have been expected, the wax-cracker product contains a high proportion of 1-alkenes, the kinetically controlled product. For the cobalt-catalyzed hydroformylation the nature of the alkene mixture is not relevant, but for other derivatizations the isomer composition is pivotal to the quality of the product. Another process involves the catalytic dehydrogenation of alkanes over a platinum catalyst.

Catalyst

The oligomerization catalyst is a homogeneous nickel complex of a phosphino-carboxylic acid. The other catalysts are fixed bed catalysts: a Co/Mo oxide for the metathesis reaction and a solid catalyst for the isomerization.

2.9 RHODIUM CATALYZED HYDROFORMYLATION OF PROPENE

Overall Reaction

Propene + H_2 + CO $\rightarrow$ butanal and 2-methylpropanal
(120°C, 30 bar, rhodium complex as the catalyst).

Feedstock

Propene from oil refinery crackers, and syngas from coal or natural gas.

Product Use

Butanal is reduced to 1-butanol, which is used as a solvent. Butanal is dimerized with base and hydrogenated to give 2-ethyl-1-hexanol which is used as the phthalate ester in PVC as a plasticizer. Butanal is used for the manufacture of a wide variety of other products.

Scale

The estimate is that 6 million tons of butanal are produced annually via the process described here.

Process Description

The low pressure oxo process was jointly developed by Union Carbide, Davy Powergas (Davy McKee), and Johnson Matthey. The latter two companies possess the rights of the Wilkinson patents. The catalyst is a rhodium complex, with a large excess of triphenylphosphine. Temperature and pressure have to be controlled very carefully because the linearity strongly depends on these parameters. Ligand and rhodium (300 ppm) are very sensitive to impurities and the feed must be very thoroughly purified. The rhodium stays in the reactor, apart from a small purification cycle, employed for the reactivation of inactive rhodium complexes, the removal of heavy ends and ligand decomposition products that slowly build up. The gases are led into the reactor at the bottom via a sprinkler (see Fig. 2.17). The alkene conversion per pass is estimated to be 30%,

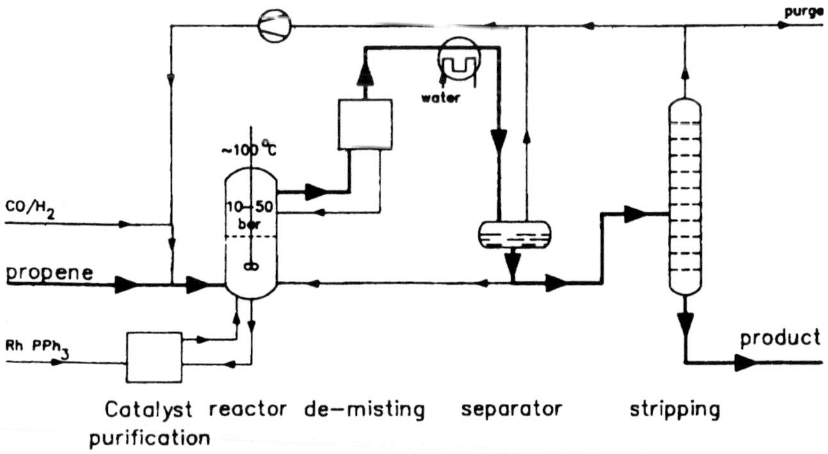

Fig. 2.17. Low pressure rhodium hydroformylation.

and the propene is recycled. Only propene is hydroformylated with the use of this process; high-boiling alkanals cannot be removed with the gases in this manner. The demister is a crucial part of the system; the fine droplets which might be present in the gas will, of course, contain rhodium. In view of the high costs this should be transferred back into the reactor very rigorously. After the product has condensed and separated from the gaseous feedstock in the separator the butanal, containing a few percent 2-methylpropanal, is led to the distillation column. The flow to the reactor from the separator must maintain the minimum liquid level in the reactor. The small catalyst recycle is necessary because slow decomposition of the ligand occurs. Triphenylphosphine decomposes in the rhodium catalyzed process to a phenyl fragment and a diphenylphosphido fragment. Diphenylphosphide forms very stable, though inactive rhodium complexes. Impurities in the feed may also cause the formation of inert rhodium complexes. A variety of reagents can be applied for the regeneration of the inactive complexes to the active $HRh(CO)(PPh_3)_3$ precursor.

Alternatives

The alternative processes include cobalt-catalyzed hydroformylation and similar rhodium-based processes. Hydroformylation with cobalt requires much higher temperatures (140–170°C) and pressures (70–200 bar). The activity ratio of rhodium and cobalt may be of the order of 1000 but the costs of the metals counterbalance this effect. The selectivity of the rhodium catalyst is of the order of 99% or higher, whereas certain cobalt catalysts may show a selectivity as low as 90%. The by-products are alkanes, through hydrogenation of the feedstock, and heavy ends through condensation of the aldehyde products. Cobalt still is the preferred catalyst for the hydroformylation of internal higher alkenes to detergent aldehydes (alcohols). In the eighties a new process has come on stream employing a two-phase system with rhodium in a water phase and the substrate and the product in an organic phase. The catalyst used is a rhodium complex with a sulphonated triarylphosphine, which is highly water soluble. The process has been commercialized by Ruhrchemie for the production of butanal from propene, after the initial work had been done by workers at Rhone-Poulenc.

Recently, Union Carbide has announced the replacement of the triphenylphosphine ligand by bulky phosphites. The new catalysts lead to higher reaction rates, selectivity and catalyst stability and may also be of interest for the hydroformylation of 2-butenes.

The Catalyst

With all the characteristics of the cobalt process in mind, it can be seen that there was a sizable incentive to improve on this performance. Fundamental work by

Wilkinson demonstrated that rhodium triphenylphosphine catalysts allowed the operation of the hydroformylation reaction at much lower pressure (1 bar was reported by Wilkinson) which consequently lead to the hope of much lower capital and operating costs. The selectivity was also reported to be considerably higher; virtually no hydrogenation was observed and the linearity was in some cases as high as 90%. A convenient catalyst precursor is $RhH(CO)(PPh_3)_3$. Under ambient conditions this will slowly convert 1-alkenes into the expected aldehydes. Internal alkenes undergo hardly any reaction and, in contrast to the cobalt catalyst, it has little isomerization activity. At higher temperatures pressures of 10 bar or more are required. Unless a large excess of ligand is present the catalyst will also have some isomerization activity, but the internal alkenes thus formed will not be hydroformylated. This makes the rhodium catalyst less suitable for the conversion of alkenes other than propene. Propene hydroformylation can be performed with a linearity ranging from 60 to 95%, depending on the phosphine concentration. At very high phosphine concentration the rate is low, but the linearity attains its maximum value. The commercial process presumably operates around 30 bar, at 120°C, at high phosphine concentrations, giving linearities around 92%. The estimated turnover frequency of mols of product per mol of rhodium complex per hour is of the order of 300. Low ligand concentrations, with concomitant low linearities (70%), will give turnover frequencies of the order of 10,000 at 10 bar and 90°C.

2.10 METHANOL SYNTHESIS

Overall Reaction

$$CO + 2\,H_2 \rightleftharpoons CH_3OH \qquad \Delta H = -90.8\ \text{kJ/mol}\ (298\ \text{K},\ 101.3\ \text{kPa})$$
$$CO_2 + 3\,H_2 \rightleftharpoons CH_3OH + H_2O \qquad \Delta H = -49.6\ \text{kJ/mol}$$

Feedstock

Synthesis gas (CO/H_2) manufactured by reforming of a hydrocarbon, usually natural gas.

Product Use

Methanol is one of the most important chemicals. The major application is in the chemical industry, where it is used as a solvent or as an intermediate. It is increasingly used in the energy sector. A survey of the most important reactions is given in Fig. 2.18.

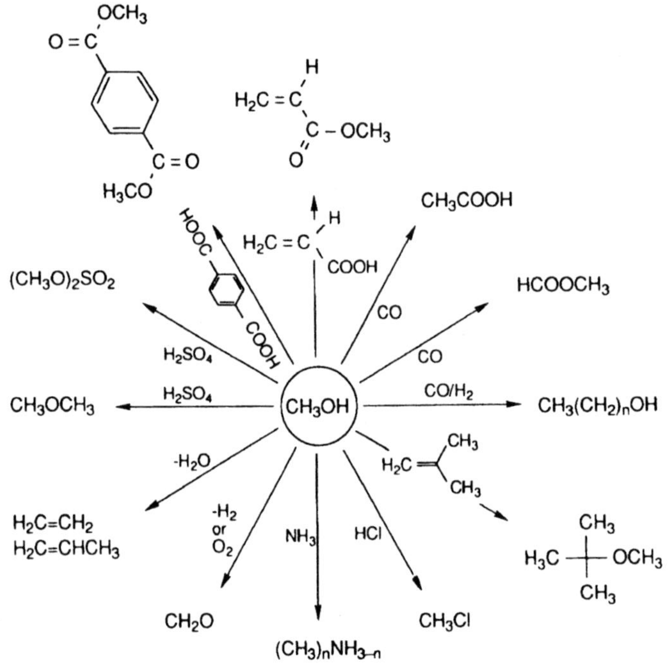

Fig. 2.18. Reactions of methanol.

Scale

Methanol is a major bulk chemical. In 1989 the production capacity exceeded 21×10^6 ton/y.

Process Description

The thermodynamic equilibrium is most favourable at high pressure and low temperature. The methanol synthesis process was developed at the same time as NH_3 synthesis. In the development of a commercial process for NH_3 synthesis it was observed that, depending on the catalyst and reaction conditions, oxygenated products were formed as well. Compared with ammonia synthesis, catalyst development for methanol synthesis was more difficult because selectivity is crucial besides activity. In the CO hydrogenation other products can be formed, such as higher alcohols and hydrocarbons that are thermodynamically favoured. Figure 2.19 illustrates this.

It is clear that methanol is less stable than many possible by-products, such as methane. The catalyst has to be selective. The selectivity of modern catalysts is above 99%. The original catalysts were active only at high temperature (300–400°C). The pressure applied was 25–35 MPa. Until the end of the 1960s basically

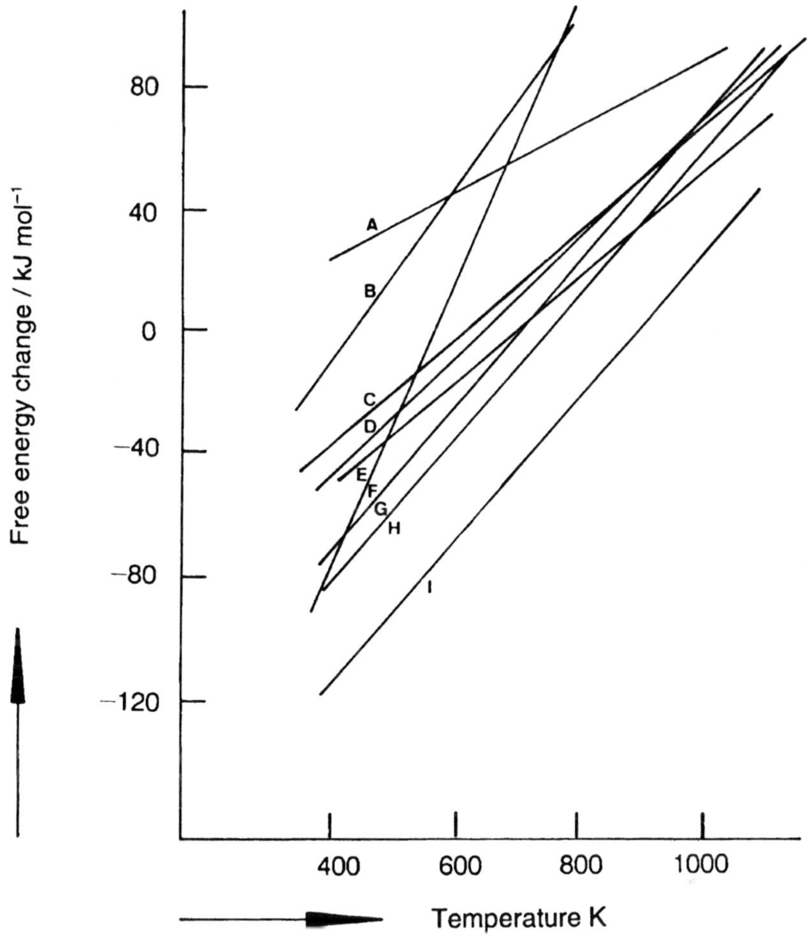

Fig. 2.19. Standard free energies of formation of hydrocarbons and alcohols from carbon monoxide and hydrogen with water as by-roduct. A, ethanol; B, methanol; C, acetylene, D, benzene; E, propylene; F, ethylene; G, propane; H, ethane; I, methane [1].

the original catalyst was used. More active catalysts were known, but they were not resistant to impurities such as sulphur. In modern plants the synthesis gas is very pure and very active catalysts can be used. This has led to 'low pressure plants' (260°C, 50–100 bar). The temperature is critical. A low temperature is favourable from a thermodynamic point of view.

Methanol production plants consist of three parts:

- production of synthesis gas;
- synthesis of methanol;
- processing of the crude methanol produced.

The composition of synthesis gas depends on the feedstock used. When naphtha is the raw material, the stoichiometry is about right. However, when CH_4 is used, H_2 is in excess. In practice the excess of H_2 is burnt as fuel or CO_2 is added to the process in order to convert all H_2 into methanol. In general the last approach is the preferred one.

Most modern processes are low pressure processes. Plant capacities range from 150–3000 tons/day. The plants differ mainly in reactor design and, interrelated with this, in the way the heat produced by the reaction is removed. In the ICI process an adiabatic reactor is used with a single catalyst bed. The reaction is quenched by adding cold reactant gas at different heights in the catalyst bed. The temperature profile in the bed has a sawtooth profile. A flow scheme of the ICI process is given in Fig. 2.20.

The product mixture leaving the reactor flows through heat exchangers in order to heat the reactant mixture to the desired temperature. Subsequently, the mixture is further cooled and raw methanol condenses while the unconverted synthesis gas is recycled to the reactor. The raw methanol is purified by distillation.

In the Lurgi process a cooled tube reactor is applied. The catalyst particles are located in the tubes and cooling takes place by boiling water. The most important difference between the two reactor types is the temperature profile. In the Lurgi reactor it is much flatter than in a quench reactor.

In the Haldor Topsoe process several reactors are used, arranged in series. The heat of reaction is removed by intermediate coolers. The synthesis gas flows radially through the catalyst beds.

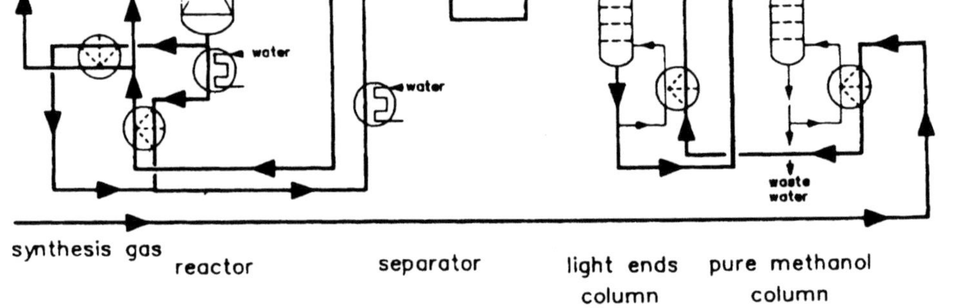

Fig. 2.20. The ICI low-pressure methanol process.

Deactivation will always occur. The usual practice (increasing temperature) is not recommended because of the sensitivity of the catalyst to sintering. Hence, the pressure is increased to compensate for the decline in activity.

Alternatives

Although a classical process, research and development are still being carried out, aiming at the realization of novel processes. These are based either upon shifting the equilibrium by removing the product or by developing a dramatically improved catalyst, permitting operating at much lower temperature.

Catalysts

Originally, a catalyst based on chromia and zinc oxide was developed. In fact, during the first 40 years this catalyst was applied without fundamental changes. In 1966 ICI introduced a $Cu/ZnO/Al_2O_3$ catalyst with much higher activity. This catalyst allowed reaction temperatures lower than 300°C, affording a commercial process at much lower pressure: 50–100 bar instead of the 250–350 bar used in the older process.

2.11 MALEIC ANHYDRIDE

Overall Reaction

$$C_4H_{10} + 3.5\ O_2 \rightarrow C_4H_2O_3 + 4\ H_2O \qquad \Delta H = -1260\ kJ/mol^{-1}$$

Feedstock

Butane and butenes from refinery crackers and natural gas. Since 1985 butane was introduced to replace benzene as the feedstock. The first commercial production, based on benzene, started in 1930.

Product Use

Maleic anhydride is an important intermediate in the chemical industry [2–3]. It is used in polycondensation and addition reactions. The end products of these reactions are polyesters, alkyd resins, lacquers, plasticizers, copolymers and lubricants. For example, the copolymer of styrene and maleic anhydride is an engineering plastic.

Scale

The typical size of the plant of the process described here is 20,000–30,000 tons per year. World production is 0.5 million tons per year.

Process Description

The commercial processes are based on fixed-bed technology. Tubular reactors are used in order to obtain good heat transfer. A fixed-bed reactor, however, is not optimal. The thermal characteristics of the reaction network are highly unfavourable; the reactions are exothermic, and intermediate products are much more reactive than butane while hot spots lead to total combustion. Low feed concentrations and an extensive heat transfer surface area are required. The butane concentration in the feed is only 1.8%, which is below the explosion limit. The butane process produces more water than the benzene process and only 35% of the maleic anhydride yield can be recovered directly from the product stream. The remainder is condensed together with water leading to maleic acid. This is then dehydrated at >130°C. The product is purified by fractional distillation. Alternatively, maleic anhydride is condensed from the product stream in a high boiling organic solvent such as phthalic acid or tetra- or hexahydrophthalic acid. The waste gases are sent to an incineration unit.

Catalysts

The basis of the commercial success was the development of highly selective catalysts [4–6], i.e. the vanadium phosphorus oxide (VPO) catalyst which is prepared from V_2O_5 digested in a refluxing acidic solution (aqueous HCl or butanol or benzyl alcohol) and orthophosphoric acid. The precipitate is filtered off and dried below 200°C. The solid is calcined in air at 400°C. A high mechanical strength is required for VPO used in a fluid-bed operation and therefore silica is also incorporated. After calcination a thin layer of silica is located at the periphery of the particles which are now very durable while the silica layer is porous to reactants and products. There are many promoters for the VPO catalysts, the most important being Zn, U, Ti, Co, Mo, Mg an Zr. There seems to be a general consensus that there is an optimum averaged valence state of vanadium, which is slightly higher than 4 (the alcohols may act as reductors during the preparation). Figure 2.21 gives a schematic view of a spray-dried particle prepared from colloidal silica and polysilicic acid.

The elucidation of the kinetic network has received a lot of attention. The following simplified scheme has been proposed:

$$C_4H_{10} \rightarrow C_4H_8 \rightarrow C_4H_6 \rightarrow \text{furan } (C_4H_4O) \rightarrow \text{MA} \rightarrow CO_2 + H_2O$$

From the extensive literature it can be concluded that a Mars–van Krevelen mechanism applies; the oxidation of the hydrocarbons is achieved using the surface oxygens and the reoxidation of the catalysts is accomplished by reduction of oxygen at separate sites.

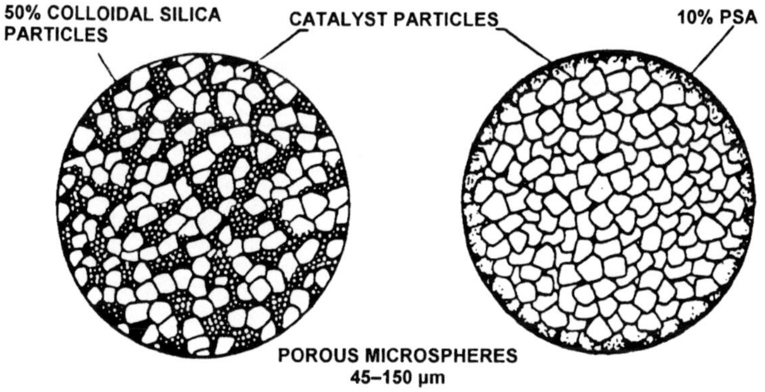

Fig. 2.21. Spray dried particles prepared from 50% colloidal silica (~20 nm diameter) and 10% silica derived from polysilicic acid [7].

Alternatives

In the past the most important process was based on the partial oxidation of benzene. Typical production units are 10,000–20,000 tons/y, which is smaller than those using butane. Benzene gives rise to smaller amounts of water and more maleic anhydride can be produced in the first condensation step.

Fluidized-bed reactors have been proposed as alternatives to circumvent the problems mentioned above (ALMA process, Alusuisse, Lummus). These allow better control of the temperature and higher feed concentrations. In addition the reactor system is less prone to explosions. A major problem in this reactor is attrition of the catalyst. In recent literature this problem has received a great deal of attention. A second drawback of the fluidized-bed reactor is the intrinsic mixing behaviour, which lowers the conversion.

An interesting option (Monsanto, Du Pont) involves the use of a riser reaction as shown in Fig. 2.22. The configuration is analogous to a modern FCC unit (see Fig. 2.3). In the riser reactor the (oxidized) catalyst transfers oxygen to the butane substrate giving maleic anhydride. The catalyst is separated from the product in a stripping zone. Subsequently the catalyst is reoxidized in the fluid-bed reactor. Data from pilot studies indicate that the catalyst is able to transfer an amount of oxygen to the substrate equivalent to a monolayer capacity, thus confirming the Mars–van Krevelen mechanism for this reaction. Separation of the two steps enables a fine tuning of the reaction conditions:

- the oxidation is performed in the absence of molecular oxygen which improves the selectivity (75% has been reported [8]). Moreover the reaction mixture is always out of the explosion zone.
- residence times in the reactor and the regenerator can be chosen independently and, as a consequence, optimal conditions for selectivity can be obtained.

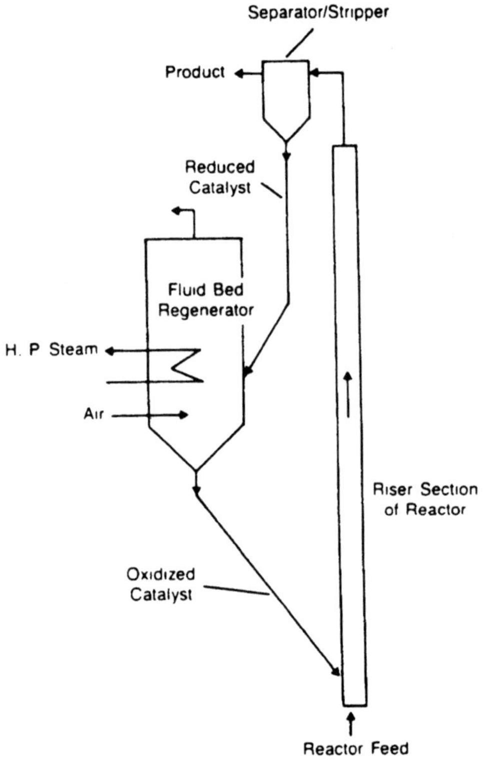

Fig. 2.22. Riser reactor for selective oxidation [4].

2.12 METHYL *t*-BUTYL ETHER (MTBE)

Overall Reaction

Addition of methanol to isobutene to yield MTBE (80–110°C, 7–20 bar, ion exchange resin in the H^+ form as the catalyst).

Feedstock

Isobutene is present in refinery streams. Especially C_4 fractions from catalytic cracking are used. Such streams consist mainly of *n*-butenes, isobutene and butadiene, and generally the butadiene is first removed by extraction. For the purpose of MTBE manufacture the amount of C_4 (and C_3) olefins in catalytic cracking can be enhanced by adding a few percent of the shape-selective, medium-pore zeolite ZSM-5 to the FCC catalyst (see Fig. 2.23), which is based on zeolite Y (large pore). Furthermore, two routes lead from *n*-butane to isobutene (see Fig. 2.24); the isomerization/dehydrogenation pathway (upper route) is

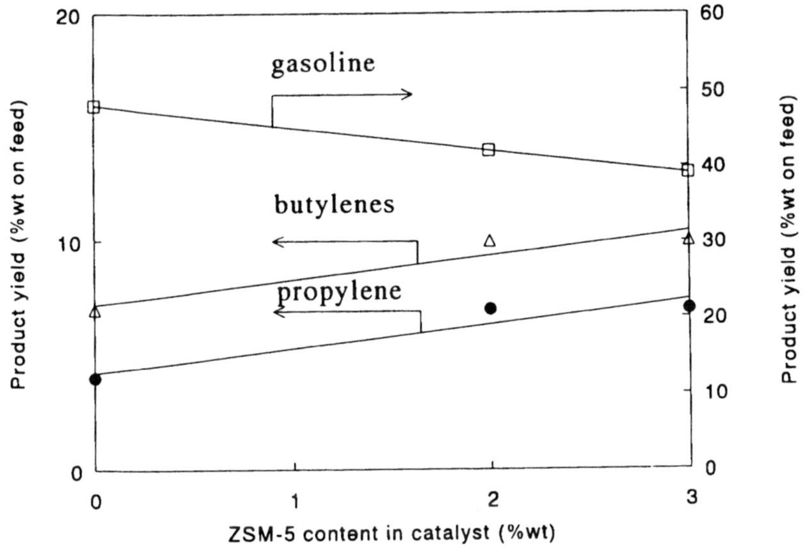

Fig. 2.23. Product yields in catalytic cracking vs. ZSM content in FCC catalyst.

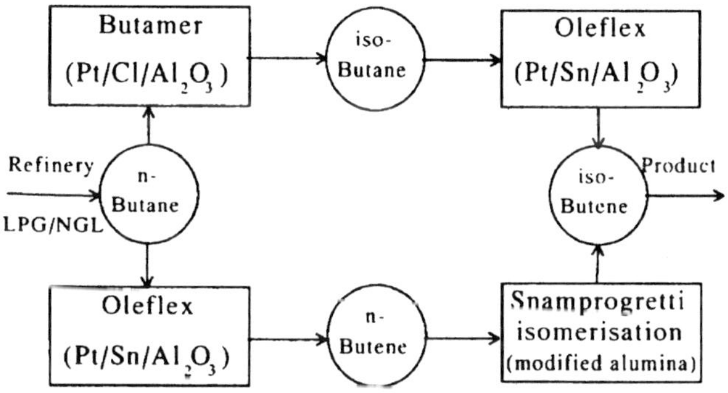

Fig. 2.24. Two routes for isobutene from butane.

industrially practised. For *n*-butene/isobutene isomerization the zeolite ferrerite is an excellent catalyst (Shell). Finally, isobutene is also industrially obtained by dehydration of *t*-butyl alcohol, formed in the Halcon process (isobutane/propene to *t*-butyl alcohol/propene oxide). The latter process has been mentioned as an alternative for the SMPO process (see Section 2.7).

Product Use

The main use of MTBE is as an octane booster in gasoline formulations. Table 2.1 (Section 2.2.1) compares octane number and boiling points of some tertiary ethers and hydrocarbons. The volatility is another important property of gasoline components. In fact, the lower volatility of ETBE is an advantage with respect to MTBE. Another (smaller scale) application of MTBE is the synthesis of high purity isobutene by cracking MTBE over amorphous silica–alumina. This isobutene serves as a monomer for polyisobutene.

Scale

MTBE is the fastest growing chemical of the last decade. Presently, over 11 million t/y capacity has been installed. Further growth is foreseen though at a lower growth rate. Essentially the success of MTBE is founded (i) on its outstanding antiknock properties as a gasoline component, and (ii) its function as a key product in the C_4 hydrocarbon processing within a refinery.

Process Description

The synthesis of MTBE is carried out in the liquid phase over a fixed bed of ion exchange resin in the H^+ form. The rate of reaction of isobutene with methanol is much higher than that of the *n*-butenes (isobutene forms a relatively stable tertiary carbenium ion in the first step), which enables the selective conversion of isobutene in the presence of the *n*-butenes. In fact, streams with an isobutene content as low as 5% can be converted.

A C_4 fraction and methyl alcohol (molar ratio methyl alcohol:*i*-C_4 = 1:1) is preheated to about 70°C before being sent to the reactor containing a fixed bed of sulphonated polystyrene resin. The reaction is exothermic ($\Delta H_{298} = -37.7$ kJ/mol) and the heat produced is removed by means of cooling jackets so as to keep the temperature below 120°C. The reaction mixture from the top of the reactor is distilled and any unreacted butene is collected overhead with the azeotropic amount of methyl alcohol. The bottom contains pure MTBE. The distillate, together with additional methyl alcohol, is passed to a second reactor. The products from the second stage are extracted with water to remove residual methyl alcohol. The water–methyl alcohol solution is distilled to recover methyl alcohol, which is recycled.

The second stage in the process is required because the MTBE formation is an equilibrium reaction. The temperature needed (~100°C) to achieve a sufficiently high *rate* of conversion means a decrease in isobutene equilibrium conversion ($X_{iB} = 0.9$ at 65°C, $X_{iB} = \sim 0.75$ at 100°C). The main side reaction in the MTBE process is the dimerization of isobutene towards di-isobutene (two isomers). Side reactions with essentially no significance are the formation of *t*-butyl alcohol (due to the presence of water as feed impurity), the formation of dimethyl ether from methyl alcohol, and the oligomerization of isobutene towards tri- and tetramers. A (three stage) process is also in operation which tolerates butadiene. The butadiene/methyl alcohol reaction is faster than that of the *n*-butenes but consider-

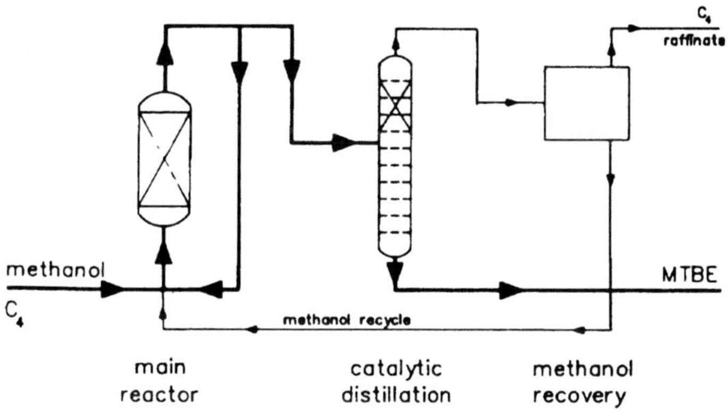

Fig. 2.25. Flow scheme of the MTBE process with catalytic distillation.

ably slower than that of isobutene, so high isobutene selectivity can still be obtained. Similar liquid phase processes, as described above for MTBE, are recommended for the reaction of isopentenes and methanol to yield TAME and for the conversion of isobutene and ethanol to give ETBE. The use of higher iso-olefins — as for TAME — leads to somewhat lower yields of methyl *t*-alkyl ethers than are obtained for MTBE. As an alternative to fixed bed catalytic conversion the formation of MTBE by 'reactive distillation' is applied. Here *in situ* separation of feed and product components is aimed at, leading to a higher isobutene conversion. A flow scheme is given in Fig. 2.25.

The Catalyst

The catalysts used industrially in the MTBE process are sulphonated polystyrene resins of the macroreticular type. These strongly acidic materials are prepared by copolymerizing styrene and *p*-divinylbenzene in the presence of an organic compound that is a good solvent for the monomers but a poor swelling agent for the polymer. Opaque round particles are obtained, which are subjected to sulphonation.

Amberlyst-15 (A15) is often used, and is prepared by using some 20% (wt.) of divinylbenzene in the formulation. The highly crosslinked product has a surface area of ~50 m^2/g and an exchange capacity of 4–5 meq/g resin.

In the preparation of MTBE it is accepted that methyl alcohol adsorbs more strongly but diffuses more slowly in the domain of the resin than the less polar isobutene. In order to keep the side reaction (isobutene to di-isobutene) at a low level, relatively small resin particles should be used.

Other catalysts for the isobutene/methyl alcohol reaction include: sulphuric acid (homogeneous), heteropolyacids (homogeneous or heterogeneous) and zeolites (heterogeneous).

2.13 CAPROLACTAM

Feedstock

Benzene or toluene via cyclohexanone (see Fig. 2.27).

Product Use

Caprolactam is the raw material for Nylon-6 (see Fig. 2.26).

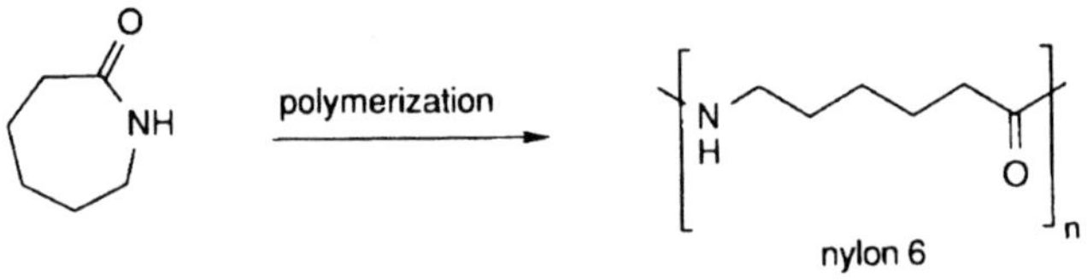

Fig. 2.26. Conversion of caprolactam into Nylon-6.

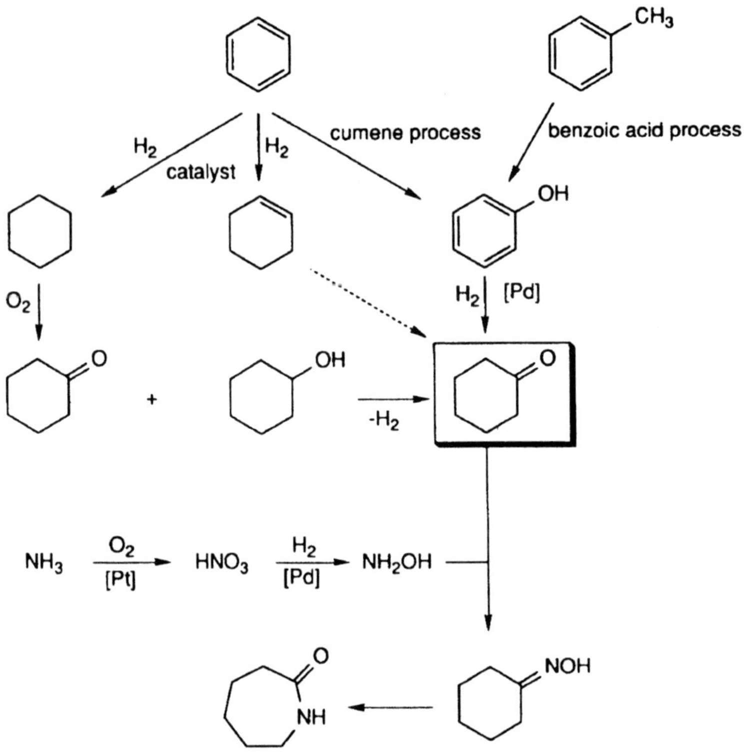

Fig. 2.27. Caprolactam manufacture.

Scale

Worldwide annual production is ca. 2 million tons.

Process Description

Several processes are used for the industrial production of caprolactam. Generally cyclohexanone is the key intermediate and it is produced by catalytic hydrogenation of phenol (*ex* benzene or toluene) or the catalytic autoxidation of cyclohexane (from benzene hydrogenation) as shown in Fig. 2.27.

In the phenol hydrogenation process phenol is fed in the gas phase with hydrogen at 140–170°C through a catalyst bed at atmospheric pressure. The catalyst generally consists of 0.2–0.5 wt.% palladium on a zeolite carrier. The yield exceeds 95% at quantitative conversion. Figure 2.28 shows the flow diagram for the process.

In the cyclohexane oxidation route cyclohexane is oxidized with air at 125–126°C and 8–15 bar in the liquid phase using Co or Mn naphthanates as the catalyst. This affords a mixture of cyclohexanol and cyclohexanone via a classical free radical autoxidation mechanism. Cyclohexane conversion is limited to 10–12% in order to minimize by-product formation via further oxidation. The selectivity to cyclohexanol/cyclohexanone is 80–85%.

The unreacted cyclohexane is distilled off and recycled. The ca. 1:1 mixture of cyclohexanol and cyclohexanone is then subjected to dehydrogenation over a palladium catalyst (the same catalyst as is used in phenol hydrogenation) to give pure cyclohexanone.

In classical processes cyclohexanone is converted to the corresponding oxime by reaction with hydroxylamine (see Fig. 2.27). The oxime subsequently affords

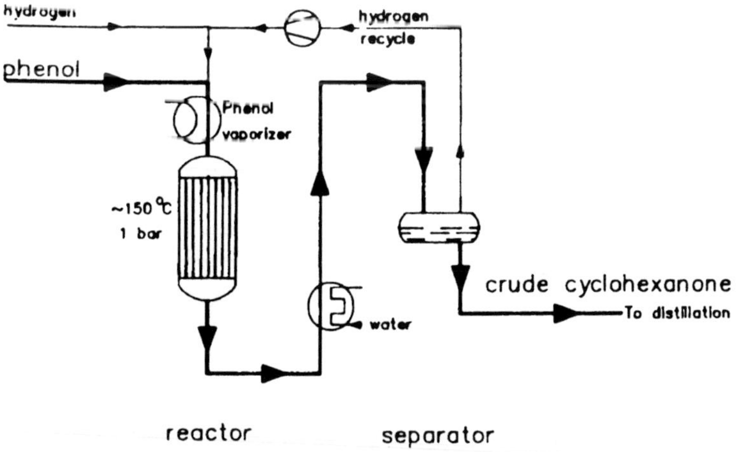

Fig. 2.28. Flow diagram for cyclohexanone production from phenol.

caprolactam via the Beckmann rearrangement with sulphuric or phosphoric acid. Alternatively, in a more recent development which has just been commercialized, a mixture of cyclohexanone, ammonia and hydrogen peroxide is directly converted to cyclohexanone oxime over a titanium(IV)–silicalite (TS-1) catalyst. Here the catalyst converts ammonia and hydrogen peroxide *in situ* to hydroxyl amine. This route is more direct than the classical route and reduces the amount of salt formation but it involves the use of a more expensive oxidant (H_2O_2 rather than O_2). Another new development comprises the methoxycarbonylation of butadiene, followed by hydroformylation of the pentenoic ester. On the basis of the cost of starting materials this route looks most promising.

2.14 VITAMIN A INTERMEDIATES

Feedstock

The major feedstocks in the modern routes to key vitamin A intermediates are butadiene, isobutene and formaldehyde.

Product Use

Vitamin A is widely used as a pharmaceutical and a food and feed additive.

Scale

Worldwide production of vitamin A is ca. 3000 tons.

Process Description

Two key intermediates in the production of vitamin A are citral and the so-called C_5 aldehyde. In the modern routes to these intermediates, developed by BASF and Hoffmann-La Roche, catalytic technologies are used (see Fig. 2.29 and 2.30). Thus, in the synthesis of citral, the key intermediate is 2-methyl-1-butene-4-ol, formed by acid-catalyzed condensation of isobutene with formaldehyde. Air oxidation of this alcohol over a silver catalyst at 500°C (the same catalyst as is used for the oxidation of methanol to formaldehyde) affords the corresponding aldehyde. Isomerization of 2-methyl-1-butene-4-ol over a palladium-on-charcoal catalyst affords 2-methyl-2-butene-4-ol. The latter is then reacted with the aldehyde from the oxidation step to form an enol ether. Thermal Claisen rearrangement of the enol ether gives citral (see Fig. 2.29).

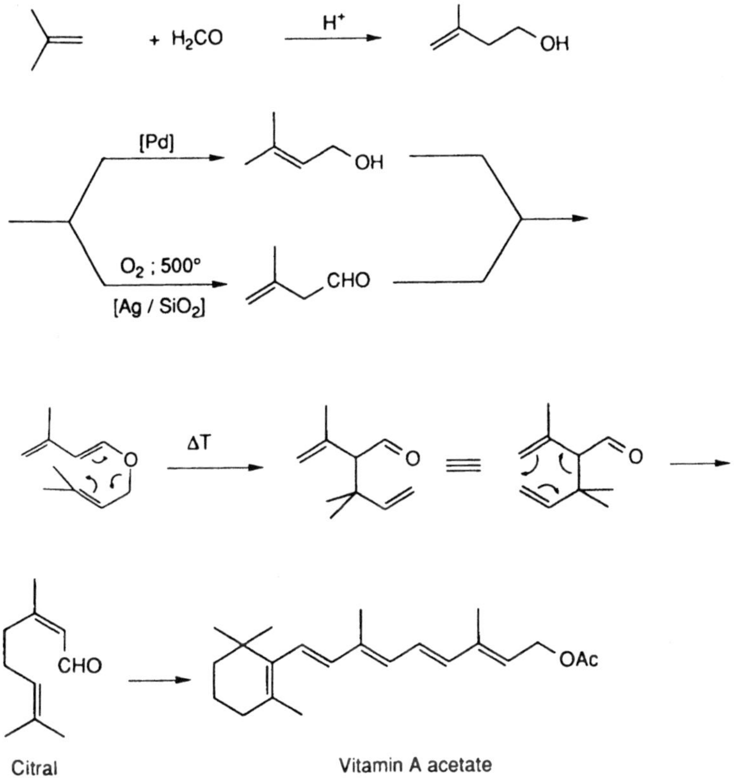

Overall: $2\ C_4H_6 + 2\ H_2CO + 1/2\ O_2 \longrightarrow C_{10}H_{16}O + 2\ H_2O$

Fig. 2.29. BASF route to citral.

The C_5 aldehyde intermediate is produced from butadiene via catalytic oxidative acetoxylation followed by rhodium-catalyzed hydroformylation (see Fig. 2.30). Two variations on this theme have been described. In the Hoffmann-La-Roche process a mixture of butadiene, acetic acid and air is passed over a palladium/tellurium catalyst. The product is a mixture of *cis*- and *trans*-1,4-diacetoxy-2-butene. The latter is then subjected to hydroformylation with a conventional catalyst, $RhH(CO)(Ph_3P)_3$, that has been pretreated with sodium borohydride. When the aldehyde product is heated with a catalytic amount of *p*-toluenesulphonic acid, acetic acid is eliminated to form an unsaturated aldehyde. Treatment with a palladium-on-charcoal catalyst causes the double bond to isomerize, forming the desired C_5-aldehyde intermediate.

In the BASF process the 1,2-diacetate is the substrate for the hydroformylation step. It can be prepared either directly via oxidative acetoxylation of butadiene

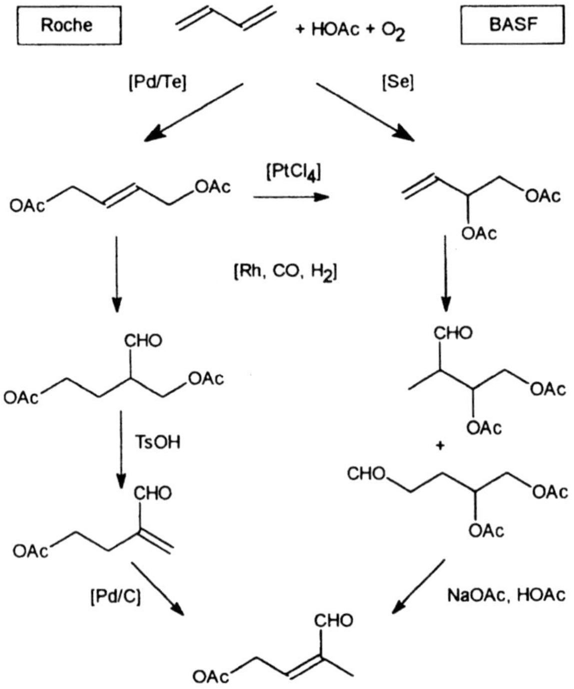

Fig. 2.30. Synthesis of vitamin A intermediate.

using a selenium catalyst or via $PtCl_4$-catalyzed isomerization of the 1,4-diacetate (see above). The latter reaction affords the 1,2-diacetate in 95% yield. The hydroformylation step is carried out with a rhodium catalyst without phosphine ligands since the branched aldehyde is the desired product (phosphine ligands promote the formation of linear aldehydes). Relatively high pressures and temperatures are used and the desired branched aldehyde predominates. The product mixture is then treated with sodium acetate in acetic acid to effect selective elimination of acetic acid from the branched aldehyde, giving the desired C_5 aldehyde.

2.15 IBUPROFEN

Feedstock

p-Isobutylbenzene (see Fig. 2.31).

Product Use

Ibuprofen is an over-the-counter drug that is widely used as an antiinflammatory and a general analgesic, where it competes with aspirin.

Scale

Worldwide annual production is of the order of 7000–8000 tons which is very large for a pharmaceutical.

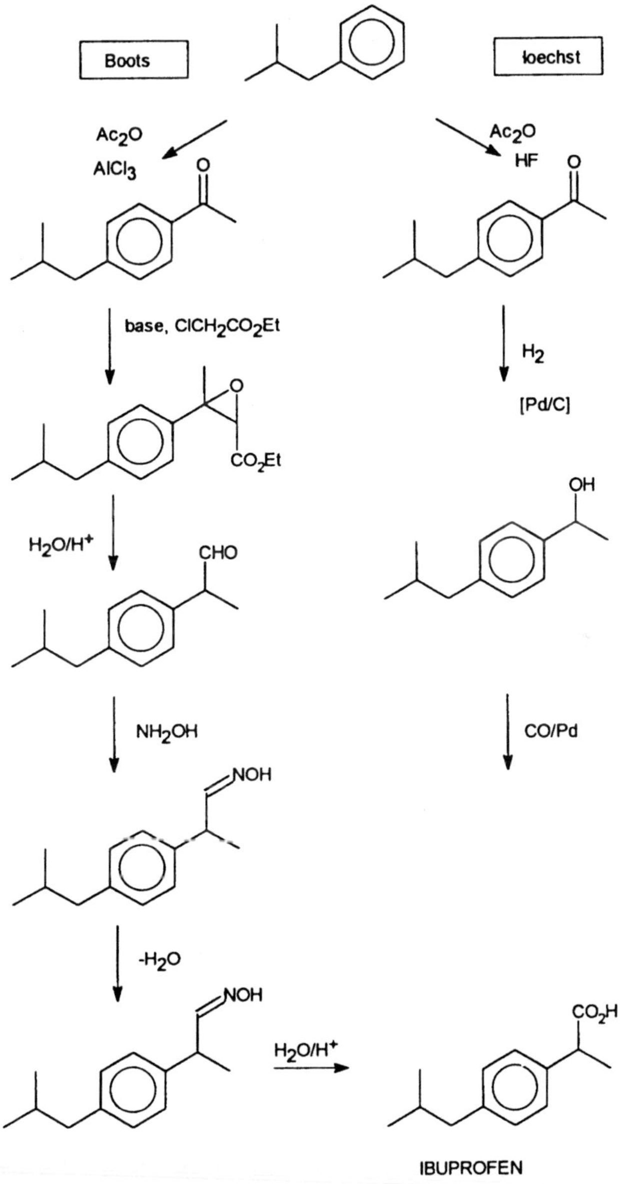

Fig. 2.31. Two routes for the production of ibuprofen.

Process Description

There are basically two processes for the production of ibuprofen (see Fig. 2.31). Both involve a Friedel–Crafts acylation as the first step. In the classical route, developed by Boots, the *p*-isobutylacetophenone is subsequently converted to ibuprofen in five steps involving conventional organic chemistry. In the catalytic route, more recently developed by Hoechst, the *p*-isobutylacetophenone is converted in two steps.

The first step is catalytic hydrogenation to the corresponding alcohol over a 5% palladium-on-charcoal catalyst in methanol solvent at 30°C and 7 bar. The selectivity is 97% at >99% conversion. In the second step the alcohol is carbonylated at 35 bar CO and 125–130°C in the presence of a $PdCl_2(Ph_3P)_2$ and HCl as catalyst. Selectivities to ibuprofen are ca. 70% at 99% conversion.

The catalytic route has the advantages of being shorter and producing considerably lower amounts of inorganic salts than the classical route.

2.16 AMINOPENICILLANIC ACID (6-APA)

Feedstock

Major feedstock is penicillin G, obtained by fermentation using *Penicillium chrysogenum* strains upon addition of phenylacetic acid (see Fig. 2.32).

Product Use

6-APA is an intermediate for the production of semi-synthetic penicillins that are widely used as antibiotics [9]. The most important derivatives are ampicillin and amoxycillin, obtained by coupling 6-APA with D-phenylglycine and D-*p*-hydroxyphenylglycine, respectively. They represent a pharmaceutical value of about $ 4 billion.

Scale

From the 25,000 tons of Penicillin G produced worldwide per year, more than 20,000 tons are converted into 6-APA, yielding ca. 10,000 tons of 6-APA.

Process Description

A major problem in the conversion of Penicillin G into 6-APA is to keep the reactive beta-lactam amide bond intact upon the desired hydrolysis of the secondary amide bond (see Fig 2.32).

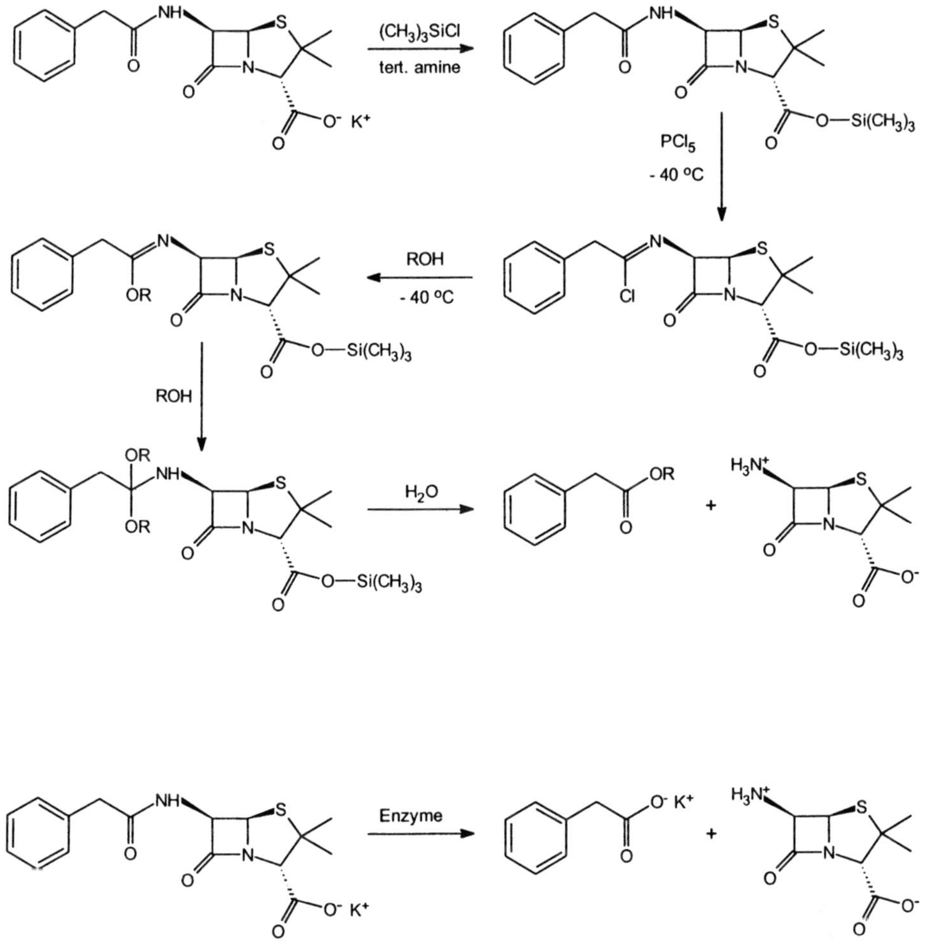

Fig. 2.32. Chemical versus enzymatic production of g-APA.

In the classical chemical process, this could be accomplished by a consecutive multistep procedure in a batch reactor, using 3 ton trimethylsilyl chloride, 8 ton N,N-dimethylaniline, 6 ton phosphorus pentachloride and 1.6 ton ammonia for the production of 5 ton 6-APA from 10 ton Penicillin G. In addition, 50 ton dichloromethane and 40 ton of butanol were used as solvents. Part of the reaction sequence was carried out at –40°C, leading to 30 MWh cooling energy costs.

In the present enzymatic process, 5 ton 6-APA is obtained using immobilized penicillin acylase (5–10 kg) as the biocatalyst in 100 ton of an aqueous solution containing 10 ton Penicillin G at ambient conditions. During the enzymatic hydrolysis the pH is kept constant by the continuous addition of ammonia (500 kg). In addition, the phenylacetic acid formed upon the hydrolysis of 6-APA can be recycled for use in the Penicillin G fermentation process.

So, apart from avoiding the use of large amounts of chlorine-containing reagents and solvents, which can only be partly recycled, the biocatalytic process offers the advantage of simplicity (one-step batch reactor) with lower energy costs and the reuse of phenylacetic acid [10].

2.17 HIGH FRUCTOSE CORN SYRUP (HFCS)

Feedstock

Corn starch (see Fig. 2.33), a polyglucose mixture consisting of both linear amylose (1,4-linkages) and branched amylopectin (both 1,4- and 1,6-linkages).

Product Use

HFCS is a concentrated (40-45% dry solids content) aqueous solution of a mixture of fructose (45%) and glucose (55%) which is used as a sweetener in liquid applications, in particular soft drinks. It is a direct replacement for, and competitor with, sucrose and has almost the same sweetening power (on a dry weight basis). Products with a further enhanced fructose content can be obtained by chromatography, usually over a sulphonic acid resin in the calcium form.

Scale

World production is 10,000,000 tons per year, on dry weight basis. The major production takes place in the USA (ca. 80%), whereas the production in Europe is still limited (5%) due to protective EC regulations in favour of sucrose. The HFCS share of the total worldwide sucrose sweetener market is less than 10%.

Process Description

The conversion of starch [11] to glucose syrups was originally based on acid catalyzed hydrolysis, giving rise to large salt loads upon neutralization during work-up and the formation of undesired coloured degradation products.

Nowadays, a two-step enzymatic hyrolysis procedure is generally applied. The first enzymatic step hydrolyzes the 1,4-linkages of starch randomly from the middle of the chain, creating oligoglucose fragments called maltodextrins. A heat-stable endo-amylase is used for 5 minutes at 105°C followed by 2 hours at 95°C, at pH 6 in a batch reactor. This starch liquefaction step is followed by a treatment with the enzyme amyloglucosidase, catalyzing the complete hydrolysis of both 1,4- and 1,6-linkages of the maltodextrins towards glucose. Process conditions of this saccharification step are 60°C, pH 4 for 1–3 days.

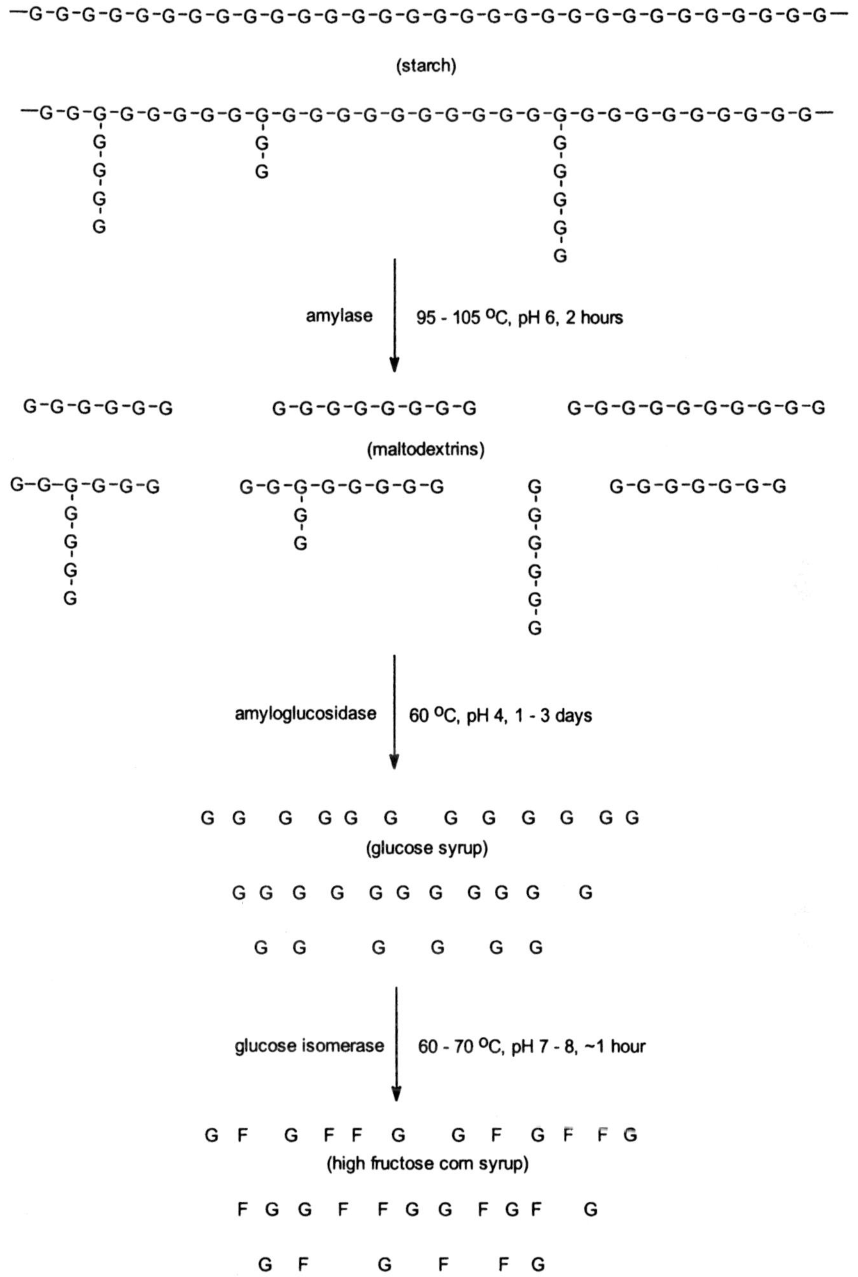

Fig. 2.33. Three-step enzymatic production of HCFS (G = glucose, F = fructose).

The isomerization of the glucose syrup thus obtained was originally also employed in batch reactors using the Mg^{++}-containing enzyme glucose isomerase in soluble form, but rapidly evolved to the use of an immobilized enzyme in downflow fixed-bed column reactors. This allowed efficient continuous processing with optimal use of catalytic activity over a long period of time (half-life of more than 50 days). Typical glucose–fructose isomerization conditions are, using glucose isomerase immobilized by occlusion in gelatin followed by cross-linking with glutaraldehyde: 60–70°C, pH 7-8, 3 mM $MgSO_4$, 40–45% glucose in water, with a flow rate of ca. 1.5 bedvolumes per hour.

It may be noted that simple alkaline-catalyzed isomerization of glucose to fructose is possible, but gives rise to serious lactic acid and coloured by-product formation. Alkaline catalysis, however, is still applied for the conversion of lactose to lactulose, used in treatment of constipation and PSE. The reason is that no enzyme has been found that is able to isomerize the glucose unit of lactose into a fructose moiety. As a consequence, a low conversion is applied or borate is used as a protecting group. In the latter case extra separation and recycle steps are required.

2.18 LOW-PHOSPHATE PIG FAECES AS FERTILIZERS

Feedstock

Feedstuffs of vegetable origin in pig fattening gives faeces slurry as a major by-product.

Product Use

Pig faeces slurry is used as fertilizer in agriculture, resulting, however, in environmental pollution because of its high phosphate content.

Scale

In the Netherlands, 15 million pigs produce ca. 5 million ton of faeces yearly (on a dry weight basis), corresponding to ca. 30,000 ton of phosphate.

Process Description

Only about 30–40% of the phosphate in feedstuffs of vegetable origin—mainly present in a bound form as phytate (see Fig. 2.34)—is digested by pigs. Traditionally, the addition of inorganic phosphate to the feedstuff was necessary to meet the required phosphate level for pig growth (bone formation).

The enzyme phytase is able to hydrolyze the phosphate ester bonds of phytate, i.e. producing a more digestible inorganic phosphate from feedstuffs of vegetable origin. This avoids part of the inorganic phosphate addition in feedstuffs and reduces the phosphate content in pig faeces slurry by 30% [12].

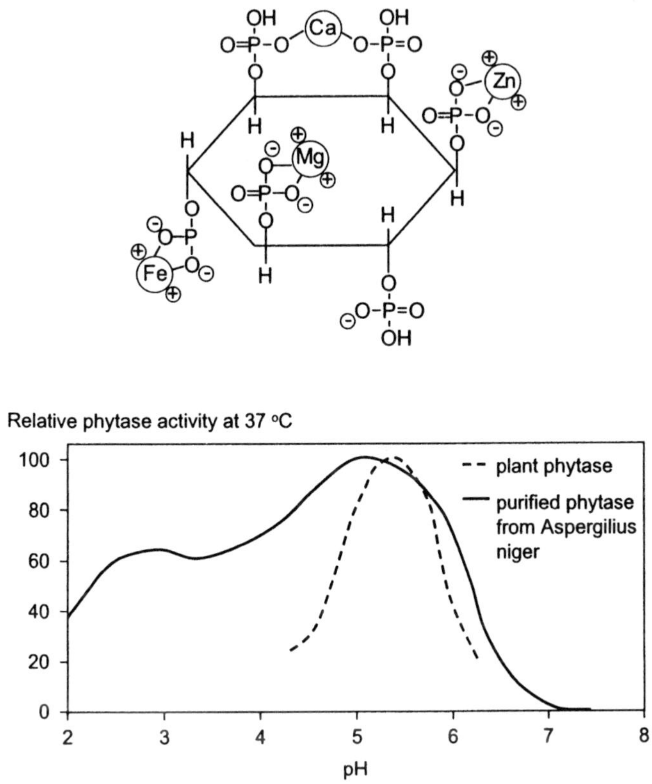

Fig. 2.34. Pig stomach 'reaction vessel': phytase (above) in the feedstuff is converted into organic phosphate by added microbial phytase (enzyme characteristics: pH dependence of microbial and plant phytase below).

Pretreating the feedstuff with phytase would, however, require a huge production facility (about 10 million tons/year on a dry weight basis for the Netherlands) with additional high energy costs due to wetting and drying of the feedstuff before and after the enzymatic conversion, respectively. Therefore, the actual enzymatic conversion step has been transferred into the stomach of the pig, which can be considered as the reaction vessel in this case. So, phytase is premixed with the feedstuff and expresses its activity in the pig stomach. As the pH in the stomach varies between 2 and 6, the phytase produced by *Aspergillus niger* fermentation also possesses sufficient activity over such a broad pH range at 37°C. Under these 'reaction' conditions, 3–4 phosphate ester moieties per phytate molecule are hydrolyzed into inorganic phosphate.

In the Netherlands, all pig feedstuff nowadays contains phytase since it reduces the costs of both inorganic phosphate addition and environmental regulations. The application of phytase in the poultry industry, where similar problems exist, is growing.

This, maybe somewhat peculiar, example demonstrates that the use of digestible catalysts, like the enzyme phytase, further broadens the field of catalysis, i.e. not in a visible production unit but in the species or product where the catalytic conversion is really needed. Other examples in the food industry are the application of the enzymes chymosin, invertase and amylase for cheese, bonbon and bread manufacturing, respectively.

REFERENCES

1 M.V.Twigg, *Catalyst Handbook*, 2nd edn., Manson Publishing, London, 1996, p.445.
2 G.J. Hutchings, *Appl. Catal.*, 72 (1991) 1–32.
3 V.A. Zazhigalov, *Katal.* 27 (1985) 373–424.
4 R.M. Contractor and A.W. Sleight, *Catal. Today*, 3 (1988) 175–184.
5 B.K. Hodnett, *Catal. Rev-Sci. Eng.*, 27 (1985) 373–424.
6 US Patent 9,931,046 (1976).
7 R.M. Contractor, H.E. Bergna, H.S. Horowitz, C.M. Blackstone, B. Malone, C.L. Torardi, B. Griffiths, U. Chowdhry and A.W. Sleight, *Catal. Today*, 1 (1987) 54.
8 *Eur. Chem. News*, 50 (Dec. 1988) 12, 36.
9 J. Verweij and E. de Vroom, *Recl. Trav. Chim. Pays-Bas*, 112 (1993) 66.
10 *BioTec Life Science* 5 (October 1991) 36.
11 C.A. Miller, in Enzymes: Tools for Success. Gist-brocades, Delft (1995), Chapter 7, p. 10.
12 A.W. Jongbloed, Z. Mroz, and P.A. Kemme, *J. Anim. Sci.*, 70 (1992) 1159.

General Literature

I. Maxwell, C. Williams, F. Muller and B. Krutzen, *Zeolite catalysis for high quality fuels*, Shell brochure, 1992.

G.A. Mills and E.E. Ecklund, Alternative Fuels, Parts 1 and 2, *ChemTech* (1989) pp. 549 and 626.

G.M. Wells, *Handbook of Petrochemicals and Processes*, 1991, p. 270.

M. Iborra, J.F. Izquirdo, J. Tejero and F. Curill, Getting the lead out with ethyl *t*-butyl ether, *ChemTech* (1988) 120.

I.E. Maxwell and W.H.J. Stork, *Stud. Surf. Sci. Catal.*, 58 (1991) 571.

D.M. Ruthven, *Chem. Eng. Progress*, Feb. (1988) 42.

Chapter 3

Chemical kinetics of catalyzed reactions

3.1 INTRODUCTION

Rate expressions are indispensable in the application of catalyzed reactions, in the design of chemical processes, in particular reactor design, and in process start-up and control. Insight into the dependence of the reaction rate on catalyst variables, temperature, and concentrations of reactants, products, and other relevant species is required to predict the size of a catalytic reactor and the optimum operating conditions.

Catalytic reaction pathwayss consist of reaction sequences formed by a series of elementary reaction steps. Hence, rate expressions are in general a function of many parameters, as depicted in Eqn. (3.1).

$$r = f(\text{catalyst}, T,\ p_i \text{ or } c_i \ldots, k_i \ldots, K_i \ldots, K_{eq}) \qquad (3.1)$$

In heterogeneously catalyzed reactions reactant molecules adsorb on the catalyst surface (characterized by equilibrium constants K_i), undergo modifications on the surface to form adsorbed products with rate constants k_i, and these products finally desorb. The surface composition and structure of the catalyst determine its overall activity and selectivity. Therefore, it is important to relate constants, such as k_i and K_i to the chemical reactivity of the catalyst surface.

Many different modes of adsorption and rearrangement of molecule fragments are possible on catalyst surfaces, so many reactions can occur in parallel. Therefore, catalysts may exhibit satisfactory activity but at the same time have a low selectivity. It is a common procedure in process development to carry out an optimization program aimed at increasing the selectivity of the catalyst system selected. Selectivity can often be increased by the use of promotors or catalyst modifiers to create particular surface sites that enhance the desired reaction but suppress undesired reaction paths.

In homogeneous catalysis ligands play a similar role as surface sites with respect to selectivity. In a solution many different catalytic complexes may be present with different catalytic activities. Types and concentrations of ligands determine the structure of the catalyst complexes to a large degree. Thus, catalytic activity and selectivity can be tuned by selecting the right ligands and reaction conditions. Similarly, in biocatalyis enzymes play the role of active sites, having a very high (shape) selectivity for specific reactions.

Besides activity and selectivity, stability is crucial in catalysis applications. Catalyst deactivation can have a kinetic origin. For instance, deactivation might occur by a serial reaction mechanism in which an intermediate can undergo a reaction to form a substance that is a poison for the active catalyst sites. Frequently encountered examples are oligomerization and coke formation.

Adsorption on a solid catalyst surface, complex formation in homogeneous catalysis with metallo-organic complexes and in biocatalysis with enzymes share the same principle, i.e. the total number of sites is constant. Therefore, the rate expressions for reactions on heterogeneous, homogeneous and biocatalysts have a similar form. The constant number of active sites results in rate expressions that differ from homogeneous gas phase kinetics. Partial pressures are usually used in rate expressions for gas-phase reactions, while concentrations are used when the reactions take place in the liquid phase. It appears that definitions and nomenclature of particular kinetics constants in the different sub-communities differ sometimes. In the following sections the expressions used by the different subdisciplines will be compared and their conceptual basis outlined.

In the approach to the derivation of rate expressions two principles are important:

- catalysts do not affect the overall equilibrium of a reaction, but only the reaction rate of the reacting molecules;
- the total number of catalytically active sites is assumed to be a constant. In heterogeneous catalysis this site density is expressed as the number of sites per unit mass of catalyst or per unit surface area; in homogeneous catalysis it is expressed as the concentration of catalytically active complex molecules, and in biocatalysis as the enzyme concentration. In fact, the above assumption is not always justified. There are examples where the number of active sites is a function of process variables such as temperature and concentration.

3.2 RATE EXPRESSION (SINGLE-SITE MODEL)

Consider the reversible reaction $A \leftrightarrow B$ (e.g. isomerization), which proceeds according to the three elemental steps represented in Fig. 3.1.

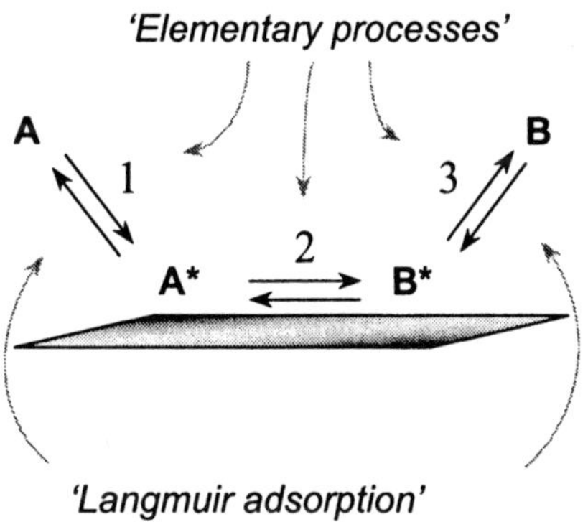

Fig. 3.1. Schematic representation of the reaction A ↔ B.

1. $A + * \underset{k_{-1}}{\overset{k_1}{\rightleftharpoons}} A*$

2. $A* \underset{k_{-2}}{\overset{k_2}{\rightleftharpoons}} B*$

3. $B* \underset{k_{-3}}{\overset{k_3}{\rightleftharpoons}} B + *$

This kinetic model can be interpreted as follows:

1. Molecule A adsorbs on an active site * at the catalyst surface with formation of an adsorbed complex A*.
2. This adsorbed complex reacts on the active site by rearrangement to an adsorbed complex B*.
3. Finally, product B desorbs from the active site, liberating the site for a new catalytic cycle.

Since these steps are considered to be elementary processes, their rates can be directly derived from their reaction rate expressions:

$$r_1 = r_{+1} - r_{-1} = k_1\, p_A\, N_T\, \theta_* - k_{-1}\, N_T\, \theta_A \qquad (3.2)$$

$$r_2 = r_{+2} - r_{-2} = k_2\, N_T\, \theta_A - k_{-2}\, N_T\, \theta_B \qquad (3.3)$$

$$r_3 = r_{+3} - r_{-3} = k_3\, N_T\, \theta_B - k_{-3}\, p_B\, N_T\, \theta_* \qquad (3.4)$$

Note that under steady-state conditions the rate of each reaction step equals the overall net rate, r. θ_*, θ_A and θ_B represent the fractions of the total number of sites that are vacant, or occupied by A and B, respectively. N_T represents the total concentration of active sites with possible dimension mol·(kg cat)$^{-1}$. For this gas-phase reaction, partial pressures of A and B are preferred in the rate equations, but for liquid-phase reactions molar concentrations should be used.

Conservation of the total number of active sites leads to the site balance expression:

$$1=\theta_{*}+\theta_{A}+\theta_{B} \tag{3.5}$$

Under steady-state conditions, which are usually satisfied in continuous-flow processes, the concentrations of the vacant sites, sites occupied by A, and sites occupied by B, do not change with time:

$$\frac{d\theta_{A}}{dt}=0=k_{1}\,p_{A}\,\theta_{*}-k_{-1}\,\theta_{A}-k_{2}\,\theta_{A}+k_{-2}\,\theta_{B} \tag{3.6}$$

$$\frac{d\theta_{B}}{dt}=0=k_{-3}\,p_{B}\,\theta_{*}-k_{3}\,\theta_{B}-k_{-2}\,\theta_{B}+k_{2}\,\theta_{A} \tag{3.7}$$

This is equivalent to the fact that under these conditions the net rates of the different steps are the same:

$$r=r_{1}=r_{2}=r_{3} \tag{3.8}$$

From Eqns. (3.5)–(3.7), the unknowns θ_{*}, θ_{A} and θ_{B} can be expressed as functions of the rate constants and partial pressures and then substituted in Eqn. (3.8). Finally, Eqn. (3.9) follows for the reaction rate expression.

$$r=\frac{N_{T}\,k_{1}\,k_{2}\,k_{3}\,(p_{A}-p_{B}\,/\,K_{eq})}{(k_{1}k_{3}+k_{1}k_{-2}+k_{1}k_{2})p_{A}+(k_{-1}k_{-3}+k_{2}k_{-3}+k_{-2}k_{-3})p_{B}+(k_{-1}k_{-2}+k_{-1}k_{3}+k_{2}k_{3})} \tag{3.9}$$

with $K_{eq} = K_1\,K_2\,K_3 (K_i = k_i\,/k_{-i})$ being the overall equilibrium constant for the reaction.

Rate expression (3.9) has been derived for a relatively simple kinetic model by application of the site balance and the steady-state hypothesis. More complex models will result in more complex expressions (see Refs. [1] and [2] for general derivation), which are hard to handle. Fortunately, some simplifications can usually be applied. The rate expression then follows by computing the net rates of intermediate formation r_i, which are the sum of the intermediate formation (forward) and intermediate removal (backward) reaction steps:

$$r_{i}=r_{+i}-r_{-i} \tag{3.10}$$

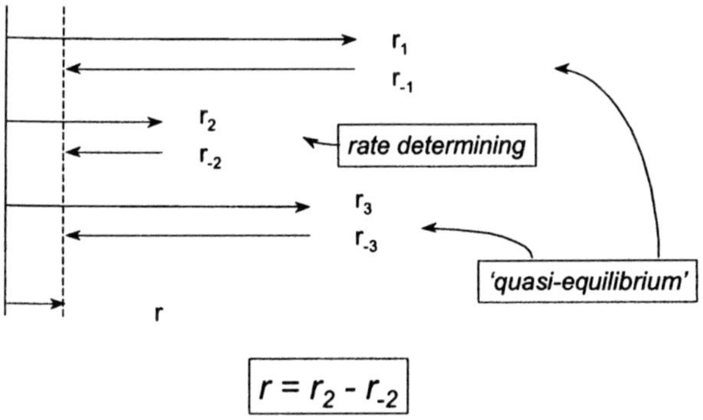

Fig. 3.2. Visualization of the *quasi*-equilibrium and rate-determining steps. The lengths of the arrows are proportional to the rates of the relevant steps.

3.3 RATE-DETERMINING STEP — *QUASI*-EQUILIBRIUM

The aforementioned reaction A↔B proceeds through three elementary processes in series. Consequently, the net rates of the individual steps are equal to the overall rate (3.8). It can be imagined that the forward and backward rates of two of these steps are large compared with the third one. They might be so large that these steps can be considered to be in *quasi*-equilibrium. The third, remaining step is then called the *rate-determining* step (r.d.s.). Figure 3.2 depicts the situation in which the surface reaction is rate determining, while the adsorption and desorption steps are in *quasi*-equilibrium.

The rate expression is now obtained as follows. Starting with the rate determining step:

$$r = r_{+2} - r_{-2} = k_2\, N_T\, \theta_A - k_{-2}\, N_T\, \theta_B \tag{3.11}$$

The *quasi*-equilibrium is then used for the other reaction steps, since:

$$\frac{r_1}{r_{+1}} = \frac{r_{+1} - r_{-1}}{r_{+1}} << 1 \tag{3.12}$$

and

$$\frac{r_3}{r_{+3}} = \frac{r_{+3} - r_{-3}}{r_{+3}} << 1 \tag{3.13}$$

Substitution of (3.2) in (3.12) gives:

$$\theta_A = \frac{k_1}{k_{-1}} p_A\, \theta_* = K_1\, p_A\, \theta_* \tag{3.14}$$

and similarly

$$\theta_B = \frac{k_{-3}}{k_{+3}} p_B \theta_* = \frac{1}{K_3} p_B \theta_* \tag{3.15}$$

This enables the elimination of the unknown quantities θ_A and θ_B from (3.11). The remaining θ_* can be eliminated by use of the site balance (3.5):

$$\theta_* = \frac{1}{1 + K_1 p_A + p_B / K_3} \tag{3.16}$$

The resulting rate expression for the case that the surface reaction is rate determining is given by Eqn. (3.17).

$$r = \frac{N_T k_2 K_1 [p_A - p_B / K_{eq}]}{1 + K_1 p_A + p_B / K_3} \tag{3.17}$$

Similar equations can be derived in case adsorption is the rate-determining step:

$$r = \frac{N_T k_1 [p_A - p_B / K_{eq}]}{1 + (1 + 1/K_2) p_B / K_3} \tag{3.18}$$

or when desorption is rate determining:

$$r = \frac{N_T k_3 K_1 K_2 [p_A - p_B / K_{eq}]}{1 + (1 + K_2) K_1 p_A} \tag{3.19}$$

If, besides reactant A and product B, other species also adsorb on the active sites, they make these sites unavailable for reaction, and hence lower the reaction rate. The effect of such inhibitors (I) should be included in the rate expression and can be summarised for the considered reaction (and surface reaction rate determining) in Eqn. (3.20).

$$r = \frac{k[p_A - p_B / K_{eq}]}{1 + K_A p_A + K_B p_B + \sum K_I p_I} \tag{3.20}$$

Here K_A, K_B and K_I represent the adsorption equilibrium constants of the components A, B and I, respectively; k represents the apparent (observed) overall reaction rate constant. It will be clear that $K_A = K_1$ and $K_B = 1/K_3$.

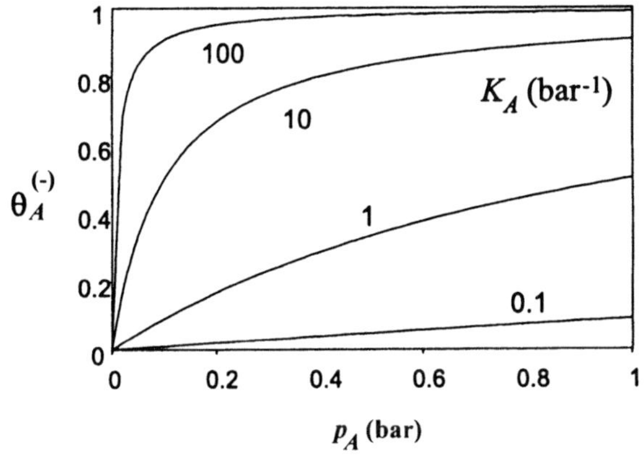

Fig. 3.3 Fractional surface coverage as a function of p_A for several values of K_A.

3.4 ADSORPTION ISOTHERMS

Reaction step (1) in Fig. 3.1 is an adsorption step. Under equilibrium conditions, the net rate is zero and the equilibrium is then described by Eqn. (3.14). This does not mean *per se* that K_1 is constant. It may vary with surface occupancy in the case of a non-uniform surface, due to the interaction of adsorbed species with each other. In conventional catalytic kinetics it is generally assumed that one is dealing with a homogeneous surface. Under the assumptions that:

- the surface contains a constant number of identical adsorption sites,
- a site can contain only one molecule; and,
- no interaction takes place between adsorbed molecules,

the so-called Langmuir adsorption is operative.

One-Component Adsorption

From the foregoing, it can be easily derived that in the case of one component adsorption, the fractional coverage of the adsorbed species A can be described by Eqn. (3.21). Figure 3.3 shows the graphical representation of equation for various values of K_A ($\equiv K_1$).

$$\theta_A = \frac{K_A\, p_A}{1 + K_A\, p_A} \tag{3.21}$$

Three regions can be distinguished for the Langmuir isotherm:

1. At low values of $K_A\, p_A$, a linear relation exists between θ_A and p_A, with slope K_A. This is a relation equivalent to Henry's law.

2. At high values of $K_A\,p_A$, θ_A approaches 1 asymptotically, i.e., nearly all sites are occupied. This can be represented by

$$1-\theta_A = \frac{1}{K_A\,p_A} \tag{3.22}$$

3. For values of $K_A\,p_A$ near 1, i.e., at moderate coverages, the complete equation (3.21) must be used. Alternatively, the Freundlich isotherm (3.23) can be used.

$$\theta_A = c\,p_A^{1/n} \quad (c,n \text{ contants, } n > 1) \tag{3.23}$$

With increasing temperature, K_A decreases and a transition from situation 2 to 1 via 3 can be expected. K_A is expressed in units of $(\text{pressure})^{-1}$. If atm^{-1} (1 atm is the standard thermodynamic reference state for gases) is used, K_A can be expressed as:

$$K_A = \exp\left(-\frac{\Delta G_A^o}{RT}\right) = \exp\left(\frac{\Delta S_A^o}{R} - \frac{\Delta H_A^o}{RT}\right) \quad \text{or} \quad \Delta H_A^0 = RT^2\frac{\partial \ln K_A}{\partial T} \tag{3.24}$$

in which ΔG^o, ΔH^o and ΔS^o represent the Gibbs free energy, enthalpy and entropy of adsorption, respectively. Both the enthalpy and entropy of adsorption is an exothermic process and the molecule loses at least translational degrees of freedom.

The assumptions for the Langmuir isotherm imply an ideal surface, but few real systems will fulfil this ideal under all conditions. Experimental determination of the heat of adsorption ($-\Delta H^o$) as a function of the surface coverage shows that the heat of adsorption usually decreases with increasing coverage. This indicates that catalyst surfaces are not uniform and/or that the adsorbed molecules exhibit a mutual interaction. Adsorption isotherms that take this coverage dependence into account are, amongst others, the Freundlich and the Temkin isotherms. In kinetic modelling practice, these isotherms are hardly used since the derivation of rate expressions becomes a cumbersome job (see e.g. Refs. [1,2]), in particular for multicomponent systems. This situation might rapidly change with the increasing use of powerful computers, but the simple physical picture of the catalyst surface will always remain the advantage of the Langmuir model.

Multicomponent Adsorption

Within the Langmuir approach, the general expression for the fractional coverage by component A in the case of multicomponent adsorption is given by equation (3.25)

$$\theta_A = \frac{K_A p_A}{1 + K_A p_A + \sum K_I p_I} \tag{3.25}$$

Dissociative Adsorption

In the foregoing, only molecular adsorption has been considered. However, some molecules (e.g. H_2, CO) can dissociate upon adsorption and, hence, two sites are required.

Consider the dissociative adsorption equilibrium of hydrogen:

$$H_2 + 2* \leftrightarrow 2H*$$

In an analogous way as in the preceding section for the molecular adsorption, the coverage with hydrogen atoms H follows (3.26). The pressure dependency now follows the square root of the partial H_2 pressure at low coverages.

$$\theta_H = \frac{\sqrt{K_{H_2} p_{H_2}}}{1 + \sqrt{K_{H_2} p_{H_2}}} \tag{3.26}$$

3.5 RATE EXPRESSIONS — OTHER MODELS AND GENERALIZATIONS

Often multiple sites are involved in a catalytic process. This is especially the case for dissociation reactions. The same procedure as applied in Section 3.1 for a single-site model can be used for the derivation of the rate expression for a dual-site model. The procedure is exemplified for the dissociation reaction A ↔ 2 B, which is thought to proceed through the following three-step sequence:

		ν_i
1.	$A + * \underset{k_{-1}}{\overset{k_1}{\rightleftharpoons}} A*$	1
2.	$A* + * \underset{k_{-2}}{\overset{k_2}{\rightleftharpoons}} 2\,B*$	1
3.	$B* \underset{k_{-3}}{\overset{k_3}{\rightleftharpoons}} B + *$	2

Since steps 1 and 2 must proceed once per overall reaction and step 3 twice, the so-called stoichiometric numbers ν_i of the steps are one, one, and two, respectively.

By application of the steady-state hypothesis (surface occupancies do not change with time), a site balance, and the assumption that the surface dissociation reaction is rate determining, while the other steps are in *quasi*-equilibrium, the following rate expression is derived:

$$r = \frac{s\,N_T\,k_2\,K_A\,[p_A - p_B^2\,/\,K_{eq}]}{(1 + K_A\,p_A + K_B\,p_B)^2} \tag{3.27}$$

in which $K_{eq} = K_1 K_2 K_3^2$. The numerator includes a parameter s, which represents the number of nearest neighbours of an active site. The necessity of this parameter in the rate equation can be understood as follows. In step 2, the rate-determining step, adsorbed A reacts with an empty site. This reaction is only possible when there is an empty site next to A. Therefore, the rate of this reaction is proportional to the concentration of adsorbed A ($N_T\theta_A$) multiplied with the number of adjacent sites s times the chance that they are empty θ^*. Using the total concentration of empty sites ($N_T\theta^*$) instead, would lead to overestimation of the reaction rate. K_A and K_B are the adsorption equilibrium constants of A and B and equal to K_1 and $1/K_3$, respectively. The denominator is now squared compared to a single-site model. This indicates that in the rate-determining step two active sites are involved.

The models described above are termed Langmuir–Hinshelwood–Hougen-Watson (LHHW) models, named after the scientists that contributed a lot to the development of these engineering models. The characteristics of these models are that adsorption follows the Langmuir isotherm, and that reaction takes place between adsorbed species. Sometimes, one distinguishes Eley Rideal models, whereby a molecule reacts directly from the gas phase with a surface complex:

$$A + B^* \leftrightarrow C^*$$

Other variants of kinetic models can be derived, of course. Froment and Bischoff [4] present an extended treatment of this approach. It follows that rate expressions based on sequences of elementary steps, one of which is rate determining, can be expressed in the general form:

$$\text{rate} = \frac{(\text{kinetic factor}) \cdot [\text{driving force}]}{(\text{adsorption term})^n} \tag{3.28}$$

The kinetic factor always contains the rate constant of the rate-determining step, together with the concentration of active sites and adsorption equilibrium constants.

The driving force represents the chemical affinity of the overall reaction to reach thermodynamic equilibrium. It is proportional to the concentration difference of the reactants with respect to their equilibrium concentrations. The driving force term does not contain parameters associated with the catalyst, consistent with the fact that the catalyst does not affect chemical equilibrium.

The adsorption term represents the reduction of the overall rate due to adsorption, with the individual terms denoting the distribution of the active sites over the different intermediate surface species and vacancies. It may contain square roots of partial pressures, indicating dissociative adsorption. The power n in the rate expression indicates the number of sites involved in the rate-determining process, and usually has a value of 0, 1, or 2. Larger values reported in literature merely represent values obtained by fitting.

In homogeneous catalysis, enzyme reactions, and reactions of microbial systems the same types of equations are used as in the LHHW models. In the latter disciplines, however, they are often referred to as Michaelis–Menten kinetics.

3.6 LIMITING CASES — REACTANT AND PRODUCT CONCENTRATIONS

In Sections 3.2 and 3.4, the approach based on one rate-determining step with the other steps in quasi-equilibrium was applied to simplify the derivation of rate expressions. A further simplification of the rate expressions is obtained when the product concentrations or partial pressures are negligibly small, i.e., at low conversions of a pure reactant feed stream. Similarly, at high conversions, when nearly only product is present, the rate expressions can also be simplified. Applying the single-site model, Eqns. (3.17)–(3.19), for the reaction $A \leftrightarrow B$ at low conversion, the following expressions (often called "initial rate expressions") are obtained, provided that the feed contains pure A.

Adsorption rate determining:

$$r_0 = N_T\, k_1\, K_A\, p_{A0} \tag{3.29}$$

Surface reaction rate determining:

$$r_0 = \frac{N_T\, k_2\, K_A\, p_{A0}}{1 + K_A\, p_{A0}} \tag{3.30}$$

In the extreme case of low $K_A\, p_{A0}$ this reduces to:

$$r_0 = N_T\, k_2\, K_A\, p_{A0} \tag{3.30b}$$

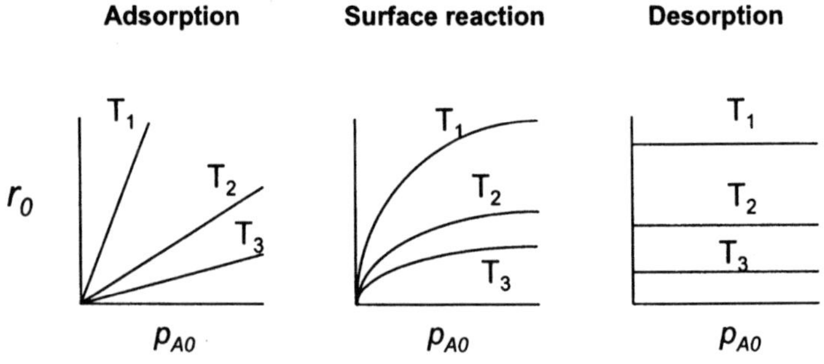

Fig. 3.4. Pressure dependency of the initial reaction rate for three different rate-determining steps and at increasing temperatures $T_1 > T_2 > T_3$.

which is a similar expression to Eqn. (3.29) and cannot be distinguished from it experimentally.

Desorption rate determining:

$$r_0 = \frac{N_T k_3 K_A K_2 p_{A0}}{1 + (1 + K_2) K_A p_{A0}} \tag{3.31}$$

When desorption is rate limiting, the surface is nearly fully occupied and the latter expression, for K_2 and $K_A p_{A0} >> 1$, reduces to:

$$r_0 = N_T k_3 \tag{3.32}$$

Figure 3.4 represents the pressure dependences of r_0 for these three cases. The effect of temperature is also shown.

Evidently, this initial rate pressure dependence gives a quick insight into which kinetic model describes best the experimental results of the reaction under consideration. Initial rate experiments are ideal for model discrimination purposes where one tries to select the best kinetic description of a process. However, two important aspects must be realised:

- The product partial pressures may be low, but when the product is strongly adsorbed not all the terms in the denominator can be neglected. In this case, only the numerator can be simplified.
- Other components in the reaction mixture may compete for adsorption sites and occupy part of the active sites. Hence, a variation of the reactant pressure will have less effect than would be expected.

In the case of the dual site model the initial rate even passes through a maximum if the surface reaction step is rate determining, which is known for e.g. the CO oxidation over noble metal catalysts.

It appears that (3.30) can be rewritten as the Michaelis–Menten expression used in biocatalysis:

$$r \equiv v = k_2 E_0 \frac{[S]}{K_m + [S]} \tag{3.33}$$

In this expression E_0 is the total concentration of enzyme catalyst present in the system. It replaces the total active site concentration N_T. The pressure is now replaced by the substrate (reactant) concentration [S] and the Michaelis constant K_m is the equivalent of $1/K_A$ in (3.48).

One notes immediately that the dependence of the rate v on [S] is similar to the Langmuir adsorption term dependence as shown in Fig. 3.3. The maximum rate V_m is obtained when all enzyme sites are occupied, leading to:

$$V_m = k_2 E_0 \tag{3.34}$$

This is exactly the same as (3.32) for strong reactant adsorption.

In biocatalysis k_2 is the rate measured when all enzyme molecules are complexed with reactant divided by the total concentration of enzyme present. This is the Turn-Over Number according to biochemists' definition. Note that this differs from the Turn-Over Frequency as defined in heterogeneous catalysis where it is simply the rate normalised to the total number of surface sites present. In the latter case it is a function of the gas phase composition.

At low substrate concentrations the expression (3.33) becomes:

$$v = \frac{k_2}{K_m} E_0 [S] \tag{3.35}$$

In the liquid phase diffusion to the catalyst may become the limiting step. Diffusion limitations provides an upper bound to the observed reaction rate (see Chapter 8). It appears that some enzyme catalytic reactions are so fast, e.g. carbonic anhydrase or acetyl cholesterase, that they exhibit this phenomenon. Catalysis under such condition is called "kinetic perfection".

The conventional way to determine the Michaelis-Menten parameters from experimental data is to invert expression (3.33) to the form:

$$\frac{1}{v} = \frac{1}{V_m} + \frac{K_m}{V_m} \cdot \frac{1}{[S]} \tag{3.36}$$

This representation yields the so-called Lineweaver–Burk plot (Fig. 3.5). This is also commonly done in heterogeneous catalysis for analysis of initial rate data [4,5]. This changes, however the error distribution and parameter values may be

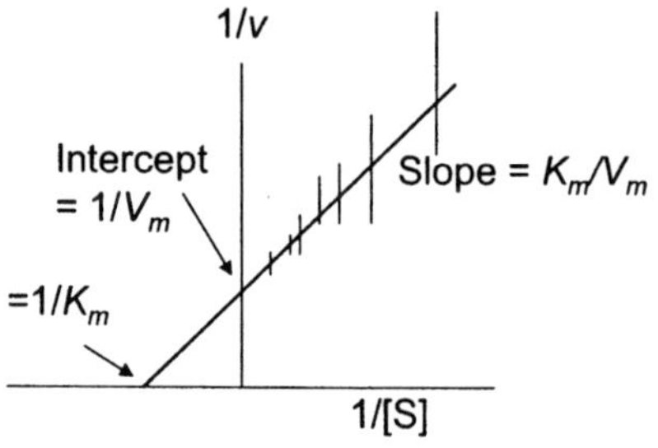

Fig. 3.5. Schematic Lineweaver–Burk plot.

less well estimated. The best way is to apply nonlinear least-squares estimation, which is nowadays a well established method.

Also in enzyme catalysis competitive adsorption effects are observed frequently and have been extensively analyzed. Three cases are distinguished:

1. Irreversible inhibition
2. Competitive inhibition
3. Non-competitive inhibition

1. Irreversible inhibition occurs when reactants adsorb, but do not desorb. Nerve gases function as such. This simply reduces the total number of available enzyme molecules and is equivalent to catalyst (self)poisoning.

2. In competitive inhibition some intermediate I competes with reactions for complexation with enzyme E. Adsorption is now reversible. The corresponding kinetic scheme is:

$$S + E \underset{k_{-1}}{\overset{k_1}{\rightleftarrows}} ES \xrightarrow{k_2} E + P \tag{3.37}$$

$$I + E \underset{k_{-1}^i}{\overset{k_1^i}{\rightleftarrows}} EI \tag{3.38}$$

the kinetic expression can be readily found using the procedure described in Section 3.2. Essential again is the constancy of total amount of catalytic centres E_0. The expression for the rate of production P becomes:

$$v = V_{\mathrm{m}} \frac{[S]}{[S] + K'_{\mathrm{m}}} \tag{3.39}$$

with

$$K'_{\mathrm{m}} = K_{\mathrm{m}}\left(1 + \frac{[I]}{K^i_{\mathrm{m}}}\right) \tag{3.40}$$

The slope of the Lineweaver–Burk plot has increased.

An interesting case of competitive adsorption is provided by the pH effect for an acidic enzyme catalyzed reaction (e.g. lipase). If the pH is too low protonation of an essential amino acid residue of the enzyme may occur with loss in activity. At very high pH again loss in the number of active enzyme molecules may occur, but now because of possible deprotonation of essential residues. Since we are dealing with competitive adsorption effects, at low substrate concentrations, where the rate is controlled by K'_m, a maximum in the rate will occur at the pH where competitive adsorption is the least. As can be deduced from (3.39) at high substrate concentration the reaction rate will not be affected.

3. In biochemistry non-competitive inhibition relates to changes in the enzyme activity due to the interaction with other inhibitory intermediates that do not adsorb to its reactive centre. Strictly speaking the assumption of no changes in reactivity of the reaction site due to co-adsorption effects is now not valid any more. The analogous situation in heterogeneous catalysis is the change in site activity due to lateral interactions with coadsorbed intermediates. It implies that k_2 in (3.37) changes into a k'_2 and is a function of intermediate concentration. Because step 2 is rate limiting the *quasi*-equilibrium assumption applies and K_m does not change. So, now only the intercept in the Lineweaver–Burk plot is changed.

3.7 TEMPERATURE AND PRESSURE DEPENDENCE

Typical behaviour can be expected with respect to the temperature dependence of catalyzed reactions, due to the effect of (competitive) adsorption. This will be demonstrated with some simplified cases of the reaction A ↔ B, starting with a consideration of the rate-determining step.

3.7.1 *Transition-State Theory*

According to the transition-state theory molecules react through unstable intermediates called transition-state complexes which then react to products. For instance, the surface reaction A* ↔ B*, which is considered to be the rate-determining step, proceeds as follows:

$$*A \underset{k_-^{\#}}{\overset{k_+^{\#}}{\rightleftarrows}} *A^{\#} \xrightarrow{k_{\text{barrier}}} *B$$

It is assumed that the reacting complex (*A) is in equilibrium with the transition-state complex (*A$^{\#}$), and that the number of molecules in the transition state that react to the product (*B) per unit of time is given by the frequency k_{barrier}. This latter step is assumed to be rate limiting. The reaction rate constant can then be computed from expression (3.41).

$$k = k_{\text{barrier}}\left(\frac{k_+^\#}{k_-^\#}\right) = k_{\text{barrier}} K^\# \tag{3.41}$$

Expression (3.41) is valid when energy exchange is fast compared to the overall reaction rate. Since $K^\#$ is an equilibrium constant, it can be written as in Eqn. (3.42), where $\Delta G^\#, \Delta H^\#$ and $\Delta S^\#$ are the Gibbs free energy, the enthalpy and the entropy differences between the transition state and the ground state, respectively.

$$K^\# = \exp\left(\frac{-\Delta G^\#}{RT}\right) = \exp\left(\frac{\Delta S^\#}{R} - \frac{\Delta H^\#}{RT}\right) \tag{3.42}$$

As long as quantum-mechanical corrections can be ignored, the rate of reaction of the transition-state complex is the same for all reactions and is given by Eqn. (3.43):

$$k_{\text{barrier}} = \frac{k_B T}{h} \tag{3.43}$$

where k_B is the Boltzmann constant and h is Planck's constant. k_{barrier} is typically $10^{12}\ \text{s}^{-1}$.

The overall rate constant then becomes:

$$k = \frac{k_B T}{h} \exp\left(\frac{\Delta S^\#}{R} - \frac{\Delta H^\#}{RT}\right) = A_0 \exp\left(\frac{-E_a}{RT}\right) \tag{3.44}$$

$\Delta S^\#/R$ may be taken as a constant, because it only varies slightly with temperature. Furthermore, since the exponential term is so much more temperature-sensitive than the pre-exponential term k_0, this latter term may also be taken as a constant, resulting in the second equality of Eqn. (3.44).

$\Delta H^\#$ can be identified with the activation energy E_a (neglecting a contribution of RT) of the reaction (see Fig. 3.6). In Fig. 3.6, ΔH represents the reaction enthalpy for the reaction A* → B*. More detailed treatments are presented by Maatman [6], Boudart, and Djégà-Mariadassou [2], and Van Santen and Niemantsverdriet [7].

3.7.2 Forward Reaction — Temperature and Pressure Dependence

For the conversion A ↔ B, the overall activation energy is a complex function of the reaction enthalpies and activation energies of the individual elementary

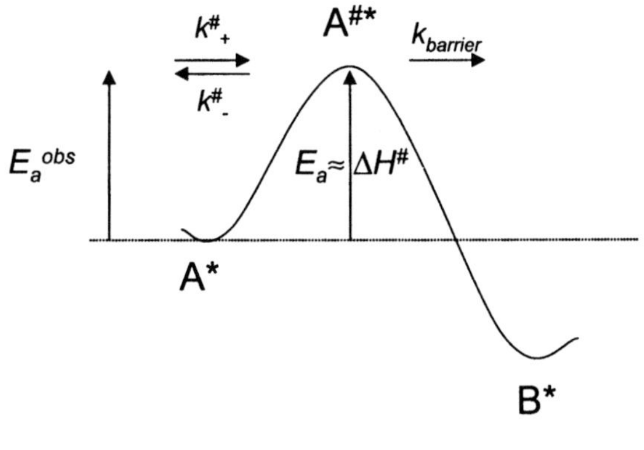

Fig. 3.6. Energy diagram for a surface reaction.

reaction steps. We will illustrate this by assuming that the surface reaction is the rate-determining step (3.17) and that the backward reaction can be neglected. The rate expression for this single-site reaction can now be written as:

$$r = \frac{k_2 N_T K_A p_A}{1 + K_A p_A + K_B p_B} \tag{3.45}$$

where $K_A = K_1$ and $K_B = 1/K_3$.

The dependence of the reaction rate (3.48) on pressure and temperature is determined by contributions of both the numerator and denominator. Often, rate expressions are presented in a power law form:

$$r = kp_A^{n_A} p_B^{n_B} \tag{3.46}$$

Comparison with Eqn. (3.45) suggests that the powers n_A and n_B will not be constants. They can be extracted as follows:

$$n_i = \frac{\partial \ln r}{\partial \ln p_i} \tag{3.47}$$

Applying Eqns. (3.45) to (3.47), and using the adsorption equilibrium relationships (Eqns. (3.14) and (3.15)), together with Eqn. (3.16), yields:

$$n_A = 1 - \theta_A \qquad \text{and} \qquad n_B = -\theta_B \tag{3.48}$$

Hence, the apparent reaction orders are related to the fractional surface coverages. From Eqn. (3.48), it follows that n_A varies from 0 to 1, and n_B from –1 to 0, depending on the conditions.

Like the pressure dependence, the temperature dependence is also often expressed in an empirical form, in which an apparent overall rate constant is used (like in Eqn. (3.44)). The observed (or apparent) activation energy can be expressed as:

$$E_{\mathrm{a}}^{\mathrm{obs}} = RT^2\left(\frac{\partial \ln r}{\partial T}\right)_p \tag{3.49}$$

The observed activation energy can now be derived from Eqn. (3.45) in a similar way as the derivation of the reaction powers as a function of surface coverage:

$$E_{\mathrm{a}}^{\mathrm{obs}} = E_{\mathrm{a2}} + (1-\theta_{\mathrm{A}})\Delta H_A - \theta_{\mathrm{B}}\Delta H_{\mathrm{B}} \tag{3.50}$$

The observed activation energy contains contributions from the rate-determining step (E_{a2}) and from the adsorption enthalpies of A and B, the latter depending on the fractional occupancies. Obviously, $E_{\mathrm{a}}^{\mathrm{obs}}$ will depend on the experimental conditions. Therefore, it is not surprising that a wide range of values have been reported for the same reaction system.

3.7.3 Forward Reaction — Limiting Cases

Based on the previous analysis of the pressure and temperature dependence of the reaction order and the observed activation energy, four different cases can be distinguished:

(i) Strong Adsorption of A

Strong adsorption of A results in $K_{\mathrm{A}}\, p_{\mathrm{A}} >> 1$ and $K_{\mathrm{B}}\, p_{\mathrm{B}}$, and hence Eqn. (3.45) reduces to:

$$r = k_2 N_{\mathrm{T}} \tag{3.51}$$

Physically this means that the whole catalyst surface is covered with A ($\theta_A \rightarrow 1$, $\theta_B \rightarrow 0$). Therefore, varying the partial pressure of A does not influence the reaction rate. The reaction is said to be zero order in A (and B). The overall activation energy is $E_{\mathrm{a}}^{(\mathrm{obs})} = E_{a2}$ (see Fig. 3.6), provided the concentration of active sites N_T is temperature independent.

(ii) Weak Adsorption of A and B

When A and B are only weakly adsorbed, $K_{\mathrm{A}}\, p_{\mathrm{A}}$ and $K_{\mathrm{B}}\, p_{\mathrm{B}} << 1$ and $\theta_A, \theta_B \rightarrow 0$, so the rate expression becomes:

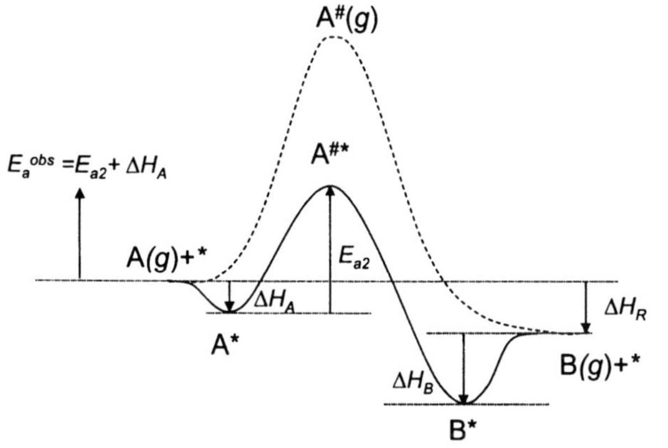

Fig. 3.7. Energy diagram for the catalytic conversion of A to B. * Denotes surface vacancy and # transition state. ΔH_R is the reaction enthalpy. Included is the energy diagram for the gas phase conversion (dashed curve).

$$r = k_2 N_T K_A p_A \tag{3.52}$$

The reaction is now first order in A and zero order in B. The observed overall activation energy will be lower than in the previous case.

$$E_a^{obs} = E_{a2} + \Delta H_A \tag{3.53}$$

since the enthalpy of adsorption of A, ΔH_A, is negative; adsorption is an exothermic process. This result can be understood from the energy diagram in Fig. 3.7. Due to the low occupancy of A, for reaction to occur, adsorption of gaseous A is required, and hence A gains adsorption enthalpy. Subsequent surface reaction requires overcoming the activation energy.

(iii) Strong Adsorption of B

In this case $K_B p_B >> 1$ and $K_A p_A$, and $\theta_A \to 0$, $\theta_B \to 1$, and the rate expression becomes:

$$r = \frac{k_2 N_T K_A p_A}{K_B p_B} \tag{3.54}$$

The reaction is first order in A and –1 in B (B decreases the reaction rate strongly by competitive adsorption). The surface is nearly completely covered with B. Therefore, the initial state for the reaction is gaseous A and adsorbed B (see Fig. 3.8). For A being able to react, firstly a molecule of B must desorb with accomp-

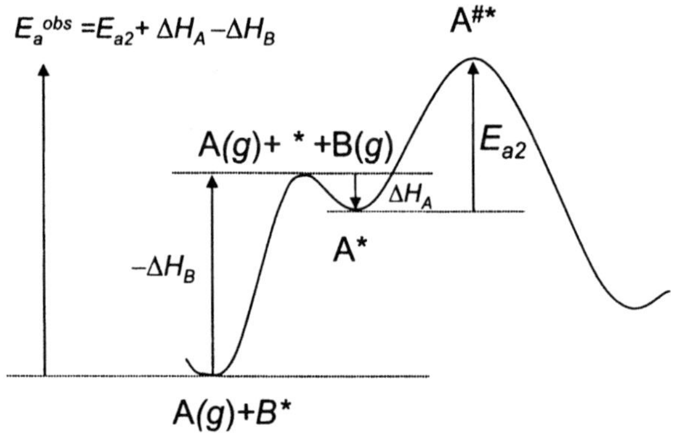

Fig. 3.8. Energy diagram for the catalyzed conversion of A to B in the case of strong adsorption of B

anying desorption enthalpy. Subsequently, A adsorbs, gaining adsorption enthalpy, and reacts through the surface reaction, where the activation energy barrier has to be overcome. Thus, the observed activation energy is higher than in the previous cases:

$$E_a^{obs} = E_{a2} + \Delta H_A - \Delta H_B \tag{3.55}$$

Note that $-\Delta H_B > -\Delta H_A$ as a consequence of the stronger adsorption of B.

(iv) Intermediate Adsorption of A and B

For intermediate values of $K_A\, p_A$ and $K_B\, p_B$expressions (3.48) and (3.50) apply for the reaction order and the observed activation energy, respectively. The reaction order will range between 0 and 1 for A, and between –1 and 0 for B. The observed overall activation energy will have intermediate values between the two extremes of cases (ii) and (iii):

$$E_{a2} + \Delta H_A < E_a^{obs} = E_{a2} + (1 - \theta_A)\, \Delta H_A - \theta_B\, \Delta H_B < E_{a2} + \Delta H_A - \Delta H_B \tag{3.56}$$

Transitions between the different situations discussed above will occur as a function of temperature, because the occupancies will vary. With increasing temperature the adsorption equilibrium constants decrease resulting in decreased occupancies. Thus, if starting with strongly adsorbing B at low temperature as example, a gradual transition can be envisaged upon increasing the temperature, during which:

- the order of A remains nearly 1
- the order of B changes from –1 to 0, and
- the observed overall activation energy will change from (3.55) to (3.53).

In addition, large changes in partial pressure can result in changes in the apparent reaction order of a component. At low p_A the reaction is first order in A, while at high p_A the rate approaches a limit, as can be expected for Langmuir adsorption, and the reaction becomes zero order in A (see Eqn. (3.48)).

Summarizing, the overall observed activation energy and the apparent reaction orders of the components depend on the degree of coverage of the active sites, which in turn depends on the temperature and partial pressures.

Analogously, limiting cases can be distinguished for the dual-site model, in which the order of A can even become negative (see for example Eqn. (3.27)). This is common for dissociation reactions.

An excellent illustration of the LHHW theory is catalytic cracking of n-alkanes over ZSM-5 [8]. For this reaction, the observed activation energy decreases from 140 to –50 (!) kJ/mol when the carbon number increases from 3 to 20. The decrease appeared to linearly depend on the carbon number as shown in Fig. 3.11. This dependence can be interpreted from a kinetic analysis that showed that the hydrocarbons (A) are adsorbed weakly under the experimental conditions. The initial rate expression for a rate-determining surface reaction applies (3.30), which in the limiting case of weak adsorption of A reduces to Eqn. (3.52). The activation energy is then represented by equation (3.53).

Measurement of the adsorption enthalpy ΔH_A revealed a linear decrease with carbon number. By applying Eqn. (3.53) the activation energy for the surface reaction, E_{a2}, was estimated. The data in Fig. 3.9 clearly shows that E_{a2} has reasonable (and positive) values, while it remains fairly constant for $n > 8$, supporting the kinetic interpretation.

Until now it has been assumed that the rate-determining step remains the same with changing reaction conditions. However, it can also change, especially under the influence of temperature [4], but also as a result of pressure changes [2].

Suppose the rate-determining step at low temperature is the desorption step. Generally, this implies that the activation energy barrier for desorption is the

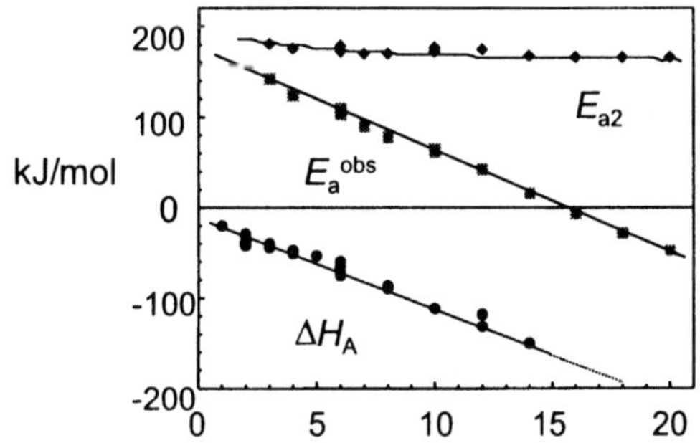

Fig. 3.9. Observed activation energies, reaction activation energies, and adsorption enthalpies in the cracking of n-alkanes over ZSM-5 (adapted from [8]).

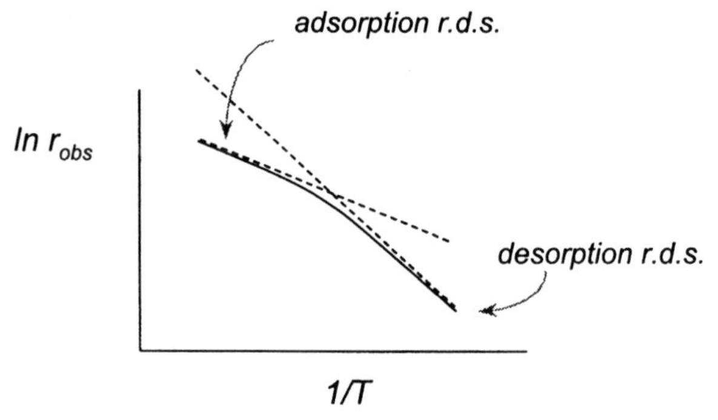

Fig. 3.10. Change of observed activation energy due to changing rate-determining step as a function of temperature.

highest. A temperature increase will enhance the rate of this step more than the other steps with lower activation energies. Then another step, for instance a dissociation step, can now become rate limiting. The temperature dependence of the overall rate will behave as depicted in Fig. 3.10. This illustrates that the observed overall activation energy decreases with increasing temperature upon a change in the rate-determining step.

In these cases one cannot comply with the assumption of only one rate determining step. To obtain an adequate rate expression, valid over the whole temperature range under consideration, two or even more steps should be assumed not to be in *quasi*-equilibrium, and are, hence, rate determining.

Energy diagram 3.7 is of course very similar in heterogeneous, homogeneous and biocatalysis, since kinetics is similar. A difference to be taken into account is that in the liquid phase adsorption is to be considered with respect to the liquid phase.

An essential difference between catalysis and gas phase kinetics is the absence of adsorption complexes in the latter case. A schematic comparison between a gas phase reaction energy diagram and one for a catalytic reaction is included in Fig. 3.7.

3.8 SABATIER PRINCIPLE — VOLCANO PLOT

According to (3.52), the larger the heat of reactant adsorption (larger K_A), the larger the overall rate of reaction. A larger heat of adsorption enhances the surface coverage (and changes the reaction order), and consequently, the reaction rate, e.g. to (3.51).

Often, a relation is found between the overall rate of reaction and the heat of adsorption. Generally, however, the rate passes through a maximum as a function of the heat of adsorption. This trend is given in Fig. 3.11.

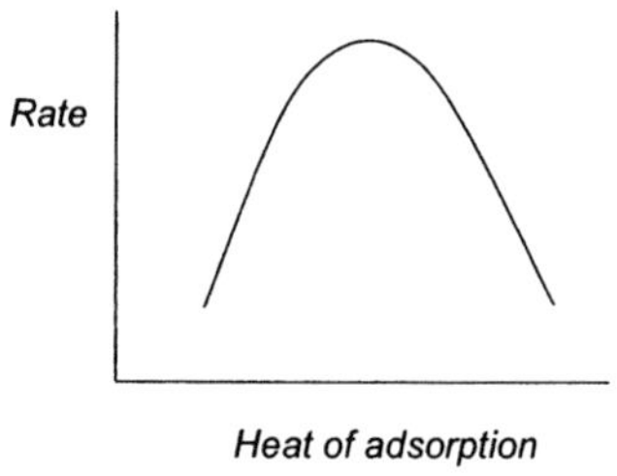

Fig. 3.11. Volcano curve for the overall reaction rate as a function of the heat of adsorption.

Volcano plots, as in this figure, have been measured for very different reactions, e.g., the formic acid decomposition, the ammonia synthesis reaction, hydrodesulfurization, or hydrodenitrogenation reactions.

As explained above, the increase of the reaction rate is due to the increased site coverage with the reactant (Eqn. (3.30)). A limit of the increase is reached once the optimum surface coverage has been reached. The rate is then controlled by the rate of product formation. The reason for the decrease in reaction rate with a further increase in adsorption strength is the increased activation energy for desorption; not only will the reactants be adsorbed more strongly, but the products will also be held more strongly (Eqn. (3.48)). Clearly, an optimum for the interaction strength between the catalytically active surface and the adsorbates exists, resulting in a maximum in the rate of reaction (the Sabatier principle). To the left of the maximum, the reaction has a positive order in the reactants, whereas to the right the order has become zero or even negative (see Eqn. (3.48). This kinetic dependence on the interaction with a catalyst can be used to test whether a Volcano plot is due to Sabatier's effect or not.

In practice, the variation of the rate of reaction with adsorption strength and the occurrence of a maximum in this rate have the following consequence for kinetic modelling of heterogeneous catalysts. Usually, the assumption of a homogeneous surface is not strictly valid. It would probably be more realistic to assume the existence of a certain distribution in the activity of the sites. However, certain sites will contribute most to the reaction, since these sites activate the reactants most. This might result in apparently uniform reaction behaviour, and can explain why Langmuir adsorption often provides a good basis for the reaction rate description. This also implies that adsorption equilibrium constants determined from separate adsorption experiments can only be used in kinetic expressions when coverage dependence is explicitly included. Otherwise, these constants have to be extracted from the rate data.

Several authors have derived rate expressions for non-uniform catalyst surfaces. Boudart and Djégà-Mariadassou [2] show that relations are obtained with a mathematical similarity to those obtained for a uniform surface. In the rate expression for ammonia production, the Temkin isotherm has been used for a long time. This isotherm accounted for a, supposedly, heterogeneous adsorption behaviour [1]. Recently, however, it has been shown that the LHHW approach can account for the data over a pressure range of ≈ 300 bar [9] without the assumption of heterogeneity. This and other examples demonstrate the usefulness of the LHHW approach for reactions of practical importance.

3.9 CONCLUDING REMARKS

The preceding sections indicate how a useful approximate reaction rate expression can be derived for catalyzed reactions, starting from an assumed kinetic model of elementary reaction steps. The derivation is based on the following assumptions:

- the reaction system is in a steady state
- the surface for adsorption and reaction is uniform;
- the number of active sites is constant, independent of reaction conditions;
- adsorbed species do not interact, apart from their reaction paths.

The form of the resulting expression differs from the gas-phase reaction rate expressions due to the presence of a denominator representing the reduction in rate due to adsorption phenomena. The individual terms of this denominator respresent the distribution of the active sites among the possible surface complexes and vacancies. Expressions of this type are termed the Langmuir–Hinshelwood–Hougen–Watson (LHHW) rate expressions in heterogeneous catalysis and Michaelis–Menten expressions in biocatalysis.

The steady-state approach generally yields complex rate expressions. A simplification is obtained by the introduction of one or several rate-determining step(s) and *quasi*-equilibrium steps, and further by the initial reaction rate approach. For complex reaction schemes, identifying the most abundant reaction intermediates ("mari") and making use of the site balance can simplify the kinetic models and rate expressions.

In practice, useful relations result even for the non-ideal heterogeneous surfaces of solid catalysts. Some reasons can be:

- similarity of mathematical relations for uniform and non-uniform adsorption models;
- Sabatier's principle of the optimum site activity. Optimum sites contribute most to the reaction, resulting in an apparently uniform behaviour.

An interesting further development in describing the kinetics of heterogeneously catalyzed reactions is the so-called microkinetics approach, whereby independent information about adsorbed species from temperature programmed desorption and spectroscopic studies are used to predetermine rate and equilibrium constants of elementary processes, thus enabling the prediction of the overall rate. Especially for metal catalyzed reactions this gives good results [9]. More information about reaction kinetics related to catalysis can be found in Refs. [10–13].

In this chapter the aspects of kinetic model selection/discrimination and parameter estimation and the experimental acquisition of kinetic data are not dealt with, since they fall outside its scope. Moreover, in interpreting the observed temperature dependence of the rate coefficients in this chapter we are assumed to be dealing with intrinsic kinetic data. As will be shown in Chapter 8, parasitic phenomena of mass and heat transfer may interfere, disguising the intrinsic kinetics. Criteria will be presented there to avoid this experimental problem.

NOTATION

A_0	preexponential factor	s^{-1}
c	constant in Freundlich isotherm	atm^n
c_i	concentration	$mol\ m^{-3}$
E_a	activation energy	$J\ mol^{-1}$
E_0	total enzyme concentration	$mol\ dm^{-3}$
h	Planck's constant	J s
k_B	Boltzmann constant	$J\ K^{-1}$
$k_{barrier}$	number of molecules reacting per unit time	s^{-1}
k_i	reaction rate constant for reaction i	s^{-1} (for first order reaction)
K_i	equilibrium constant of reaction i	atm^{-1} or $m^3\ mol^{-1}$
K_m	Michaelis constant	$mol\ dm^{-3}$
K'_m	Modified Michaelis constant (inhibition)	$mol\ dm^{-3}$
K^i_m	Michaelis constant inhibitor	$mol\ dm^{-3}$
n	constant in Freundlich isotherm	–
n_i	reaction order in i	–
N_T	total concentration of active sites	mol $(g\ cat)^{-1}$ or mol $(m^2\ cat)^{-1}$
p_i	partial pressure of component i	atm, kPa
r	reaction rate (overall)	$atm\ s^{-1}$ or $mol\ m^{-3}\ s^{-1}$
r_i	reaction rate of reaction i	$atm\ s^{-1}$ or $mol\ m^{-3}\ s^{-1}$
R	ideal gas constant	$J\ mol^{-1}\ K^{-1}$
s	number of nearest neighbours of active site	–
t	time	s
T	temperature	K
v	reaction rate (biocatalysis)	$mol\ s^{-1}\ dm^{-3}$
V_m	maximum reaction rate (biocatalysis)	$mol\ s^{-1}\ dm^{-3}$

Greek

α	constant in Temkin isotherm	-
ΔG	Gibbs free energy change	$J\ mol^{-1}$

ΔH	Enthalpy change	J mol^{-1}
ΔS	Entropy change	J mol^{-1} K^{-1}
θ_i	fraction of total number of sites occupied by i	–
ν_i	stoichiometric number of elementary step i	–

Subscripts

0	initial or at zero coverage
+	forward reaction
–	backward reaction
eq	equilibrium
g	gas phase
obs	observed / apparent

Superscripts

0	standard conditions
#	transition state
*	surface species

REFERENCES

1 M.I. Temkin, *'The Kinetics of Some Industrial Heterogeneous Catalytic Reactions'* in *Advances in Catalysis*, Vol. 28, Academic Press, New York, 1979, p. 173.

2 M. Boudart and G. Djégà-Mariadassou, *Kinetics of Heterogeneous Catalytic Reactions*, Princeton University Press, Princeton, NY, 1984.

3 C.N. Satterfield, *Heterogeneous Catalysis in Industrial Practice*, 2nd ed., McGraw-Hill, New York, 1991, p.61.

4 G.F. Froment and K.B. Bischoff, *Chemical Reactor Analysis and Design* 2nd ed., Wiley, New York, 1991.

5 O.A. Hougen and K.M. Watson, *Chemical Process Principles*, Vol. III, Wiley, New York, 1947.

6 R.W. Maatman, Site Density and Entropy Criteria in Identifying Rate-Determining Steps in Solid-Catalyzed Reactions, in *Advances in Catalysis*, Vol. 29, Academic Press, New York, 1980, p. 97.

7 R.A. van Santen and H. Niemantsverdriet, *Chemical Kinetics and Catalysis*, Plenum, 1995.

8 J. Wei, *Ind. Chem.Eng. Res.*, 33 (1994) 2467–2472.

9 P. Stoltze and J.K. Nørskov, An interpretation of the high-pressure kinetics of ammonia synthesis based on a microscopic model. *J. Catal.* 110, 1 (1988).

10 J.M. Thomas and W.J. Thomas, *Introduction to the Principles of Heterogeneous Catalysis*, Academic Press, London, 1967.

11 R. Mezaki and H. Inoue, *Rate Equations of Solid-Catalysed Reactions*, University of Tokyo Press, Tokyo, 1991.

12 F. Kapteijn and J.A. Moulijn, Kinetics and transport processes, in: *Handbook of Heterogeneous Catalysis* Vol. 3, Ch. 6.1, G. Ertl, H. Knözinger and J. Weitkamp (Eds.), VCH, Weinheim, p. 1189 (1997).

13 L. Stryer, *Biochemistry*, W.H. Freeman, New York, Ch. 8, 1995.

Fundamental catalysis

Chapter 4

Bonding and elementary steps in catalysis

4.1 INTRODUCTION

In this chapter we discuss bonding to transition metal elements as a basis for our understanding of the elementary reactions that can occur in transition metal complexes and on surfaces of solids. The nature of the chemical bond will be discussed in Section 4.2 in terms that are applicable to both homogeneous and heterogeneous catalysts, the emphasis being on what they have in common and what distinguishes them. Section 4.3 gives a summary of the organometallic reactions that are important as elementary steps in homogeneous catalysts; working with this 'basis set' it is in many cases possible for the reader to write down a likely mechanism for a reaction for which only the overall formula is known. In part these reactions are also relevant to heterogenous catalysis and this section forms a good introduction to the next, Section 4.4, which deals with reactions on metal surfaces and solid acids. It will be very instructive to compare the steps of the organometallic complexes with the new possibilities created by the metal and oxide surfaces. In passing, we have also tried to point out the differences existing in the vocabulary between the homogeneous schools and the heterogeneous schools in catalysis. We do not propose that unique meanings can be attached to the words used, but it is necessary, and hopefully sufficient, that the reader is made aware of these minor obstacles. The chapter will conclude with a section on elementary reaction steps in biocatalytic reactions (4.6) and biocatalytic oxidation reactions (4.7); many of the concepts presented in earlier sections will be recognized also in the biocatalytic context but of course the detailed chemistry will be different.

4.2 BONDING

4.2.1 General Introduction

This introductory section intends to present the basic bonding concepts, necessary to understand chemical bonding to transition metal complexes, clusters and transition metal surfaces.

The valence electrons in transition metals are distributed over $d(n-1)$ and $s(n)$ and $p(n)$ electrons, n labeling the principle quantum-numbers.

The number of d electrons can vary between 0 and 10, distributed over five d-atomic orbitals. The maximum number of s electrons is 2 and p electrons is 6. The latter are distributed over the three p atomic orbitals. Varying the position of the metal element from left to right through a row of the periodic systems shows an increase of the number of valence electrons.

The d-valence electron occupation reaches saturation for the elements Cu, Ag and Au.

Whereas the distribution of electrons over the d and s electrons is irregular for the transition metal atoms to the left of Cu, Ag and Au, this is much less the case for the transition metals in their solid state. Here the electron occupation of the s, p valence electronband is approximately constant and equal to one electron per atom, whereas the number of d-valence electrons steadily increases moving from the left to the right through a row of the periodic system.

The spatial extension of the $s(n)$ and $p(n)$ atomic orbitals is significantly larger than the more contracted $d(n-1)$ atomic orbitals. Therefore the contribution to the chemical bond from the interaction with the s and p atomic orbitals is sometimes dominant and always substantial.

4.2.1.1 The Covalent Bond: Homonuclear Systems

For clarity we will limit the introductory discussion to chemical bonding to interacting atoms that only contribute an s-valence atomic orbital to the chemical bond.

To illustrate relevant aspects of the electronic factors that control the difference in bonding to a single atom and bonding to several atoms, i.e. coordination changes, Fig. 4.1 gives the orbital energies that correspond to the two different situations. At the left side of the scheme (Fig. 4.1) for two atoms of equal energy, the energies of the bonding and antibonding orbitals are given, as computed within the Extended Hückel Approximation. In this method [1] the atomic orbital energies are denoted as α, the overlap energy by β' and non-orthogonality of orbitals is accounted for by the overlap S. β' leads to bond formation and has a negative value. It is due to the attraction of an electron on one atom by the nuclear charge on the other atom. The overlap S, as we will see, is

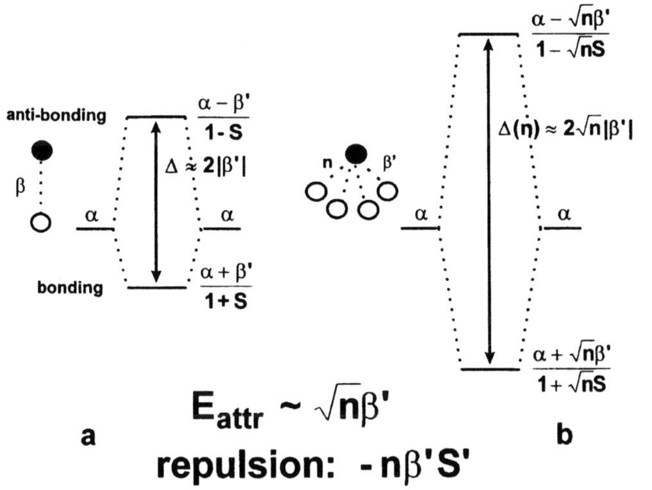

Fig. 4.1. Interaction scheme between two atoms and one atom in a cluster.

responsible for the repulsive part of the potential energy curve. The bond energy follows by occupying the molecular orbitals with electrons, (maximum 2 per orbital), adding the orbital energies and subtracting the atom energies of the system before interaction.

In the interacting system one orbital energy is lower than that of the free atoms and it is called the bonding orbital. The other orbital is destabilized; the anti-bonding orbital. For finite overlap S ($|S|$ is always smaller than 1, the normalization constant), the stabilization of the bonding orbital is less than the destabilization of the antibonding orbital. One notes that the difference in the energies of the bonding and antibonding orbitals $\Delta \approx -2\beta'$. The exact expression for $\Delta = \dfrac{-2(\beta' - \alpha S)}{1 - S^2}$. Δ is a measure of the covalent interaction strength.

When the system contains 2 electrons they will occupy the bonding orbital and the bond energy is equal to:

$$E_b^{at} = -\Delta(1 - S) \tag{4.1a}$$

$$\approx 2\beta' \qquad (S << 1) \tag{4.1b}$$

When 4 electrons are present in the system, the bonding as well as antibonding orbitals become occupied. The bonding energy ΔE_b now becomes:

$$E_b^r = -2\Delta S \tag{4.2a}$$

$$\approx -4\beta' S \qquad (S << 1) \tag{4.2b}$$

Since β' is a negative, attractive energy, it follows that the interaction is attractive when the total number of orbitals is half occupied, but is repulsive when all orbitals are doubly occupied.

The attractive interaction, approximately proportional to β' is the covalent attractive contribution to the chemical bond energy. The repulsive part, approximately proportional to $-\beta' S$ depends explicitly on the overlap S. It can be understood as the Pauli repulsion, due to repulsive interaction of overlapping orbitals with electrons of equal spin.

As we will see later, orbital interactions between molecules with partially occupied orbitals will be predominantly attractive. However the interactions between the deeper doubly occupied orbitals are repulsive. The latter are responsible for steric repulsive interactions.

The right part of Fig. 4.1 gives the changes in electronic structure when an atom interacts with n neighboring atoms. Again the free atom energies are assumed equal. The n cluster atoms are assumed not to interact. The consequences of this interaction will be discussed later in this section.

The most important change when co-ordination number n changes is the value of $\Delta(n)$, the difference between bonding and antibonding orbital energy values:

$$\Delta(n) = \frac{-2\sqrt{n}(\beta' - \alpha S)}{1 - nS^2} \tag{4.3a}$$

$$= \sqrt{n}\,\Delta(1) \qquad (nS^2 << 1) \tag{4.3b}$$

The attractive interaction when the atomic orbitals are half filled becomes:

$$E_b^{at}(n) = \sqrt{n}\, E_b^{at}(1) \tag{4.4}$$

The repulsive Pauli interaction, that one finds for the system with all orbitals doubly occupied becomes:

$$E_b^{r}(n) = 2n\,\Delta(1) S \tag{4.5}$$

Comparison of expression (4.5) with (4.1) shows that the attractive interaction becomes approximately proportional to the square of the number of coordinating atoms, but the Pauli-repulsion term is linear in the number of coordinating atoms.

As we will see later this difference in the dependence of the attractive and repulsive interaction on co-ordination number will lead to a preference for low coordination numbers for systems that have significant Pauli-repulsive interactions.

4.2.1.2 The Covalent Bond; Heteronuclear System: The Lewis Acid—Lewis Base Interaction

Aspects of bonding between non-equal atoms are illustrated by Fig. 4.2, which gives the orbital energy levels for a two atom system with free atom energies α_0 and α_1. Again bonding (ε_b) and antibonding (ε_a) levels are formed.

When the energy difference $|\alpha_0 - \alpha_1| >> |\beta'|$, their respective energies are:

$$\delta\varepsilon_b = \varepsilon_b - \alpha_1 = -\frac{\beta'^2}{\alpha_0 - \alpha_1} + \frac{1}{2}\Delta^1 S \tag{4.6a}$$

$$\delta\varepsilon_a = \varepsilon_a - \alpha_0 = +\frac{\beta'^2}{\alpha_0 - \alpha_1} + \frac{1}{2}\Delta^0 S \tag{4.6b}$$

Two terms contribute to the shifted orbital energies. The first terms represent the stabilization and destabilization of the initially non-interacting orbitals due to covalent bonding. Their form, proportional to the square of the overlap energy and inverse to the energy difference, is the usual one from Frontier Orbital Theory. The second term is the Pauli-repulsion. The expressions for Δ^1 and Δ^0 are slightly different from Δ.

$$\Delta^1 = -2\left(\frac{\beta' - \alpha_0 S}{1 - S^2}\right) \tag{4.7a}$$

$$\Delta^0 = -2\left(\frac{\beta' - \alpha_1 S}{1 - S^2}\right) \tag{4.7b}$$

One notes that if the two orbitals become doubly occupied the two fragments have a repulsive interaction equal to:

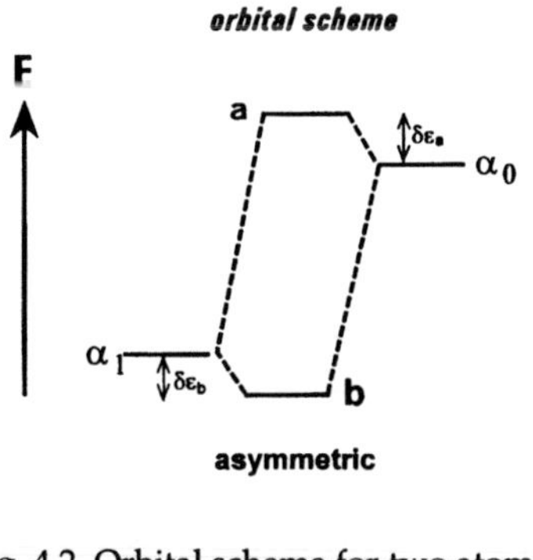

Fig. 4.2. Orbital scheme for two atoms.

$$E^{\text{rep}} = (\Delta^1 + \Delta^0)S \tag{4.8}$$

When each of the two atoms has initially one electron, the bond energy becomes:

$$E_b^{\text{cov}} = -\frac{2\beta'^2}{\alpha_0 - \alpha_1} + \Delta^1 S + (\alpha_1 - \alpha_0) \tag{4.9}$$

The third term arises from electron transport from atom (0) to atom (1). (α_0-α_1) is equal to the Ionization Potential(I.P.) of atom (0) minus the Electron Affinity (E.A.) of atom (1). This electron transfer results in a positive charge on atom (0) and a negative charge on atom 1.

Therefore the total bond energy is equal to:

$$E_b(\text{total}) = E_b^{\text{cov}} + E_{\text{electrostatic}} \tag{4.10}$$

To the covalent contribution an electrostatic term has to be added.

$$E_{\text{electrostatic}} = -\frac{|q|^2}{|r_0 - r_1|} \tag{4.11a}$$

with

$$q = \pm\left\{1 - 2\left|\frac{\beta'}{\alpha_0 - \alpha_1}\right|^2\right\} \tag{4.11b}$$

$$= \pm\{1 - \Delta e\} \tag{4.11c}$$

The charge q is reduced due to backdonation of electrons from the doubly occupied orbital on atom (1) by an amount Δe. Backdonation reduces the electron density on atom 1, but enhances the electron density on atom 0. The corresponding energy contribution equals:

$$E_{\text{backdonation}} = -\frac{2\beta'^2}{\alpha_0 - \alpha_1} \tag{4.12a}$$

$$= -(\alpha_0 - \alpha_1)\cdot\Delta e \tag{4.12b}$$

Expressions (4.12) can be considered the interaction-energy between a Lewis acid (empty orbital, α_0) and Lewis base (doubly filled orbital, α_1).

Since the disturbance of the original orbitals with energies α_1 and α_0 is small, the orbital φ_{α_1} becomes mainly part of the bonding orbital ψ_b and atomic orbital φ_{α_0} becomes mainly part of the antibonding orbital ψ_a. The admixture of the respective other atomic orbitals in the bonding and antibonding molecular orbitals gives electron transfer Δe.

4.2.1.3 The Embedded System

Figure 4.3 schematically illustrates electronic structure changes when a cluster becomes embedded in a surface. The first scheme (Fig. 4.3a) is similar to scheme 4.1b. When an atom interacts with *n* neighbours (that have no interaction between themselves) the covalent interaction is $\sim \sqrt{n}\,\beta'$.

Figure 4.3b shows the electron distribution $\rho(E)$ as a function of electron energy *E* for a surface cluster of $(z + 1)$ atoms embedded in the surface of a metal atom interacting with overlap energy β. Because the bulk system terminated with a surface is semi-infinite, the energies of an infinite number of metal orbitals are continuously distributed between an upper and lower energy value determined by lattice topology and interaction energy parameter β. At the surface the width of energy-density on the surface atoms is proportional to $\sqrt{z}\,\beta$ [1] (*z* is the number of nearest neighbour metal atoms of a surface atom, β the overlap energy of the metal atoms).

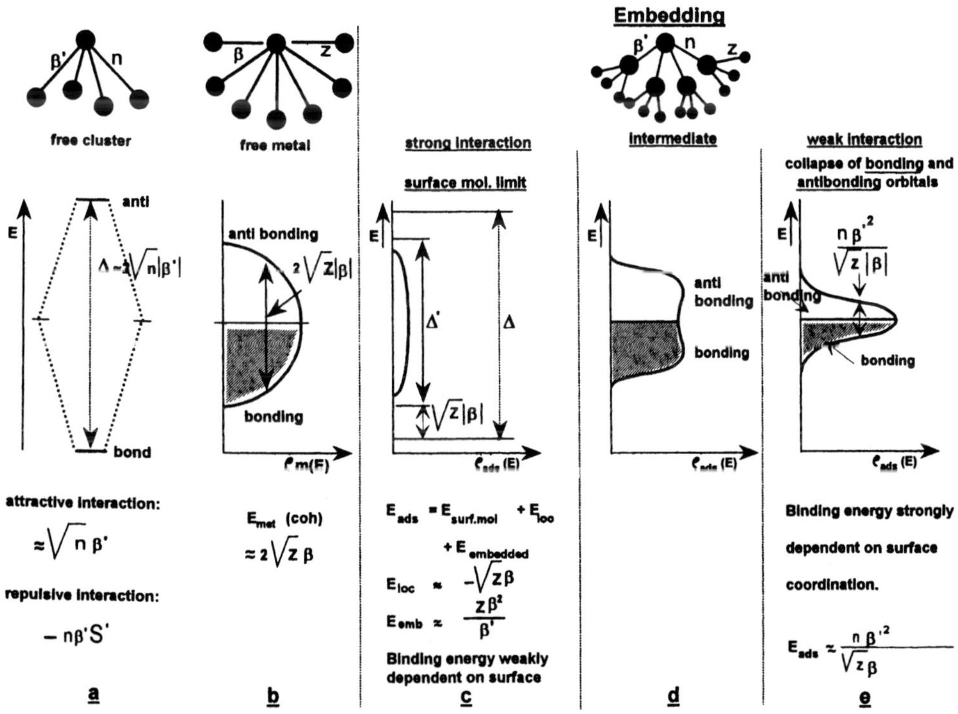

Fig. 4.3. Electronic structure changes when a cluster is embedded in a surface.

The attractive contribution to the bond energy of the surface metal atoms has the same co-ordination dependence.

Electron density at the bottom of the metal electron energy belongs to bonding-orbitals between the metal atoms, electron density at the top of this band belongs to antibonding orbitals.

Figures 4.3c, d and e illustrate changes in electronic structure of cluster 3a when the cluster atoms become part of a surface. Three physically different situations have to be distinguished: surface molecule limit, intermediate adsorption and weak adsorption. Figure 4.3c schematically presents the electronic structure in the surface molecule limit. Then the interaction between adatom and surface atoms β' is significantly larger than β:

$$\frac{\sqrt{n}\beta'}{\sqrt{z}\beta} >> 1 \tag{4.13}$$

Whereas in the free adatom-cluster the difference between bonding and antibonding orbitals is Δ, this splitting decreases for the embedded cluster. Embedding delocalizes the cluster atom electrons and the effective interaction between adatom and cluster-atoms decreases. This is reflected in the smaller bond splitting Δ' compared to Δ. The decrease in bandsplitting is proportional to $\sqrt{z}\beta$, the localization energy of the surface electrons.

The adsorption energy can be approximately written as the sum of three terms:

$$E_{ads} = E_{surface\ mol} + E_{loc} + E_{emb} \tag{4.14a}$$

$$= (2\sqrt{n}\beta' - n\beta' S) - 2\sqrt{z}\beta - \frac{4z\beta^2}{\sqrt{n}|\beta'|} \tag{4.14b}$$

The first term is the interaction energy within the free cluster; the second term the localization (or decoupling) energy of the surface cluster electrons, the third term the interaction due to the restoration of bonding between cluster (now in interaction with adsorbate) and surface. Remember that the overlap energies are negative.

In the surface molecule limit, the first term dominates and the last term is small compared to the localization-energy.

Hence in the surface molecule limit the interaction-energy between adsorbate and surface decreases when the number of surface neighbour atoms (z) of the interacting surface atom increases.

This is a result often found in chemisorption. The chemisorption strength to a dense surface (surface atoms have high co-ordination, the coordination number is proportional to the number of nearest-neighbour atoms) is small compared to the adsorption energy to an open surface (surface atoms have low co-ordination).

Note that the adsorption energy dependence on surface atom co-ordination relates to the weakening of the bond energies between the surface atoms (due to electron localization, only partially being restored by E_{emb}).

When intermediate surface bonding ($\sqrt{n}\beta' \approx \sqrt{z}\beta$) occurs the adatom bonding and antibonding orbitals are not any more separate, but then overlap as broadened bands in the surface electron density regime (Fig. 4.3d).

In the weak interaction limit (Fig. 4.3e) the orbitals collapse. The localization energy cannot be overcome and the surface electron structure is only weakly disturbed.

The band width of the electron distribution on the surface atoms (the analogue of Δ) has now become very small and equal to:

$$\Delta_w = \frac{2n|\beta'|^2}{\sqrt{z}|\beta|} = 2\frac{\sqrt{n}\beta'}{\sqrt{z}\beta}\cdot(\sqrt{n}|\beta'|) \tag{4.15a}$$

$$<< \sqrt{n}|\beta'| \tag{4.15b}$$

The covalent contribution to the bond energy has now become very much less than the free cluster value $\sqrt{n}|\beta'|$. The bond energy decreases ~$(\sqrt{z})^{-1}$ when the surface co-ordination of surface atoms increases.

4.2.2 *Bonding in Transition Metal Complexes*

In transition metal complexes the distribution of electrons over the *d*-atomic orbitals determines complex geometry. We will illustrate this for chemical bonding in carbonyl complexes.

Oxidative addition or reductive elimination reactions (see Section 4.3) require electron occupation or depletion of antibonding orbitals since bonds are formed or broken. The presence of occupied or empty metal *d*-orbitals is essential for such processes. This will be illustrated for dissociative adsorption H_2 and the formation of the formyl bond from adsorbed H and CO (see Ref. [2] for more detail).

Also, the similarity in electronic structure of catalytically important inorganic surfaces as NiO or MoS_2 to that of co-ordination complexes will be pointed out.

Figure 4.4a illustrates the geometry of tetrahedrally co-ordinated Ni in the $Ni(CO)_4$ complex.

The valence electronic structure of CO is sketched in Fig. 4.5. The valence orbitals consist of the doubly occupied σ-symmetric lone pair orbital, directed away from the carbon atom (5σ), and two empty asymmetric $2\pi^*$ orbitals. Their energy separation is approximately 7 eV.

The relative energies of CO valence orbitals and Ni atomic orbitals are sketched in Fig. 4.4b. The Ni atom has eight occupied 3*d* and two occupied 4*s* orbitals.

Because of the difference in spatial extension, one can ignore hybridization between *d* and the *s*, *p* electrons. As in tetrahedrally co-ordinated carbon atoms, the 4*s* and three 3*p* orbitals can be hybridized to four sp_3 orbitals (Fig. 4.4a). This

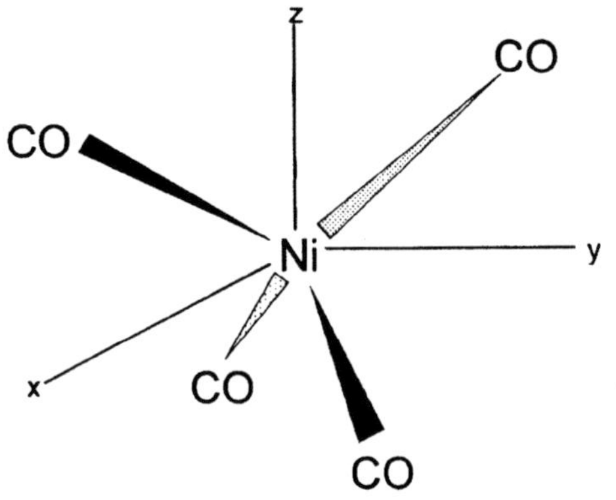

Fig. 4.4a. Geometry of $Ni(CO)_4$.

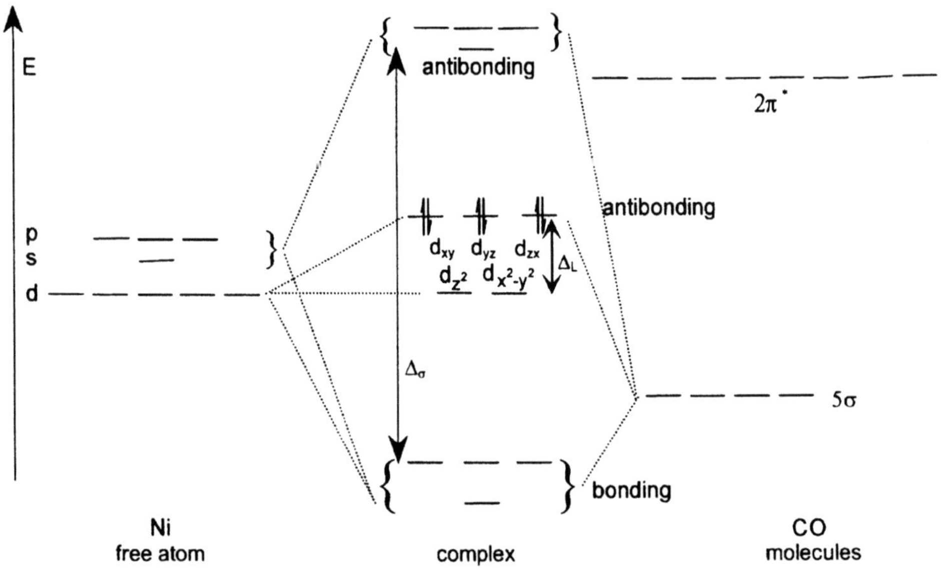

Fig. 4.4b. σ-interaction in $Ni(CO)_4$.

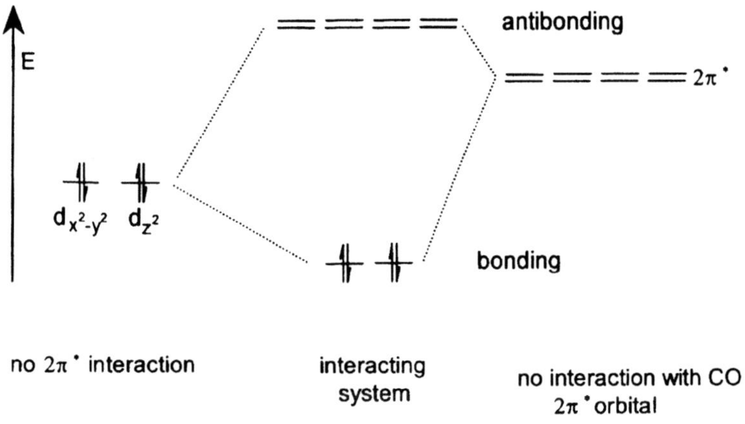

Fig. 4.4c. π-interaction in $Ni(CO)_4$.

will push their energy upwards with respect to the 4*s* orbital of atomic Ni. Therefore the two 4*s* electrons present in the free Ni atom will transfer to the empty 4*d* orbital and the Ni 4*d* electron system becomes d_{10}.

In the $Ni(CO)_4$ complex the four empty Ni sp_3 orbitals are directed towards the four doubly occupied 5σ CO orbitals, with which they form four occupied bonding and four unoccupied antibonding orbitals, separated by bandgap parameter Δ_σ. Because of the large overlap between the Ni sp_3 orbitals and the CO 5σ orbitals the resulting covalent bond will be strong and dominate the Ni–CO bond interaction. Because the CO 5σ orbitals are doubly occupied and the Ni sp_3 orbitals are empty, this interaction is called *donative* with respect to the CO ligand. As in a Lewis base–Lewis acid interaction, a fraction of electronic charge is donated into the empty Ni orbitals from the doubly occupied CO 5σ orbitals.

Weak covalent interactions with the *d* atomic orbitals are also present, leading to the electronic energy level scheme as shown in Fig. 4.4b. The total number of valence electrons is 18. The complex is stable at this electron count because of the large energy gap between highest occupied *d* levels and the antibonding σ orbitals. This corresponds to the so-called 18 electron rule.

Compounds are stable as long as no significantly destabilized antibonding orbitals become occupied.

The d_{xy}, d_{zx} and d_{zy} atomic orbitals on the Ni atom become part of the weakly destabilized antibonding levels due to interaction with the 5σ CO levels. Since all of the *d* orbitals are doubly occupied, the overall 5σ CO 4*d* Ni interaction is repulsive.

So far we have not discussed the interaction of the Ni electrons with the empty CO 2π* orbitals. Because of symmetry constraints, these orbitals interact with the doubly occupied $d_{x^2-y^2}$ and d_{z^2} orbitals (Figs. 4.4b and 4.4c). Bonding and antibonding orbitals are formed as a linear combination of the respective Ni *d* and

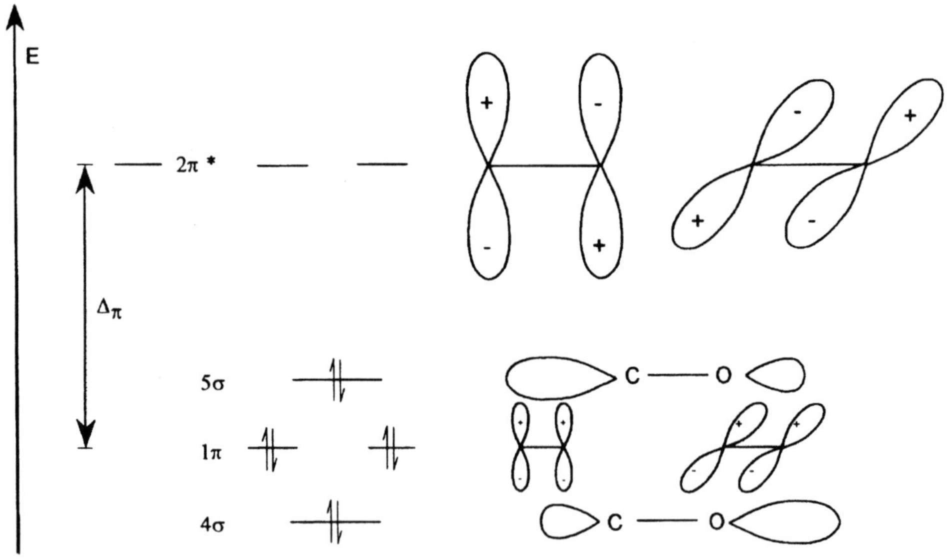

Fig. 4.5. Molecular orbital scheme of the electronic structure of CO.

CO 2π* orbitals. The bonding orbitals become doubly occupied and have mainly a d character. They are downwards shifted with respect to the Ni $d_{x^2-y^2}$ and d_{z^2} orbitals.

The antibonding orbitals remain empty and have mainly a 2π* character. This π interaction system is called the backdonation part of the CO–metal interaction, because the bonding interaction results in electron transfer from the Ni 4d atomic orbitals to the CO 2π* orbitals.

Since the 2π* orbitals of CO are antibonding with respect to the CO bond, population of the 2π* orbitals weakens the CO bond. This can be observed using infrared spectroscopy, where one usually finds a lowering of the CO stretching frequency when CO becomes part of co-ordination complexes or adsorbs to surfaces. It implies significant backdonation.

From the above it may be clear that the magnitude of this shift has no immediate relation to the strength of the metal to carbon bond. For instance, in $Ni(CO)_4$ the bondstrength is dominated by the donative σ-type interaction between the CO 5σ orbitals and Ni 4s and 4p atomic orbitals.

The orbital interaction scheme that describes bonding of ligands as CO, but also of other ligands, in terms of the sum of donative and backdonative interactions is called the Chatt–Dewar–Duncanson picture of the chemical bond. Historically it was first used to describe bonding of ethylene. As we will see later, bonding to surfaces (Section 4.4.1.2) is quite analogous and there it is called the Blyholder model.

Within the classical Chatt–Dewar–Duncanson and Blyholder descriptions of chemical bonding as a sum of donative and backdonative interactions, only the attractive interactions (see Eqn. (4.12a)) between doubly occupied and empty valence orbitals are considered.

The case of $Ni(CO)_4$ illustrates that repulsive interactions between doubly occupied orbitals, due to filling of antibonding orbitals are also important. The need to reduce this repulsive interaction often controls the particular bonding configuration of adatoms to surfaces or in cluster compounds.

For Co, having one electron less than Ni, one finds that $Co(CO)_4$ can accept one additional electron. This is the reason why dimerization occurs giving the stable $[Co(CO)_4]_2$ complex, but bonding of $Co(CO)_4$ with an hydrogen atom can also occur. Because of the stability of the $Co(CO)_4$ anion (18 electron rule), the hydrogen atom in the hydrogen cobalt carbonyl complex has an acidic character:

$$HCo(CO)_4 \rightarrow H^+ + Co(CO)_4^- \qquad (4.16)$$

In contrast, a covalent metal hydride bond is formed in $HMn(CO)_5$. Its bonding scheme is shown in Fig. 4.6. Manganese has seven valence electrons: the H atom donates one electron. Again, in the octahedral complex bonding with CO and H is dominated by the donative interaction between the doubly occupied CO 5σ orbitals, the H s orbital and empty Mn 4*s*, 4*p* and $d_{x^2-y^2}$ and the metal d_{z^2} orbitals. Similar to the situation as described in Section 4.2.1.2, bonding between H and Mn can be best described as electron transfer from Mn to H, to give Mn^{+1} and H^- and a donative interaction between H^- and empty Mn^{+1} orbitals. For simplicity in Fig. 4.6 the H(1s) orbital energy and Co 5σ orbitals have been treated as equivalent. One notes that $HMn(CO)_5$ again is an 18 electron system. Now the antibonding $d_{x^2-y^2}$ and d_{z^2} type orbitals are also non-occupied and the interaction between the *d*- as well as *s*, *p* type orbitals of Mn^+ and CO and H^- is only of donative and attractive nature. It results in a stable complex, having stronger bonds than in $Ni(CO)_4$.

Electron backdonation occurs between the doubly occupied d_{xy}, d_{yz} and d_{zx} orbitals and empty CO 2π* orbitals.

As a prelude to oxide or sulfide surfaces, it is useful to consider the electronic structure of $Mn(CO)_5$. The removal of H changes octahedral co-ordination to one of fewer neighbours.

The resulting orbital scheme is shown in Fig. 4.7. Now only five occupied, bonding σ type orbitals are formed and electrons redistribute over the non- and anti-bonding *d*-character orbitals of Mn. The Mn atom is neutral. The non-shared electron occupies an antibonding d_{z^2} type orbital (Fig. 4.7b) that has hybridized along the *z*-axis (the direction of the empty site), with *s* and p_z atomic orbitals. A dangling outwards directed orbital is formed containing one electron. Another d_z^2, *s*, p_z orbital combination is directed inwards and forms a strong bond with a CO 5σ orbital.

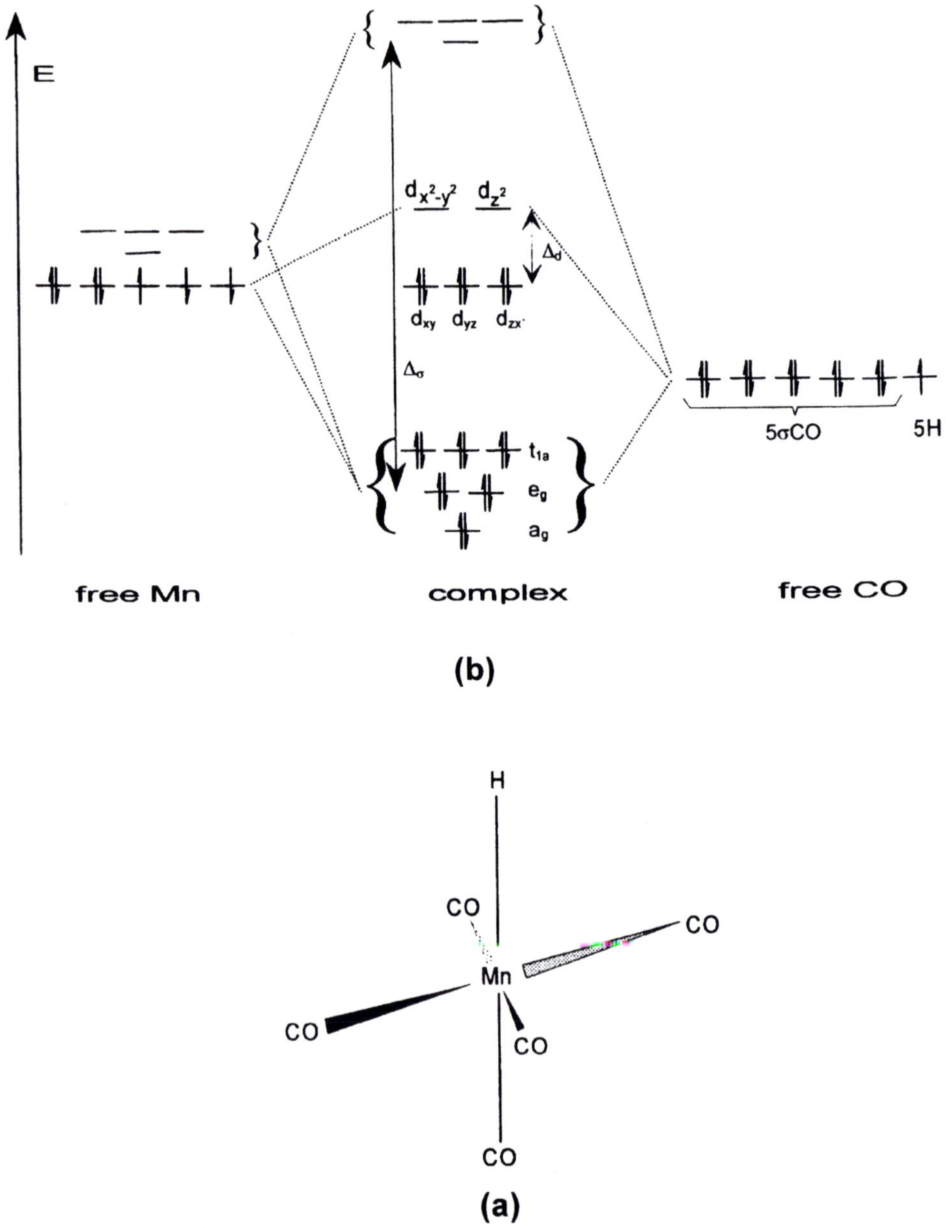

Fig. 4.6. (a) Structure of the $HMn(CO)_5$ complex. (b) The molecular orbital scheme of octahedral $HMn(CO)_5$ complex; CO $2\pi^*$ interaction not included.

We have seen that complexes become stabilized for particular electron counts. When an increase in electron count results in the population of strongly anti-bonding orbitals, no stable compounds are formed. An analogous rule has been

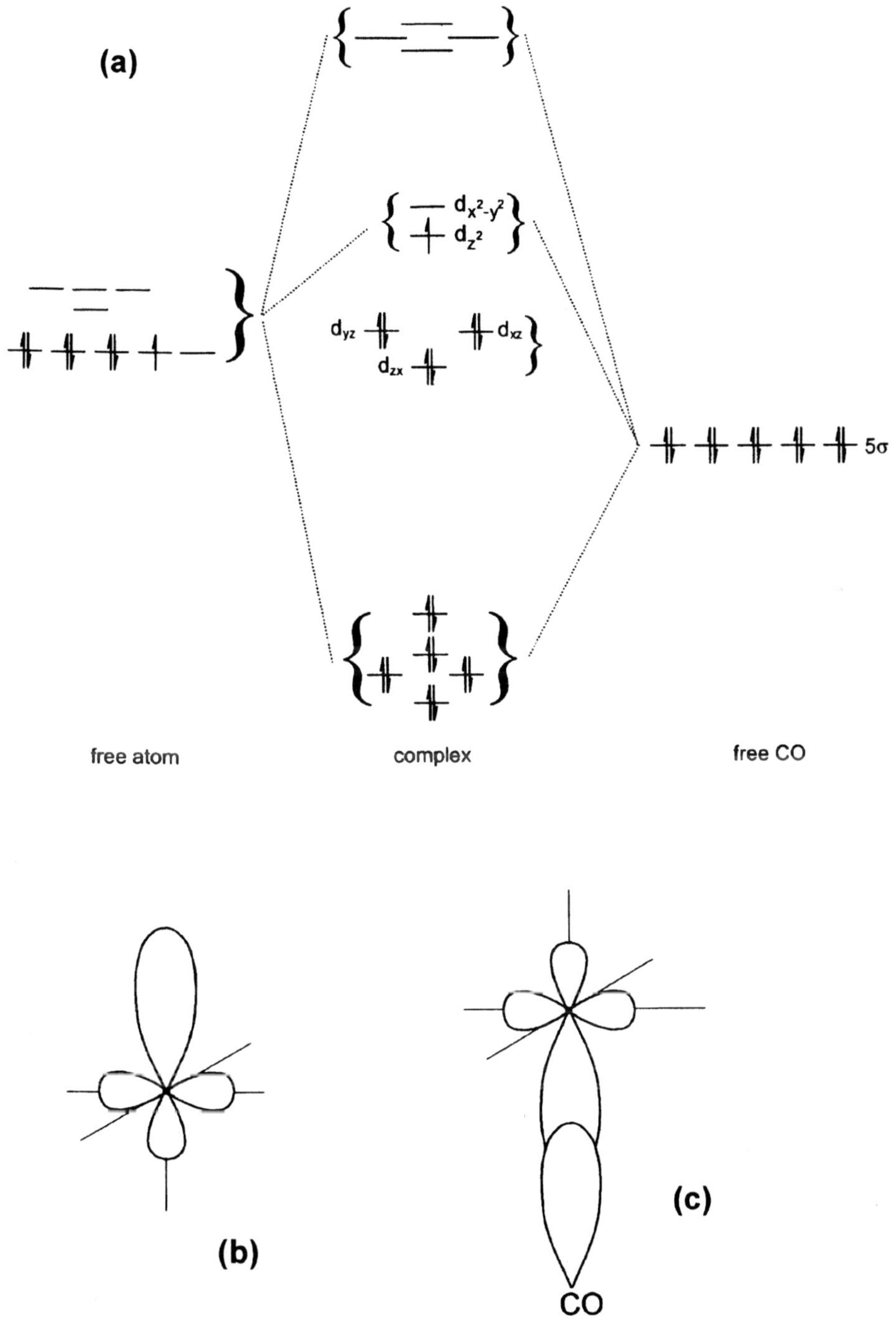

Fig. 4.7. (a) Molecular orbital scheme of Mn $(CO)_5$. (b) The antibonding $d_z{}^2$ type dangling bond orbital. (c) The bonding $d_z{}^2$–5σ orbital system.

found for the stability and hence reactivity of metal clusters. It stems from the large overlap of the *s* and *p* atomic orbitals and the resulting importance of the *s* and *p* valence electrons to the cohesive energy of the metal clusters. Each atom can be considered to contribute one electron per atom to the *s*, *p* valence electron bonds of the cluster. In a closed packing geometry the clusters are approximately spherical, and hence, a stable cluster will be one with $n = 2, 8, 18$, etc., n being the number of atoms that are part of the cluster. Indeed, beam experiments indicate exceptional stability or nonreactivity for such cluster atom numbers for many metals. The shell rule, however, is very approximate and in most cases it is better to relate surface atom co-ordination numbers with cluster reactivity.

The description of the bonding of transition metal cations in oxide or sulfide surfaces, is very similar to that of the carbonyl complexes. The chemistry of these ions is related to that in co-ordination complexes when cation charge and local co-ordination are similar. We will illustrate this point of view for NiO and MoS_2. The local co-ordination of the cations is illustrated in Fig. 4.8.

In oxides or sulfides the anionic electrons are considered to contribute one doubly occupied σ-type atomic orbital to the anion–cation bond and two doubly occupied π-type orbitals.

The approach we follow is justified, as long as the anionic orbitals remain doubly occupied. Then the electron distribution around the anions is spheric. Using this procedure the orbital schemes for bonding of octahedral Ni^{2+} or prismatic Mo^{4+} can be readily deduced. The corresponding orbital schemes are shown in Fig. 4.9.

The Molecular Orbital scheme for NiO is very similar to that of $HMn(CO)_5$. The six bonding σ type orbitals are mainly localized on oxygen or sulfur and doubly occupied. In NiO, as in $HM(CO)_4$, the $d_{x^2-y^2}$ and d_{z^2} atomic orbitals are weakly antibonding. In MoS_2 the splitting of the antibonding and non-bonding *d*-orbitals is different. Therefore at high *d*-electron-occupation the prismatic structure becomes less stable than octahedral co-ordination.

The stability of the compound increases when, compared to Ni^{2+}, the number of electrons on the central M^{n+} ion decreases because less antibonding *d*-orbitals become occupied. Co^{3+} or Fe^{2+} should give optimum bonding, since with less

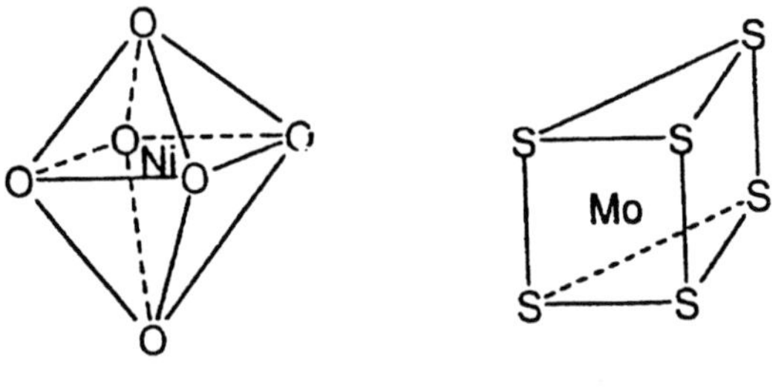

Fig. 4.8. Local cation coordination in NiO and MoS_2.

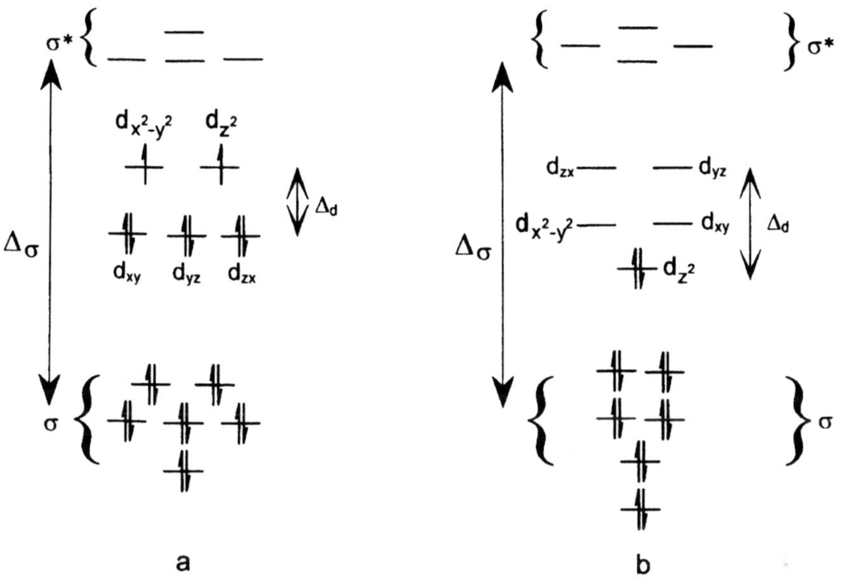

Fig. 4.9. (a) Molecular Orbital scheme of octahedrally coordinated Ni^{2+} in NiO. (b) Molecular Orbital Scheme of prismatically coordinated Mo^{4+} in MoS_2.

electrons the non-bonding or weakly bonding *d*-electrons become non occupied.

At the surface the cations and anions have fewer neighbours. The resulting electronic reorganization of valence electrons is quite analogous to that previously discussed for $Mn(CO)_5$.

In the (100) surface of NiO, Ni^{2+} becomes five co-ordinated to oxygen. The outward-directed orbital is the rehybridized d_{z^2} orbital, which is doubly occupied and hence will have a strong repulsive interaction with occupied adsorbate valence orbitals of similar symmetry.

Again Co^{3+} or Fe^{2+}, in similar co-ordination will have a strong interaction, because of the absence of this repulsive interaction.

Four co-ordinated Ni^{2+} on edges of NiO will become more reactive because now also the occupied outward-directed d_{yz} orbitals can backdonate electrons into empty asymmetric adsorbate orbitals. As we will see later, this interaction is responsible for dissociative adsorption reactions. The corresponding orbital scheme and structure is shown in Fig. 4.10.

This very short intermezzo on the electronic structure of transition metal compounds had as its main purpose to illustrate that bonding is very similar to that in molecular co-ordination compounds. Hence their chemical reactivity can be understood using the same concepts.

The presence of Lewis basic oxygen or sulfur anions around the positively charged cation provides additional bonding positions for protons generated in heterolytic dissociation reactions (Fig. 4.11). The cation acts in essence as a Lewis

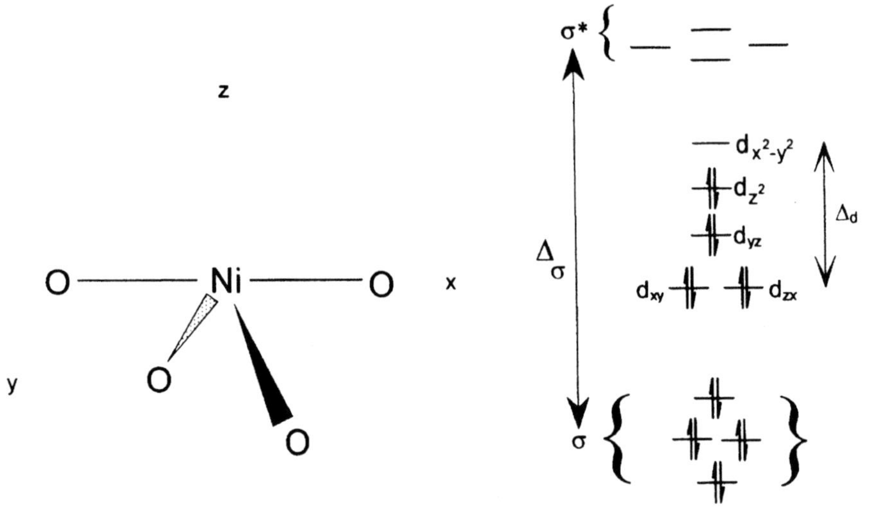

Fig. 4.10. (a) Local structure of 4 coordinated Ni^{2+} in NiO surface. (b) Orbital scheme of Ni^{2+} corresponding to structure 4.10a.

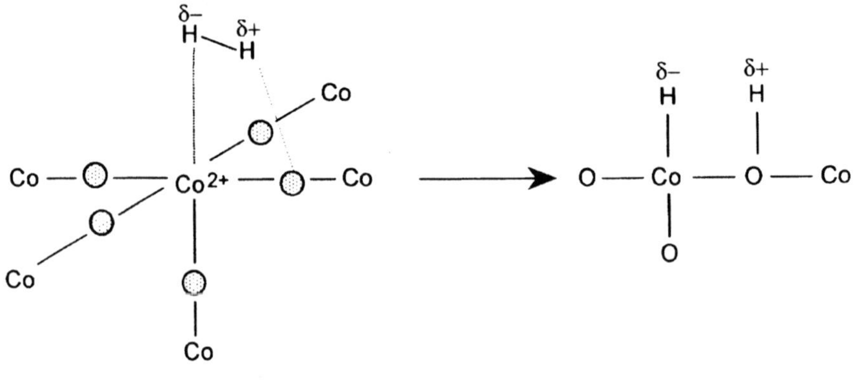

Fig. 4.11. Heterolytic dissociation of H_2 on a Co^{2+} centre.

acid. The hydrogen binds to it as an anion, as in $HM_n(CO)_5$. Its acidity depends strongly on the electrostatic field generated by its neighbour anions.

To illustrate oxidative addition of H_2 to a metal centre and the importance of the rehybridization due to the interaction with metal atom ligands, we will compare the reaction of H_2 by Ni atoms and the $Ni(PH_3)_2$ complex [3]. Oxidative addition is the analogue of dissociative adsorption of a molecule to a catalyst centre. A terminology used in surface science and heterogeneous catalysis.

In the Ni atom, the 4*s* orbital contains two electrons. This spatially extended, doubly occupied orbital experiences a large repulsive interaction with the doubly occupied σ orbital of H_2 of the same symmetry. Dissociation of H_2 has a large activation energy because electron promotion has to occur from 4*s* to 3*d* atomic orbitals in order to obtain dissociation. Co-ordination of ligands, as in $Ni(PH_3)_2$, decreases the promotion energy. The $Ni(PH_3)_2$ complex has initially a linear geometry (Fig. 4.12).

Directed occupied bonding orbitals and non-occupied antibonding orbitals between the Ni atom and the two phosphorus ligands are formed when two electrons of the Ni 4*s* orbitals are promoted into the Ni 3*d* orbitals. The Ni 4*s* and 4*p* atomic orbitals hybridize into two empty atomic orbitals directed to the occupied lone pair on the phosphorus atom. When the hydrogen molecule approaches this complex, the interaction is again weak. The empty, spatially extended Ni *p* orbitals required for bonding have been pushed upwards to a high

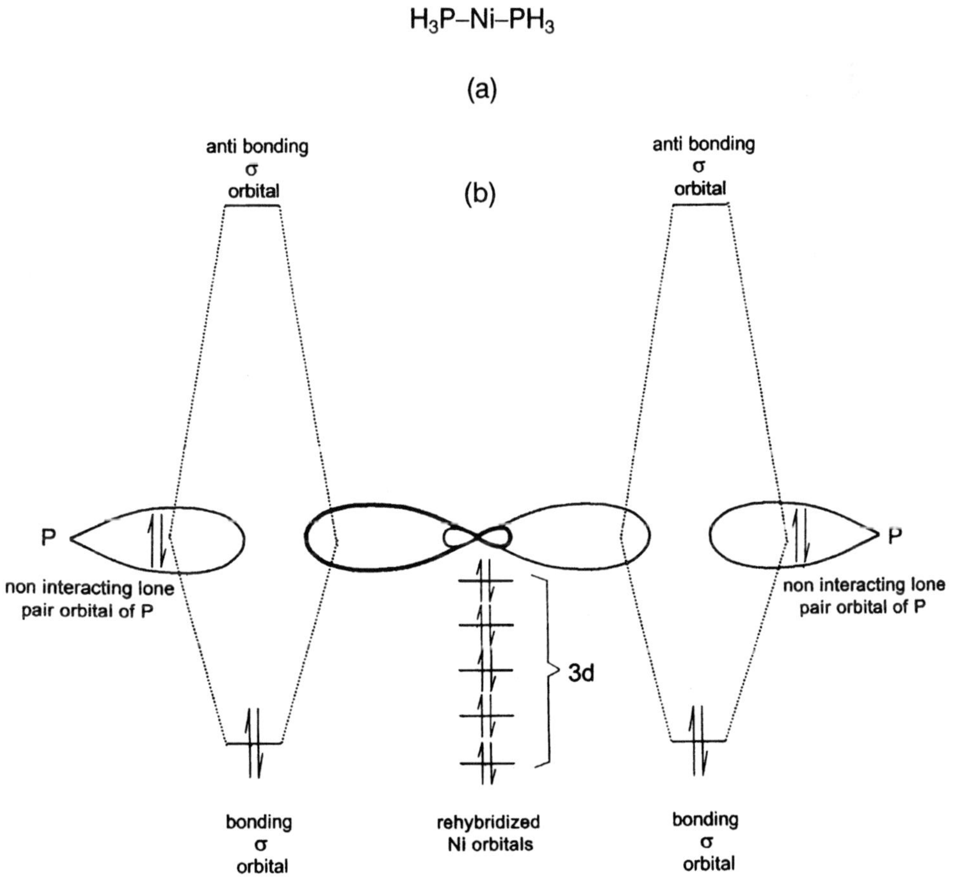

Fig. 4.12. The interaction between a Ni atom and a hydrogen molecule. (a) Geometry of $H_3P–Ni–PH_3$. (b) The orbital scheme of $H_3P–Ni–PH_3$.

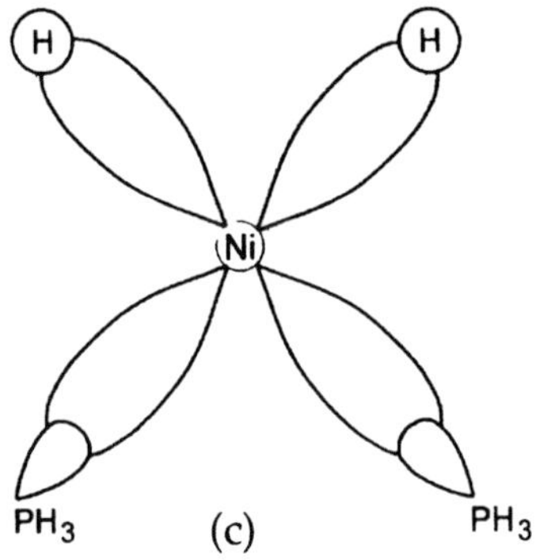

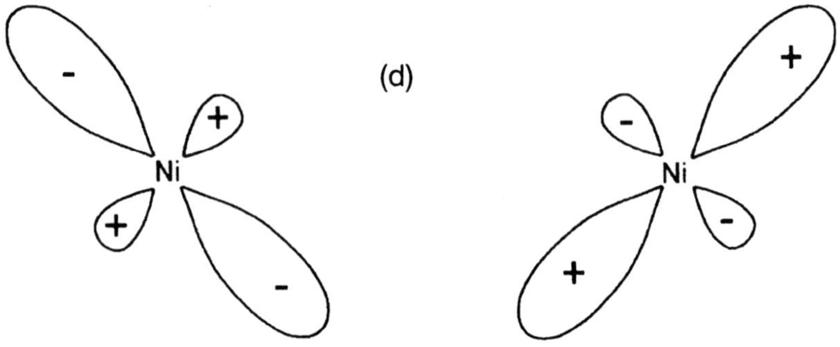

Fig. 4.12 continued. (c) Orbitals in the square-planar complex $H_2Ni(PH_3)_2$. (d) Rehybridized orbitals in $H_2Ni(PH_3)_2$.

energy. Dissociation occurs only when the $Ni(PH_3)_2$ molecule is deformed in such a way that empty spatially extended orbitals become available.

When the H_3P–Ni–PH_3 angle becomes 90°, hybridization between the d_{xz} and s atomic orbitals, rather than s–p hybridization, results in directed orbitals suitable for co-ordination with the PH_3 ligands. In the process, empty antibonding orbitals having an s and d_{xz} nature are created, as sketched in Fig. 4.12d. Their orientation and symmetry are suitable for interaction with the hydrogen molecule. Now the antibonding H_2 σ orbital can be populated and the molecule will dissociate. Note that hybridization on the Ni atom is now d^9s^1, which is very similar to the hybridization of a Ni atom in the Ni metal surface.

For dissociative adsorption of H_2 we pointed out the importance of reducing repulsive interactions between doubly occupied orbitals and the need for easily accessible occupied orbitals, asymmetric with respect to the reaction co-ordinate, that backdonate electrons into antibonding adsorbate orbitals. Occupation of the latter weakens the adsorbate bonds and reduces the activation energy for dissociation.

In organometallic chemistry dissociative adsorption is called oxidative addition. The hydrogen atoms formally become negatively charged. In the

$Ni(PH_3)_2$ complex, the valence of Ni changes from zero to 2+ in the $H_2Ni(PH_3)_2$ complex.

Whereas in oxidative addition electrons are used to form new bonds between metal centre and adsorbate fragment, in the reverse reaction reductive elimination of the adsorbate donates electrons to the metal centre. The H-atoms recombine to form H_2 that desorbs.

Recombination reactions in which bonds are formed between fragments around a metal centre, even when no desorption occurs, also require the availability of empty transition metal orbitals, asymmetric with respect to the reaction co-ordinate.

We will illustrate this for the insertion reaction of CO into a CH_3 group co-ordinated to Pd^{2+}. This reaction can be considered as a prototype for C–C bond formation catalyzed by a metal centre (see Fig. 4.13).

The formal charge on the methyl group is –1, which implies that 2 electrons occupy the σ-symmetric lone pair orbital. When the CH_3 fragment migrates to CO and a carbon–carbon bond is formed, the doubly occupied 5σ CO orbital interacts with the doubly occupied σ–CH_3 orbitals resulting in doubly occupied bonding and antibonding orbitals, giving a repulsive interaction (Fig. 4.13).

The empty $d_{x^2-y^2}$ orbital on the Pd atom can accept electrons from the antibonding C–C fragment orbital, thus reducing the repulsive interaction [4,5].

As we will also see later, the possibility of stabilizing surface interactions with occupied or unoccupied orbitals, asymmetric with respect to the reaction co-ordinate, will also be essential to enable surface associative or dissociative reactions.

4.3 ELEMENTARY STEPS IN ORGANOMETALLIC COMPLEXES

The following paragraphs deal with the elementary steps that occur at a metal centre in a homogeneous catalytic reaction. Although the greater part of the elementary steps described have been studied in detail in model reactions of homogeneous organometallic complexes [6–10], there is general consensus that many of them can also take place at metal ions on surfaces [11]. At metallic surfaces, however, several additional reactions can take place that, by their nature, do not occur with low molecular weight, soluble organometallic complexes. The latter will be discussed in Section 4.4.

4.3.1 *Creation of a Vacant Site*

The function of a catalytic centre is to bring reactants together and to lower the activation barrier of their reaction. In homogeneous catalysis the catalytic centre comprises a metal centre and surrounding organic molecules. The latter mole-

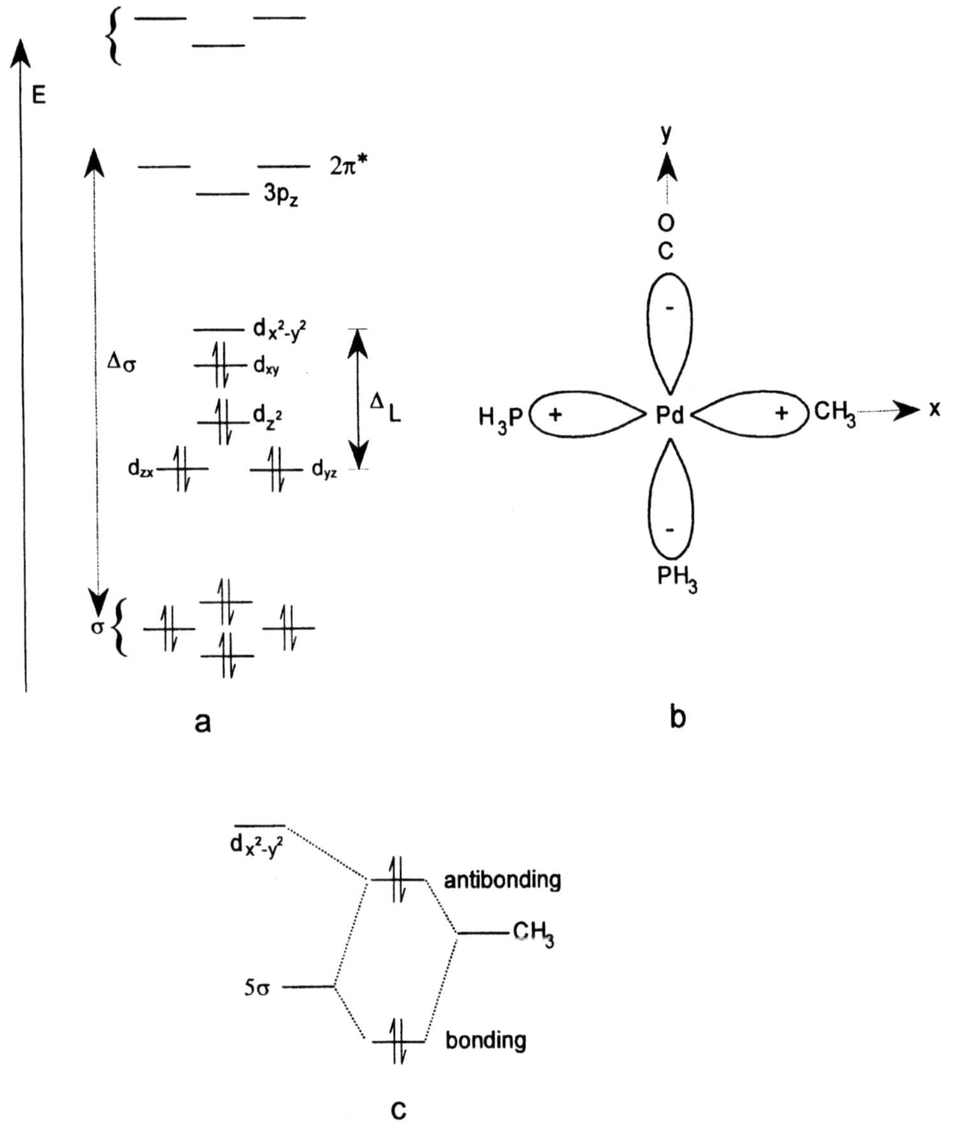

Fig. 4.13. (a) Molecular orbital energy scheme of $Pd(CO)(CH_3)(PH_3)_2$ complex.(b) The Pd d_{x2-y2} orbital in the $Pd(CO)(CH_3)(PH_3)_2$ complex. (c) Release of the repulsive interaction between CH_3 and CO by backdonation into the empty d_{x2-y2} orbital.

cules are called ligands and they can be neutral molecules (solvent, phosphines, CO, alkenes) or anions (e.g. halides). The ligands are usually σ donors, bringing the electron count of the complex to 18 or 16 or, especially on the left-hand-side of the periodic table, to 14 electrons, so that no electrons are forced into antibonding orbitals (see Section 4.2.2). To bring the reactants together, the metal centre must have vacant sites, i.e. it must be coordinatively unsaturated. Metal catalysis begins, we could say, with the creation of these vacant sites. In the condensed phase solvent molecules will always interact with the metal ion. The reactant molecules are present in excess, and so may the ligands be. Therefore, a competition exists in complex formation between the desired substrate and other potential ligands present in the solution. With heterogeneous catalysts the presence of competing molecules in the gas phase controls the state of the surface. A negative order in one of the concentrations of the reactants can often be found in the expression for the rate of product formation (see the chapter on kinetics). In homogeneous catalysis this order can also be negative in ligand concentration. When the substrate coordinates strongly to the metal centre this may give rise to a zeroth order in the concentration of the substrate, i.e. Michaelis–Menten kinetics. Furthermore, coordination of the product formed during the reaction may slow down or inhibit the catalytic process. As an illustration, we may mention the cobalt-catalyzed hydroformylation reaction [12,13]:

$$HCo(CO)_4 \rightarrow HCo(CO)_3 + CO \quad (4.17)$$

$$HCo(CO)_3 + C_2H_4 \rightarrow HCo(CO)_3(C_2H_4)$$

$$HCo(CO)_3(C_2H_4) \rightarrow Co(CO)_3(C_2H_5)$$

$$Co(CO)_3(C_2H_5) + CO \rightarrow Co(CO)_4(C_2H_5)$$

$$Co(CO)_4(C_2H_5) \rightarrow Co(CO)_3(C(O)C_2H_5)$$

$$Co(CO)_3(C(O)C_2H_5) + H_2 \rightarrow O{=}C(H)C_2H_5 + HCo(CO)_3$$

The catalyst is $HCo(CO)_4$, which must lose one molecule of CO in order to make room for an alkene molecule (Fig. 4.14).

A high CO pressure would shift equilibrium (4.17) to the left and the catalytic reaction would become slower. In this complex CO is a far better ligand than an alkene. On the other hand the reaction uses CO as a substrate, so it cannot be omitted. Furthermore, low pressures of CO may lead to decomposition of the

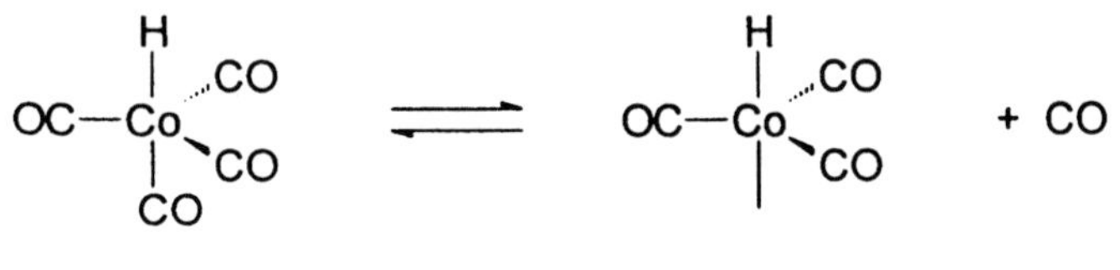

Fig. 4.14. Desorption of carbon monoxide.

cobalt carbonyl complexes to metallic cobalt and CO, which is also undesirable. Finally, the product alcohol may stabilize divalent cobalt species which are not active as a catalyst:

$$6\ ROH + 3\ HCo(CO)_4 \rightarrow Co(ROH)_6^{2+}\{Co(CO)_4^-\}_2 + 4\ CO + 1\tfrac{1}{2}\ H_2$$

In a nutshell, we see several counteracting factors.

4.3.2 *Coordination of the Substrate*

The vacant site created in the required initial step of the catalytic process at the metal centre can now be occupied by the incoming substrate. Another way of looking at the question of 'creation of a vacant site' and 'coordination of the substrate' is the classical way of studying substitution reactions [14]. This is particularly useful for homogeneous complexes. Two extreme mechanisms are distinguished: a (complex) associative and a (complex) dissociative one. The discussion above concurs with the dissociative mechanism, that is to say the rate-controlling step is the breaking of the bond between the metal and the leaving ligand. In the associative process the displacement is a bimolecular process with simultaneous bond breaking and bond formation. The bond being broken or formed denotes here the bond between the reactant *molecule* and the *metal*. Note the difference in terminology between heterogeneous catalysis and coordination chemistry, the terminology adopted in homogeneous catalysis:

Homogeneous catalysis:	Meaning:	Heterogeneous catalysis:
dissociation	metal ligand bond breaking	desorption
association	metal ligand bond forming	adsorption
oxidative addition	fission of bond in substrate	dissociation
reductive elimination	making of bond giving substrate	association

In square-planar, 16-electron complexes (i.e. coordinatively unsaturated) as found for many group 9 (Co, Rh, Ir) and 10 (Ni, Pd, Pt) metals, the associative process is most common. The *trans* effect and *trans* influence ligand series [6–10] are also useful measures in the study of homogeneous catalysis. Apart from very small ligands, such as CO, H_2, and NO, steric repulsion between ligands, as well as complexes and incoming substrates, plays a dominant role in determining the kind of intermediates and complexes formed and their equilibria in solution.

Carbon monoxide and ethylene are common substrates involved in homogeneous catalysis. The bonding of carbon monoxide to a transition metal has been depicted in Fig. 4.4. The bonding of alkenes to transition metals is described by the Chatt–Dewar–Duncanson scheme involving σ donation by the filled π orbital of the alkene, and π back donation from the metal into the π^* orbital of the alkene (see Fig. 4.41).

4.3.3 *Insertions and Migrations*

Elementary steps involving insertion or migration reactions are of prime importance for catalysis employing alkenes and carbon monoxide. Two examples taking place on a platinum centre have been depicted in Figs. 4.15 and 4.16. In these reactions an acetyl fragment is formed on the platinum centre from a coordinated CO and a methyl group. The important mechanistic difference between insertion and migration is that, in the insertion mechanism, CO inserts into the metal methyl bond and consequently the acyl bond formed takes the position of the former methyl group, i.e. the σ bonded fragment retains its position *trans* to P* (Fig. 4.15). In the migration mechanism the methyl group migrates to the coordinated CO, and now the resulting acetyl group occupies the position *cis* to P* (Fig. 4.16). There is convincing experimental evidence [15,16] (see below) in support of the migration mechanism versus the insertion mechanism. Results of theoretical calculations [17–19] are also in better agreement with the migration mechanism, i.e. with the anionic methyl group moving to the positively charged carbon atom of carbon monoxide. Hence the most accurate description for this process is migration. In the literature, however, the reaction is usually referred to as 'insertion', perhaps mainly because the reactive un-

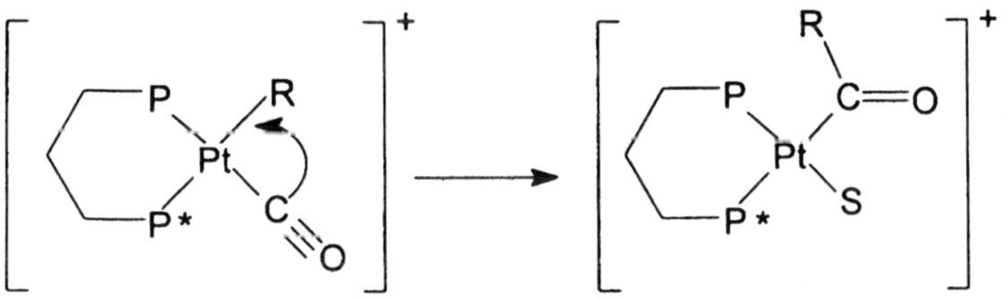

Fig. 4.15. Insertion reaction.

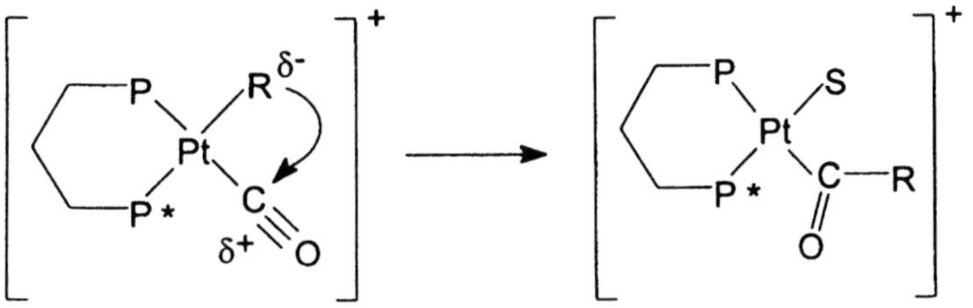

Fig. 4.16. Migratory insertion reaction.

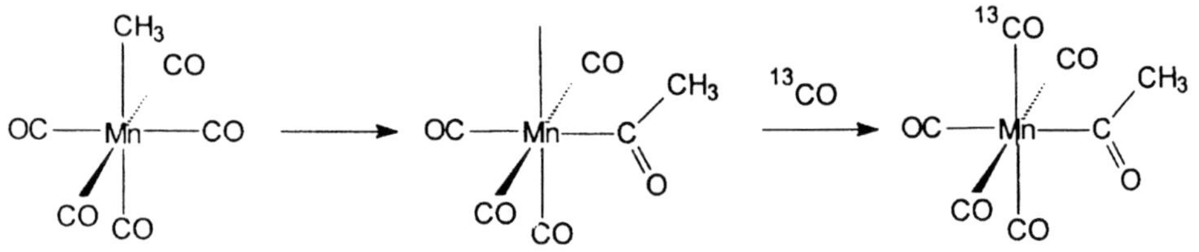

Fig. 4.17. Migratory insertion to pre-co-ordinated CO.

saturated species (CO or ethene) is chosen as the linguistic subject of our sentences: 'carbon monoxide inserts giving an acyl complex'. To do justice to the intimate details of this reaction one also writes 'migratory insertion'.

So far we have automatically presumed that the reacting carbon monoxide is coordinated to the metal. For the platinum example shown that might not be an easy thing to prove, but there is experimental evidence for complexes in which methyl migration to coordinated CO occurs. The classic proof stems from a relatively inert complex in which both the migration and the exchange of co-ordinated CO with free CO is slow. The reaction of $CH_3Mn(CO)_5$ in the presence of ^{13}C labelled free ^{13}CO results in the formation of $CH_3(CO)Mn(CO)_4(^{13}CO)$ in which the labelled CO is present as carbon monoxide and not in the acetyl group (Fig. 4.17).

Hence there is no direct reaction between the methyl manganese unit and the newly incoming carbon monoxide. Later, when we discuss the activation of a coordinated substrate molecule toward nucleophilic attack, the latter will turn out to be an alternative for the insertion/migration process. In the reaction involving nucleophilic attack at the coordinated, activated, unsaturated substrate, the anionic fragment is an uncomplexed species. There is, as yet, no proven example of an insertion of an uncomplexed unsaturated substrate into a metal–carbon σ bond.

A second important migration reaction takes place with alkenes, rather than carbon monoxide. Figure 4.18 gives a schematic representation of a hydride that migrates to a coordinated ethene molecule *cis* to the hydride. The figure shows the hydride migration which would leave an empty space in the coordination sphere of the metal. This coordinative unsaturation can be lifted in two ways: firstly, an agostic [20] interaction with the β-hydrogens may occur; secondly, an incoming ligand may occupy the vacant site [21]. How and whether activation of a coordinated alkene takes place prior to migration is not always clear-cut. Coordinated alkenes are subject to σ donation and π back donation, and the overall outcome of the shift of electron density from and to the coordinated alkene cannot be predicted. Molecular orbital calculations [22] show that, to a simple first approximation, the coordinated alkene is not activated towards

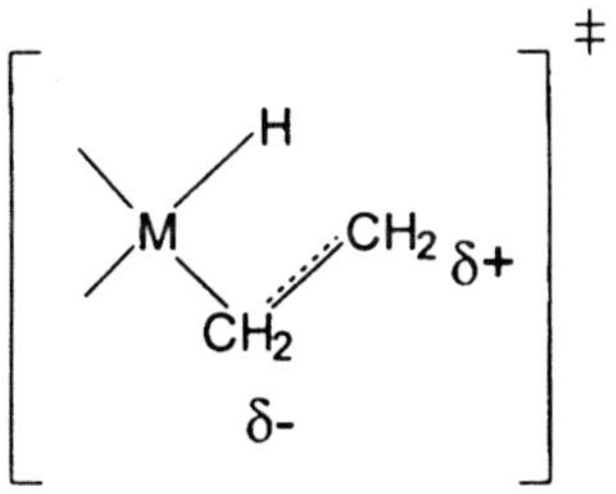

Fig. 4.18. Hydride migration to ethene.

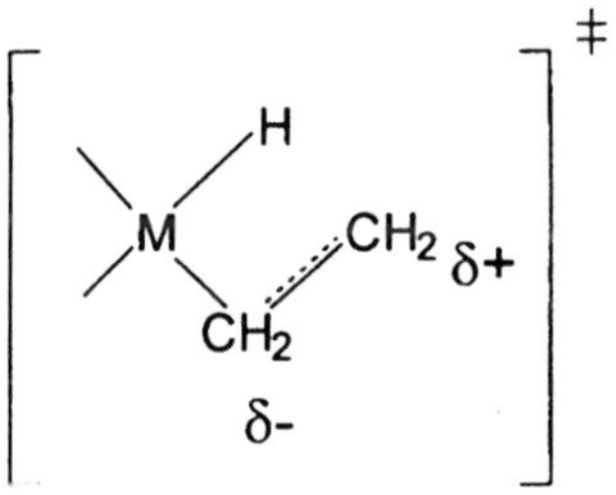

Fig. 4.19. Asymmetric co-ordination giving activation for nucleophilic attack.

migration (i.e. nucleophilic attack) and it cannot be predicted *a priori* whether rapid migration of the hydride will occur. However, a strong polarization of the alkene will be achieved when an asymmetric bonding of the alkene is arranged, as shown in Fig. 4.19.

The migration reaction takes place in a *syn* fashion with respect to the alkene; the two atoms M and H add on the same face of the alkene (2+2 *syn*-addition). This has been unequivocally established by experiments. Later we will see reactions where this is not the case, although the overall stoichiometry is the same for both types.

Thermodynamically the insertion of an alkene into a metal–hydride bond is much more favourable than the insertion of carbon monoxide into a metal–methyl bond. The latter reaction is more or less thermoneutral and the equilibrium constant is near unity under standard conditions. The metal– hydride bond is stronger than a metal–carbon bond and the insertion of carbon monoxide into a metal hydride is thermodynamically most often uphill. Insertion of alkenes is also a reversible process, but slightly more favourable than CO insertion. Formation of new σ bonds at the cost of the loss of the π bond of the alkene during alkene hydrogenation etc., makes the overall processes of alkenes thermodynamically exothermic, especially for early transition metals.

Other insertions may involve isonitriles, alkynes, alkadienes, CO_2, and SO_2. They will not be dealt with here.

4.3.4 β-Elimination and Deinsertion

The reverse reaction of the migration of η^1-bonded anionic groups to co-ordinated alkenes is named β-elimination (compare Figs. 4.20 and 4.21).

The migration reaction diminishes the total electron count of the complex by two, and creates a vacant site at the metal; β-elimination does the opposite. β-Elimination requires a vacant site at the metal centre, and the electron count of the complex increases by two electrons during the process. The reaction resembles the β-elimination reaction occurring in many organic processes, but the difference lies in the intramolecular nature of the present process, as the eliminated alkene may be retained in the complex. In organic chemistry the reaction may well be a two-step process, e.g. proton elimination with a base followed by the leaving of the anion. In transition metal chemistry, however, it is the availability of *d* orbitals that greatly facilitates a concerted *syn* β-elimination.

Suppression of β-elimination in catalytic processes is often a desirable feature. It can be achieved in several ways, although these recommendations in many instances have limited practical value. Rule number one would be to avoid β-hydrogens! When carrying out a polymerization reaction with a specific alkene carrying three hydrogens this will indeed be impractical. Rule number two is to maintain coordinative saturation. Again, in a catalytic cycle this may be a counter-productive suggestion, since the next reaction in the catalytic cycle will also require a vacant site in order to bring the next substrate molecule into the

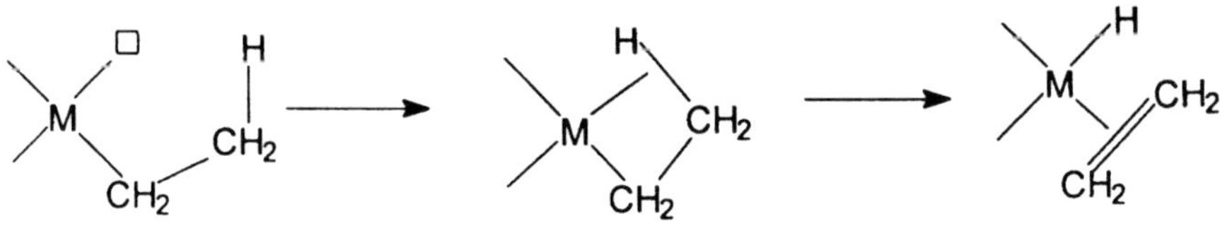

Fig. 4.20. β-Elimination.

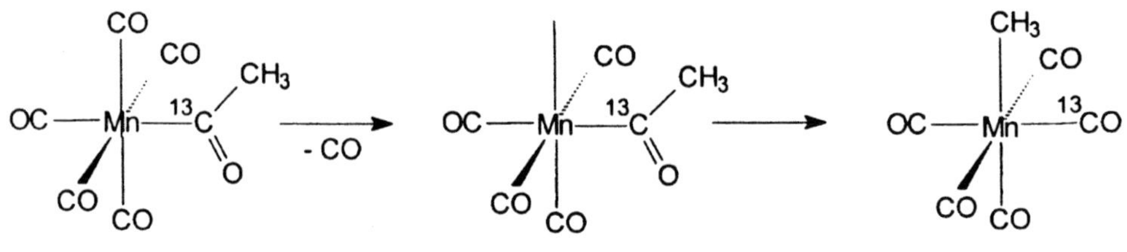

Fig. 4.21. *Cis*-deinsertion.

coordination sphere of the metal. Thirdly, steric hindrance may hamper the correct stereochemistry required for β-elimination, and perhaps this can be used to stabilize our metal alkyl complex. The last recommendation is to seek the metals for which the metal alkyl complexes are stable with respect to the corresponding hydride/alkene metal complexes. The relative stability of metal alkyls with respect to hydrides increases for the metals on the left-hand side of the periodic table: the early transition metals and the lanthanides [23–25]. It is not surprising that the best alkene polymerization catalysts are found amongst these metals.

For metal hydride elimination from a metal alkyl (to produce an alkene) one will also find the terms 'deinsertion' and 'extrusion' rather than β-elimination. For alkenes 'β-elimination' is correct, but for CO the deinsertion nomenclature is obligatory. We have seen that insertion takes place between a σ-bonded, anionic fragment and a neutral ligand in mutual *cis* positions, as was described for the manganese complex in Fig. 4.17. By the same token, the deinsertion reaction can only proceed if there is a vacant site *cis* to the acyl group. A further modification of the experiment outlined in Fig. 4.17 proves this point. A manganese acetyl complex which is ^{13}C labelled at the acyl carbonyl group was synthesized and heated to give deinsertion of CO. The result was that the only product formed contained the methyl substituent in a position *cis* to the labelled ^{13}C, see Fig. 4.21.

4.3.5 *Oxidative Addition*

In an oxidative addition reaction an electrophilic compound XY adds to a metal complex during which the XY bond is broken and two new bonds are formed: MX and MY. X and Y are reduced, and will at least *formally* have a minus one charge and hence the *formal* oxidation state of the metal is raised by two. The coordination number of the metal also increases by two. The electron count of the metal complex increases by two, while the electron count of the metal decreases by two. Figure 4.22 gives a formal representation. Clearly, the oxidative addition will be faster when the starting metal centre is more electron rich and/or more nucleophilic and hence the reaction is promoted by electron-releasing donor atoms at the metal. The simplest form of an oxidative addition has been presented in Section 4.2.2 (Fig. 4.12), the reaction of nickel with dihydrogen. It has already been mentioned that both the electron count and the geometry of the coordination complex play an important role in determining the energetics of such a reaction. In the following, only one popular example will be described.

The oxidative addition has been most extensively studied on iridium complexes, particularly Vaska's complex. The latter is a square planar complex, *trans*-$L_2IrCl(CO)$, with a d^8 electron count containing iridium(I). After the oxidative addition we formally obtain iridium(III), an octahedral complex, with a d^6 electron configuration; i.e. the 16-electron square-planar complex is converted into

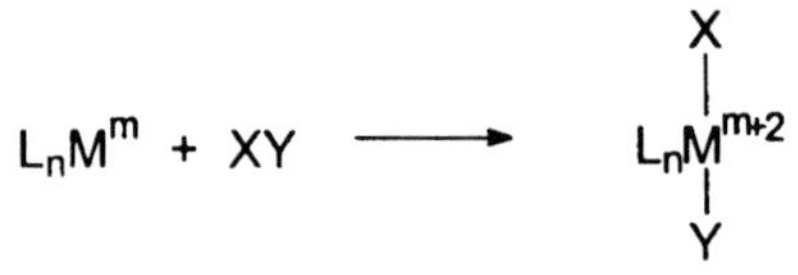

Fig. 4.22. Oxidative addition.

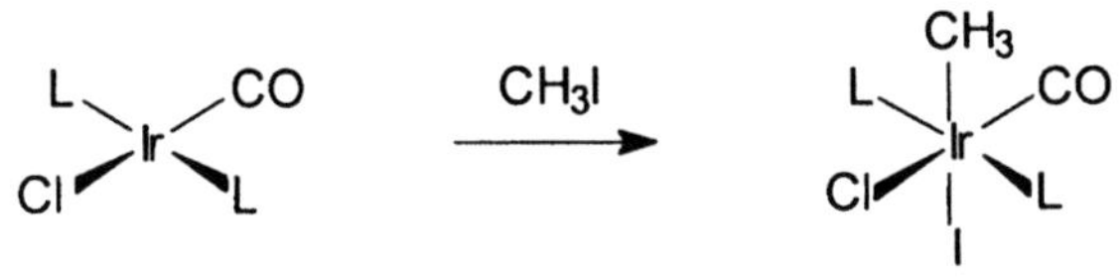

Fig. 4.23. Oxidative addition to Vaska's complex.

an octahedral 18-electron complex. In Fig. 4.23 we have depicted the oxidative addition of methyl iodide to Vaska's complex (L = phosphine) [26]. A large series of organic and inorganic molecules have been added to d^8-iridium and d^8-rhodium centred complexes in this manner.

The oxidative addition of alkyl halides can proceed in different ways, although the result is usually a *trans* addition independent of the mechanism [27]. In certain cases the reaction proceeds as an S_N2 reaction, as in organic chemistry. That is to say that nucleophilic attack is carried out by the electron-rich metal at the carbon atom attached to the halide, the halide being the leaving group. This process leads to inversion (Walden inversion) of the stereochemistry of the carbon atom (this can be observed only when the carbon atom is asymmetric). Note that the effect of steric hindrance in the electrophile RX on the relative rates of the reaction is the same as in organic chemistry, i.e. MeI >> EtI >> PrI >> iPrI. There are also examples in which racemization occurs. In some cases this has been explained on the basis of a radical chain mechanism. Indeed, radical scavengers have proved the presence of radicals by slowing the reaction down, and radical traps have demonstrated the expected ESR signals. The reaction sequence for the radical chain process reads as follows:

$$L_nM + R\bullet \rightarrow L_nRM\bullet$$

$$L_nRM\bullet + RX \rightarrow L_nRMX + R\bullet$$

The oxidative addition of dihydrogen to low-valent metal complexes is a common reaction in many catalytic cycles. Common electronic configuration changes involved are $d^2 \rightarrow d^0$, $d^8 \rightarrow d^6$, and $d^{10} \rightarrow d^8$. In spite of the high strength of the dihydrogen bond the reaction proceeds smoothly to afford *cis* dihydrido complexes. The bond energy of a metal–hydrogen bond is in the order of 240 ± 40 kJ mol^{-1} which is sufficient to compensate for the loss of the H–H bond (436 kJ mol^{-1}). The hydride anion is formally charged with a minus one charge and this electron count gives dihydrogen the role of an oxidizing agent!

In general, oxidative addition reactions occur at mononuclear complexes as discussed above. Oxidative addition of dihydrogen often occurs at binuclear complexes. The reaction of dicobalt octacarbonyl may illustrate this:

$$Co_2(CO)_8 + H_2 \rightarrow 2\ HCo(CO)_4$$

The electronic configuration of the metals changes from d^7 to d^6. This reaction has also been referred to as the homolytic cleavage of dihydrogen, although no hydrogen radicals are involved in the cleavage reaction itself. We will call it an 'oxidative addition' to a binuclear metal centre. One may have some doubt about the hydridic nature of the hydrides formed, because in the hydrido cobalt carbonyl the hydrogen atom shows an 'amphoteric' character: sometimes it reacts as an H^- anion, and in other cases its reactions are best described as an H^+ cation (it is reported to be as acidic as sulphuric acid). Note that in the former case the Co centre is formally monovalent, whereas in the proton case its valency is minus one.

The oxidative addition of acids is another instructive example. It resembles the reactions with alkyl halides and may result in another 'amphoteric' hydride:

$$L_4Ni + H^+ \rightarrow L_4HNi^+$$

The starting material is an 18-electron nickel(0) complex which is protonated forming a divalent five-coordinate nickel(II) hydride [28]. This can react further with alkenes to give alkyl groups, but it can also react as an acid with hard bases to regenerate the nickel(0) complex. Similar oxidative addition reactions have been recorded for phenols, water, amines, carboxylic acids, mineral acids (e.g. HCN), etc.

Intermolecular oxidative additions involving C–H bond breaking is a topic which has been extensively studied recently. It is usually referred to as C–H activation; the idea is that the M–H and M–hydrocarbyl bonds formed will be much more prone to functionalization than the unreactive C–H bond [29–31]. Intramolecular oxidative additions of C–H bonds have been known for quite some time [32,33] (see Fig. 4.24). This process is termed orthometallation. It occurs frequently in metal complexes, and is not restricted to 'ortho' protons. It has considerable importance in metal-mediated synthesis.

4.3.6 *Reductive Elimination*

Reductive elimination is simply the reverse reaction of oxidative addition: the formal oxidation state of the metal is reduced by two (or one in a bimetallic reaction), and the total electron count of the complex is reduced by two. While

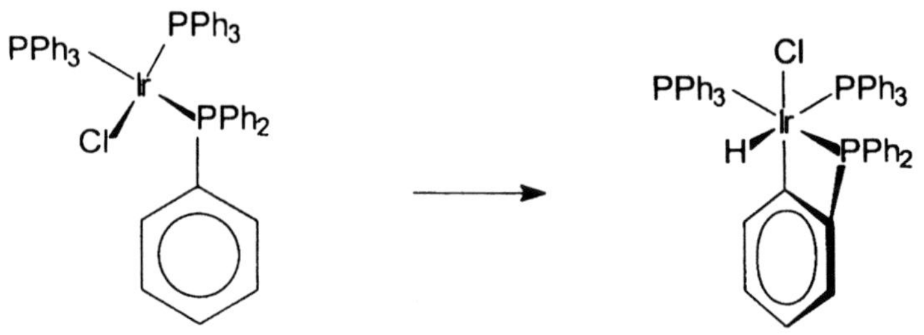

Fig. 4.24. Orthometallation.

oxidative addition can also be observed for main group elements, this reaction is more typical of the transition elements in particular the electronegative, noble metals. In a catalytic cycle the two reactions occur pairwise. At one stage the oxidative addition occurs, followed by, for example, insertion reactions; and then the cycle is completed by a reductive elimination of the product. Reductive elimination can be enhanced by electron-withdrawing ligands on the metal, i.e. when the state of lower valency (electron-rich metal centre) is stabilized. Chemical oxidation (one or two electron) of a complex may also initiate reductive elimination. Only one example will be discussed here, as most reactions are directly the reverse of the ones discussed above under oxidative additions.

Reductive elimination of molecules with carbon–carbon bonds has no counterpart in oxidative addition reactions because the metal–carbon bond energies (120–240 kJ mol^{-1}) may not always be large enough to compensate for the energy of the carbon–carbon bond (370 kJ mol^{-1}), and secondly, the carbon–carbon bond is much less reactive than a carbon–hydrogen bond or a dihydrogen bond due to repulsive interactions between the attacking substituted C–C bond and the ligands surrounding the metal centre.

Examples of reductive elimination are present in the literature mostly for group 10 metals, e.g. Fig. 4.25. As outlined above, the reaction can be induced by the addition of electron-withdrawing ligands; in this case electron-poor alkenes are very effective as ligands.

In practice the reactions are less straightforward than is shown in Fig. 4.25, and several other decomposition pathways are available (see α-elimination below).

4.3.7 α-*Elimination Reactions*

α-Elimination reactions have been the subject of much study since the mid-seventies, mainly due to the pioneering work of Schrock [34,35]. The early transition metals are most prone to α-elimination, but the number of examples of the later elements is growing. A classic example is shown in Fig. 4.26.

The process is called α-elimination because a hydrogen at the alpha carbon at the metal is being eliminated (if we regard the product as a metalla-alkene the

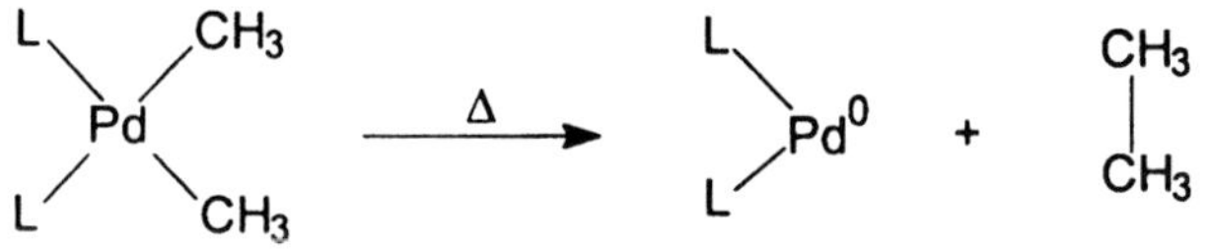

Fig. 4.25. Reductive elimination of a C–C bond.

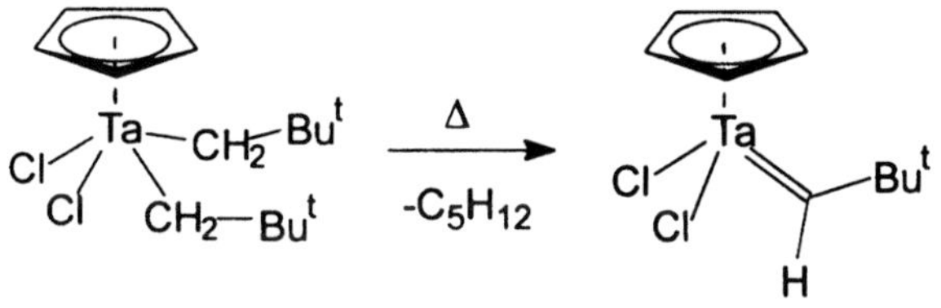

Fig. 4.26. α-Elimination.

process should be called β-elimination, but this only adds to the confusion). The electron-deficient tantalum atom forms 'agostic' bonds with hydrogens in the α-positions. (Agostic interaction implies electron donation by nearby atoms or bonds to highly electrophilic metal centres which is observed spectroscopically or as a distortion in the crystal structure). In the starting material this activates one hydrogen that will then leave with the neopentyl anion as neopentane. In the alkylidene complex formed there is a strong interaction between tantalum and the electrons of the α-hydrogen of the alkylidene. Consequently, in suitable complexes α-elimination can occur twice, yielding alkylidyne complexes. See Fig. 4.27 for a simplified example with tungsten.

In decomposition reactions of dimethyl–metal complexes of palladium(II) and nickel(II) one finds the formation of only traces of methane [36] which may also be attributed to an α-elimination process. In regard to the valence state, note that, formally, the alkylidene ligand is considered as a neutral ligand and therefore, in the tantalum–alkylidene complex in Fig. 4.26, tantalum is trivalent. The electronic structure of the alkylidene is of course reminiscent of the corresponding oxide $CpTa(Cl)_2O$, which we would definitely call pentavalent. All that matters is that there should be a sufficient number of electrons for the multiple bonds which we draw.

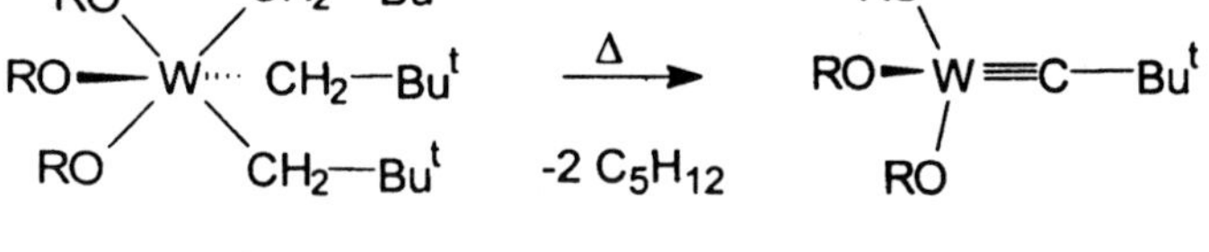

Fig. 4.27. α-Elimination giving alkylidynes.

4.3.8. Cyclometallation

Cyclometallation refers to a process in which unsaturated moieties form a metallacyclic compound. It is sometimes categorized under oxidative additions, but we prefer this separate listing. Examples of the process are presented in Fig. 4.28. Metal complexes which actually have displayed these reactions are M = L_2Ni for reaction **a**, M = Cp_2Ti for reactions **b** and **c**, M = Ta for **d**, and M = $(RO)_3W$ for **e**. The latter examples involving metal–carbon multiple bonds, have only been observed for early transition metal complexes, the same ones mentioned under the α-elimination heading.

In examples **a** and **b** in Fig. 4.28, the metals increase their valency by two, and this is not just a formalism as the titanium(II) and the nickel(0) are indeed very electron-rich metal centres. During the reaction a flow of electrons takes place from the metal to the organic fragments which end up as anions. In these two reactions the metal provides two electrons for the process, as in oxidative addition reactions. The difference between cyclometallation and oxidative addition is that, during oxidative addition, a bond in the adding molecule is being broken, whereas in cyclometallation, fragments are added together. For reactions **c** and **d** the situation is slightly different. The formal valency of the metal increases in **c** and **d** with two units again when the alkylidene is considered as a two-electron

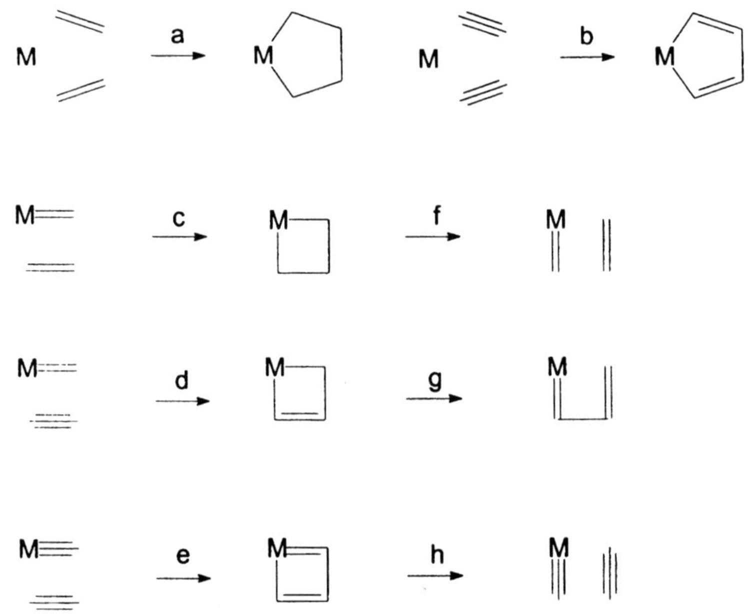

Fig. 4.28. Cyclometallations and retro-cyclometallations.

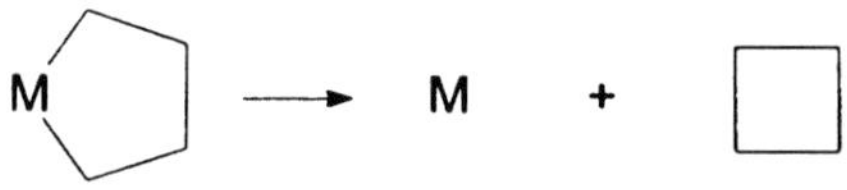

Fig. 4.29. Cyclisation through reductive elimination.

neutral ligand. The two electrons donated by the metal in the 'oxidative addition' process were already involved in the bonding of the alkylidene, and not present at the metal centre as a surplus nonbonding lone pair (cf. square planar d^8 complexes, or the above d^2 Ti and d^{10} Ni cases). The reaction of the alkylidyne **e** is a cyclometallation reaction for which other descriptions make little sense.

Three mechanisms can be proposed for the intimate reaction mechanism for **c–e**, analogous to the organic 2+2 cycloadditions: a pericyclic (concerted) mechanism, a diradical mechanism, and a diion mechanism. In view of the polarization of the metal(+) carbon(–) bond, an ionic intermediate may be expected. The retention of stereochemistry, if sometimes only temporary, points to a concerted mechanism.

The reverse reaction of a cyclometallation is of importance for the construction of catalytic cycles. The simple retrocyclometallations of **a** and **b** are not productive, since they would again lead to ethene and ethyne, and additional reactions have to be invoked for a productive catalytic cycle. For **c–e** the retro reactions **f–g** can be envisaged, leading to new products: see Fig. 4.28.

Reaction **a** in Fig. 4.28 may be succeeded by various other reactions such as insertions, β-eliminations or regular reductive eliminations (see Fig. 4.29).

The reductive elimination reaction is governed by the common rules given in Section 4.3.6. Nickel-catalyzed cyclodimerization of butadiene [37] affords 1,5-cyclooctadiene which is used as such in a variety of chemical applications or is reduced to cyclooctene which is, *inter alia*, used for the production of cyclooctenemers, a ring-opening metathesis product. The first step involves the coordination of a diene molecules to zerovalent nickel (and an additional ligand L). Then a cyclometallation takes place during which the nickel is transferred into divalent nickel. A reductive elimination completes the cycle regenerating nickel(0) and the two carbon–carbon bonds have been formed. Figure 4.30 shows the essential details.

4.3.9 Activation of a Substrate toward Nucleophilic Attack

Alkenes

Coordination of an alkene to a relatively electronegative metal (it may often carry a positive charge) activates the alkene toward attack of nucleophiles. After the nucleophilic attack the alkene complex has been converted into a σ-bonded alkyl

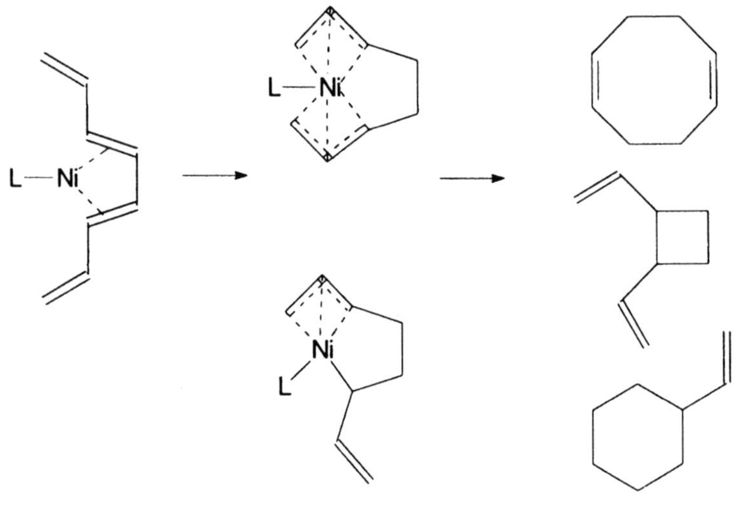

Fig. 4.30. Cyclodimerisation mechanism.

complex with the nucleophile at the β position. With respect to the alkene (in the 'organic' terminology) the alkene has undergone *anti*-addition of M and the nucleophile Nu, see Figs. 4.31 and 4.32.

As indicated in Fig. 4.31, the overall result is the same as that of an insertion reaction, the difference being that insertion gives rise to a *syn*-addition and formation of a vacant site, and nucleophilic attack to an *anti*-addition. There is ample proof for the *anti* fashion; the organic fragment can be freed from the complex by treatment with protic acids and the organic product can be analyzed. Suitably substituted alkenes will show the *anti* or *syn* fashion of the addition. Hydrides can be added *anti* to an alkene by taking a borohydride anion as the nucleophile and adding that to a coordinated, and thus activated, alkene.

In Fig. 4.31 the nucleophile depicted is anionic, but Nu may also be a neutral nucleophile, such as an amine or H_2O. There are many alkene complexes of middle and late transition elements which undergo this type of reaction, e.g. M = Pd^{2+}, Pt^{2+}, Hg^{2+}, Zn^{2+}, $FeCp(CO)^{2+}$. The addition reaction of this type is the key step in the Wacker-type processes catalyzed by palladium.

Alkynes

Alkynes show the same reaction, and the metal–vinyl product obtained is the *trans* isomer. After a suitable elimination from the metal the alkene obtained is the product of the *trans*-addition. Earlier we have seen that insertion into a metal–hydride bond and subsequent hydrogenolysis of the M–C bond will afford the *cis*-alkene product. Thus, with the borohydride methodology and the hydrogenation route, both isomers can be prepared selectively.

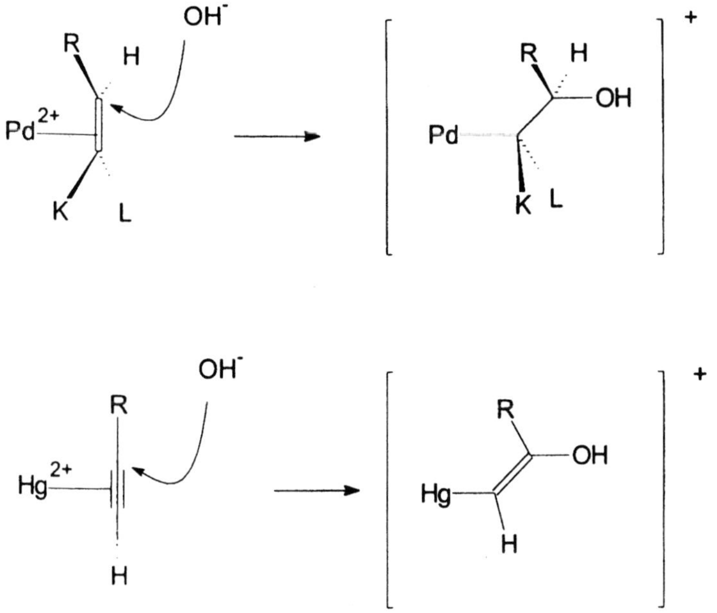

Fig. 4.31. Nucleophilic *anti*-attack on a co-ordinated alkene and alkyne.

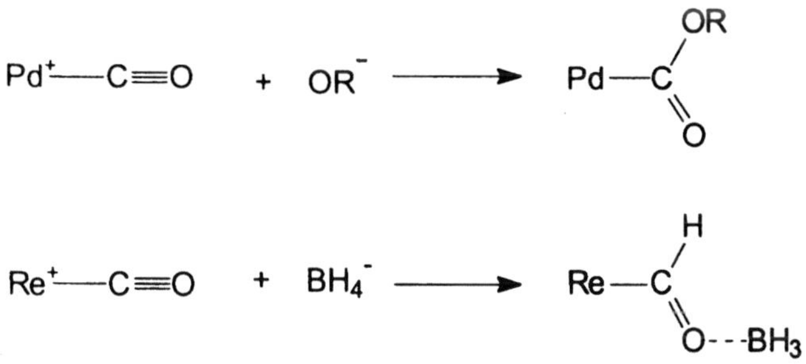

Fig. 4.32. Nucleophilic attack at co-ordinated CO.

Carbon monoxide

Carbon monoxide, when coordinated to a metal centre, is subject to activation toward nucleophilic attack. Through σ donation and π back donation into the antibonding CO π^* orbitals the carbon atom has obtained a positive character. This makes it not only more susceptible to a migrating anion at the metal centre, but also to a nucleophile attacking from outside the coordination sphere. In this instance it is more difficult to differentiate between the two pathways. There are examples showing that the electrophilicity of the carbon atom can be further increased by the action of Lewis acids complexing to the oxygen atom of the coordinated CO.

Figure 4.32 shows an alkoxide attack at coordinated CO giving a carboalkoxy complex, and a borohydride attack at coordinated CO in which the boron simultaneously acts as a Lewis acid in the final product [38]. The BH_3 complexation now stabilizes the formyl complex that would otherwise be thermodynamically inaccessible. The latter reaction has so far only been of academic interest in homogeneous systems (it may be relevant to heterogeneous systems though proof is lacking). The nucleophilic attack by alkoxides, amines, and water is of great interest to homogeneous catalysis. A dominant reaction in syn-gas systems is the conversion of carbonyls with water to metal hydrides and carbon dioxide ('shift reaction'), see Fig. 4.33.

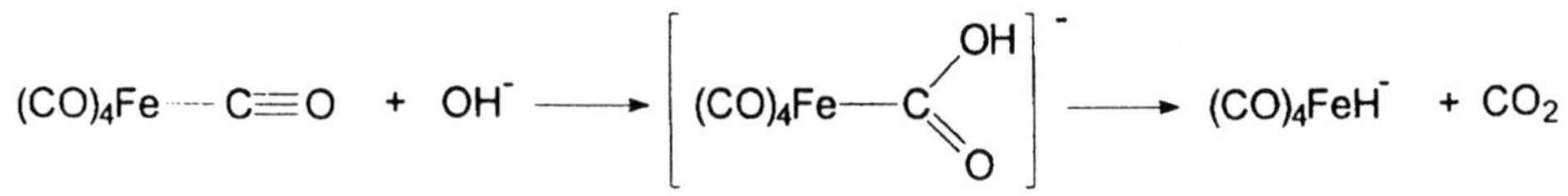

Fig. 4.33. Nucleophilic attack at $Fe(CO)_5$.

4.3.10 σ-Bond Metathesis

A reaction which is relatively new is the so-called σ-bond metathesis [39,40]. The word metathesis is used because of the now well-known metathesis of alkenes and alkynes. It is a concerted 2+2 reaction immediately followed by its retrograde reaction giving metathesis. The transition state is strongly polarized, such that in the reaction of M–R with H–H (Fig. 4.34), the transition state contains a $[H...H...R]^{-}M^{+}$ unit with large negative charges at the terminal groups.

The highly electronegative metal ion affords the appropriate empty orbitals for the stabilization of the complex anion in the transition state. Both late and early transition metal alkyls are prone to this reaction, but its occurrence had to be particularly invoked in the case of the early transition metals. Many similar reactions, such as the reaction of metal alkyls with other HX compounds, could be described as if they followed this pathway, but the use of the term σ-bond metathesis is restricted to those reactions in which one reacting species is a metal hydrocarbyl or metal hydride and the other reactant is a hydrocarbon or dihydrogen. Two reactions have been depicted in Fig. 4.34. There are, of course, borderline cases; when the reacting hydrocarbon is acidic, as in the case of 1-alkynes, a direct attack of the proton at the carbanion can be envisaged. It has been proposed that acyl metal complexes of the late transition metals may also

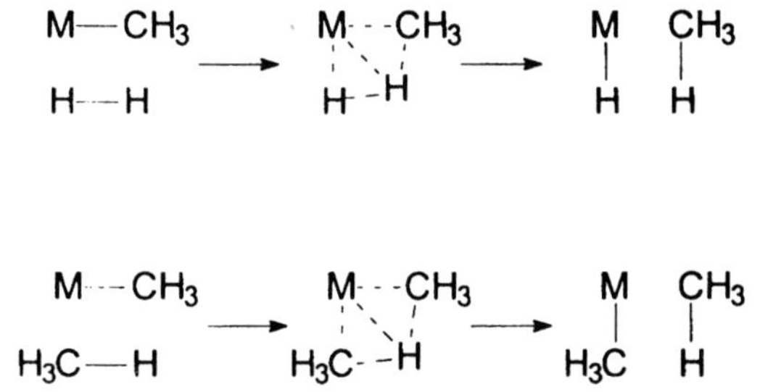

Fig. 4.34. σ-Bond metathesis.

react with dihydrogen according to a σ-bond metathesis mechanism. However, an alternative exists for the late elements in the form of an oxidative addition reaction. This alternative does not exist for d^0 complexes such as Sc(III), Ti(IV), Ta(V), W(VI), etc. and in such cases σ-bond metathesis is the most plausible mechanism. σ-Bond metathesis can also be regarded as a special case of heterolytic cleavage (see Section 4.3.11), but the latter reaction has only been observed for dihydrogen, so far, and not for alkanes.

4.3.11 Heterolytic Cleavage of Dihydrogen

Dihydrogen is said to be heterolytically cleaved when it is dissociated into a proton and a metal bonded hydride. It has been the topic of much study and discussion, and the evidence for its occurrence has been growing steadily [41,42]. In the ideal case the heterolytic splitting is catalyzed by the metal ion and a base which assists in the abstraction of the proton. In this reaction there is no formal change in the oxidation state of the metal. The mechanism has been proposed for Ru(II) complexes which can react with dihydrogen according to:

$$RuCl_2(PPh_3)_3 + H_2 \rightarrow RuHCl(PPh_3)_3 + HCl$$

Note, however, that ruthenium has a sufficient number of *d* electrons to allow for oxidative addition of dihydrogen, which could then be quickly followed by reductive elimination of HCl. It is difficult to distinguish between the two pathways experimentally. Recent observations on dihydrogen complexes of ruthenium have thrown a new light on the heterolytic splitting of dihydrogen. $CpRu(L)(L')(\eta^2\text{–}H_2)^+$ reacts rapidly with NEt_3, as can be deduced from the dynamic ^1H NMR spectra, which indicate a rapid exchange of the dihydrogen complex with its conjugate base, CpRu(L)(L')H (Fig. 4.35) [41]. This reaction is much faster than the exchange with the corresponding dihydride complex. The present studies on dihydrogen complexes may lead to a better understanding of the heterolytic splitting of dihydrogen, which is now shown to be activated towards reaction with a base through complexation to a cationic complex.

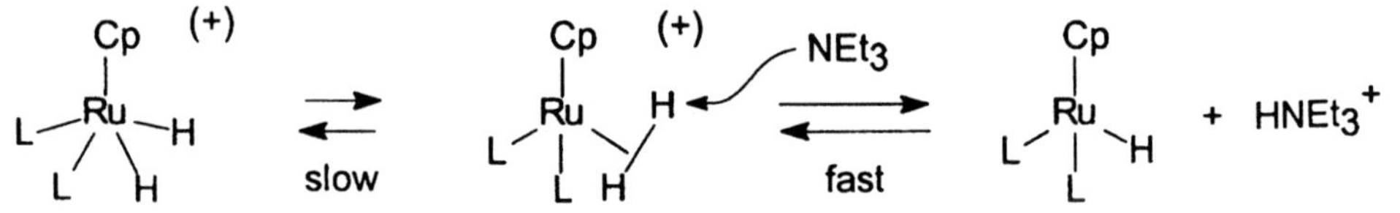

Fig. 4.35. Heterolytic cleavage of dihydrogen.

4.4 ELEMENTARY SURFACE REACTION STEPS

4.4.1 Elementary Surface Reaction Steps at Transition Metal Surfaces

4.4.1.1 Introduction

The catalytic cycle of elementary reaction steps on a transition metal surface consists of the following reactions:
- Adsorption/desorption
- Surface reconstruction/diffusion
- Surface dissociation/recombination
- Surface transfer reactions

These will be discussed in more detail in the next sections. The energies of reaction intermediates and activation energies of the subsequent reaction steps can be summarized in a reaction energy diagram. A proposed reaction energy diagram for the ammonia synthesis reaction is given in Fig. 4.36 [43]. The energetics of such a diagram depends on surface geometry and composition and on a chosen overall reaction mechanism. It can be constructed from experiment and complemented by theoretical calculations.

It is important to realize that on surfaces the adsorbate energies can be strongly concentration dependent and hence the reaction diagram should in principle be constructed as a function of surface concentration.

Usually only the low surface coverage values are given. Of additional importance is the realization that surface diffusion and reconstruction have important kinetic consequences.

Surface reconstruction usually leads to the formation of stable overlayers. This cannot occur without some mobility of catalyst substrate atoms. The reconstructed phase often has a different surface atom density from the non-reconstructed surface. For this reason the macroscopic consequence may be facetting of the catalyst along particular preferred crystallographic directions. This explains the often observed phenomenon that a heterogeneous catalyst often shows only stable performance after some initiation period. In a reactive system the surface composition of the adlayer may strongly vary with conditions and hence the details of surface facetting and surface reactivity.

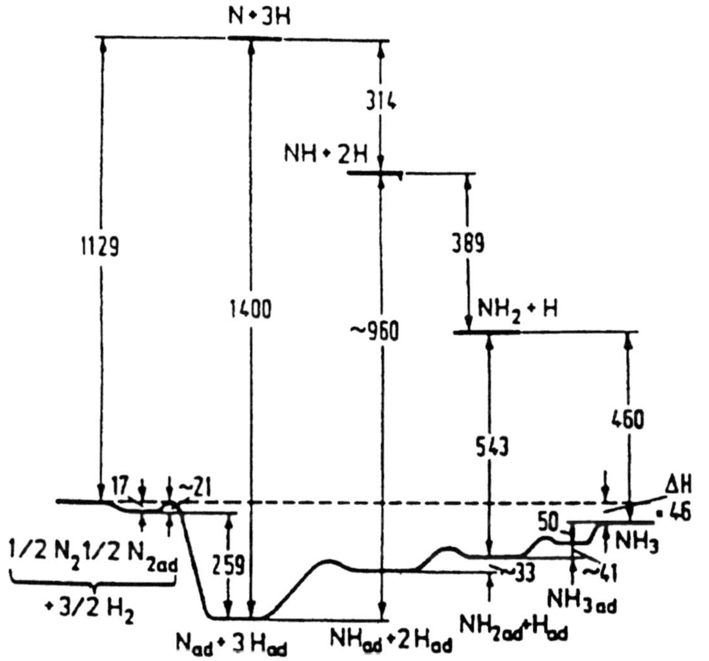

Fig. 4.36. Scheme for ammonia synthesis on Fe(III) at low nitrogen coverages from the work of Ertl (Ref. 43).

An adlayer with a variety of adspecies may become covered with patches of surface islands.

Only when diffusion is fast compared to rate limiting elementary reaction steps, these surface composition inhomogeneities may not have kinetic consequences and the adsorbate layer may be considered ideally mixed.

The reaction energy diagram (Fig. 4.36) illustrates competition between dissociative adsorption of the nitrogen molecule and desorption of ammonia. This confirms the conceptual basis of the Sabatier effect (3.7) that predicts that the rate is maximum at optimum interaction with the catalyst surface: the interaction energy where the rate of dissociative adsorption or another surface reaction and desorption rate are in balance.

The reaction energy diagram is temperature independent because it refers to Helmholtz energies and not Gibbs free energies. The latter are temperature dependent because they also contain the entropy dependence.

Generally changes in entropy of surface reactions are small, but between gasphase and surface, changes in entropy are large. This results in the general phenomenon that, at low temperature, reactions become inhibited because products do not desorb. At higher temperature the desorption equilibrium shifts to the gasphase and surface reactions tend to become rate limiting.

When reactions take place at the gas–solid interphase no additional interactions with medium molecules, as solvent molecules present at the solid-liquid interphase, can take place. Solvation effects are absent. Therefore reactions that require charge-separation are rare and if they occur need charge stabilization by strong electrostatic interactions with the surface. Examples of such reactions are acid-base reactions that occur on oxidic surfaces.

4.4.1.2 Adsorption of Atoms and Molecules

In this section we will discuss main features of atoms and molecules chemisorbed to transition-metal surfaces. Adsorption induced reconstruction will be included.

The interaction of adsorbed molecules with a surface is usually considerably less than that of atoms. For example the heat of adsorption of CO is of the order of 150 kJ/mole, whereas the energy of adatom adsorption of oxygen or carbon on a comparable surface is of the order of 400–500 kJ/mole. It is this difference in energy that provides the driving force for dissociative adsorption.

$$E_{diss} = -E_{molecule}(\text{gasphase}) - E_{ads}(\text{mol}) + \sum_i \{E_{\text{fragment } i}(\text{gasphase}) + E_{ads}(\text{fragment } i)\}$$

The gain in energy due to the strong adsorption of the fragment intermediates formed by dissociation has to overcome the energy cost of the cleavage of an atom–atom bond in the molecule. Differences in the reactivity of a surface are due to the changed balance between adatom (fragment) interactions with the surface and that of the non-reacted adsorbed molecule. When one compares chemisorption to different surfaces or metals, it appears that differences in the adatom energies are significantly larger than those of adsorbed molecules. This changes E_{diss} and hence the reactivity of a surface. This follows from application of the Brønsted–Polanyi rule.

Scheme 4.1 illustrates this relation between change in activation energy δE_{act} and change in reaction energy δE_R. According to the Brønsted–Polanyi rule, when there is no change in the reaction path

$$\delta E_{act} \sim 0.5\, \delta E_R$$

The different dependence of molecular versus atomic chemisorption has its electronic basis in the very different balance of donating and backdonating interactions. We will illustrate this for chemical bonding of chemisorbed CO versus chemisorbed C or O.

The valence electron orbitals and energies of CO and C are sketched in Fig. 4.37. The valence electrons of CO have been briefly discussed earlier in Section 4.2.2. In a molecule, atomic orbitals form bonding and antibonding orbitals,

Δ ≡ Activation Energy
d ≡ Reaction Energy

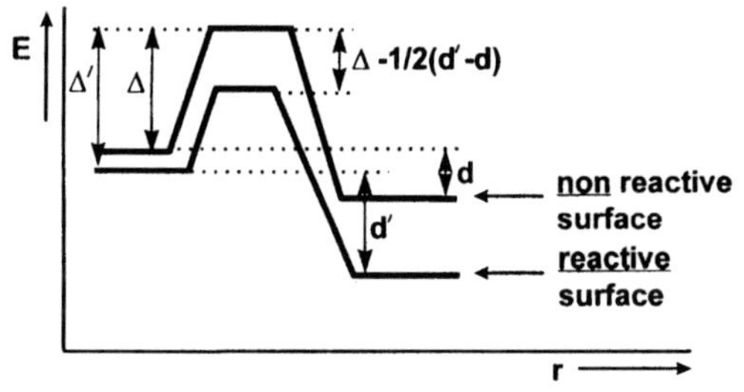

Illustration of the
Brønsted-Polanyi relation: $\Delta' = \Delta - 1/2(d'-d)$

Scheme 4.1.

separated by an energy gap. In CO the highest occupied molecular orbital, the 5σ orbital is σ symmetric with respect to the molecule's axis. It is separated by approximately 7 eV from the lowest two unoccupied degenerate $2\pi^*$ orbitals, of π symmetry with respect to the molecule's axis. The unoccupied $2\pi^*$ orbitals are antibonding and result from the interaction between the C_{p_x} and O_{p_x} and C_{p_y} and O_{p_y} orbitals. In the atom, atomic p orbitals are partially occupied, separated by approximately 20 eV from the doubly occupied $2s$ atomic orbital. CO adsorbs perpendicular to the transition metal surface, attached via its carbon atom. When adsorbed atop the surface valence s-electrons will interact with the 5σ orbital, but their interaction is symmetry forbidden with the p-symmetric $2\pi^*$ orbitals.

Interaction of π-type CO orbitals with the s valence atomic orbitals is only possible in high coordination sites (in the organometallic nomenclature these are called bridging sites, denoted μ (or μ_2), μ_3, and μ_4 for sites involving 2, 3 and 4 metal atoms, respectively). As illustrated in Fig. 4.38 they can then interact with asymmetric group orbitals that are linear combinations of atomic s orbitals of the surface atoms.

When CO is adsorbed atop (monometallic, end-on), the π-type CO $2\pi^*$ orbitals will interact with the symmetric metal d atomic orbitals as well as p_x or p_y orbitals. The 5σ orbital will interact with d_{z^2} as well as s and p_z metal atomic orbitals. To illustrate the formation of the surface chemical bond, the interaction scheme of bond formation between the CO valence orbitals (5σ and $2\pi^*$) and metal d valence electrons is sketched in Fig. 4.39. Both bonding and antibonding orbital fragments are formed. Bonding π-type fragment orbitals are occupied. They are mainly located on the metal atom. Bonding as well as antibonding σ-type surface fragment orbitals become occupied. The σ-type bonding (σ_b) orbital fragments

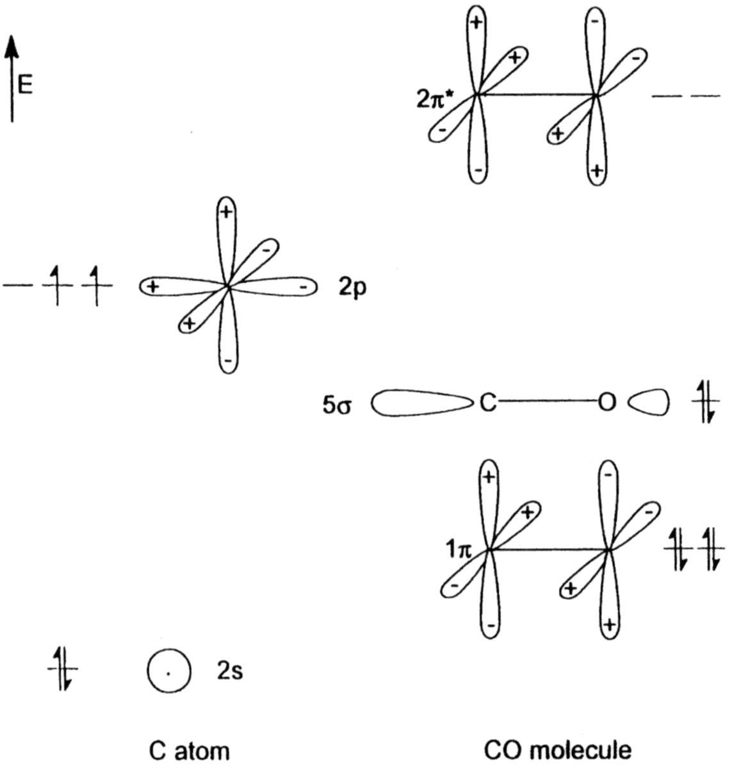

Fig. 4.37. Atomic and molecular valence electron levels.

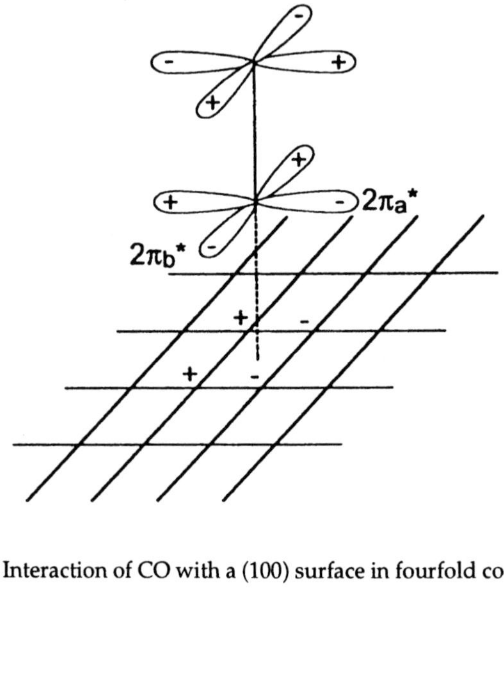

Fig. 4.38. Interaction of CO with a (100) surface in fourfold coordination.

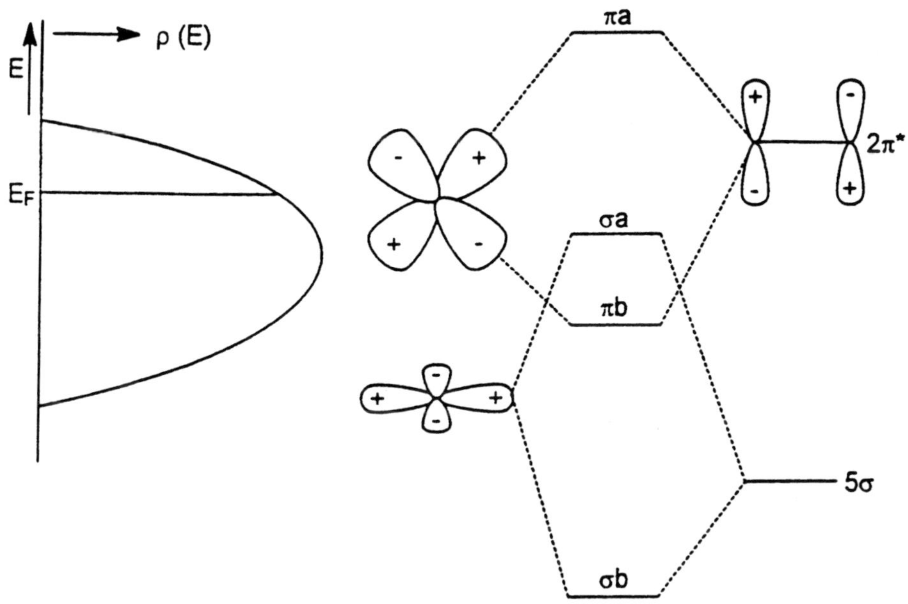

Fig. 4.39. Quantum chemistry of the CO *d* valence electron interaction.

are mainly located on CO, the σ-type antibonding (σ_a) orbital fragments are mainly located on the metal atom. This bonding scheme corresponds to the earlier mentioned Blyholder model.

When the *d* valence electron band is nearly completely filled, interaction with the doubly occupied CO 5σ orbital, leading to a significant fraction of occupied antibonding orbital fragments σ_a between adsorbate and surface atoms, will be repulsive. This Pauli repulsion is proportional to the number of surface atom neighbours and hence is a minimum in atop coordination. This counteracts the interaction with the $2\pi^*$-CO molecular orbitals, which favours high coordination sites. Different from the CO 5σ orbital, the energy of the unoccupied $2\pi^*$ orbitals is higher than the highest occupied metal orbital, the Fermi energy. The occupied orbital fragments resulting from the interaction with the metal surface are mainly of bonding nature and are located on the metal surface atoms. Occupied bonding orbital fragments favour adsorption in high coordination sites.

A decrease in the *d* valence electron occupation decreases the occupation of the bonding π-type orbitals. This decreases the interaction energy. For the σ-type symmetric orbital fragments a decrease in electron occupation depletes antibonding orbital fragments, which results in an increase of the interaction energy. Changes in the interaction of π- and σ-type orbitals give a counteracting contribution to the overall bond strength.

Bonding of the C atom to the transition metal surface is dominated by interaction with the three low-lying $2p$ C atomic orbitals. The p atomic orbitals parallel to the metal surface can only interact with the s valence orbitals in high-coordination sites. This interaction is much stronger than the corresponding CO interaction (the $2\pi^*$ orbitals should be compared), because of the lower energy of the atomic orbitals. Also these C p atomic orbitals are not part of a molecular orbital as in CO, which increases orbital overlap significantly. The overlap of the $2\pi^*$ orbital with a surface group orbital ψ_g is given by:

$$S_{2\pi^*,\psi_g} = \int (c_1 \psi_{p_x}^{C} - c_2 \psi_{p_x}^{O}) \cdot \psi_g \, d\tau \quad (4.18a)$$

$$= c_1 \int \psi_{p_x}^{C} \cdot \psi_g \, d\tau \quad (4.18b)$$

$$= c_1 S_{p_x,\psi_g} \quad (4.18c)$$

Equation (4.18c) relates the carbon p_x overlap to the overlap with a CO $2\pi^*$ orbital. The value of $|c_1|^2$ is significantly less than 1. Whereas the interaction with a molecular $2\pi^*$ orbital results only in occupied bonding orbital fragments, interaction with the atomic p_x, p_y or p_z orbitals results in a bonding orbital fragment as well as antibonding ones. This is due to the lower energy of the atomic $2p$ orbitals, as compared with molecular $2\pi^*$ orbitals.

In the atom three $2p$ orbitals are involved in the surface chemical bond. The interaction with the p_z orbital has σ symmetry with respect to the surface. Apart from an interaction with different surface symmetry orbitals and a larger overlap energy with the surface orbitals, its dependence on electron occupation is qualitatively similar to that of the p_x and p_y orbitals. Because the p_x and p_y orbitals can only interact with the surface s atomic orbitals in high coordination sites, and this bonding interaction contributes significantly to the total surface bond energy, bonding to high coordination sites is favoured.

The covalent interaction with the adatom $2p$ orbitals increases with decreasing surface d valence electron occupation, because fewer antibonding orbital fragments then become occupied. The adsorption energy of atoms varies much more strongly with d valence electron occupation than that of molecules because compensating effects occur in the surface chemical bond of molecules. As we discussed above for CO, variation in the interaction with the 5σ orbital is partially compensated for by changes in the interaction with the $2\pi^*$ orbital.

In general, therefore, high coordination sites are favoured when adsorbates interact that have partially occupied p or π type orbitals [1]. This is typically the case for most adatoms. The interaction of s or σ type orbitals with partially filled metal valence orbitals usually also shows a preference for high coordination sites. When the interaction with d valence orbitals dominates atop adsorption is

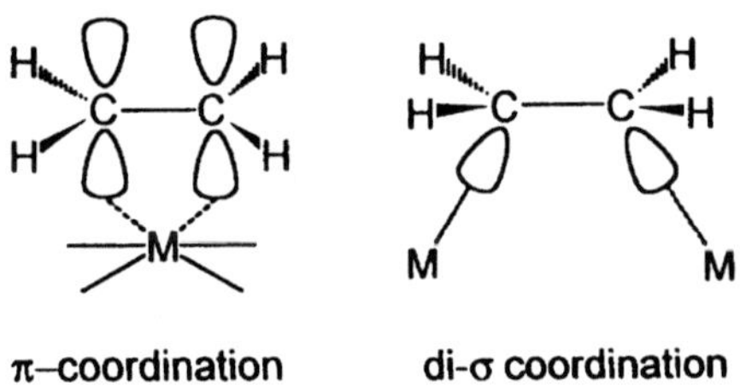

Fig. 4.40. Coordination of ethylene.

favoured in group 8–11 metals. Hydrogen atoms will adsorb in high coordination sites as long as their interaction with the *d* valence electrons is small. This generally appears to be the case. Methyl fragments adsorbed to the Ni surface tend to favour high coordination sites [42]. The CH_3 fragment mainly interacts via its σ-symmetric lone-pair orbital. For a transition metal like Pt, with spatially more extended *d* atomic orbitals, one expects the methyl radical to show less preference for high coordination sites and to adsorb strongly atop [6,7]. Chemisorption of ammonia to a metal surface also occurs mainly by way of its σ type lone-pair orbital. Calculations show that on Cu, with a completely filled *d* valence electron band, NH_3 adsorbs preferentially atop. Bonding with the CH_3 radical or ammonia occurs mainly by the electron donating interaction with their highest occupied molecular orbital (HOMO), the σ orbital, because of the weak interaction with their lowest unoccupied molecular orbital (LUMO), which has an unfavourably high energy.

Often, as we saw for CO, interaction with HOMO as well as LUMO type adsorbate orbitals is important. This is also the case for a molecule like ethylene. This can coordinate in two ways to the transition metal surface, namely in π coordination or di-σ coordination (Fig. 4.40) (η^2 and μ-η^1:η^1, respectively). (η^n denotes a ligand bonded with *n* atoms to the metal.)

The chemical bonding of π coordinated ethylene is very similar to the Chatt–Dewar–Duncanson picture of coordination (Fig. 4.41). The donating orbital is the doubly occupied π orbital that is σ-symmetric with respect to the normal to the surface. When adsorbed atop it interacts with the highly occupied d_{z^2} surface atomic orbital and the partially filled *s* and p_z orbitals. The ethylene LUMO is the empty asymmetric π^* orbital, which interacts with the surface d_{xz} and p_x orbitals. The corresponding overall interaction is relatively weak and no hybridization on ethylene is assumed to occur. The orbital interaction diagram of the occupied ethylene π orbital with surface atom d_{z^2} orbital is analogous to that sketched for the CO 5σ orbital in Fig. 4.39. When this d_{z^2} becomes nearly completely occupied, as occurs, for instance, for Pd or Pt, the ethylene-π surface-atom-d_{z^2} interaction becomes repulsive. Because the spatial extension of Pt *d* atomic orbitals is larger than those of Pd or Ni, the repulsive interaction is largest for Pt. In order to

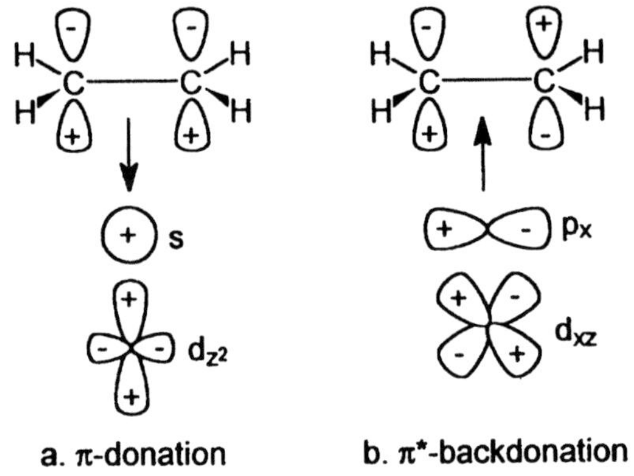

Fig. 4.41. Chatt-Dewar-Duncanson chemical bonding scheme of chemisorbed ethylene.

overcome this repulsive interaction ethylene rehybridizes on Pt and adsorbs in di-σ coordination, as sketched in Fig. 4.40b (μ–η^1:η^1). The orbitals on the C atom of ethylene rehybridize from sp_2 to sp_3 hybridization, as in ethane. The CH bonds bend away from the surface and two lone pair orbitals, one in each carbon atom, develop along the missing CH bonds. They are directed towards the surface metal atoms. Ethylene now prefers coordination to two surface metal atoms [8]. Chemisorption controlled by ad-molecule hybridization generally appears to be important when the interaction with highly occupied surface *d* valence atomic orbitals dominates. On Pt CH_3 adsorbs atop but CH_2 adsorbs in twofold position (μ, without the subscript 2) and CH in threefold position (μ_3). On Ni, with a relatively unimportant *d* valence orbital interaction, all three molecule fragments prefer high coordination [44].

When surface atoms participate in more open surfaces, their average co-ordination number decreases, resulting in an increase of coordinative unsaturation. The width of the valence band electron density is related to the coordination number of the metal atoms. The larger the coordination number, the larger is the number of metal–metal bonds per atom, and the electrons have to be shared over more bonds. This increases the degree of delocalization of the surface electrons, which is reflected in an increased valence electron bandwidth. Because the total electron density per atom remains constant, an increased bandwidth implies a decreased average electron energy density. As is illustrated in Fig. 4.42, the electron density decreases with decreasing coordinative unsaturation in the centre of the valence electron band. This is because the integrated density is constant.

As a result, the interaction of adsorbate orbital and surface orbitals is largest for surface atoms that have the lowest surface metal atom coordination number. The bonding orbital surface fragment has a larger downward shift and the antibonding orbital fragments shift more upward. For the situation sketched in Fig. 4.42, fewer antibonding orbital fragments become occupied for the co-

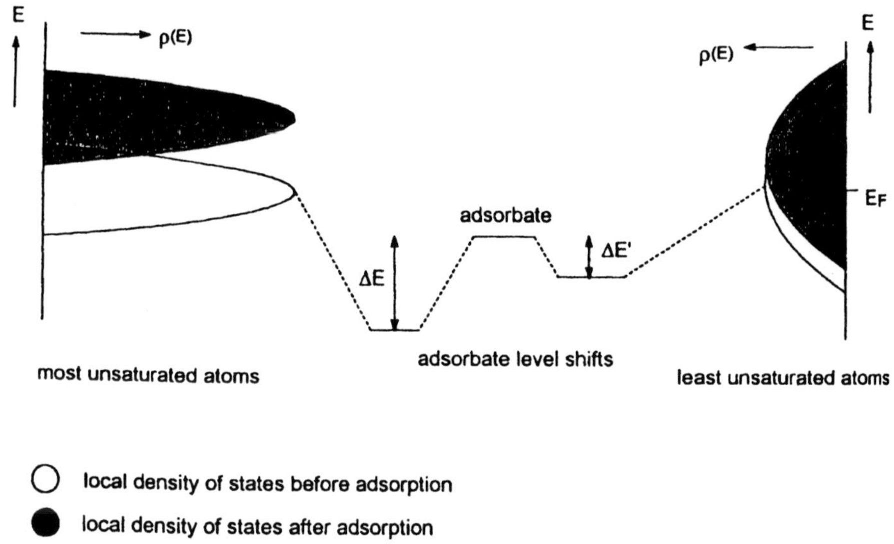

Fig. 4.42. Interaction of adsorbate orbital with surface atoms of different coordination number.

ordinatively most unsaturated atoms. This, in combination with the larger downward shift of the bonding orbital fragments, results in a stronger interaction with the coordinatively most unsaturated surface atoms. A similar overall result can be derived when empty adsorbate orbitals interact, forming only occupied bonding surface orbital fragments.

Changes in bond strength with variation of coordinative unsaturation of surface atoms are very similar to the effects of coadsorption. When two adsorbate atoms share a bond with the same surface atoms, the surface atom effectively has an increased coordination number. Therefore the adatom bond strength is weaker than when there is no sharing between metal surface atoms. The heat of adsorption generally decreases at high adsorbate surface coverage as a result of this effect.

On the adsorption of strongly interacting surface atoms, in particular, the neighbouring metal–metal bonds between surface metal atoms weaken strongly. The surface–adatom interaction enhances the total number of surface atom neighbour atoms. The delocalization of the surface electrons increases, which results in a decrease of bond strength between the other atoms. This also holds for the interaction between the surface metal atoms. An alternative way of viewing this is the concept of conservation of bond order [45]. The bond order of a chemical bond is a measure of its bond strength. To a very good approximation the sum of the bond orders of the chemical bonds with a single atom is a constant. Because the total bond order per metal surface atom is conserved, an increase in

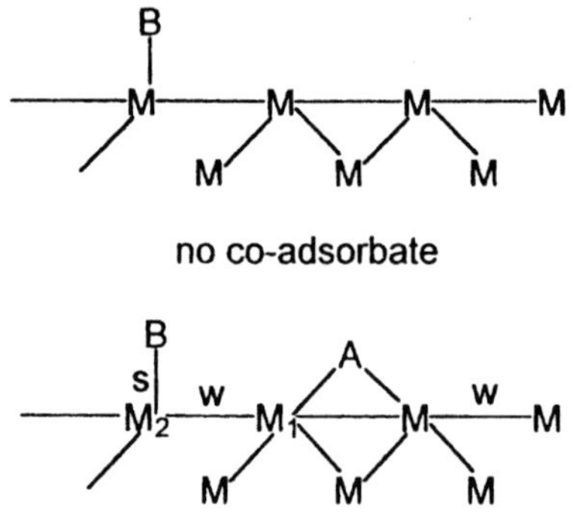

Fig. 4.43. Change in bond strength with a surface atom by a coadsorbate in the second coordination shell with respect to adsorbate (schematic).

coordination number decreases the bond order per bond, and hence the bond strength of the atom bonds.

The effect on the bond strength of a coadsorbate adatom B, adsorbed in the second coordination shell with respect to adatom A, is an increase of the bond energy (see Fig. 4.43). The weakened interaction between metal atoms 1 and 2, results in an increase of the M–B bond strength.

The weakening of the surface metal atom bonds by adsorbed atoms may lead to surface reconstruction. Secondly reconstructed phases become thermodynamically preferred, when the cost of reconstruction is overcome by the gain in interaction with adsorbed atoms. As an example we will take the reconstruction of the Cu(110) surface upon adsorption by oxygen atoms.

Chemisorbed oxygen leads to the so-called added row reconstruction of Cu(110). Since the average number of Cu surface atom neighbours decreases, this reconstruction costs energy in the absence of adsorbed oxygen atoms. Computation [46] indicates an energy cost of 16 kJ/mol. On the other hand the lower surface co-ordination of the surface Cu atoms leads to an enhanced interaction with oxygen. Compared to bonding to the non-reconstructed surface, the gain in energy is 52 kJ/mol (see Fig. 4.44).

The phenomenon of surface reconstruction is quite general and is one of the causes of surface facetting observed in working catalysts.

The surface reactivity is enhanced by increased reactivity of surface atoms, resulting from lower surface co-ordination. Such effects are also basic to the enhanced reactivity of small metal particles, containing atoms of lower co-ordination, as in large particles. In practice small particles may appear less reactive, because due to their high reactivity they are rapidly deactivated in a catalytic reaction.

The dependence of surface reactivity as a function of metal in the periodic system can be best understood by summarizing the general features of the

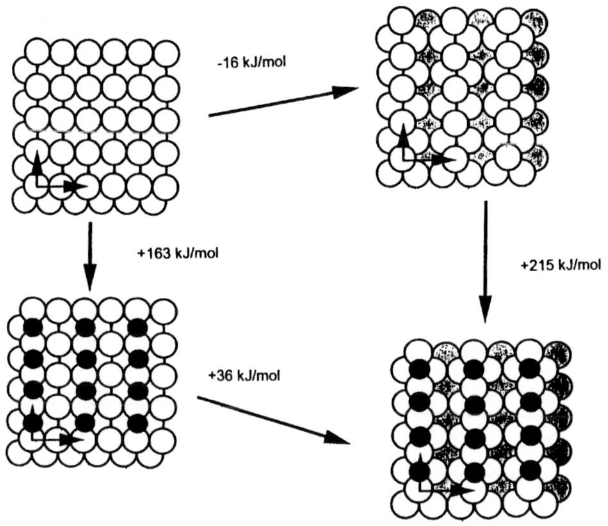

Fig. 4.44. Energy changes in kJ/mole of oxygen adsorption to non-reconstructed Cu(110) surface (163 kJ/mole) compared to oxygen adsorption to added-row reconstructed Cu(110) surface (215 kJ/mole). Energy cost of surface reconstruction is 16 kJ/mole.

transition metal valence electron bond structure.

The valence electron distribution of a transition metal surface is sketched in Fig. 4.45. A narrow *d* valence electron band is overlapped by a broad *s–p* valence electron band. Per atom, the number of electrons in the *s,p* valence electron band can be considered to be constant and equal to 1. The *s* character of the atomic orbitals dominates in this valence band. The *d* valence electron content varies. For the metals Cu, Ag and Au the *d* valence electron band is completely occupied. When one moves from right to left in a row of the periodic system, the *d* valence electron occupation of the metal decreases. In Ni, Pd and Pt the *d* valence electron occupation is 9, etc. The width of the *d* valence electron band increases when moving down in a column of the periodic system or to the left, moving along a row. For a metal (i.e. the solid) the ionization potential is less than for an atom (the resulting positive charge is screened in the bulk metal) and is called the work function. The work function (E_F) decreases when moving from right to left along a row, and increases when moving downward along a column in the periodic system. The *d* atomic orbitals have a smaller spatial expansion than the *s* atomic orbitals, which explains the smaller width of the *d* valence atomic orbitals.

The bond strength to transition metal surfaces decreases with increasing *d*-valence electron occupation. This is because more anti-bonding adsorbate-surface orbital interactions occur. It tends to decrease when moving downward along a column of the periodic table.

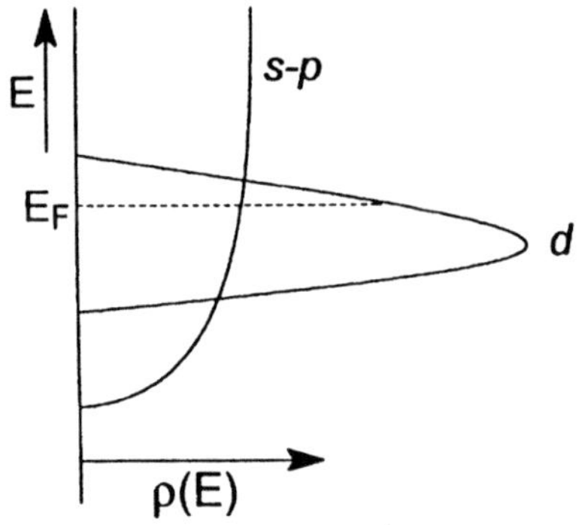

Fig. 4.45. The valence electron distribution at a transition metal surface (schematic).

The metal–metal bond strength becomes larger because the metal electrons become more delocalized due to the larger orbital overlap. Upon chemisorption electrons have to localize on the surface involved in the chemisorptive bond. The larger the delocalization of the metal electrons, the higher is the cost involved in localization of the electrons and the smaller the chemisorptive bond energy. In addition the work function tends to increase, resulting in a decrease of polarity of the adatom (adsorbed atom) surface chemical bond. For instance carbon, oxygen and hydrogen atoms are bonded more strongly to Ni than to Pt. Exceptions to this rule exist, however. For example CO binds more strongly to Pt than to Ni, but O_2 and N_2 show the expected, reverse behaviour.

These trends explain the important fact that H_2 dissociates on the group VIII metals, hydrogen molecule dissociation is activated on Cu, but will not dissociate on Ag or Au.

Au is the most unreactive of the $d^{10}s^1$ metals. It is the only metal on which oxygen will not dissociate.

The activation energies for adatom or admolecule diffusion are significantly less than the heats of adsorption. In order to diffuse a molecule or atom has to change co-ordination (e.g. one-fold to two-fold, to one-fold etc.) These energy differences are typically 10% of the heats of adsorption.

If one dilutes a surface with reactive surface atoms, by alloying with non-reactive atoms (e.g. Au) or with atoms of a different reactivity, the heat of adsorption of adatoms will be affected because in high co-ordination sites re-active atoms become replaced by other weaker bonding atoms. The change in heat of adsorption then reflects this altered chemical environment of the adatoms.

The decreased heat of dissociative adsorption of O_2 to Pt by alloying with Au has this as an explanation. Also the weaker interaction energy of C atoms to such alloys compared to non-alloyed Pt is a consequence of this ensemble type effect. It is called the *secondary ensemble* effect.

4.4.1.3 Surface reactions: dissociative, associative and transfer reactions; selectivity.

The elementary reaction of surface dissociation will occur when this reaction becomes thermodynamically feasible. This has been extensively discussed in the previous section.

We learned in Section 4.2.1 that the surface adatom bond energy varies more than the molecular adsorption bond with variation of metal. Adsorption of atoms is favoured by metals with a partially filled *d* valence electron band or metals of a low work function. CO will dissociate on the first row transition metals (except on Cu), it will barely dissociate on Rh, and not at all on Pt or Ir. O_2 will dissociate on all transition metals, including Cu and Ag, but not on Au.

On a transition metal surface the dissociating molecule prefers a specific reaction-path [47].

The surface atoms generated by dissociation prefer high co-ordination sites. According to the nomenclature of organometallic chemistry such sites are denoted μ_n sites, *n* being the number of metal atoms involved in the chemisorptive bond. CO bonded with one molecular atom to the surface is η^1 bonded. Side-on adsorbed CO is η^2 bonded. When the atoms are adsorbed in neighbouring adsorption sites they experience a large repulsive interaction. The effective number of neighbours of the surface metal atom has increased by adsorption, which reduces the bond strength of the atoms that are co-ordinated to this surface-metal atom. When one of the atoms adsorbs in site a, the other metal surface sites, c or d, are positions where the dissociated atoms share the least number of surface atoms and hence have the smallest repulsive interaction.

The activation energy is least when the CO bond is weakened because of its interaction with the metal surface. Electron transfer from the metal surface into antibonding molecular orbitals weakens the CO bond. This is a very general feature of dissociative adsorption. The dissociation energy of adsorbed molecules decreases when they are adsorbed with considerable electron transfer from the metal surface to the adsorbate.

For instance, O_2 molecules readily dissociate on the Ag (110) surface. The molecule is adsorbed as O_2, with a charge of the order of ~−1.6 a.u. [48]. Chemisorbed CO undergoes considerable CO bond weakening, because electron donation to CO populates its antibonding $2\pi^*$ orbitals (see Fig. 4.39). The antibonding orbitals (i.e. weakening orbitals) are usually antisymmetric with respect to a molecule or bond symmetry axis. The dissociating molecule, parallel to the surface, has to interact with surface orbital fragments of proper symmetry in order that electron donation may occur. As sketched in Fig. 4.45 d_{xz} or d_{yz} orbitals have the proper symmetry. So, for a low dissociation energy barrier, interaction with occupied *d* surface atomic orbitals asymmetric with respect to the surface is essential. Whereas, for a dissociation reaction, electron transfer between surface and adsorbate is favourable, the reverse is the case for the association reaction.

The activation barriers of surface dissociation reactions are low compared to the energies of the adsorbate bonds that are to be broken. Often one finds a value ~10% of the bond energy. This arises from the stabilization of the stretched bond by the stabilizing interaction with the metal surface.

Adsorption followed by a surface dissociation step is the analogue of the oxidative addition reaction we have met earlier. Associative recombination reactions become favoured when the adatom energies become destabilized, i.e. to the right of the periodic system, when the *d*-valence electrons have become highly occupied. The associative recombination reaction followed by desorption is the analogue of the reductive elimination reaction.

Since dissociation generates from one molecule two fragments that can be adsorbed in high co-ordination, dissociative adsorption requires a surface ensemble of atom. This is called the *primary ensemble* effect. The presence of large surface ensembles with a large number of surface atoms distinguishes surfaces from organometallic complex and small transition metal cluster from heterogeneous catalysts. Blocking of adsorption sites by alloying with inert atoms (Sn, Au) or by coadsorption with non-reactive adsorbates will suppress dissociation reactions.

In the activation of larger molecules similar principles apply, but now steric effects play an additional role.

For instance the activation of ethane proceeds after initial cleavage of the CH bonds, whereas the C–C bond is weaker than the CH bond. This is because the C–C bond can only be activated once removal of some hydrogen atoms has made both C atoms accessible for activation. When, for instance, due to alloying the ensemble of atoms is not large enough to accommodate the fragments of dissociated ethane only CH activation occurs and the preferred product will be ethylene, a molecule in which the number of C–C bonds is maintained.

The activation of a CH bond requires a smaller ensemble than C–C bond cleavage, because the alkyl radical and hydrogen can adsorb at low co-ordination sites with small cost.

Surface coverage not only affects the selectivity by varying ensemble size, but can also influence the electrophilic nature of reacting atoms. The reactivity of coadsorbed surface atoms provides an example.

Oxygen adatoms adsorbed to group IB or adsorbed to group VIII metal as Pt can activate CH, OH or NH bonds. This can be especially important when activation of such bonds is endothermic. An interesting example is the dissociative adsorption of NH_3 on a Cu(111) surface [49]:

Compare:

$$NH_{3ads} \rightarrow NH_{2ads} + H_{ads} \qquad \Delta H = +176 \text{ kJ/mol} \tag{a}$$

$$NH_{3ads} + O_{ads} \rightarrow NH_{2ads} + OH_{ads} \quad \Delta H = +86 \text{ kJ/mol} \tag{b}$$

Both elementary reaction steps are endothermic, but the hydrogen transfer reaction with coadsorbed oxygen is considerably less endothermic than without coadsorbed oxygen.

To Ni, oxygen as well as hydrogen is much more strongly bonded than to Cu. Therefore, as follows from Bond Order Conservation, the (Ni)OH bond is much weaker than the (Cu)OH bond. In the case of N_i oxygen adatom coadsorption will suppress dissociative adsorption of NH_3 to Ni.

Chemically oxygen adatoms adsorbed to Cu or Ag behave as Lewis basic reagents.

Heterolytic splitting of H_2 as a ZnO or MoS_2 is analogous to the hydrogen transfer reaction of CH, NH to form OH. However the electropositive nature of oxygen adatom may be affected at high surface coverage. For instance, on a silver surface in the oxidized state, with oxygen adsorbed in the external surface layer as well as in the subsurface layer the electropositive reactivity of oxygen adatoms is changed into an electronegative one. When exposed to ethylene, CH activation occurs at low oxygen coverage, but the surface in the highly oxidized state will give insertion of oxygen into the π bond of ethylene leading to ethylene epoxide. It is this state of the surface that is responsible for the use of Ag as a catalyst for the epoxidation of ethylene (see also Section 5.5.2).

Activation of the CH bond leads to total combustion. Whereas this reaction can be suppressed for ethylene, allyl formation of molecules as propylene occurs with such a low activation energy that total combustion remains the dominating reaction even at high surface coverage.

Interestingly when one alloys silver with gold, and the surface oxygen concentration decreases, the alloy appears to be a selective catalyst for the production of acrolein, requiring electropositive oxygen so that aldehyde formation can occur.

The MoO_3 based catalysts used commercially for the production of acrolein from propylene, have a reduced total oxidation activity because of the strong Mo–O bonds. The electropositive reactivity of lattice oxygen results in selective acrolein production, by attachment of lattice oxygen to the end C-atoms of the allylic species generated by hydrogen transfer of propylene hydrogen atoms to catalyst oxygen atoms.

The weakening of adatom surface bonds with increased coverage, may also result in enhanced rates of surface association reactions and desorption. An example is the oxidation of NH_3 catalyzed by Pt at low temperatures. The peak maximum for desorption of N_2 from Pt(111) is at 400 K at low coverage. At higher coverage the temperature programmed desorption profile broadens with a width of over 100 K. As a consequence low temperature oxidation of NH_3 requires high surface coverages.

So far we discussed how selectivity is influenced by altered surface compositions. We identified the ensemble effect as a geometric control factor and competitive adatom reactivity as an electronic control factor.

A third, statistical control factor that affects the selectivity of a surface reaction is the probability for adsorbed intermediates to have a particular co-reactant in a neighbouring site, so that an associative surface reaction may occur. An interesting example is the selectivity to produce nitrogen by oxidation of NH_3. The low reactivity of NH_3, causes the copper surface to be predominantly covered with oxygen, with nitrogen atoms dispersed in between. At low N coverage the probability of having oxygen atoms as neighbours is high and hence the selectivity to NO is enhanced. The selectivity to N_2 increases at higher NH_3 partial pressures because then the N_{ad} coverage increases and hence the probability of having N_{ad} as a reactive neighbour is enhanced.

4.5 ELEMENTARY REACTION STEPS ON SOLID ACIDS

4.5.1 General introduction

Oxidic surfaces in particular develop acid or basic properties which are important in catalysis. We will approach this subject first by taking as a starting point the ionic bond model [52]. The lattice is considered to consist of cations and anions held together by electrostatic interactions. Later we will discuss a more balanced theory that also accounts for covalent bonding aspects.

Surface Lewis acidic or basic sites are created on ionic surfaces because surface atoms have a lower coordination number compared to their bulk values. A condition for surface stability is that the sum of the charges over the surface unit cell is zero and the surface is overall neutral. A useful method to estimate the degree of coordinative unsaturation on an ionic surface is to use Pauling valencies and to compute charge excesses. The Pauling valency, v, is defined as:

$$v = Q / C \tag{4.19}$$

Where Q is the formal ion charge and C its coordination number.

For the Mg^{2+} ion, the Pauling valency equals

$$v^{+} = \frac{2}{6} = \frac{1}{3} \tag{4.20a}$$

and for the O^{2-} ion, the valency becomes

$$v^{-} = -\frac{2}{6} = -\frac{1}{3} \tag{4.20b}$$

On the surface the charge excess of an ion is defined as the formal charge Q of the

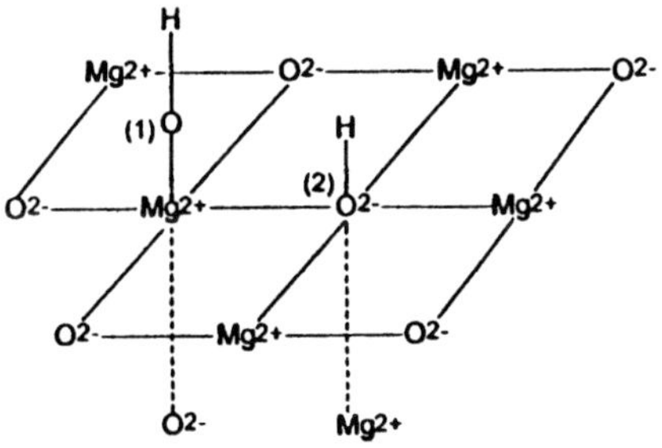

Fig. 4.46. H_2O dissociation on the $M_gO(100)$ surface.

ion considered plus the sum of the Pauling valencies, computed with respect to the other ion:

$$e^+ = Q^+ + \sum_i v_i^- \tag{4.21}$$

On the MgO(100) surface one finds for e^+

$$e^+ = 2 - \frac{5}{3} = +\frac{1}{3} \tag{4.22}$$

A negative value is computed for e^- on the oxygen anion. A positive charge excess implies Lewis acidity and a negative charge excess implies Lewis basicity. When the surface becomes exposed to NH_3, ammonia will be adsorbed to the Mg^{2+} ion. Similarly, H_2O will also bind through its oxygen atom to the surface Mg^{2+} ion.

Brønsted acidity or basicity develop when the H_2O molecule dissociates. Two different hydroxyl coordinations can develop on the MgO surface. As shown in Fig. 4.46, one hydroxyl becomes end-on coordinated to Mg^{2+}, the proton coordinates to the basic bridging oxygen anion.

To determine acidity or basicity it is useful to compute the charge excesses on oxygen atoms O(1) and O(2).

$$e(O(1)) = -2 + 1 + \frac{1}{3} = -\frac{2}{3} \tag{4.23a}$$

$$e(O(2)) = -2 + 1 + \frac{5}{3} = +\frac{2}{3} \tag{4.23b}$$

A negative excess charge implies incomplete coordination with positive charge carriers. The end-on hydroxyl group can be considered as being Brønsted basic. Oxygen atom O(2) has excess positive charge, the proton coordinated to this

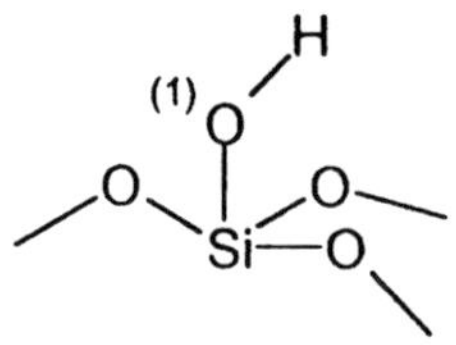

Fig. 4.47. Silanol group.

bridging oxygen atom behaves as a Brønsted acid. This is an important and very general conclusion.

When oxidic surfaces become exposed to water, basic hydroxyls, as well as Brønsted acidic protons, are generated upon dissociation of the water. This also occurs on the surfaces of basic oxides. There is an abundance of infrared spectroscopic information confirming the appearance of different OH groups by adsorption of HO on oxidic surfaces. Catalytic reactions induced by Brønsted acidic sites take place on such surfaces. The acidic proton is located on bridging oxygen sites.

Ammonia interacting with a surface as sketched in Fig. 4.46, will form NH_4^+ ions coordinated to the bridging oxygen atom. Most oxide surfaces behave very similarly to the MgO surfaces. An exception is the surface generated on silica. In SiO_2 the Si cations are tetrahedrally coordinated. This coordination can be restored on the surface when the SiO_2 surface becomes exposed to H_2O. Silanol groups develop, as sketched in Fig. 4.47.

One can easily see that the excess charge on O(1) is zero. So no decision on Brønsted acidity or basicity can be made on the basis of Pauling excess charges. Chemically, the silica–silanol group is slightly acidic, due to the high charge of the small Si^{4+} ion to which the hydroxyl is coordinated.

Whereas the surfaces discussed so far have been generated from the bulk by a simple cut, leading to a decrease in the coordination number of the surface atoms, catalytically important acidic surfaces can also be generated in microporous or layered materials by isomorphous substitution of lattice cations. This occurs in zeolites and smectite clays. Zeolites and clays can be considered as aluminosilicates. Their lattice compositions can vary significantly. In zeolites the Al^{3+} ion can be substituted by many other trivalent cations. Si^{4+} can be partially substituted by Ti^{4+} or Ge^{4+}.

In the zeolites each lattice cation is tetrahedrally coordinated to four oxygen anions. Each oxygen anion shares two lattice cations. In smectite clays an octahedral layer, usually containing Al^{3+} or Mg^{2+}, connects two tetrahedral layers. The tetrahedral layer can be considered neutral when the tetrahedral site contains a four valent cation. This is usually a Si^{4+} ion. A side view is shown in Fig. 4.48.

Isomorphous substitution of Si^{4+} by a trivalent ion (this is often an Al^{3+} ion) results in a negative lattice charge. This negative charge can be compensated by a

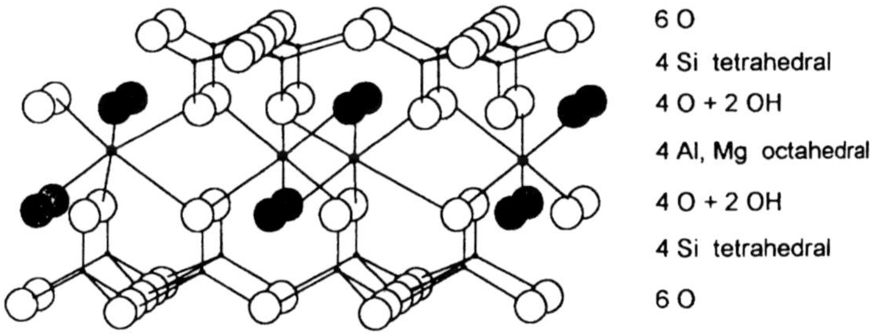

Fig. 4.48. Smectite clay (schematic).

cation located in the zeolite cage or micropore or on the clay layer. When a cation is exchanged by an ammonia ion and NH_3 is subsequently desorbed, protons are generated, located on bridging oxygen atoms as sketched in Fig. 4.49.

It is interesting to compute the excess charge on the bridging oxygenation to which protons are attached:

$$e(O(1)) = -2 + 1 + \frac{3}{4} + 1 = +\frac{3}{4} \tag{4.24}$$

Compared with (4.20b) this excess charge is quite high and indicates a strong Brønsted acidity of the attached protons.

As we saw earlier when we discussed the acidity of the silanol group, computation of charge excesses provides only a very crude estimate of acidity. As we will see for an explanation of differences in the Brønsted acidity of different zeolites, Pauling's valency rule is too crude.

The tetrahedral unit that forms the basis of the network of the zeolite makes many different structures possible. α-Quartz, the low temperature form of SiO_2, is one of the dense polymorphs. Many low-density structures can be formed, which are microporous and contain interconnected channels, bound by oxygen atoms that connect the cation-containing lattice tetrahedra. Approximately 60 different structures exist; two examples of are given in Fig. 4.50a and 4.50b. The oxygen atoms are located in the middle of the lines connecting tetrahedrally

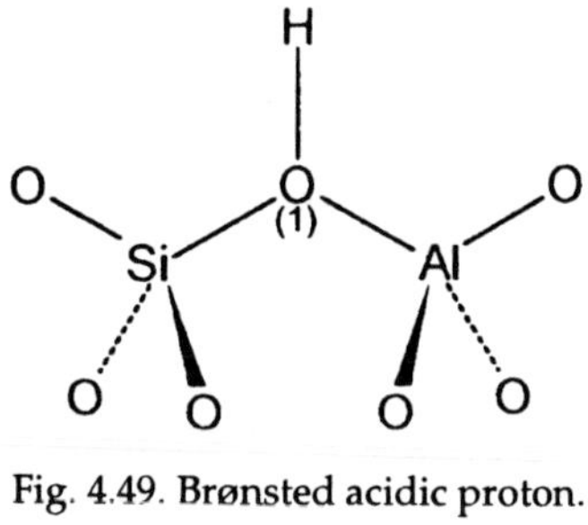

Fig. 4.49. Brønsted acidic proton.

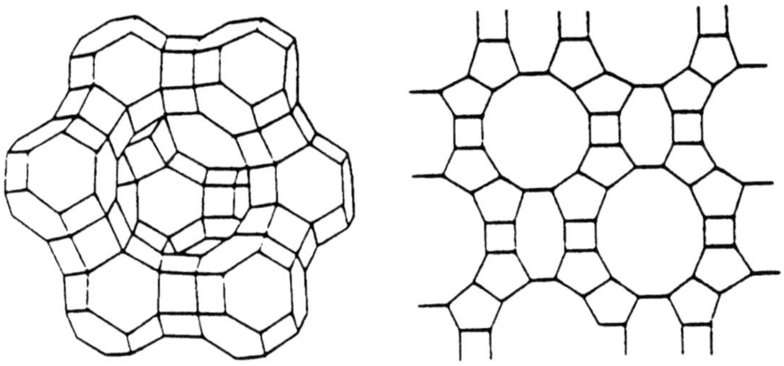

a b

Fig. 4.50. (a) Structure of faujasite. (b) Structure of mordenite.

coordinated atoms. The 12- or 10-ring channel openings enable organic molecules to enter the zeolite cavities.

An important factor that controls the Brønsted acidity of these catalysts is the lattice composition. For a zeolite to be Brønsted acidic, a proton has to connect two tetrahedra containing one tetravalent cation (often Si^{4+}) or a trivalent cation (often Al^{3+}). It appears that within a particular concentration region the deprotonation energy is a strong function of the Al/Si ratio.

To understand this one has to consider the cation composition of tetrahedra in the second coordination shell with respect to the proton in a zeolite (see Fig. 4.51). Because a $[TO_2]$ tetrahedron containing an Al^{3+} ion carries an overall negative charge, in zeolites no neighbouring tetrahedra can exist which both contain an Al^{3+} ion. For the [Si–O–Al] unit this implies that Al ions can only occupy the three tetrahedra next to a given silicon containing tetrahedron. The deprotonation energy is lower when the occupation by Al of the tetrahedra next to Si decreases, that is when Al is replaced by Si. Once the Al concentration has decreased so that no two [Si–O–Al] dimer pairs occur, the deprotonation energy does not vary any further with the Al content. Because of the lower charge on Al^{3+} its interaction with oxygen atoms as O(1) is less than that of a Si^{4+} ion, so that the Si atom (2) becomes more strongly bound and the bond strength between this Si atom and O(3) decreases. This strengthens the OH bond and hence increases the deprotonation energy. The generally found dependence of proton catalyzed reaction on the (Al/Si) content is sketched in Fig. 4.52.

Brønsted acid-catalyzed reactions will be proportional to the proton content of a zeolite as long as the intrinsic acidity of a proton remains unchanged, since the proton concentration is proportional to the Al/Si ratio. Once the Al/Si ratio exceeds a critical value $(Al/Si)_c$, the intrinsic acidity of a zeolitic proton starts to decrease. This effect dominates over the increased number of protons and, as a consequence, the overall rate of the reaction decreases.

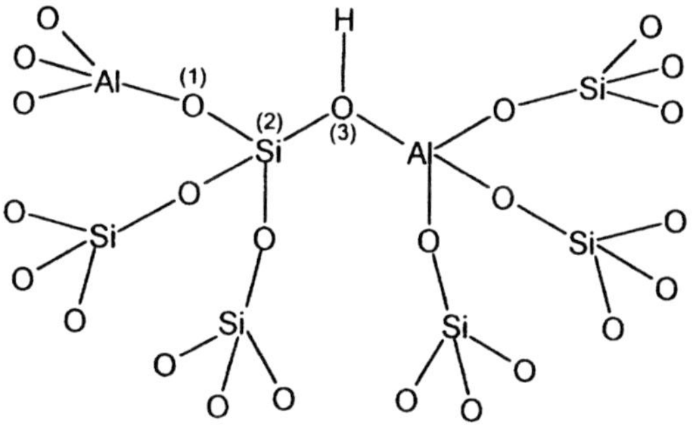

Fig. 4.51. The second coordination shell tetrahedra with respect to a proton in a zeolite.

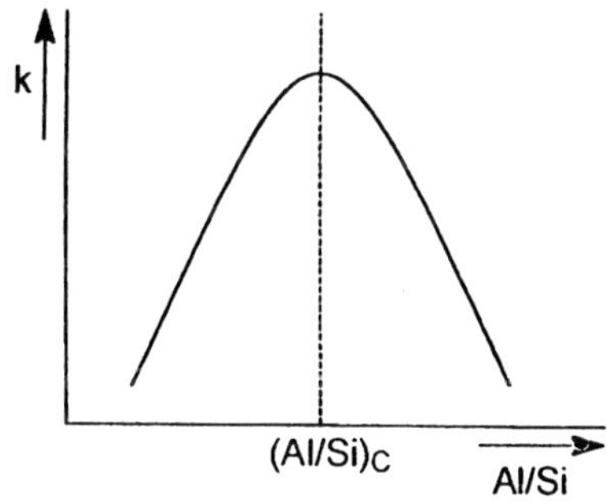

Fig. 4.52. Dependence of acid catalyzed reaction on zeolite Al/Si ratio.

4.5.2 Mechanism of protonation

In order to protonate a molecule the acidic OH band has to be heterolytically broken. Not withstanding its acidic character, the energy cost to deprotonate a zeolite side is of the order of 1250 kJ/gat. This is one of the main reasons why energetically protonation reactions in zeolites are quite different from reactions in superacids.

Secondly, many zeolite catalyzed reactions occur in apolar media and stabilization of the charged state by hydration of solvent complexation is absent. The high energy cost of charge separation involved with the cleavage of the zeolite OH bond is partially compensated for in zeolites by the electrostatic interaction between cation and negatively charged zeolite wall.

The essential features of protonation are [53]:

- formation of transition states that are carbenium- or carbonium-ion like
- formation of stable intermediates as alkoxy species.

This is illustrated in Fig. 4.53 for the protonation of an ethylene molecule. When an ethylene molecule approaches a proton, a weak π hydrogen bond is formed initially. The energy gain is typically of the order of 30 kJ/mole. The stable

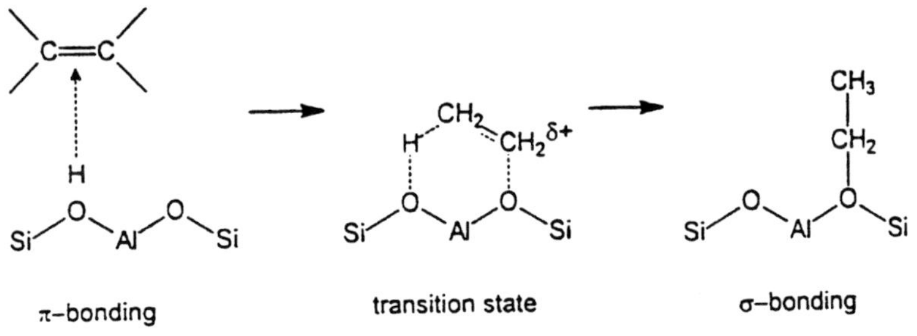

Fig. 4.53. Protonation of ethylene; o- and ρ-bonded ethylene.

intermediate that results from protonation is a stable σ ethoxy species. With respect to the π-bonded state its energy is of the order of 50 kJ/mole [54]. In between the π- and σ-bonded state there is the transition state with an energy upwards compared to the π-bonded state of 100 kJ/mole.

The structure of the protonated species in the transition state is similar to that of a carbenium-ion. The stability of a cation depends on the type of carbon atom that is protonated:

$$H_3C{-}C^+H_2 < CH_3{-}C^+H{-}CH_3 < CH_3{-}\overset{\displaystyle CH_3}{\overset{|}{C_+}}{-}CH_3$$

A primary carbenium-ion is less stable than a secondary and a tertiary ion is most stable. As a consequence the activation energies for protonation increase in the reverse order [55].

To understand differences in acidity of zeolites one has to consider not only changes in the bond strength of the OH bond, but also zeolite 'solvation' effects on the positively charged carbenium or other ions generated by protonation.

Figure 4.54 illustrates protonation of NH_3 [56,57]. Protonation proceeds via the initial formation of hydrogen-bonded NH_3, the analogue of the π complex formed with ethylene. Proton transfer only proceeds when the NH_4+ ion can become stabilized by the negative charge on the zeolite wall. This occurs when it binds as a tri-dentate to the three negatively charged basic oxygen anions around Al^{3+}.

We will illustrate this also for water adsorption. The low dielectric constant of the zeolite leads to the unexpected result that contact of a single water molecule with a proton does not lead to its protonation. Bonding of a water molecule can be considered as Brønsted acid–Lewis base assisted adsorption. Only when two water molecules adsorb around a proton, protonation occurs. The hydronium-ion now is stabilized by the second polar water molecule (compare Figs. 4.55a and 4.55b).

Reaction of methanol with a zeolitic proton readily gives dimethyl ether. Whereas the coadsorbed state of two methanol molecules is very similar to that

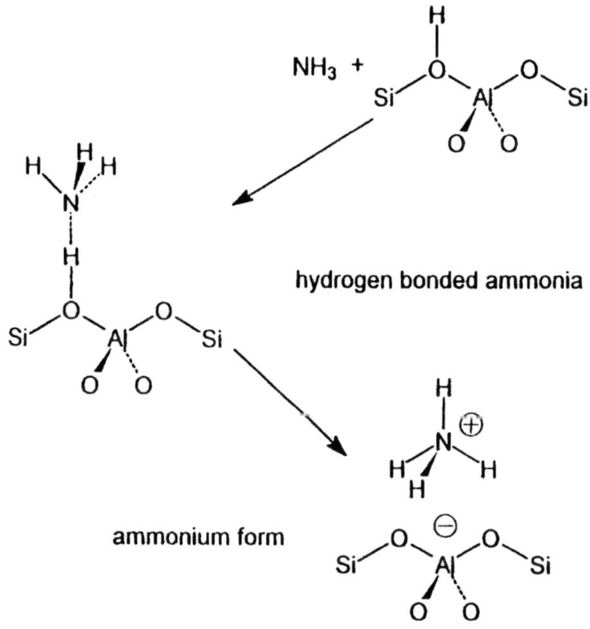

Fig. 4.54. Protonation of ammonia.

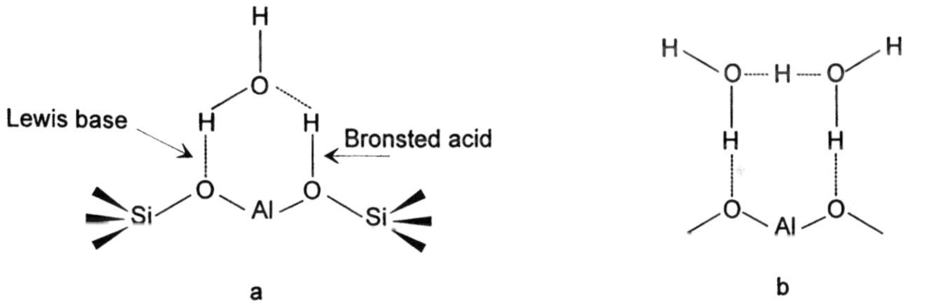

Fig. 4.55. (a) Adsorption of single H_2O molecule. (b) Proton transfer in di- H_2O complex.

of water, carbon–oxygen bond formation and cleavage only occurs in the transition state [58]. It is schematically shown in Fig. 4.56.

Whereas trimethyloxonium formation occurs in zeolites, they have been shown not to be the intermediates for C–C bond formation from methanol as has been believed for a long time. In the presence of coadsorbed methanol C–C bond formation occurs in a direct reaction between adsorbed methoxy and methanol [59].

The corresponding transition state is schematically shown in Fig. 4.57. Two methanol molecules react with a surface methoxy, formed by dissociation of methanol. After reaction ethanol is formed and protonated methanol.

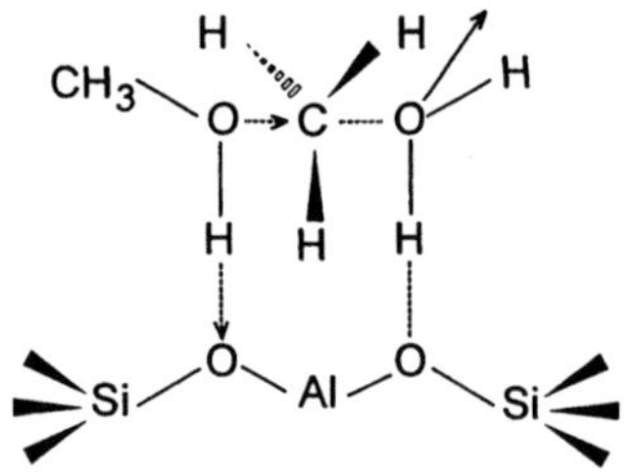

Fig. 4.56. Transition state for methanol formation. The transition state analogue of the S_N2 reaction.

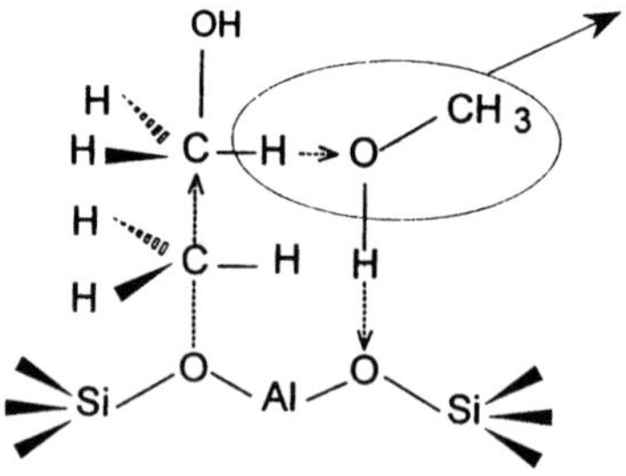

Fig. 4.57. The transition state for C–C bond formation.

4.5.3 Brønsted acid-catalyzed hydrocarbon activation reactions

This section describes elementary reaction steps and reaction chemistry of proton activated alkane reactions as understood mainly from studies in superacids. The reaction steps and reaction intermediates are also useful to consider in zeolite catalysis. However there is an important difference. Whereas carbonium-ion and carbenium-ion in superacids are usually stable intermediates, in zeolites they are highly activated states often corresponding to transition states [53].

Stable states of carbenium-ions in zeolites correspond often to covalently bonded alkoxy species. Carbenium ions behave freely only when part of transition states.

The difference between liquid superacid catalysis and zeolite catalysis relates essentially to the fact that in liquids ionic protonated intermediates are formed, e.g.

$$2H_2SO_4 \rightarrow HSO_4^- + H_3SO_4^+$$

The reactivity of the positively charged protonating agents is substantially higher than that of the neutral zeolite OH groups [60].

Proton activation of alkanes only occurs at relatively high temperature, even with strongly acidic protons, as present in the zeolite. It proceeds via the formation of carbonium ions (Fig. 4.58).

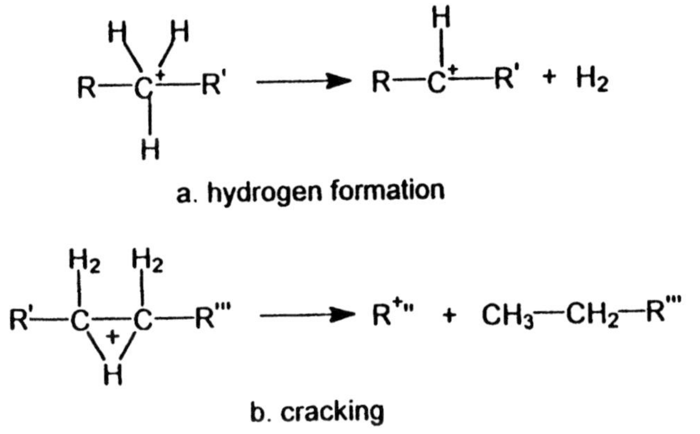

Fig. 4.58. Carbenium ion formation from carbonium ion.

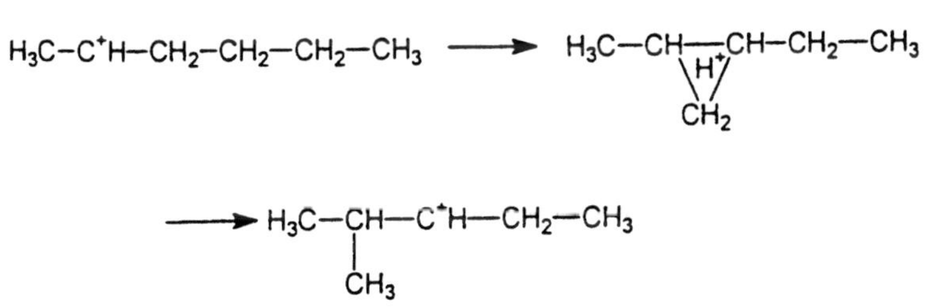

Fig. 4.59. Isomerization of carbenium ions via a triangular intermediate.

The carbonium ion is a nonclassical molecule, because it contains a penta-coordinated carbon atom. It is very unstable and is therefore difficult to form. It is formed at temperatures of about 500°C. Once formed it readily decomposes to a carbenium ion by H_2 formation or cracking (Fig. 4.58).

Carbenium ions have been shown to isomerize readily (Fig. 4.59). As long as no primary carbenium ions are involved, isomerization of carbenium ions occurs with low activation energy and at low temperatures. Isoalkane formation occurs by reaction with another alkane and transfer of a hydride ion (Fig. 4.60).

Because the activation energy of the hydride transfer reaction is lower than that of carbonium ion formation, after a short initiation period the hydride transfer reaction will maintain the carbenium ion concentration at a steady-state

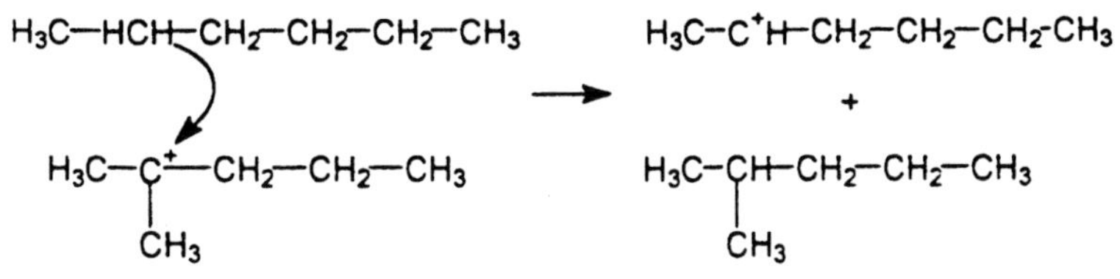

Fig. 4.60. The hydride transfer reaction.

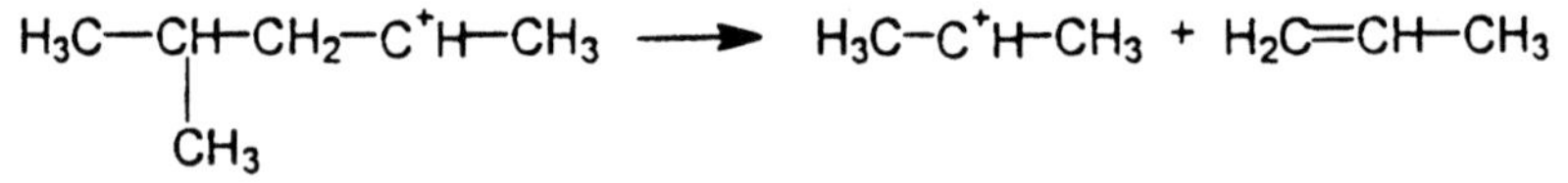

Fig. 4.61. β-Carbon–carbon cleavage leading to cracking.

level. However, secondary reactions occur that give rise to short catalyst lifetimes. Apart from the hydride transfer reaction the carbenium ion can undergo two other reactions. One is the cracking reaction (Fig. 4.61).

In the cracking reaction, transfer of a hydrogen atom on a β-carbon atom with respect to the ionic atom results in C–C splitting to give a smaller carbenium ion, as well as alkenes. The alkenes can undergo secondary oligomerization and polymerization reactions deactivating the catalysts. When one wishes to reduce the size of hydrocarbon molecules, the cracking reactions of Fig. 4.58 are desired. Catalytic cracking is performed on a large scale using acidic zeolites. Zeolites are preferred because the micropores of the zeolite have a shape-selective effect which prevents the alkenes from polymerizing and forming heavy molecules of large size thus rapidly deactivating the catalyst. Nonetheless the zeolites can also only be applied with short residence times and need to be regenerated frequently.

Whereas catalytic cracking is a useful reaction that requires high temperatures, a low reaction temperature is required in order to produce branched isomers at a high yield in reactions of the type shown in Fig. 4.62. The equilibrium shifts to the *n*-alkanes at high temperature.

In order to reduce the reaction temperature of acid catalyzed alkane conversion reactions one can reduce the temperature by replacing carbonium formation by a route via the carbenium ion by protonation of alkenes generated by metal-catalyzed (group 8–10 metals, e.g. Pt, Pd) dehydrogenation of alkanes. The metals can be readily dispersed in the micropores of a zeolite. A lowering of the reaction temperature is especially useful for alkane isomerization. A low temperature favours the branched product and inhibits consecutive reactions.

The reaction scheme is shown in Fig. 4.62. The reaction is performed under a high H_2 pressure in order to maintain a low alkene concentration. This is important to reduce the steady-state carbenium ion concentration on the catalyst. Once the residence time of carbenium ions becomes too long, not only iso-

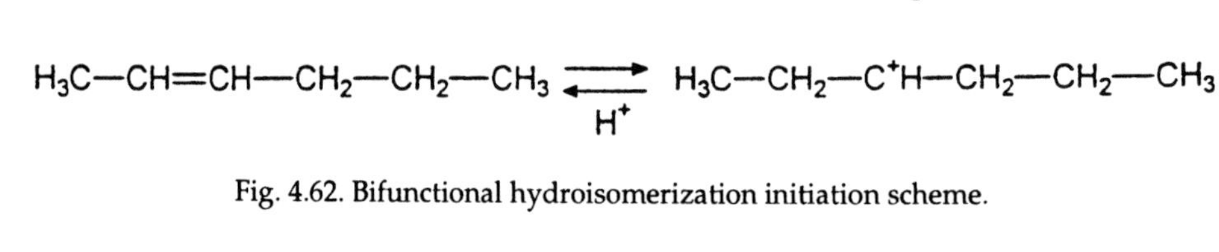

Fig. 4.62. Bifunctional hydroisomerization initiation scheme.

merization reaction will occur, but also the now undesirable cracking reaction, which is a consecutive reaction. The low alkene partial pressure reduces the carbenium ion concentration. The catalyst now performs two functions: metal-catalyzed CH activation and protonation of alkenes. For this reason the hydro-isomerization reaction according to Fig. 4.62 is called a bifunctional reaction.

4.6 ELEMENTARY STEPS IN BIOCATALYTIC REACTIONS

4.6.1 *Introduction*

All reactions in biochemical systems are carried out by enzymes. This is true even for such a simple reaction as the hydration of CO_2, which is catalyzed by the enzyme carbonic anhydrase:

$$CO_2 + H_2O \rightleftharpoons H^+ + HCO_3^-$$

The reaction rate is 10^5 CO_2 molecules per second per enzyme molecule, which is ca. 10^7 x greater than the rate without enzyme. This illustrates a key point: enzymes are catalysts that give tremendous enhancement of slow reactions (up to 10^{14} times that of the uncatalyzed reaction!) Compared to zeolites or organometallic catalysts, which typically have turnover numbers of the order of 0.1–1 sec^{-1}, enzymes are incredibly efficient catalysts, with turnover numbers of $\geq 10^6$ sec^{-1} known. In addition, just as with the heterogeneous and homogeneous catalysts discussed in the previous chapters, enzymes can catalyze reactions that are so slow that they are in fact never observed in the absence of the catalyst, making it impossible to calculate a rate enhancement. Enzymes are usually proteins, although examples of catalysis by RNA molecules are known, and the number is increasing steadily. At the moment, however, all practical applications of biocatalysis utilize proteins as catalysts.

4.6.2 *Classification of Enzymes*

Six classes of enzyme-catalyzed reaction are recognized in systematic nomenclature [61]. Their names and the type of chemical reaction catalyzed by each are indicated in Table 4.1. All enzymes have systematic names based on the above, but many are known by historically important trivial names: e.g. trypsin, chymotrypsin, pepsin, lysozyme, catalase.

4.6.3 *General Features of Enzymes*

4.6.3.1 *Specificity*

Enzymes are generally highly specific; this specificity extends to the identity of the substrates that they will transform, to the portion of the substrate molecule

TABLE 4.1

The six categories of enzymes and the type of reaction catalyzed by each

Name	Reaction catalyzed
1. Oxidoreductase	oxidation–reduction reactions
2. Transferases	transfer of chemical group from one substrate to another or from one part of substrate to another
3. Hydrolases	hydrolysis reactions
4. Lyases	elimination of groups from adjacent atoms or addition of groups to double bonds
5. Isomerases	rearrangements (isomerizations)
6. Ligases	formation of bonds to groups with hydrolysis of ATP, etc.

that is transformed, and to the identity of the products when more than one possible species can be produced [62,63].

Consider, for example, the proteases pronase, trypsin, and thrombin. All are endopeptidases, meaning that they hydrolytically cleave internal peptide bonds in a polypeptide chain. Although the same chemical reaction is catalyzed in all three cases, the three enzymes exhibit very different specificities. Pronase is a bacterial protease whose function is to hydrolyze proteins in order to provide amino acids as nutrients for growth; hence, it is essentially non-specific, and hydrolyzes virtually all peptide bonds at reasonable rates. In contrast, trypsin is a component of the mammalian digestive system and is also the key enzyme involved in activating other proteolytic enzymes by hydrolysis of specific peptide bonds in inactive precursor molecules; trypsin will only hydrolyze internal peptide bonds in which the carbonyl-containing amino acid bears a positively-charged (basic) side chain, i.e., arginine or lysine. Finally, thrombin is a component of the blood-clotting cascade, and its substrate is a protease that is the next member of the cascade and which is activated by proteolysis; as a result, thrombin is absolutely specific for cleaving an arginine– glycine peptide bond. These specificities are summarized in Fig. 4.63.

As an example of enzyme specificity in distinguishing between chemically identical portions of a substrate molecule, consider the enzyme alcohol dehydrogenase. This enzyme converts ethanol to acetaldehyde in the liver:

$$CH_3CH_2OH + NAD^+ \rightarrow CH_3CHO + NADH$$

Superficially, this looks like a simple reaction, but in fact it is completely stereospecific. The enzyme can and does distinguish between the two chemically

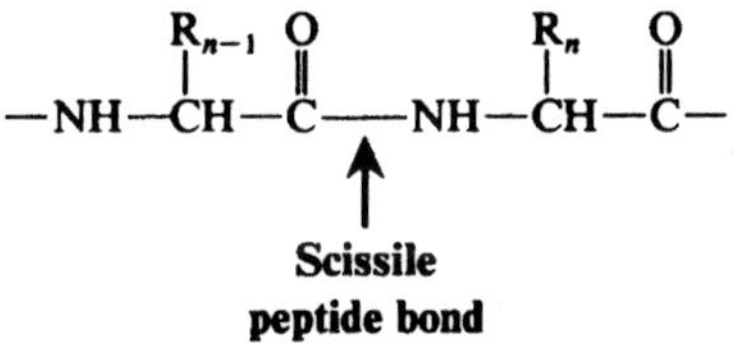

pronase: $R_{n-1} = R_n$ = anything (almost)
trypsin: R_{n-1} = Arg, Lys only
thrombin: R_{n-1} = Arg, R_n = Gly only

Fig. 4.63. Specificity of selected endopeptidases for peptide bond cleavage.

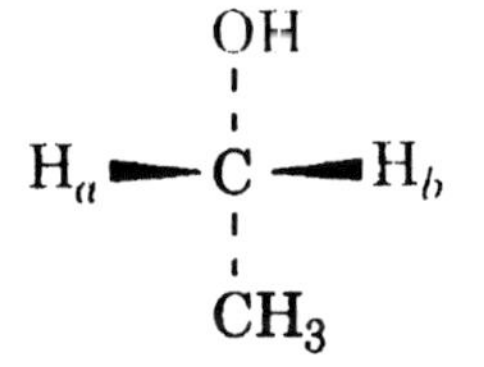

Fig. 4.64. The prochiral methylene hydrogen atoms of ethanol.

identical but prochiral methylene hydrogen atoms of ethanol, and *only* the H-atom labelled H_a in Fig. 4.64 is transferred to the NAD^+ coenzyme, with H_b remaining in the acetaldehyde product. This illustrates a general and potentially very useful point: enzymes, because they are polymers of chiral building blocks, are intrinsically chiral and will *always* distinguish between prochiral groups of a substrate molecule.

As an additional example of enzymatic specificity, consider the enzyme DNA polymerase, which synthesizes a polymer of the four basic nucleic acid units (adenosine, guanosine, thymosine, and cytosine) on a template strand of DNA. This enzyme exhibits an extraordinarily high degree of fidelity in the copies it produces: typically <1 error per 10^6 nucleotides polymerized.

A final example of the kind of specificity exhibited by enzymes is provided by squalene oxidocyclase. As shown in Fig. 4.65, this enzyme catalyzes the conversion of the 30-carbon isoprene derivative squalene 2,3 epoxide, which contains only one asymmetric carbon atom, to the steroid lanosterol in a single step that forms *four* new carbon–carbon bonds and *seven* new chiral centres in 100% yield and with no isomeric impurities.

The aspects of enzyme-catalyzed reactions cited above are so far beyond what is normally achieved in chemical reactions that there is a tendency to believe that unusual principles must be invoked to explain the results. This is absolutely untrue: just like any other kind of catalysts, *enzymes do chemistry, not magic!* The goal of this and the next section is to illustrate how enzymatic reactions are controlled by familiar chemical principles of structure and reactivity.

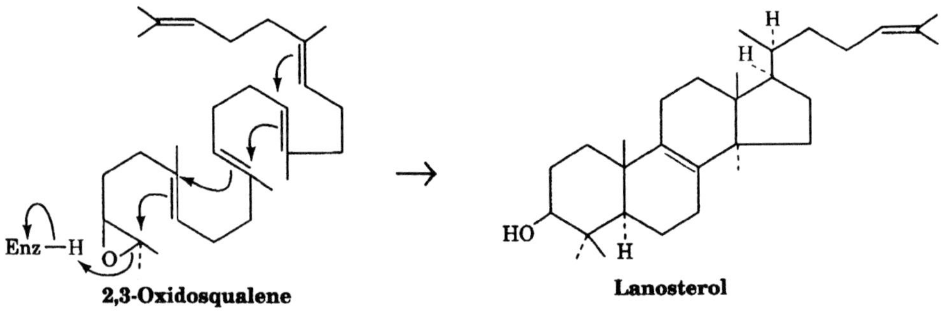

Fig. 4.65. The conversion of squalene-2,3-epoxide to lanosterol by squalene oxidocyclase.

Enzymes are capable of the kind of selectivity and rate enhancements discussed above because their active sites exhibit a number of distinctive features compared to the active sites employed by soluble transition metal complexes and solid state catalysts: multi-point contact with the substrate, which is very hard to engineer in a synthetic catalyst; the structural flexibility to undergo collective and rapid changes in structure to facilitate catalysis of a reaction; and a unique ability to combine apparently incompatible features in catalysis, such as simultaneous acid and base catalysis and hydrophobic/hydrophilic interactions [62,63]. These points are discussed in more detail in the following sections.

4.6.3.2 Regulation

The activity of many enzymes is regulated physiologically in one or more of several ways. Any or all of these can become important in the design of a process that attempts to take advantage of the many desirable features exhibited by enzymes as catalysts, discussed in the previous section. At the molecular level, regulation of enzyme activity occurs in at least three distinct ways: feedback inhibition, allosteric regulation, and covalent modification. A detailed discussion of these phenomena is outside the scope of this text, but the key points of each are summarized very briefly below. Although some of these features are similar to those exhibited by organometallic and solid state catalysts, biocatalytic systems are generally more complex, in that interactions remote from the active site are important in dictating the activity of the catalyst. The reader will recognize many features in common with the organometallic and solid state catalysts discussed in previous chapters, but biocatalytic systems are unique in the way in which complex reaction systems are coupled.

(a) *Feedback inhibition* — As shown schematically in Fig. 4.66, often the product of the *last* enzyme will inhibit the *first* in a sequence of several biosynthetic enzymes [64]. This can be an important consideration in applied biocatalysis if

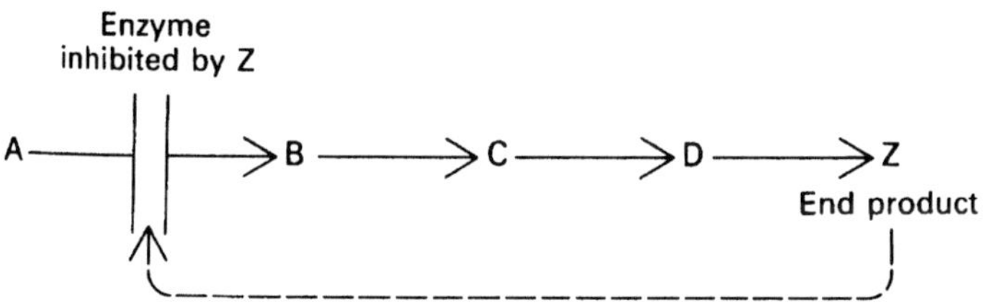

Fig. 4.66. Schematic illustration of the fact that the last product in a pathway often inhibits the first enzyme in that pathway (*feedback inhibition*).

the aim is to accumulate high concentrations of the last product of the pathway! Feedback inhibition in this sense is a phenomenon that is distinct from the related phenomenon of product inhibition, which is generally exhibited by all catalysts, including enzymes. In product inhibition, the product that has been formed in the reaction simply competes with the substrate for the active site. In the case of feedback inhibition, the inhibitor is usually a molecule that may not be a close structural analog of the substrate or product. In addition, the inhibitor usually binds at a site that is distinct from and often remote from the active site, and exerts its effect on the enzyme's activity via allosteric interactions.

The term *allosteric interactions* refers to the ability of a substance to affect the catalytic activity of an enzyme by binding at a remote site, thereby inducing subtle structural changes that are transmitted via the protein structure to the active site. Such substances are referred to as *allosteric effectors,* and can act as activators, to increase the activity of an enzyme under specified conditions, or as inhibitors, which decrease the activity of the enzyme.

(b) *Covalent modification* — Many enzymes can be turned on or off by addition or removal of a small group, often a phosphate, or by cleavage of the peptide chain [65]. Covalent modification by phosphorylation is illustrated schematically in Fig. 4.67, and is very common. For example, the enzymes that metabolize glycogen (a polymeric form of glucose used for short-term energy storage) are regulated by phosphorylation of a single serine residue in the protein. Reversible phosphorylation of the serine $-CH_2OH$ group is used to turn on glycogen

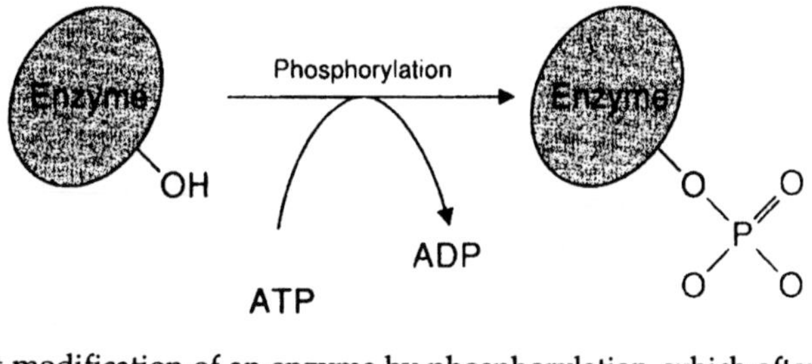

Fig. 4.67. Covalent modification of an enzyme by phosphorylation, which often results in altered catalytic properties.

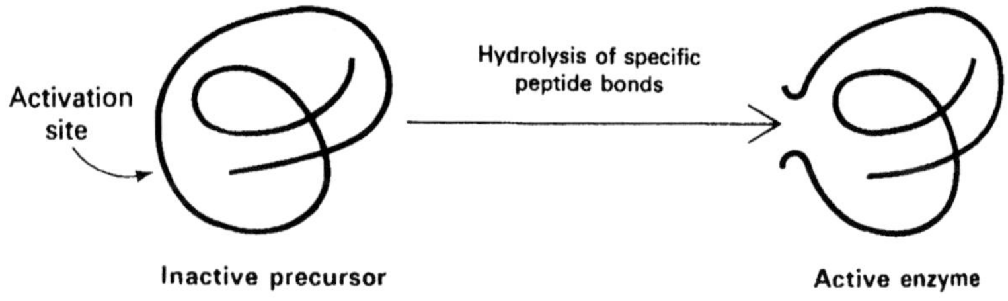

Fig. 4.68. Schematic drawing illustrating the selective cleavage of a peptide bond in a protein, which can result in altered enzymatic properties.

phosphorylase and to turn off glycogen synthetase in an inverse manner, ensuring that both enzymes are never fully functional at the same time. This phenomenon is distinct from activation of organometallic or solid state catalysts by ligand or solvent desorption, because it generally involves a modification of the structure that is remote from the active site, which results in subtle structural changes that are transmitted to the active site, much as is the case with allosteric interactions.

Some enzymes are synthesized as inactive precursors (*zymogens*) that are activated by cleavage of one or more peptide bonds, as shown schematically in Fig. 4.68. Probably the best-characterized examples are found among digestive enzymes such as chymotrypsin, which is synthesized as an inactive precursor that consists of a single polypeptide chain containing 245 amino acids. It is activated by proteolytic cleavage of the peptide chain in two places and removal of amino acids number 14, 15, 147 and 148.

Again, these processes are distinct from those typically observed in solid or organometallic complexes, where precursors are often activated by ligand dissociation or solvent desorption, in that the covalent modification occurs at a site that is remote from the active site.

4.6.3.3 *Active sites*

The fact that enzymes form stoichiometric enzyme-substrate (E·S) complexes (Chapter 3) suggests that substrates bind at a specific site on the enzyme, which presumably contains the functional groups that interact directly with the substrate during catalysis. Although the details of active structure vary greatly from one enzyme to another, the following general statements can reliably be made.

(a) *Active sites are a small part of the overall enzyme molecule.* Most enzymes consist of >100 amino acids and are roughly globular proteins that are >25 Å in diameter. In contrast, the active site of the enzyme is of roughly the same size as the substrate or, in the case of polymeric substrates or products such as glycogen or nucleic acids, the monomeric unit of the polymer.

(b) *Active sites are larger than the substrate.* The enzyme at least partially surrounds the substrate during catalysis, providing a sheltered microenvironment that contains the right combination of functional groups in the proper location to effect catalysis.

(c) *Active sites tend to be clefts or crevices in an enzyme, lined with appropriate amino acid side chains.* Polar groups are used for both acid-base catalysis and binding, and non-polar groups are used for hydrophobic binding. Water is generally excluded from the active site unless it is a reactant, and catalysis therefore occurs in a non-aqueous local environment. The amino acid side chains at the active site usually come from residues that are widely separated in the amino acid sequence (e.g., lysozyme uses amino acids # 35, 52, 62, 63, 101 at its active site). The complex three-dimensional structure of the protein results in the juxtaposition of these groups.

(d) *Substrate binding to the active site tends to be relatively weak.* The equilibrium constant for formation of the E·S or E·P complex is typically only ca. 10^{-2} to 10^{-8} M, corresponding to a value of $\Delta G = -12$ to -50 kJ/mol. This is an important point because an enzyme must be able to release its product readily in order to turn over rapidly. If the E·S complex is too stable, then the depth of the valleys in the energy vs. reaction coordinate profile can become so great that the activation energy is actually *greater* than for the uncatalyzed reaction. Thus, excessively high affinity of the enzyme for the substrate *by itself* is counterproductive, and would lead to an anti-catalyst!

(e) *The specificity of an enzyme depends on complementary structures of substrate and active site.* This is the so-called *lock and key* model, shown schematically on the left in Fig. 4.69. Although the drawing shows only complementarity in shape between the enzyme and the substrate, it is important to note that complementarity in both *shape* and *charge* is usually observed. In fact, the lock and key model is oversimplified: most enzymes have less rigid active sites that fold around the substrate during formation of the E·S complex. This is the *induced-fit* model, shown schematically on the right in Fig. 4.69 for comparison.

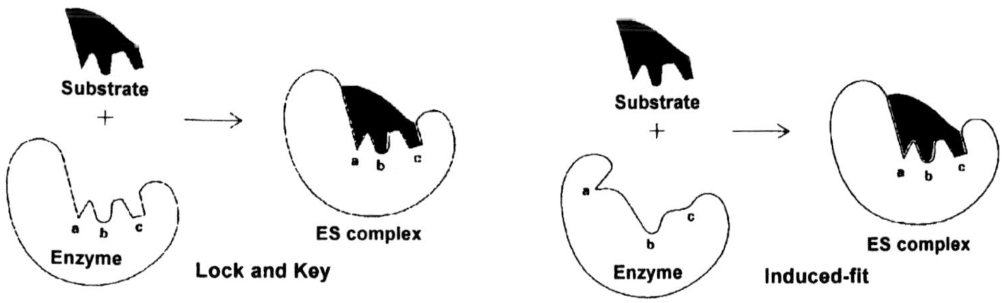

Fig. 4.69. Comparison of the lock and key vs. the induced-fit models for enzyme–substrate interaction.

In summary, enzyme active sites exhibit a number of distinctive features that differ from those of the active sites employed by soluble transition metal complexes and solid state catalysts: multi-point contact with the substrate, which is very hard to engineer in a synthetic catalyst; the structural flexibility to undergo collective and rapid changes in structure to facilitate catalysis of a reaction; and a unique ability to combine apparently incompatible features in catalysis, such as simultaneous acid and base catalysis and hydrophobic/hydrophilic interactions.

4.6.4 Factors Important In Enzymatic Catalysis

4.6.4.1 Entropic Factors

Binding of substrate(s) to an enzyme has two important effects. (i) The substrate is positioned or oriented properly for reaction, both with respect to functional groups in the active site and to other substrates. (ii) In cases where more than one substrate is involved (e.g. $A + B \rightleftharpoons C + D$ or $A + B \rightleftharpoons C$), the enzyme can assemble or collect the substrates from solution and place them in close proximity. Studies on cleverly-designed synthetic systems have clearly demonstrated that proximity effects alone can result in tremendous rate enhancements, as shown in Figure 4.70 for selected systems. Formation of an enzyme-substrate complex is, however, entropically unfavourable ($\Delta S < 0$) and therefore costs energy. Typically, enzymes use the favourable enthalpic change ($\Delta H < 0$) of substrate binding to overcome the unfavourable entropic term and to place the substrate in an environment in which it can be transformed, a phenomenon sometimes referred to as the "Circe effect" [66].

4.6.4.2 Acid-Base Catalysis

Most enzymatic reactions are sensitive to the pH of the reaction medium, as shown for several enzymes in Fig. 4.71. In fact, most enzyme reactions are both acid-and base-catalyzed. As a result, they usually exhibit an optimum pH, resulting in the typical bell-shaped rate vs. pH profile shown at the upper left for chymotrypsin in Fig. 4.71.

Examples of acid- and base-catalyzed reactions are common in organic chemistry: e.g. *ester hydrolysis* (Fig. 4.72). Since ester hydrolysis can be catalyzed by either acids or bases, clearly such a reaction would be even faster if both modes of catalysis could be employed simultaneously. Although this is not generally possible in solution, the complexity of the structure of an enzyme allows it to position both proton donors and acceptors properly to utilize both modes of catalysis simultaneously.

	Reaction	Relative rate
1.	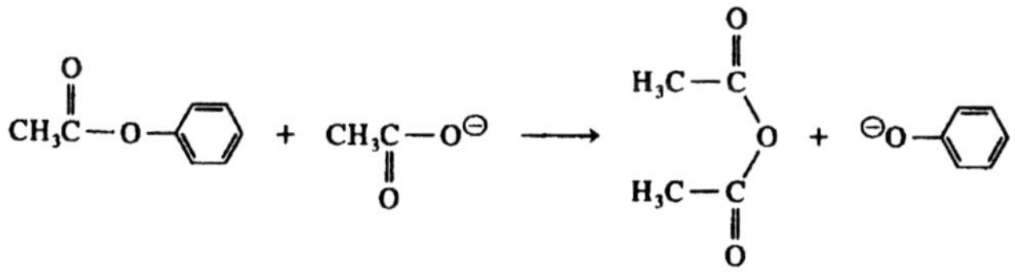	1
2.	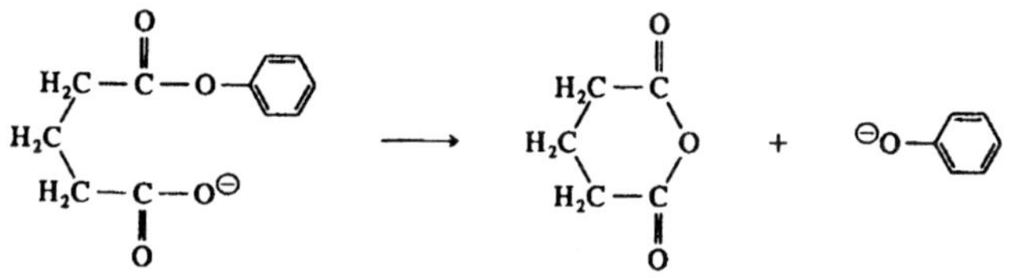	6.0×10^2
3.	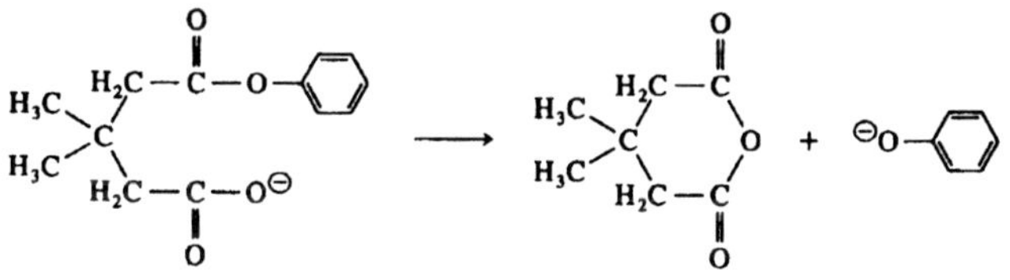	1.2×10^3
4.	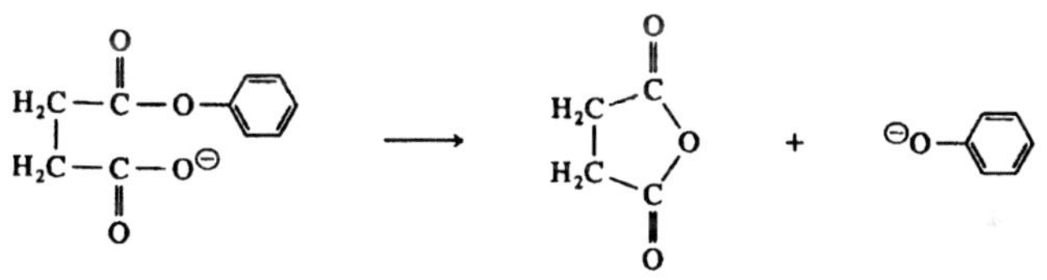	1.4×10^4
5.	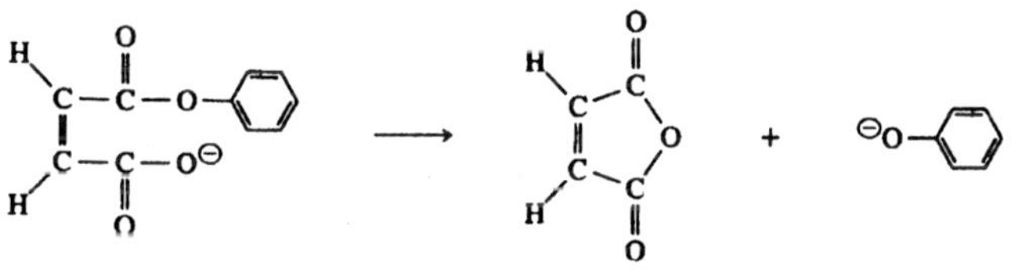	6.0×10^6
6.	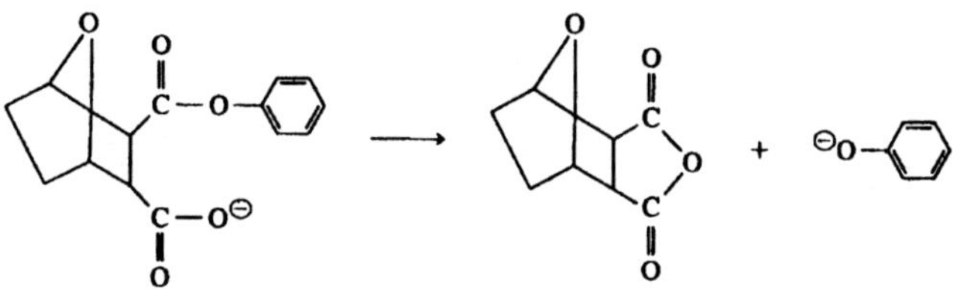	3.0×10^7

Fig. 4.70. Simple chemical systems illustrating rate enhancements due to proximity effects.

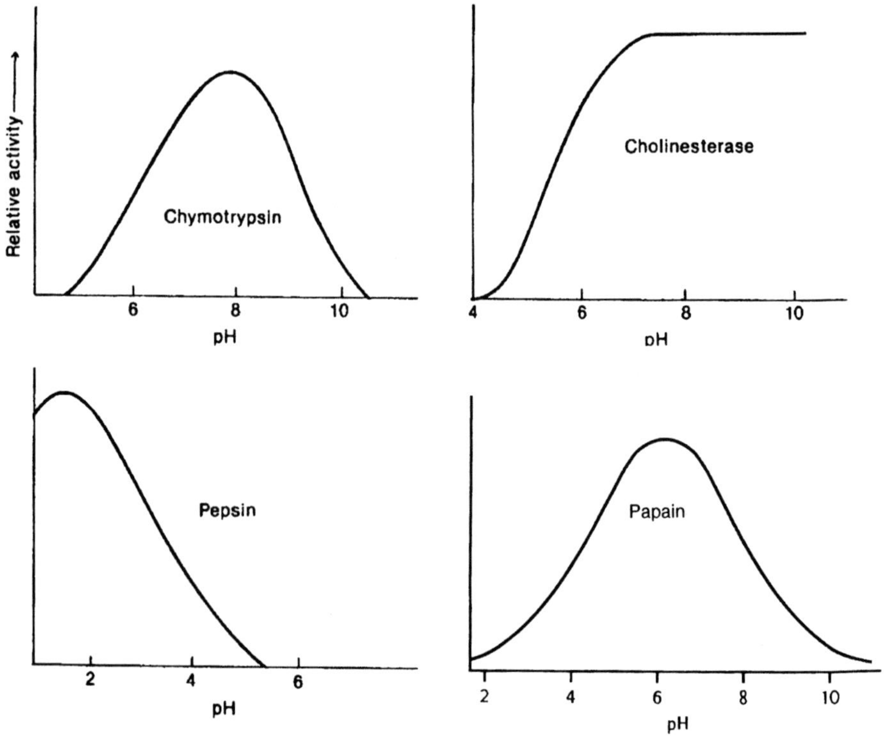

Fig. 4.71. Rate vs. pH profiles for selected enzymes.

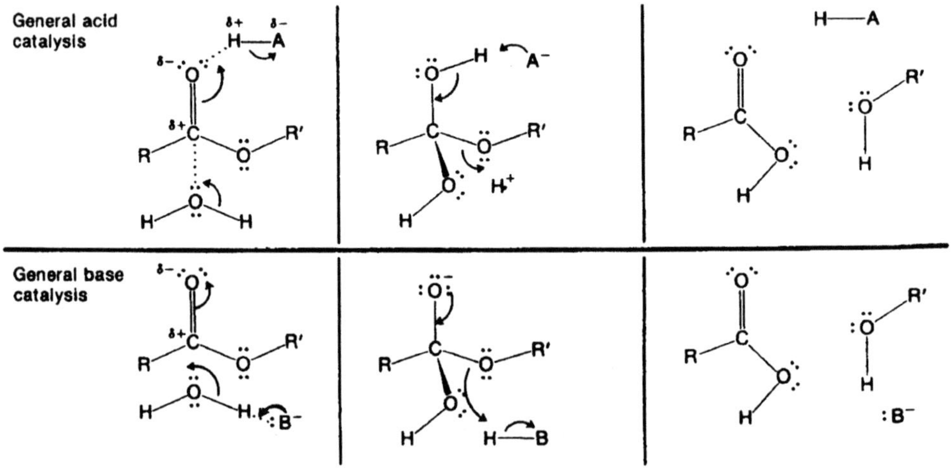

Fig. 4.72. The mechanisms for general acid and general base-catalyzed ester hydrolysis.

As an example, consider papain, a protease from papaya juice that is used to treat meat to make it more tender and easier to chew. The pH optimum of the enzyme is 6.2, and analysis of the pH dependence indicates that two groups, with pK_a's of 4.2 and 8.2, are required for catalysis. One group must be protonated and the other deprotonated for catalysis to occur. The active site of papain contains a cysteine thiol and a histidine imidazole group. There are two possible ways to write the protonation state of the active site:

$$\text{Cys S}^- \ldots \text{HisImH}^+ \quad \text{vs.} \quad \text{CysSH} \ldots \text{HisIm}$$

In other words, either the cysteine thiol or the protonated imidazole could be the acidic group needed for catalysis, and either the thiolate or the imidazole group could be the basic group. These two alternatives cannot be distinguished by kinetics measurements, but spectroscopic studies have shown that the CysSH has an unusually low pK_a of 4.2, while the pK_a of the $HisImH^+$ is rather high at 8.2. Hence, the form that is responsible for catalysis is $CysS^-$ ... $HisImH^+$. This example illustrates an important point: the microscopic environment around a particular acidic or basic group within an enzyme can differ substantially (in dielectric constant, presence of charged or polar groups, etc.) from that in simple solutions, leading to pK_a values that differ greatly from those expected for a similar functional group in a small molecule.

4.6.4.3 *Metal Ion Catalysis*

Metal ion catalysis is simply a special case of acid catalysis using a metal ion, M^{+n}, as a Lewis acid rather than H^+. The mechanism of metal ion-catalyzed hydrolysis of an ester is shown in Fig. 4.73. Note that the presence of the M^{+n} ion polarizes the carbonyl group and stabilizes the negative charge on the carbonyl oxygen atom that develops during the reaction. This is an example of *electrophilic attack* of the metal on the carbonyl, which is a mode of catalysis that is used by many enzymes. For example, carboxypeptidase is a protease that contains a Zn^{2+} ion, which coordinates to the carbonyl oxygen of the amide group that is hydrolyzed.

4.6.4.4 *Nucleophilic Catalysis*

In addition to the above factors, catalysis by many enzymes proceeds via formation of *covalent adducts* between the substrate and a group on the enzyme. In many cases, these adducts are formed by *nucleophilic attack* on the substrate. Typical nucleophiles include the –OH side chains of serine and threonine, the –SH of cysteine, the $-CO_2^-$ of aspartate and glutamate, the $-NH_2$ of lysine, and the imidazole of histidine. An example of an enzyme that utilizes nucleophilic catalysis is papain, discussed above, where the $CysS^-$ forms a covalent adduct with the carbonyl carbon of the amide to be cleaved, as shown in Fig. 4.74.

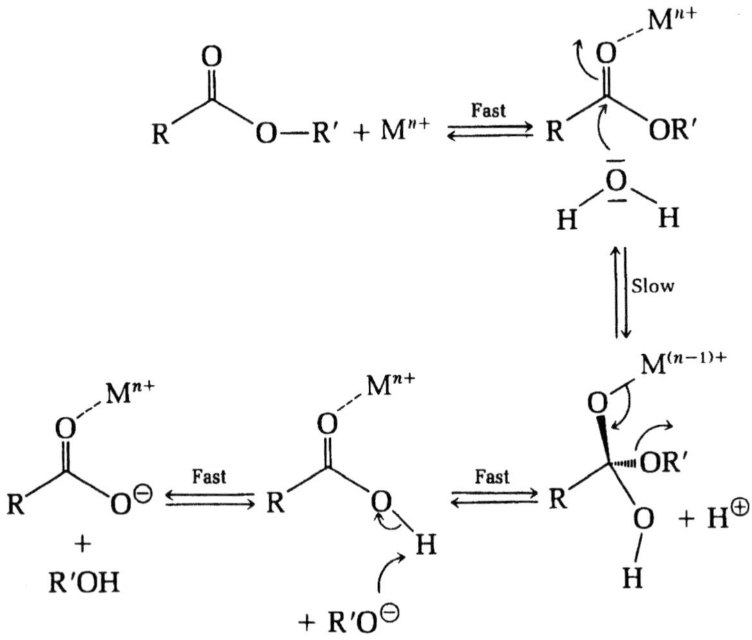

Fig. 4.73. The mechanism of metal ion-catalyzed hydrolysis of an ester.

4.7 ELEMENTARY STEPS IN BIOCATALYTIC OXIDATION REACTIONS

4.7.1 Introduction

Oxidation is an important catalytic phenomenon, and many industrially and scientifically interesting reactions can be characterized as oxidations. Section 5.5 focuses on chemically catalyzed oxidation reactions. The present section, however, focuses on elementary steps in biocatalytic oxidation reactions. Thus, special emphasis is placed on elementary principles used by biocatalysts to generate powerful and reactive oxidizing species capable of performing oxidation reactions under mild reaction conditions. Biocatalytic oxidation reactions can vary from simple electron transfer redox reactions to peroxidase-, monooxygenase-, dioxygenase- and oxidase-type of reactions, the basic principles of which will be outlined in this section.

The same factors discussed in the previous section as being important for biocatalytic reactions in general apply to biocatalytic oxidation reactions as well. In addition, however, another aspect of biocatalysts becomes important: the use of cofactors or coenzymes. Many enzymes function with the help of a specific cofactor, a small non-protein organic or metallo-organic group that is capable of facilitating the reaction to be performed. Enzymes that catalyze oxidation re-

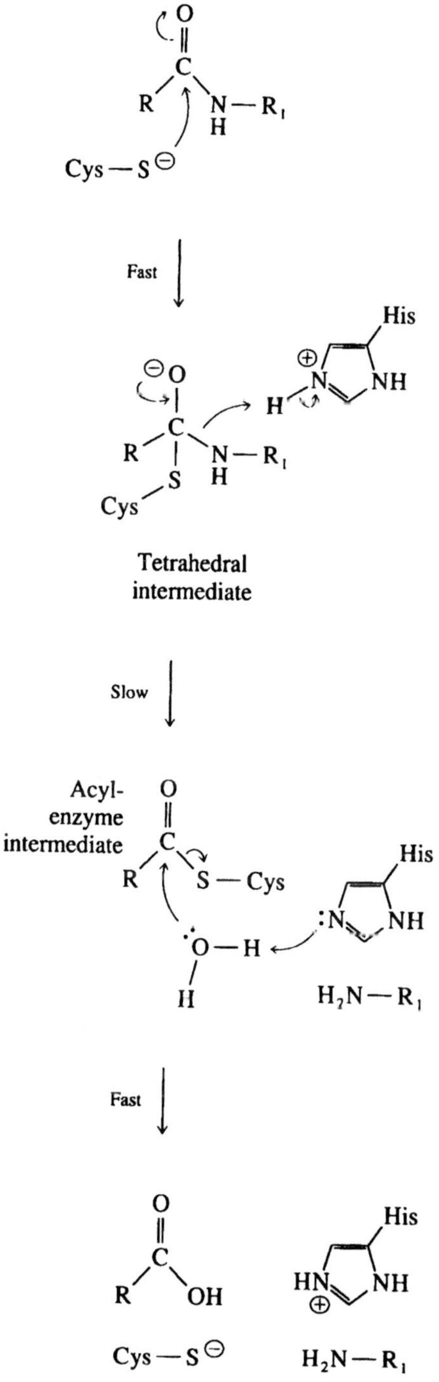

Fig. 4.74. Nucleophilic catalysis during amide hydrolysis by papain.

TABLE 4.2

Examples of some cofactors (coenzymes) frequently used in biocatalytic oxidation reactions and enzymes using these cofactors

Type of oxidation reaction	Cofactor involved	Example of enzyme or protein
Electron transfer	heme	cytochromes b, c, a
	flavin (FMN)	flavodoxins
	FeS clusters	ferredoxins
	copper	plastocyanin
Monooxygenase	heme	cytochromes P450
	flavin (FAD)	phenol hydroxylase, para-hydroxybenzoate hydroxylase
	metal clusters	methane monooxygenase, tyrosinase
Peroxidases	heme, vanadate	horseradish peroxidase, soybean peroxidase, chloroperoxidase
Dioxygenases	metal centre (Fe)	catechol dioxygenases
Oxidases	flavin (FAD)	glucose oxidase

actions generally do so with the intermediate help of different types of cofactors. Table 4.2 summarizes the types of oxidation reactions known in biocatalysis and the relevant cofactor(s) that are involved in these reactions. Figure 4.75 presents the structure of two of these cofactors in their inactive resting state, a flavin and a heme [67].

In the following sections some general features of special interest regarding elementary steps in oxidations in biocatalysis will be presented.

4.7.2 Electron transfer reactions

Without biological electron transfer reactions (also called reduction/oxidation or redox reactions) life would not exist. Well-organized electron transfer reactions in a series of membrane-bound redox proteins form the basis for energy conservation in photosynthesis and respiration. The basic reaction is simply the transfer of electrons from the donor to the final electron acceptor. Perhaps the best example of these redox reactions, their importance for living organisms, and the nature of the different type of biocatalysts that are involved is the respiration chain present in the membranes of mitochondria. The membrane-bound nature of this electron transport chain, supporting electron transfer from NADH to O_2 as

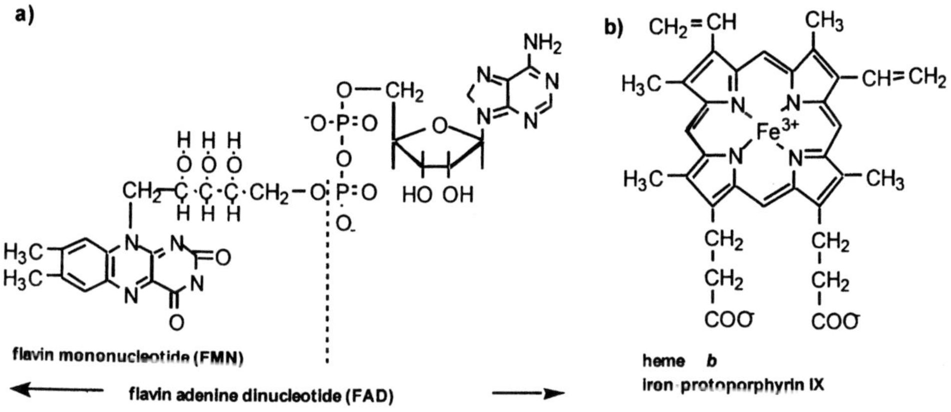

Fig. 4.75. Schematic drawings of the structures of the (a) flavin (FMN and FAD) and (b) heme cofactor for biocatalytic oxidations in their inactive resting states.

the terminal electron acceptor, is essential. This is true because protons are translocated across the membrane as electrons are passed along the electron transfer chain, generating a transmembrane proton gradient and an electrostatic potential gradient, which can subsequently be used to generate ATP, the immediate energy source for constructive cellular processes. Figure 4.76 presents an overview of the respiratory chain, and illustrates especially the nature of the electron-carrying redox proteins involved, which contain a variety of different types of cofactors.

The main point to stress from a biocatalytic perspective is the unique nature of the flavin cofactor in these types of electron transfer reactions. Whereas all other

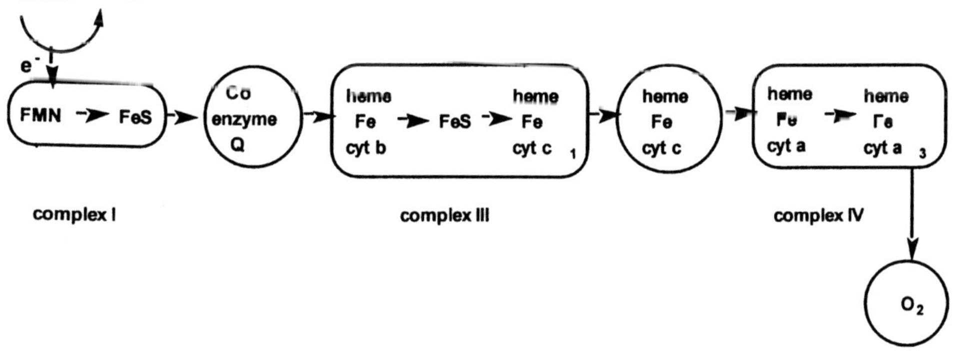

Fig. 4.76. Schematic representation of the respiratory electron transport chain. Note the presence of a variety of different cofactors.

cofactors exclusively accept and donate either only one (e.g., hemes, FeS clusters) or two (NAD(P)H) electrons at a time, the flavin cofactor can do both. Thus, flavin-containing electron transfer proteins provide the natural link between one- and two-electron transfer proteins. This is also elegantly illustrated by the electron transport processes in the respiratory chain, where an FMN-type flavin cofactor forms the link between NADH and an FeS cluster.

Finally, it is also important to mention that electron transfer by proteins is a process that has been studied in great detail recently. The prevailing hypothesis at present is that electron transport "through bonds" in the protein is more efficient than electron transport "through space". Regardless of the precise mechanism, however, electrons can be transported at reasonable rates over relatively long distances (up to 15 to 20 Å), and thus from one protein cofactor to another.

4.7.3 Heme-based peroxidases

Heme-based peroxidases are biocatalysts that perform the oxidative conversion of a wide range of substrates by means of formation of a remarkably reactive electrophilic species; the general catalytic cycle is schematically depicted in Figure 4.77.

Upon interaction of hydrogen peroxide with the Fe^{3+} metal centre that is present in the heme cofactor of the enzyme in the resting state, cleavage of the

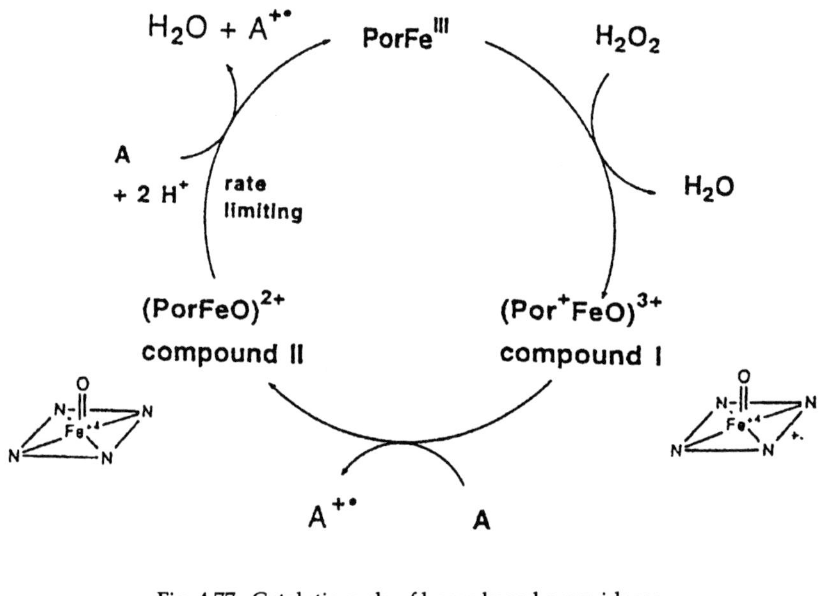

Fig. 4.77. Catalytic cycle of heme-based peroxidases.

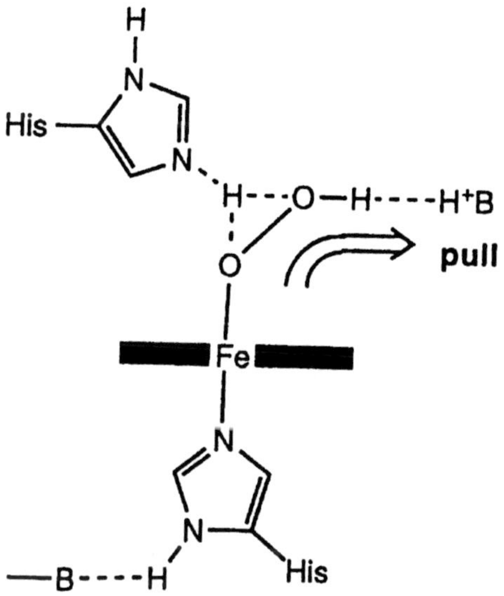

Fig. 4.78. Role of the distal histidine in the "pull" mechanism for the cleavage of the dioxygen bond and creation of the high-valent iron-oxo porphyrin π cation radical (Compound I) in peroxidases.

dioxygen bond is achieved. This leads to formation of the so-called compound I. The exact nature of this compound I has been deduced from studies with simple chemical model systems, and its existence has also been proven directly in several heme-based peroxidases, such as for example the most widely studied peroxidase, horseradish peroxidase. Compound I can best be described as an Fe(IV)-oxo porphyrin π cation radical, in which the Fe is referred to as high-valent state because it is in an oxidation state above the normal Fe(II) or Fe(III) valency. Originally the Fe might even be present in an Fe(V) state, but this has never been observed. The supposed Fe(V) (if formed initially) instead seems to abstract an electron from the porphyrin ring system of the heme, thus leading to formation of the Fe(IV)-oxo porphyrin π cation radical intermediate depicted schematically in the catalytic cycle of Fig. 4.77.

This high-valent iron-oxo species is among the most powerful oxidizing species known in biology. Formation of Compound I from the Fe^{3+}-heme containing resting state and H_2O_2 in the active site of peroxidases is assisted by the so-called distal histidine, as depicted in Fig. 4.78. This mechanism is often referred to as the "pull" mechanism for dioxygen bond cleavage by peroxidases. The proximal histidine, acting as an axial ligand to the Fe centre, is not involved in this mechanism for Compound I formation.

The highly electrophilic Compound I intermediate is able to catalyze the one-electron oxidation of a wide variety of substrates, leading to a wide range of different types of conversions. Examples of reactions catalyzed are the dimerization of phenols through formation of phenolic radicals, formation of coloured dimeric reaction products, polymerization reactions and/or the formation of products resulting from the reaction of substrate radicals with solvent molecules such as H_2O, or from dismutation of radicals to give two-electron oxidized reaction products. In addition, the strong oxidizing power of peroxidase Compound I intermediates might be useful for bleaching reactions of interest for the paper and detergent industries. Since peroxidases show a wide substrate specificity and use H_2O_2 as a clean oxidant to create the reactive Compound I species, industrial interest in these peroxidases is no surprise. In addition, their reactions may be of interest commercially, since they may affect taste and result in browning of fruits and vegetables through polymerization reactions.

Once Compound I has catalyzed the one-electron oxidation of a substrate, the porphyrin cation radical is "neutralized" and the so-called Compound II is formed (Fig. 4.77). This Compound II, perhaps upon protonation of its oxo atom, is still a powerful oxidizing species, and can catalyze the one electron oxidation of a second substrate molecule, leading also to formation of a molecule of water and conversion of the heme moiety to its Fe^{3+} resting state. Because the oxidative power of Compound II is lower than that of Compound I, substrate oxidation by Compound II is the rate-limiting step in peroxidase biocatalysis.

4.7.4 *Monooxygenases*

The biocatalysts capable of performing monooxygenation reactions are of special interest from a catalytic point of view, since in these reactions transfer of one of the oxygen atoms from molecular oxygen to the substrate occurs. The overall reaction catalyzed by monooxygenases is

$$S + O_2 + NAD(P)H + H^+ \rightarrow SO + H_2O + NAD(P)^+$$

Monooxygenase reactions can lead to a wide range of conversions, such as hydroxylations, epoxidations, heteroatom oxygenations, heteroatom dealkylations, and oxidative dehalogenations. Some of the monooxygenases not only catalyze a wide range of reactions under relatively mild conditions, but they also do so in a stereo- and/or regiospecific way. In addition, in some cases these monooxygenases show an extremely broad substrate specificity. For example the mammalian cytochromes P450 are a family of enzymes able to convert over 300,000 different substrates. For all these reasons, the enzymes capable of catalyzing monooxygenase reactions are of special industrial interest.

From a biocatalytic point of view the most interesting question to be answered is what enables these enzymes to catalyze the monooxygenation of even relatively inert chemical substrates, such as benzene derivatives, under mild conditions. To answer this question several examples of monooxygenases, with different cofactors used to perform the job, will be discussed.

4.7.4.1 Flavin-dependent monooxygenases

Flavin-dependent monooxygenases function through a reaction cycle that is schematically presented in Fig. 4.79. These enzymes perform their catalytic reaction through the generation of a reactive form of the flavin cofactor, formed by 2-electron reduction of the flavin followed by its reaction with molecular oxygen. As a result, the so-called C(4a)-peroxyflavin intermediate is formed, whose electrophilic reactivity is further increased by protonation of the distal oxygen of the peroxide to yield the C(4a)-hydroperoxyflavin form of the cofactor (Fig. 4.80).

This C(4a)-hydroperoxyflavin intermediate is the active form of the cofactor in a monooxygenation reaction, leading to the hydroxylation of aromatic substrates. The reaction mechanism is schematically depicted in Fig. 4.81. Table 4.3

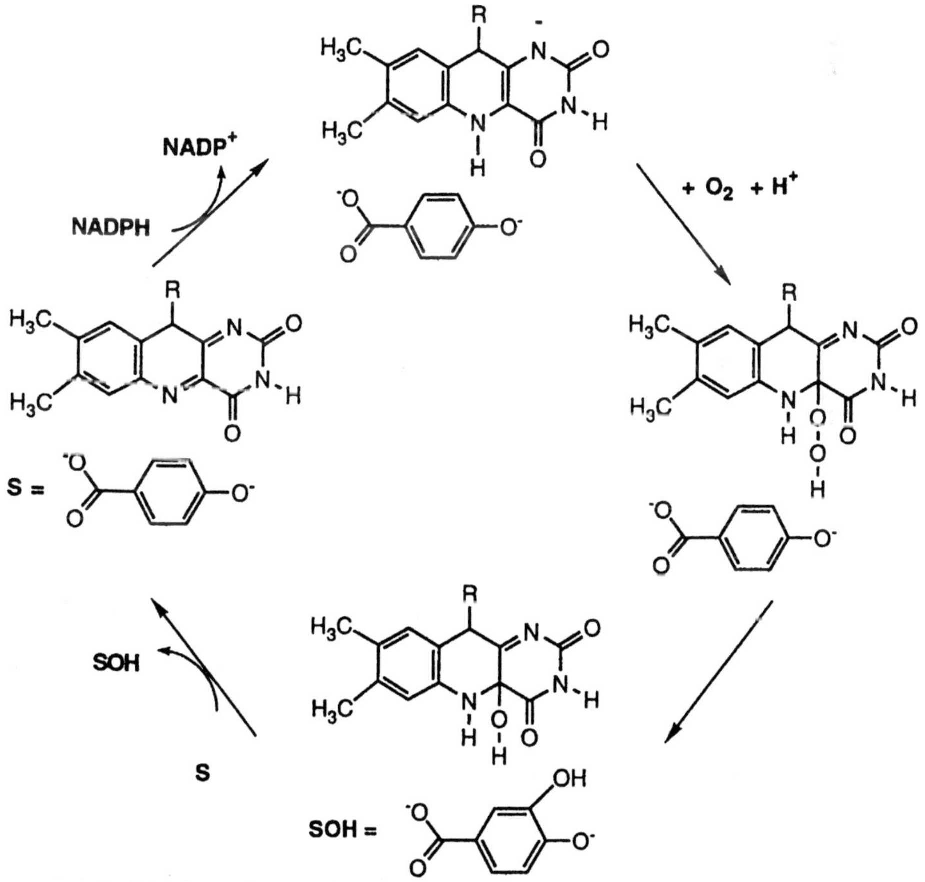

Fig. 4.79. Reaction cycle of flavin-dependent monooxygenases.

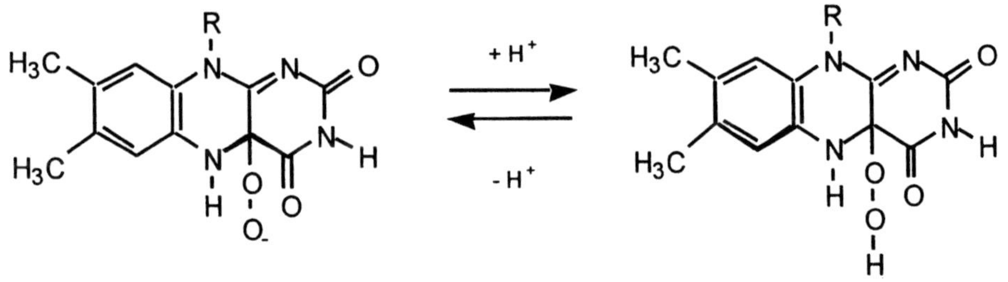

C(4a)peroxyflavin **C(4a)hydroperoxyflavin**

Fig. 4.80. Structure of the activated C(4a)-(hydro)peroxyflavin formed upon 2-electron reduction of the oxidized flavin and its subsequent reaction with molecular oxygen.

TABLE 4.3

Some flavin-dependent aromatic hydroxylases and the monooxygenase reactions catalyzed

Enzyme	Reaction catalyzed
4-hydroxybenzoate-3-hydroxylase	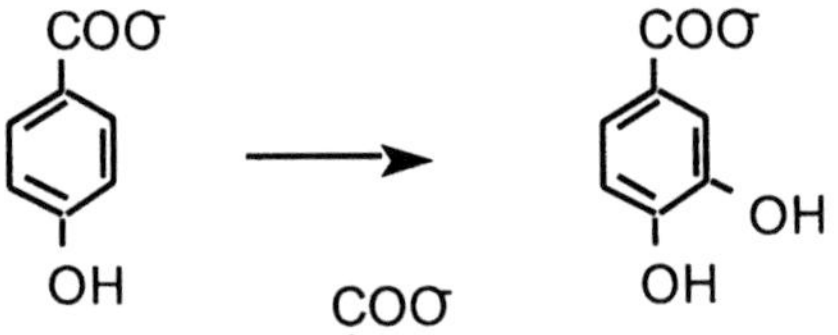
3-hydroxybenzoate-4-hydroxylase	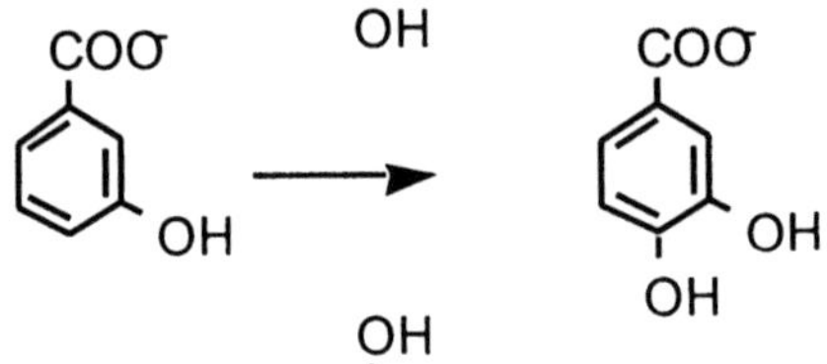
phenol hydroxylase	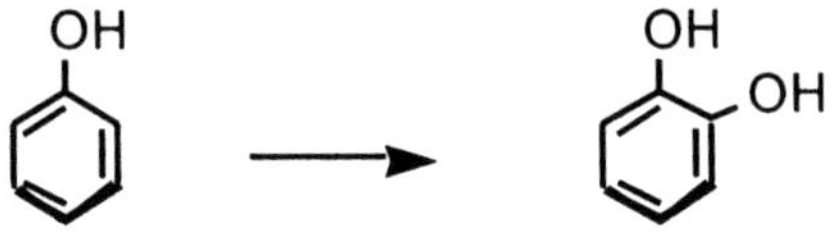
4-hydroxyphenylacetate-3-hydroxylase	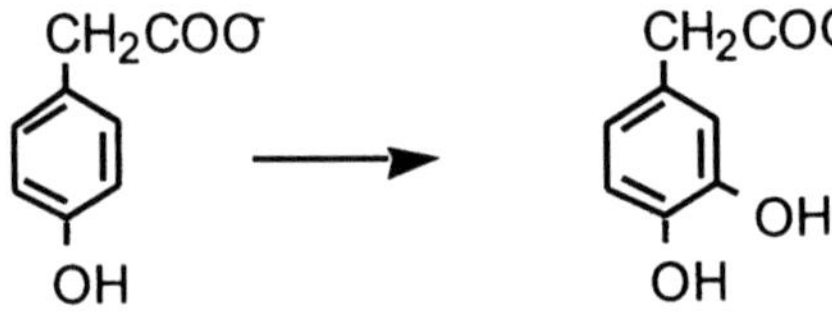
resorcinol hydroxylase	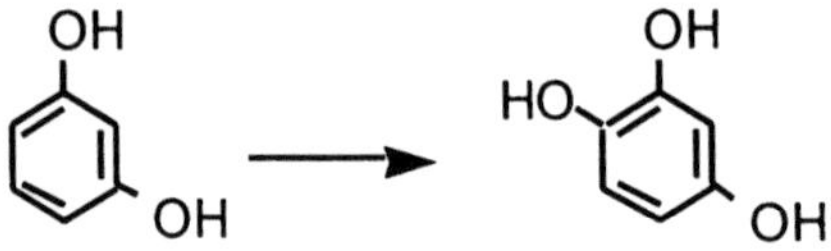

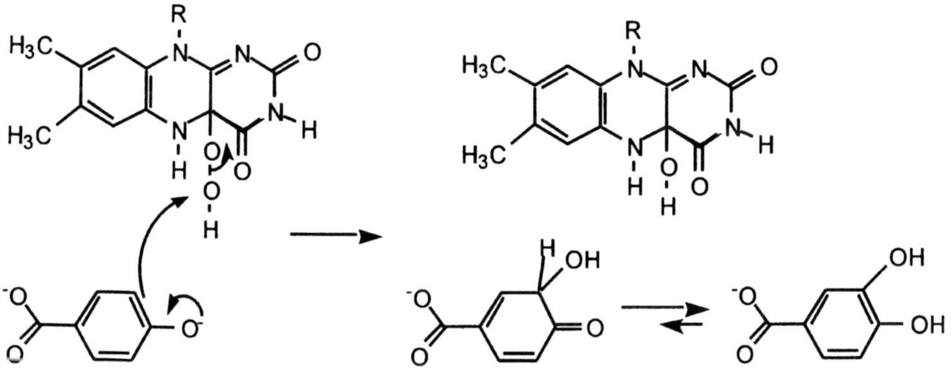

Fig. 4.81. Schematic presentation of the reaction mechanism of aromatic hydroxylation by a C(4a)-hydroperoxyflavin intermediate.

presents the reactions catalyzed by some of the many flavin dependent aromatic hydroxylases identified so far.

However, the activated, protonated C(4a)-hydroperoxyflavin is not by itself a powerful oxidizing species. From detailed biochemical studies on the catalytic mechanism of such enzymes, especially para-hydroxybenzoate hydroxylase (PHBH) and phenol hydroxylase (PH), it appears that not only the peroxyflavin needs to be activated through protonation of the distal oxygen, but also the phenolic substrates require activation in order to obtain substrate conversion. This substrate activation is achieved through deprotonation of the hydroxyl moiety of the phenolic substrate. The active site of PHBH, for example, shows a tyrosine network consisting of tyrosines 385 and 201 (Fig. 4.82) responsible for this deprotonation and activation of the substrate.

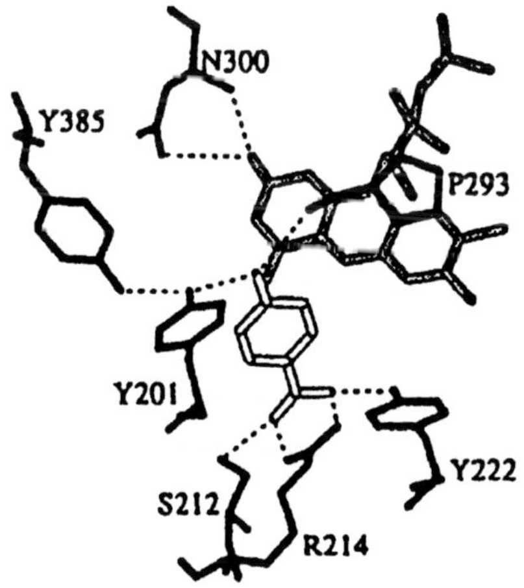

Fig. 4.82. Activation of para-hydroxybenzoate in the active site of para-hydroxybenzoate hydroxylase through deprotonation of its hydroxyl moiety by tyrosine residues 201 and 385.

Thus, these flavin-dependent monooxygenases demonstrate the capacity of enzymes to modify the pK_a of reactants through specific interactions of amino acid residue side-chains with the substrate. In principle, the catalytic process depicted in Fig. 4.81 is impossible outside the active site of the enzyme, since in solution at any given pH the simultaneous generation of both the protonated C(4a)-hydroperoxyflavin and the deprotonated para-hydroxybenzoate substrate would be impossible due to the pK_a values of these two reactants. The enzyme provides a way for simultaneous generation of the activated protonated C(4a)-hydroperoxyflavin cofactor and the activated deprotonated substrate.

In contrast to the situation for these flavin-dependent monooxygenases, other monooxygenases are able to generate very powerful oxidizing cofactor intermediates, which are capable of attacking a wide variety of substrates without the requirement for substrate activation. The cytochromes P450 and methane monooxygenase discussed below are good examples of such biocatalytic monooxygenases.

4.7.4.2 Cytochromes P450

The cytochromes P450 are an extensive group of hemoproteins that are found in all forms of aerobic life and are among the most versatile enzymes known, since they are able to convert over 300,000 different substrates. These enzymes are involved in the conversion of both endogenous substrates and of xenobiotic compounds, such as those derived from pollutants, foods and drugs. The cytochromes P450 are generally believed to catalyze monooxygenase reactions through the intermediate formation of an activated high-valent iron-oxo-porphyrin form of their heme cofactor [68]. Figure 4.83 presents the catalytic cycle involved.

This catalytic cycle proceeds via sequential binding of substrate, reduction of the heme-Fe^{3+} to heme-Fe^{2+}, binding of molecular oxygen, and, upon reduction of the heme-Fe^{2+}–O_2 complex by a second electron and protonation of the distal oxygen of the peroxide moiety, heterolytic cleavage of the oxygen–oxygen bond leading to formation of H_2O and the so-called high-valent iron-oxo porphyrin intermediate. In cytochromes P450 no distal histidine is present to facilitate the oxygen–oxygen bond cleavage, as in heme-based peroxidases (Fig. 4.77). The cytochromes P450, however, contain an axial cysteinate instead of an axial histidine ligand to the Fe. This cysteinate ligand is postulated to assist the heterolytic oxygen–oxygen bond cleavage due to its electron donating capacity. Figure 4.84 schematically presents this so-called "push" mechanism for oxygen–oxygen bond cleavage in cytochromes P450.

The exact nature of the high-valent iron-oxo intermediate that is formed in cytochromes P450 is still not unambiguously proven, but it is generally assumed that its structure is comparable to that of the reactive high-valent iron-oxo

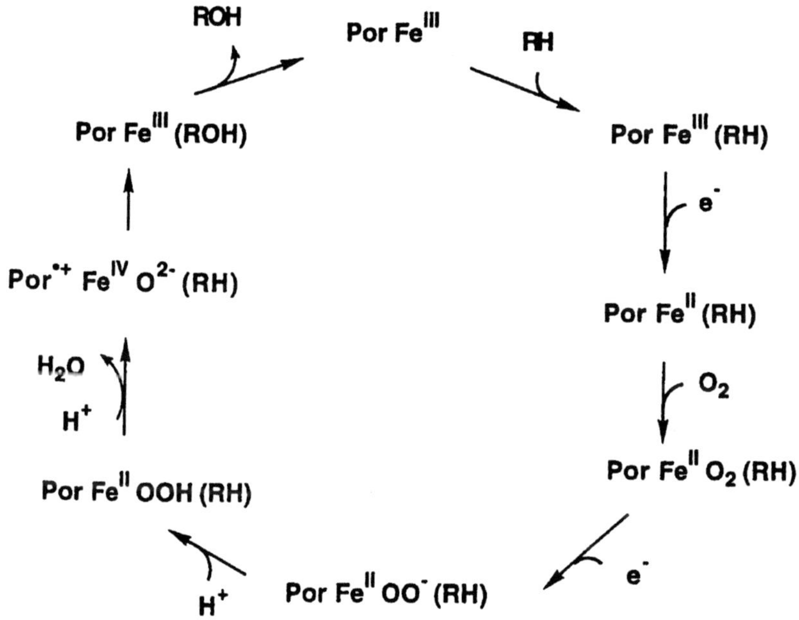

Fig. 4.83. Catalytic cycle of cytochromes P450.

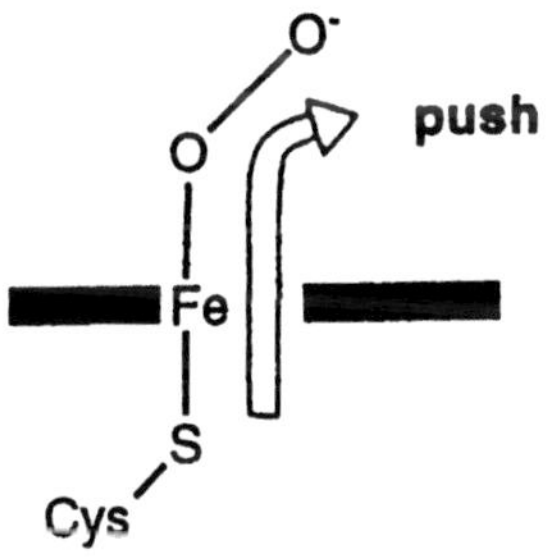

Fig. 4.84. Schematic presentation of the "push" mechanism for oxygen–oxygen heterolytic bond cleavage in cytochromes P450.

porphyrin π cation radical formed in heme-based peroxidases, i.e., compound I (Fig. 4.77). The oxidative power of these high-valent iron-oxo porphyrin π cation radical complexes is extremely high, and explains why the enzymes do not require further activation of the substrate in order for electrophilic attack by the high-valent iron-oxo intermediate on the substrate to occur, leading to substrate monooxygenation.

Not only is the number of substrates that can be converted high, but in addition many different types of oxidative monooxygenase-type reaction can be catalyzed. The most important of these are presented in Fig. 4.85.

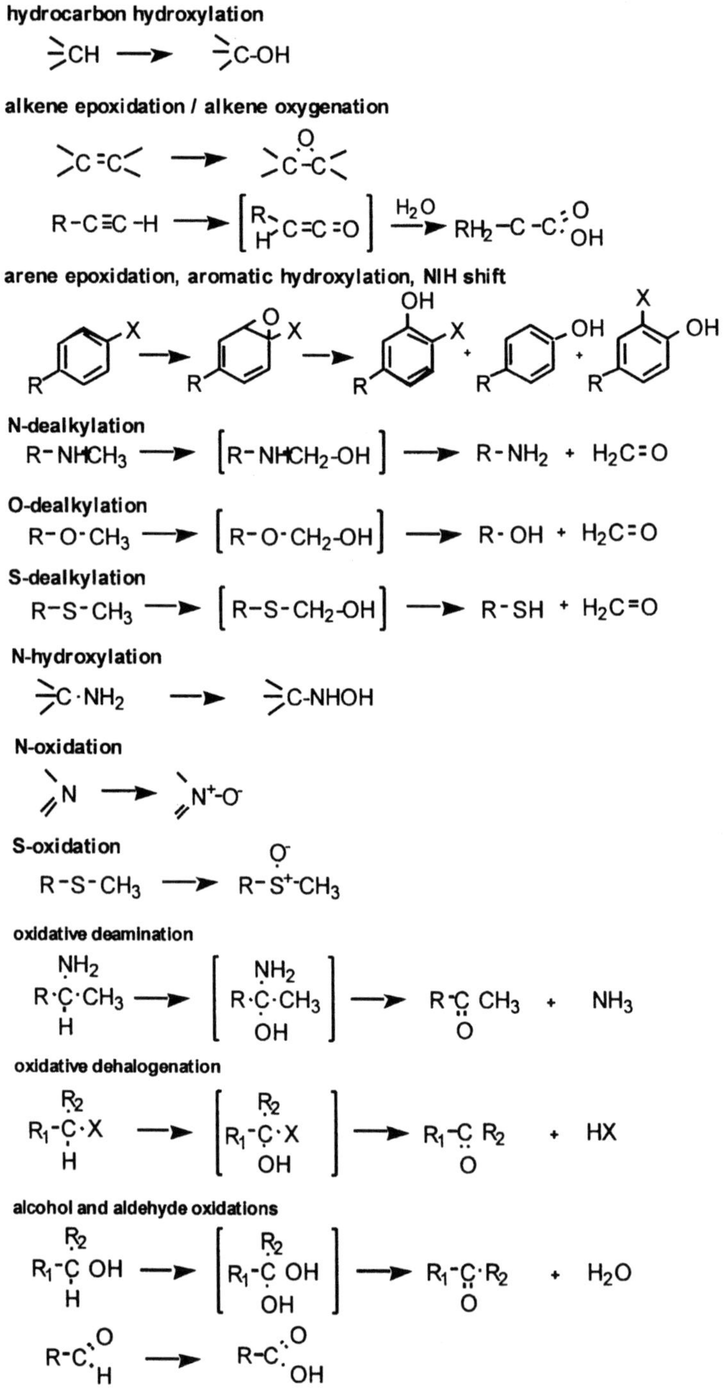

Fig. 4.85. Types of monooxygenase reactions catalyzed by cytochromes P450.

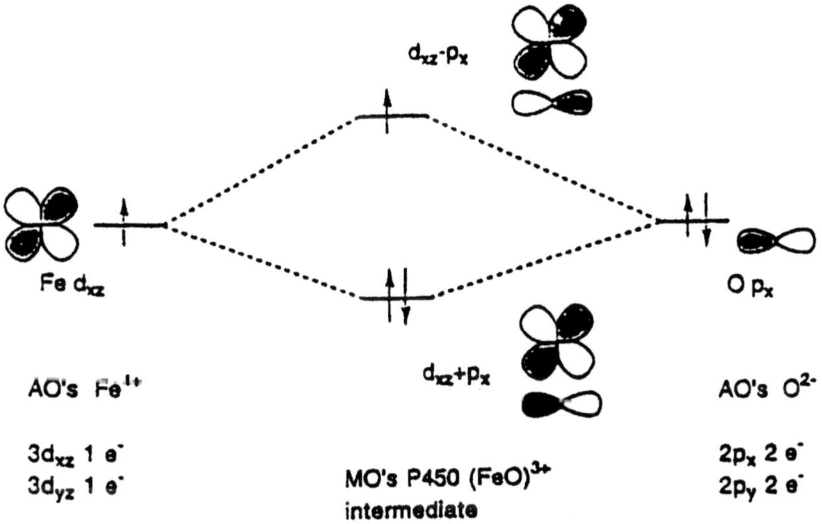

Fig. 4.86. Schematic picture of the MO interactions within the high-valent iron-oxo intermediate, illustrating the presence of unpaired spin density on the oxygen atom. Only the d_{xz} plane is illustrated, but the same situation exists in the d_{yz} plane.

The mechanisms involved in all these different catalytic conversions have one feature in common: they are generally believed to proceed via radical intermediates. This can be understood from the simple MO picture of the FeO species (Fig. 4.86), demonstrating that part of the unpaired spin density on the high-valent iron can be transferred to the oxygen, generating possibilities for radical-type interactions of the ferryl-oxo with the substrate. Computer calculations on spin distributions in model complexes confirm this picture (Table 4.4), and also indicate that the sulphur of the cysteinate ligand may participate in resonance interactions within the high-valent iron-oxo species, thereby stabilizing these intermediates (Fig. 4.87).

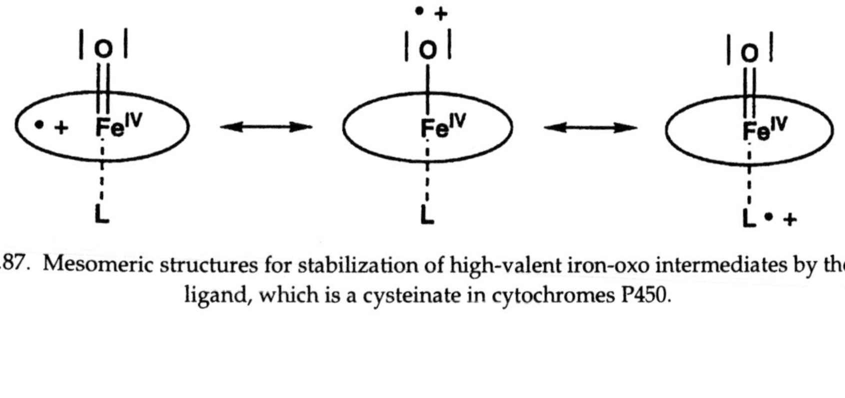

Fig. 4.87. Mesomeric structures for stabilization of high-valent iron-oxo intermediates by the axial ligand, which is a cysteinate in cytochromes P450.

TABLE 4.4

Calculated spin distributions in high-valent iron-oxo porphyrin species.

Axial ligand:	Imidazole (histidine model)		CH_3S^- (cysteinate model)	
Calculation method:	SCC-Xα	INDO	SCC-Xα	INDO
Spin density on				
Fe	1.21	1.36	1.28	1.47
O	0.86	0.60	0.83	0.50
Porphyrin	0.93	1.02	0.73	0.88
Axial ligand	0.00	0.00	0.16	0.15

Finally, Fig. 4.88 presents two examples of such radical-type mechanisms catalyzed by cytochromes P450 and leading to aliphatic hydroxylation, through the so-called oxygen-rebound mechanism, and epoxidation, respectively.

Thus, both the heme-based peroxidases and the cytochromes P450 function in biological oxidation reactions through the formation of electrophilic high-valent iron-oxo porphyrin complexes. The question that arises is why peroxidases catalyze one-electron oxidations of their substrates, whereas cytochromes P450 catalyze monooxygenation, i.e., oxygen transfer reactions. One of the possible factors causing this difference might be the different nature of their axial ligand

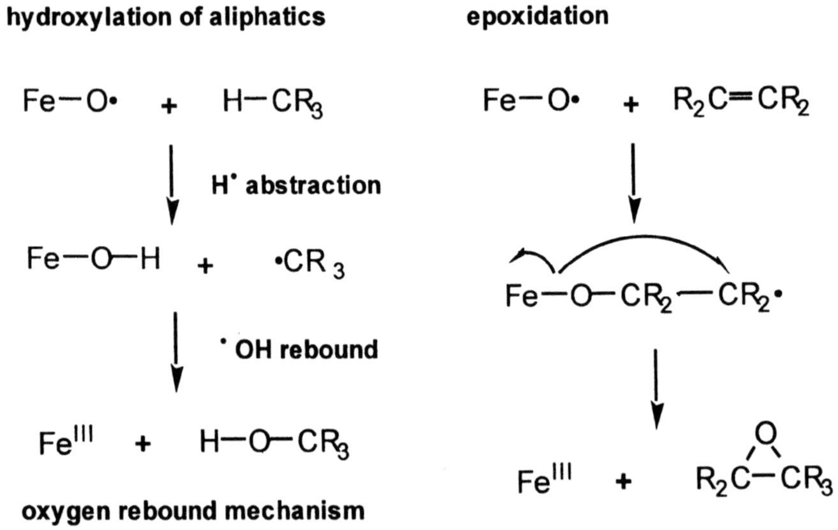

Fig. 4.88. Proposed radical-type reaction mechanisms for (a) aliphatic hydroxylation and (b) epoxidation by the high-valent iron-oxo porphyrin intermediate of cytochromes P450. The mechanism depicted in (a) is called the "oxygen-rebound" mechanism.

to the heme Fe. The cysteinate ligand of cytochromes P450 may stabilize oxygen transfer reactions. In addition, an important factor may be the (in)accessibility of the oxygen atom of the high-valent iron-oxo species. In cytochromes P450 this oxygen is freely accessible, providing possibilities for oxygen transfer to a substrate, whereas in peroxidases access to the oxygen is sterically hindered by the enzyme structure, leaving only the heme edge of the activated high-valent iron-oxo porphyrin intermediate free to interact with the substrate.

Finally, it is of interest to note that, due to the high oxidative power of these activated porphyrins, many chemists are currently focusing on the development of analogous porphyrin-based synthetic catalysts to perform oxidations of relatively inert compounds.

4.7.4.3 Methane monooxygenase

During the past decade, the oxygen-activating properties leading to monooxygenase catalysis by non-heme-iron-containing metalloenzymes has been the focus of a great deal of attention [69]. These enzymes use an oxygen-bridged binuclear non-heme-iron centre as the oxidizing species. Methane monooxygenase (MMO) is the classical example of this class of enzymes; it catalyzes the conversion of methane to methanol via the monooxygenase reaction shown below:

$$CH_4 + O_2 + NAD(P)H + H^+ \rightarrow CH_3OH + NAD(P)^+ + H_2O$$

Methane is produced in enormous quantities in the environment and assimilated as biomass by methanotrophs. The key step in this process is monooxygenation to give methanol, as catalyzed by MMO. Although cytochromes P450 catalyze similar reactions with the same overall stoichiometry, only MMO is capable of catalyzing the efficient monooxygenation of methane. MMO also catalyzes the oxidation of other hydrocarbons, including saturated and unsaturated linear, branched and cyclic hydrocarbons up to about C_8 in size, halogenated alkenes and alkanes, one- and two-ring aromatics, and heterocycles.

During catalysis a very powerful, relatively non-specific oxidizing species is generated, which is responsible for the wide range of substrate conversions. From the crystal structures of several MMOs , the structure of the binuclear iron site has been determined; the structure of the oxidized form is presented in Fig. 4.89.

The two Fe(III) atoms are linked by two bridging oxygen atoms. This so-called 'diamond core' may play a role in the oxygen activation chemistry of this class of enzymes. Upon two-electron reduction of the Fe(III) atoms to give two Fe(II) sites, an O_2 molecule binds and bridges the Fe centres. This leads to the subsequent formation of various intermediates, generally denoted compounds O, P

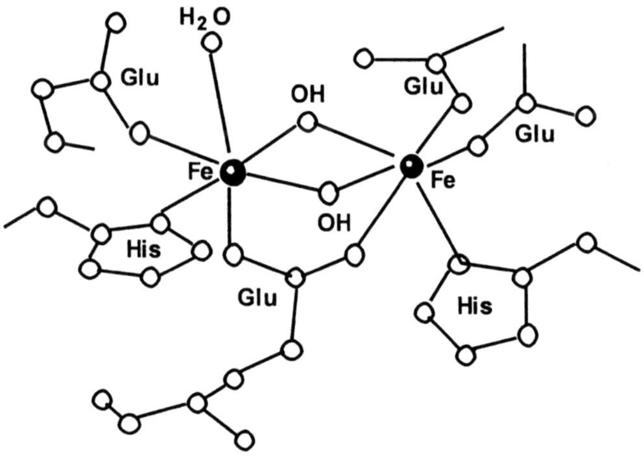

Fig. 4.89. Schematic presentation of the structure of the oxidized binuclear non-heme iron "diamond core" of MMO.

and Q, the latter of which is supposed to be involved in the conversion of methane to methanol. Figure 4.90 presents the catalytic cycle of MMO, taking into account the possible nature of these reactive intermediates as derived from extensive spectroscopic studies using inorganic model systems. From this scheme, it can be deduced that the binuclear 'diamond core' indeed plays an essential role in catalysis. In this scheme, the catalytically active form of the enzyme, compound Q, is a symmetrical analogue of an [Fe(IV)–Fe(IV)]=O species that is the electronic equivalent of the heme-Fe(IV)=O π cation radical species of cytochromes P450. In essence, this model proposes that the second iron of the cluster fulfils the role of the porphyrin in supplying the electron that is required to create the Fe(IV)=O species with a filled octet at the oxygen atom. Compound Q would then act to abstract a hydrogen atom from the substrate to yield the substrate radical, followed by hydroxyl rebound from the cluster to the substrate radical, generating methanol. This oxygen-rebound mechanism is analogous to the oxygen-rebound mechanism discussed above for cytochrome P450-catalyzed aliphatic hydroxylation (Fig. 4.88).

4.7.4.4 Tyrosinases

Figure 4.91 presents the catalytic cycle of tyrosinase, an enzyme presently of commercial and industrial interest since it is involved in the browning of fruits in the presence of oxygen, due to the oxidation of the phenols catalyzed by this enzyme. The most important feature to note is that the enzyme uses a binuclear Cu site able to bind a molecule of oxygen in a bridging mode. In contrast to the other monooxygenases, the two electrons required to split the molecule of

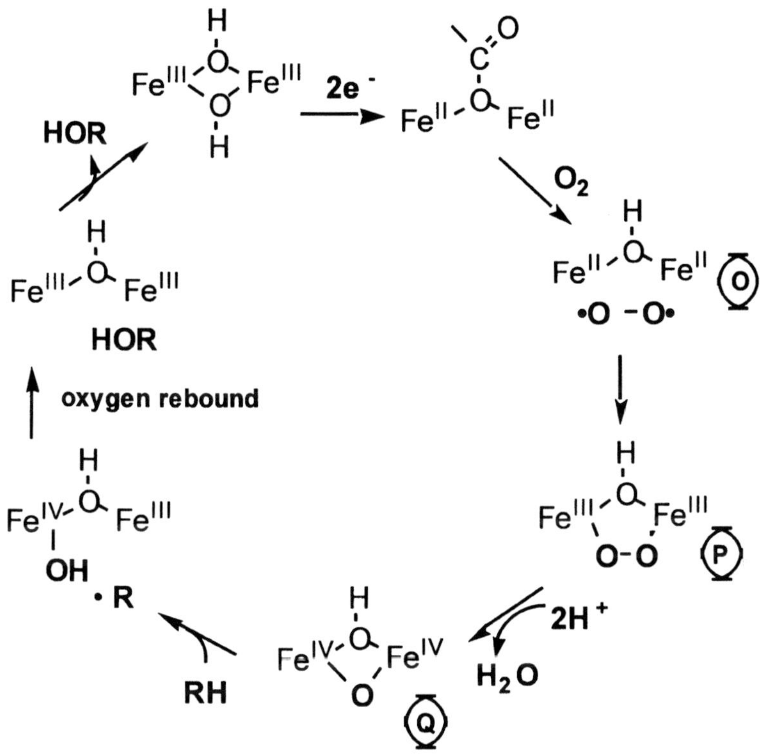

Fig. 4.90. Catalytic cycle of MMO.

oxygen to allow monooxygenase type of catalysis and formation of H_2O, come from a catechol, which is converted to its benzoquinone analogue. Thus, in addition to Fe, other metals, in this case Cu, can be involved in biological monooxygenation reactions by biocatalysts [70].

4.7.5 *Dioxygenases*

In contrast to the enzymatic splitting of the dioxygen bond of molecular oxygen by biocatalysts such as the cytochromes P450, methane monooxygenase, and the flavin-dependent monooxygenases, other biocatalysts are able to activate molecular oxygen and/or substrates in a way that leads to incorporation of both oxygen atoms into the final product. They are therefore named dioxygenases, in contrast to monooxygenases.

The intra- and extradiol ring-cleaving biocatalysts are especially good examples of this group of dioxygenases. The reactions catalyzed by the intra- and extra-diol dioxygenases are presented in Fig. 4.92.

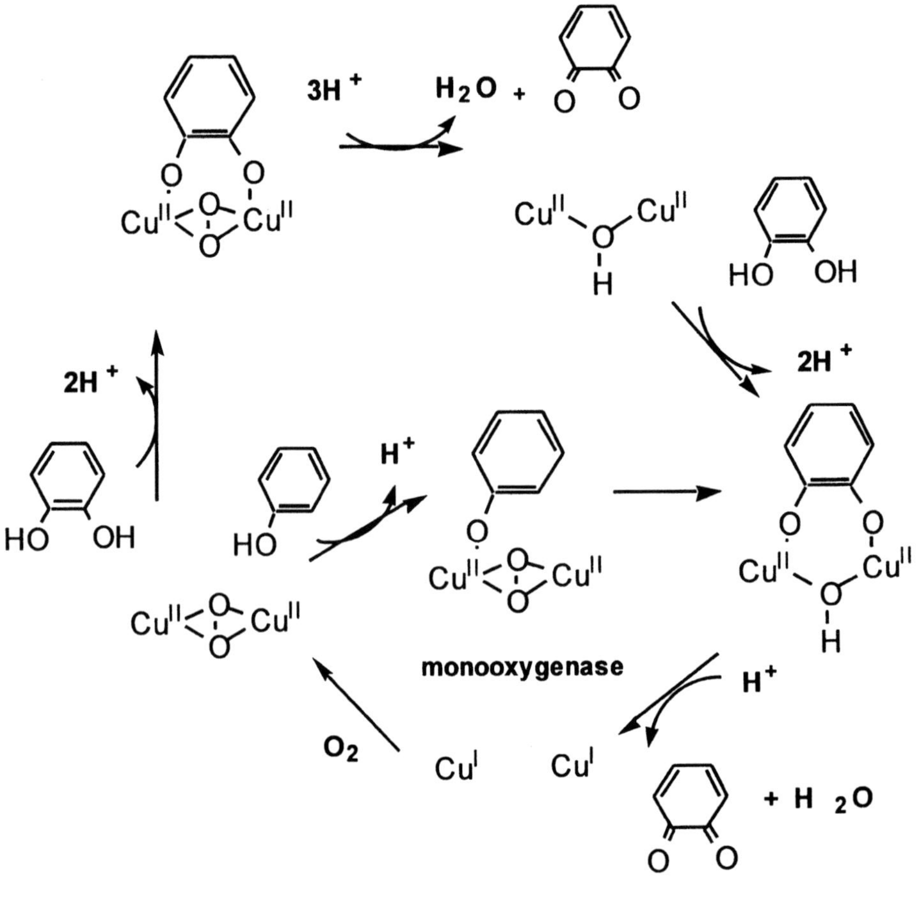

Fig. 4.91. Catalytic cycle of tyrosinase.

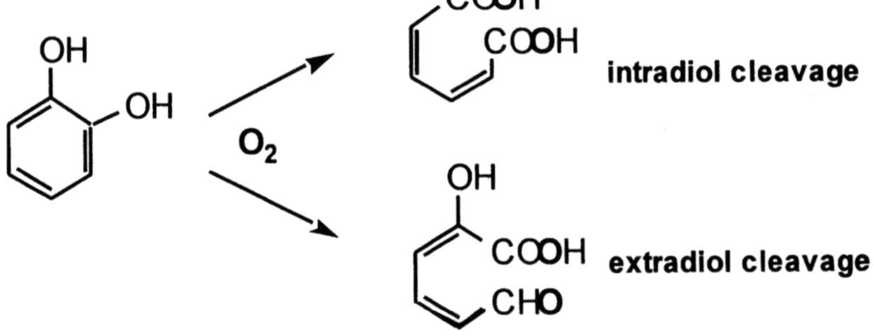

Fig. 4.92. Reactions catalyzed by intra- and extradiol ring cleaving dioxygenases.

These intra- and extradiol dioxygenases use a mononuclear metal centre to perform catalysis. In intradiol dioxygenases, the catalytic iron is bound to the active site through ligation by two tyrosinate and two histidine side chains. Upon binding of the catecholate substrate in an asymmetrical bidentate mode, one of these tyrosines is replaced by an oxygen of the deprotonated catechol; the other oxygen substituent of the catecholate replaces a water ligand molecule. In extradiol dioxygenases, the Fe ligands are known to be different, namely two histidines, one glutamate, and two water molecules. Another major difference between the intra- and extradiol dioxygenases is the different charge of the catalytic Fe centre. Intradiol dioxygenases contain and use an Fe^{3+} centre, which is unable to bind molecular oxygen. Instead it binds the catecholate substrate, thereby activating it so that direct electrophilic attack by O_2 can occur. In contrast, in extradiol dioxygenases the metal centre is Fe^{2+}, which is capable of oxygen binding and is thought to activate the O_2 for attack on the catecholate substrate. Figures 4.93 and 4.94 present in schematic form the suggested catalytic cycles of these enzymes [71].

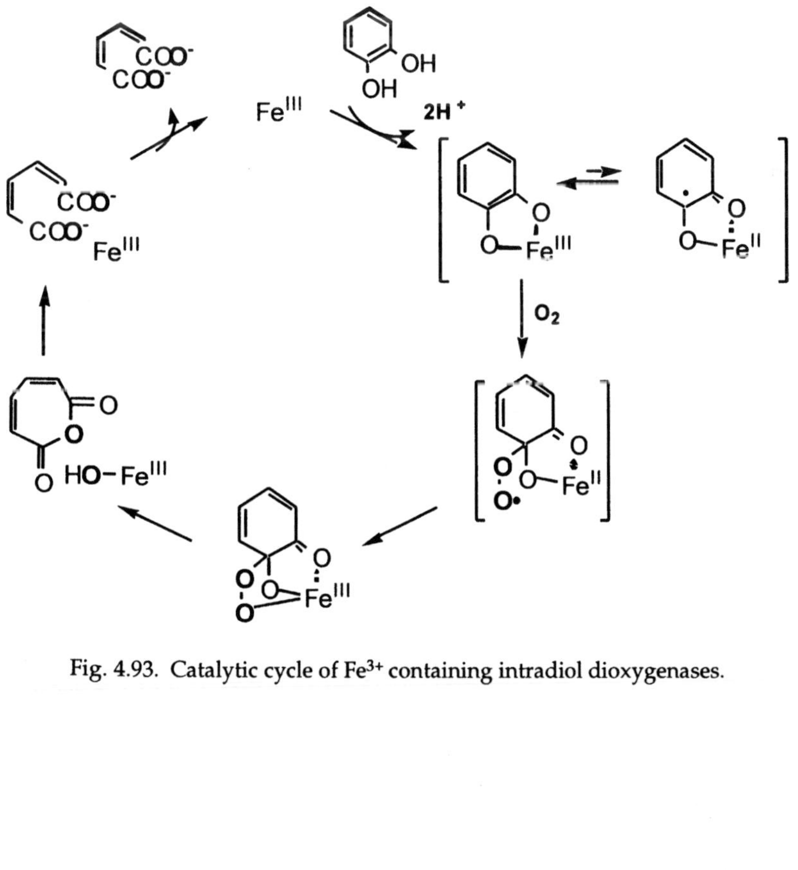

Fig. 4.93. Catalytic cycle of Fe^{3+} containing intradiol dioxygenases.

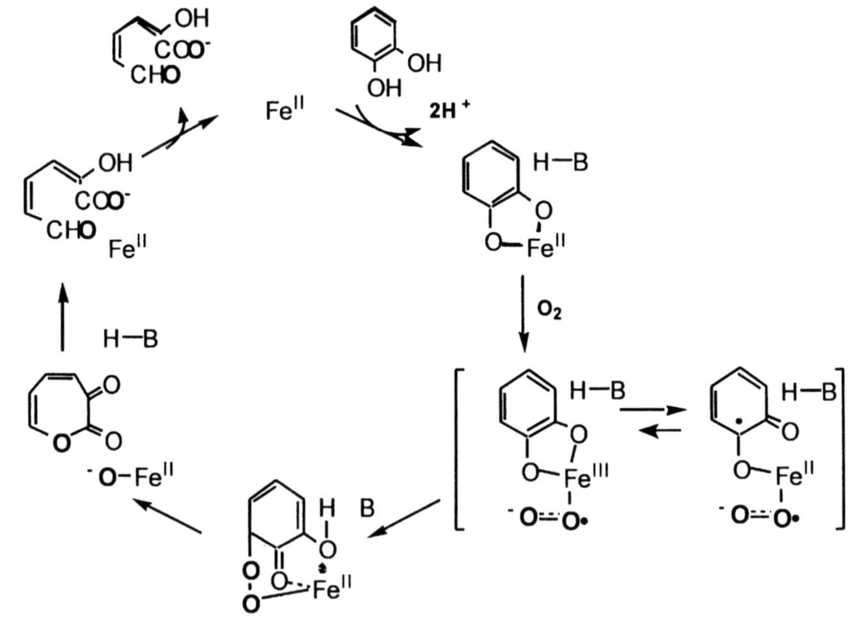

Fig. 4.94. Catalytic cycle of Fe^{2+} containing extradiol dioxygenases.

REFERENCES

1 R. Hoffmann, *Solids and Surfaces*, VCH, 1988; R.A. van Santen, *Theoretical Heterogeneous Catalysis*, World Scientific, Singapore, 1991.

2 R.A. van Santen, E.J. Baerends, in Z.B. Maksic (Editor), *Orbital Interaction and chemical reactivity of metal particles and metal surfaces*, Theoretical treatment of large molecules and their interactions, part 4, Springer Verlag, 1991, pp. 323–390.

3 M.R.A. Blomberg and P.E.M. Siegbahn, *J. Chem. Phys.*, 78 (1983) 5689; U.B. Brandemark, M.R.A. Blomberg and P.E.M. Siegbahn, *J. Phys. Chem.*, 88 (1984) 4617; J.J. Low and W.A. Goddard, *J. Am. Chem. Soc.*, 106 (1984) 6928; J.J. Low and W.A. Goddard, *J. Am. Chem. Soc.*, 106 (1984) 8321.

4 D.L. Thorn and R. Hoffmann, *J. Am. Chem. Soc.*, 100 (1984) 2079.

5 H. Berke and R. Hoffmann, *J. Am. Chem. Soc.*, 105 (1983) 427.

6 J.P. Collmann, L.S. Hegedus, J.R. Norton and R.G. Finke, *Principles and Applications of Organotransition Metal Chemistry*, University Science Books, Mill Valley, 1987.

7 P. Powell, *Principles of Organometallic Chemistry*, 2nd Edn., Chapman and Hall, London, 1988.

8 A. Yamamoto, *Organotransition Metal Chemistry*, Wiley, New York, 1986.

9 R. Crabtree, *The Organometallic Chemistry of the Transition Metals*, 1988.

10 C. Elschenbroich and A. Salzer, *Organometallics: A Concise Introduction*, VCH Publishers, New York, 1989.

11 J.M. Basset, J.P. Candy, A. Choplin, M. Lecomte, and A. Théolier, in R. Ugo (Editor), *Aspects of Homogeneous Catalysis*, Vol. 7, p. 87, Kluwer Academic Publishers, Dordrecht, 1990.

12 J. Falbe (Editor), *New Synthesis with Carbon Monoxide*, Springer Verlag, New York, 1980.

13 R.F. Heck and D.S. Breslow, *J. Am. Chem. Soc.*, 83 (1961) 4023.
14 F. Basolo and R.G. Pearson, *Mechanisms of Inorganic Reactions*, 2nd Edn., Wiley, New York, 1967.
15 K. Noack and F. Calderazzo, *J. Organomet. Chem.*, 10 (1967) 101.
16 P.W.N.M. van Leeuwen, C.F. Roobeek, and H.J. van der Heijden, *J. Am. Chem. Soc.* 116, (1994), 12117.
17 M.J. Calhorda, J.M. Brown and N.A.Cooley, *Organometallics*, 11 (1991) 1431.
18 N. Koga and K. Morokuma, *J. Am. Chem. Soc.*, 108 (1986), 6136.
19 J.R. Rogers, O. Kwon, and D.S. Maryneck, *Organometallics*, 10 (1991) 2816.
20 M. Brookhart and M.L.H. Green, *J. Organomet. Chem.*, 250 (1983) 395.
21 For an example of agosting interaction in the resting state of the catalyst see: M. Brookhart, A.F. Volpe, D.M. Lincoln, I.T. Horváth and J.M. Miller, *J. Am. Chem. Soc.*, 112 (1990) 5634.
22 O. Eisenstein and R. Hoffmann, *J. Am. Chem. Soc.*, 103 (1981) 4308.
23 J.W. Bruno, T.J. Marks and L.R. Morss, *J. Am. Chem. Soc.*, 105 (1983) 6842.
24 H.E. Bryndza, L.K. Fong, R.A. Paciello, W. Tam and J.E. Bercaw, *J. Am. Chem. Soc.*, 109 (1987) 1444.
25 T. Ziegler, W. Cheng, E.J. Baerends and W. Ravenek, *Inorg. Chem.*, 27 (1988) 3458.
26 L. Vaska, *Acc. Chem. Res.*, 1 (1968) 335.
27 J.K. Stille and K.S.Y. Lau, *Acc. Chem. Res.*, 10 (1977) 434.
28 C.A. Tolman, *J. Am. Chem. Soc.*, 92 (1970) 4217, 6777, 6785.
29 R.H. Crabtree, *Chem. Rev.*, 85 (1985) 245.
30 W.D. Jones and F.J. Feher, *Acc. Chem. Res.*, 22 (1989) 91.
31 G.W. Parshall, *Acc. Chem. Res.*, 8 (1975) 113.
32 G.W. Parshall, *Acc. Chem. Res.*, 3 (1970) 139.
33 G.W. Parshall, W.H. Knoth and R.A. Schunn, *J. Am. Chem. Soc.*, 91 (1969) 4990.
34 R.R. Schrock, *Acc. Chem. Res.*, 12 (1975) 98.
35 R.R. Schrock, *Organometallics*, 9 (1990) 2262.
36 T. Ito, H. Tsuchiya and A. Yamamoto, *Bull. Chem. Soc. Jpn.*, 50 (1977) 1319.
37 P.W. Jolly and G. Wilke, *The Organic Chemistry of Nickel*, Vol. II, Academic Press, New York, 1975.
38 F. Correa, R. Nakamura, R.E. Stimson, R.L. Burwell Jr. and D.F. Shriver, *J. Am. Chem. Soc.*, 102 (1980) 5112.
39 P.L. Watson and G.W. Parshall, *Acc. Chem. Res.*, 18 (1985) 51.
40 J.E. Bercaw et al. *J. Am. Chem. Soc.*, 109 (1987) 203.
41 M.S. Chinn and D.M. Heinekey, *J. Am. Chem. Soc.*, 112 (1990) 5166.
42 B.R. James, *Homogeneous Hydrogenation*, Wiley, New York, 1973.
43 G. Ertl, *Catal. Rev. Sci. Techn.*, 21 (1988) 201.
44 Y. Schule, P. Siegbahn and U. Wahlgren, *J. Chem. Phys.*, 89 (1988) 6982.
45 P. Sautet and J.-F. Paul, *Catal. Lett.*, 9 (1991) 3.
46 R.A. van Santen, *Progr. Surf. Sci.*, 25 (1987) 253.
47 E. Shustorovich, *Adv. Catal.*, 37 (1990) 101.
48 F. Frechard and R.A. van Santen, *Surf. Sci.*, 407 (1998) 200.
49 A. de Koster and R.A. van Santen, *Surf. Sci.*, 233 (1990) 366.
50 C. Backx, C.P.M. de Groot and P. Biloen, *Surf. Sci.*, 104 (1981) 300; M.A. Barteau and R.J. Madix, *Surf. Sci.*, 97 (1980) 101.
51 M. Neurock and R.A. van Santen, *J. Am. Chem. Soc.*, 116 (1994) 4427.
52 B. Grzybonska, J. Haber and J. Janas, *J. Catal.*, 49 (1977) 150.
53 R.A. van Santen and G.J. Kramer, *Chem. Rev.*, 95 (1995) 637.

54 V.B. Kazansky and I.W. Senchenya, *J. Catal.*, 119 (1989) 108.
55 D.M. Brouwer in R. Prins and G.C.A Schuit (Editors), *Chemistry and Chemical Engineering of Catalytic Processes*, NATO-ASIE: 39, Sijthoff and Noordholt, Alphen aan de Rijn, 1980, p. 137.
56 E.H. Teunissen, F.B. van Duyneveldt and R.A. van Santen, *J. Phys. Chem.*, 96 (1992) 366.
57 E.H. Teunissen, R.A. van Santen, A.P.J. Jansen and F.B. van Duyneveldt, *J. Phys. Chem.*, 97 (1993) 203.
58 S.R. Blaszkowski and R.A. van Santen, *J. Phys. Chem.*, 99 (1995) 11728.
59 S.R. Blaszkowski and R.A. van Santen, *J. Am. Chem. Soc.*, 119 (1997) 5020
60 V. Kazansky and R.A. van Santen, *Catal. Lett.*, 38 (1996) 115
61 M. Dixon and E.C. Webb, *Enzymes*, (3rd. ed.), Academic Press, 1979, Ch. V.
62 A. Fersht, *Enzyme Structure and Mechanism*, (2nd. ed.), W.H. Freeman, 1983.
63 C. Walsh, *Enzymatic Reaction Mechanisms*, W. H. Freeman, 1979.
64 L. Stryer, *Biochemistry*, (4th ed.), W.H. Freeman, 1995, Ch. 10.
65 H. Neurath, *Trends Biochem. Sci.*, 14 (1989) 268–271.
66 W.P. Jencks, *Cold Spring Harbor Symp. Quant. Biol.*, 52 (1987) 65-73.
67 M.A. Ator and P.R. Ortiz de Montellano, *J. Biol. Chem.*, 262 (1987) 1542-1551.
68 M. Sono, M.P. Roach, E.D. Coulter and J.H. Dawson, *Chem. Rev.*, 96 (1996) 2841–2887.
69 B.J. Wallar and J.D. Lipscomb, *Chem. Rev.*, 96 (1996) 2625–2657.
70 E.I. Solomon, U.M. Sundaram and T. Machonkin, *Chem. Rev.*, 96 (1996) 2563–2605.
71 L. Que Jr and R.Y.N. Ho, *Chem. Rev.*, 96 (1996) 2607–2624.

Chapter 5

Heterogeneous catalysis

5.1 INTRODUCTION

Three important classes of heterogeneous catalytic reactions have been selected for a more detailed description:
- reduction reactions,
- oxidation reactions,
- electrocatalytic reactions.

Two types of reduction reactions will be treated. The synthesis gas reaction (Section 5.2) in which CO is reduced by hydrogen shows a rich spectrum of products and is becoming of increasing importance as a reaction step in the transformation of natural gas to chemical products.

As the other key molecule for reduction NO has been chosen. Environmental constraints require its removal in automotive exhaust (Section 5.3) as well as for stack gas (Section 5.4). In the latter case ammonia as reductant is to be chosen, because of the high oxygen concentration in stack gas.

Several oxidation reactions (Section 5.5) have been selected. Selective oxidation of propene Section 5.5.1) is a key example of the use of oxides in oxidation catalysis.

Epoxidation reactions proceed via unique mechanisms and therefore require separate discussion (Section 5.5.2).

For completeness a final section on electrocatalysis has been included. The metal electrode shows reactivity similar to that of the shown by the metal-gas interphase. However the water environment in combination with the presence of electric gradients provided leads to another unique class of catalysis.

5.2 SYNTHESIS GAS CONVERSION

5.2.1 The Fischer–Tropsch Mechanism and its Consequences for the Technology

Hydrogenation of CO to CH_4 was first observed by Sabatier [1] while Orlov was the first to describe [2] the formation of higher hydrocarbons. Around 1925 Fischer and Tropsch published several patents on industrial processes producing hydrocarbons with a commercially interesting high fraction of C_5–C_{10} hydrocarbons. Catalysts were promoted iron and cobalt, and the processes ran at medium (or low) pressures [3]. Fischer and Tropsch also suggested a mechanism for the synthesis which nowadays bears their name. Carbon monoxide dissociates and a carbide and an oxide are formed. The oxide is reduced and the carbide, which has the carbon atoms at just the right distances, forms hydrocarbons [3,4]. It appeared later that, in its original form, the suggested mechanism was not correct. First, very few branched hydrocarbons are formed, while a random cross linking in three or two-dimensional carbides should lead to the formation of iso-products. Second, pre-carbidization of the catalysts did not lead to an enhanced rate of synthesis [5]. Therefore, several authors suggested a new mechanism [6]. The initiation step of the synthesis is then the formation of hydroxycarbene:

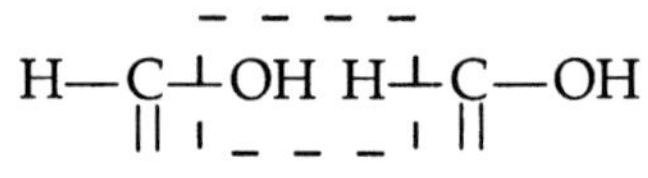

and the propagation step is dehydrocondensation, as indicated above. This mechanism was generally accepted for more than twenty years.

Pichler and Schulz [7] suggested a third mechanism, which is worth considering and seeking to verify. According to this suggestion, a carbonyl of a metal in question is formed first, which is followed by the formation of formyl (in later stages, acyl, etc.). Formyl (or a higher analogue) is subsequently hydrogenated, so that hydroxyl containing species are formed which, by a hydrogenolytic splitting, produce methyl or methylene (ethyl, etc.).

There are several other mechanisms to be found in the literature, but the three just mentioned seem to be the most realistic and best supported.

In 1974, the oil supply crisis stimulated research throughout the world on the Fischer–Tropsch Synthesis (FTS) of fuels. Surprisingly, the first result of this was evidence concerning the mechanism: with typical FTS and methanation catalysts — Fe, Co, Ni (Ru) — the initiation step is the dissociation of CO [8] and not the formation of hydroxycarbene.

Let us consider an experiment, shown schematically in Fig. 5.1 [8c]. The surface of a metal is partially covered by disproportionation of ^{13}CO:

$$2\,^{13}CO \rightarrow {}^{13}C + {}^{13}CO_2$$

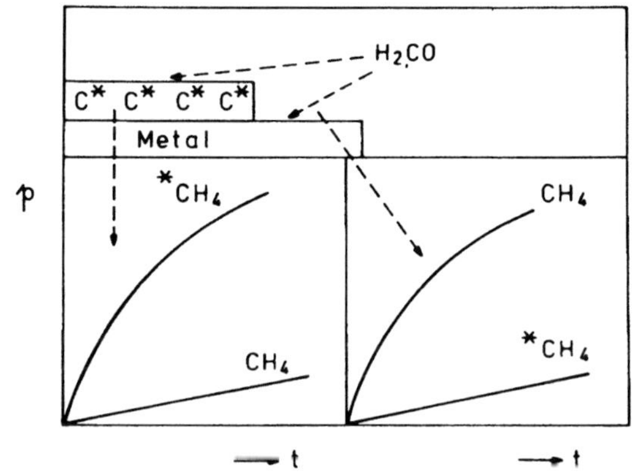

Fig. 5.1. Two possible pathways for the course of methanation. A metal surface partially covered by ^{13}C is offered to the CO/H_2 reaction mixture. *Left*: The result if C *ex* CO dissociation must be formed first. *Right*: Reaction runs via oxygen-containing intermediates.

This can be done easily with Fe, Co and Ni. Subsequently, this surface is exposed to the $^{12}CO/H_2$ mixture. If the reaction sequence is: hydrogenation of CO (to hydroxycarbene or formyl) → hydrogenolytic C–O bond splitting, then initially the production of $^{12}CH_4$ should prevail, as shown on the right-hand side of Fig. 5.1. If the sequence is dissociation → hydrogenation of C, $^*CH_4$ would prevail as on the left-hand side of Fig. 5.1. The experiments revealed that the second mechanism operated [8c]. It also appeared that the dissociation of CO requires several contiguous surface metal atoms (large ensemble) to occur [8c]. Since FTS was also found to be a large ensemble reaction, the conclusion could be drawn that the initiation of the FTS and the methanation reaction both comprise a CO dissociation [8c,9].

FTS is a process which is plagued by the self-poisoning of the catalyst, essentially by coking. A homogeneous catalyst would in principle be free of such troubles and therefore, several years after 1974, an intensive search was conducted for homogeneous FTS catalysts. However, when it appeared that the most efficient synthesis pathway comprised initiation by CO dissociation, which required large ensembles of active atoms [8], the research in this direction was stopped.

The next problem was whether or not the hydrocarbon chain growth requires oxygen-containing intermediates (formyl, acyl, hydrocycarbene, etc.). During a study of this problem, it appeared that the chain growth is promoted when the reaction mixture contains molecules (di-, tri- or tetrachlormethane, azomethane), which can generate CH_x species *in situ* [9–11]. Furthermore, it has been observed that, when a surface of a metal is covered by disproportionation of ^{13}CO, higher hydrocarbons (for example, propane or butane) contain more than one ^{13}C [8d]. In other words, it has been clearly proved that oxygen-free intermediates CH_x can polymerize into longer hydrocarbon chains. Moreover, no oxygen-containing intermediates were ever unambiguously detected (IR, EELS, XPS) on the metal surface of working FTS catalysts.

The step-by-step growth (adding CH_x groups) implies that the product molecules form a distribution, derived for polymers by Schulz and Flory and for FTS by Harrington, and Anderson et al. [12–14]. The unavoidable consequence of the Schulz–Flory–Anderson distribution is that, of all the hydrocarbons, only CH_4 can be formed with a selectivity nearing 100%. However, when producing fuels (the C_5–C_{10} fraction) both the low and the very high molecular weight hydrocarbons are undesired products. The Shell middle distillate synthesis (SMDS) solved this selectivity problem in a very ingenious way: the reaction is first driven up to high molecular weight molecules (waxes), by applying conditions most suitable for just that process (promoted Co catalyst, low temperatures). Then, in the second step, waxes are selectively cracked by zeolites, thus producing the optimum yield of hydrocarbons in the desired molecular weight region. The resulting gas-oil (diesel) is free of aromatics and has excellent properties as a fuel. The FTS is less suitable for the production of a high-quality gasoline. This is due to the fact that branched and aromatic compounds are not formed in amounts sufficient to achieve a high octane number in the product.

5.2.2 *Kinetics of the FTS and Methanation Reaction*

For a reaction of such complexity as methanation (or FTS) an exact kinetic theory is actually out of the question. One has to introduce one or more approximations. The usual assumption made is that one reaction step is rate determining (r.d.s.) and other steps are in equilibrium or steady state. Adsorption equilibria are described by Langmuir formulas (Langmuir–Hinshelwood, Hougen–Watson approach) [15] and the approach is sometimes made simpler by using so-called virtual pressures [16] (cf. Chapter 3).

People approaching the mechanism from the kinetic point of view tend to work on the basis of first principles, i.e. ignoring the sometimes already abundant non-kinetic experimental information on the mechanism. Thus, in formulating the kinetic equations, they consider, for example, two alternative pathways as equally possible: (i) the r.d.s. involves an oxygen-containing complex; (ii) the r.d.s. involves C or CH_x ($x = 1$–3), i.e. an oxygen-free complex. Subsequently, they try to describe, on the basis of the kinetics only, which alternative is the correct one. With regard to (i), again three cases can be discerned: (a) r.d.s. is a hydrogen-assisted dissociation of CO [17]; (b) the r.d.s. is the formation of an oxygen-containing intermediate [18]; (c) the r.d.s. is hydrogenolysis of the oxygen-containing intermediate [19]. All the kinetic models just mentioned lead to an approximate power rate law:

$$\text{rate}(r) = k_{\text{tot}} \cdot p^{\alpha}_{H_2} p^{\beta}_{CO}$$

where α is positive and β negative. This has all been done for methanation. However, if the overall FTS has its r.d.s. in one of the above mentioned reactions (a)–(c), it would follow the same kinetics as the methanation. Such an assumption is frequently made.

With the rate determining step involving C or CH_x, the sequence of reaction steps starts with the relevant dissociations

$$H_2 \rightarrow 2H_{ads}$$

$$CO \rightarrow C + O_{ads}$$

followed by an appropriate number of hydrogenation steps:

$$C + H_{ads} \rightarrow CH_{ads}$$

$$CH_{ads} + H_{ads} \rightarrow CH_{2ads}$$

$$CH_{2ads} + H_{ads} \rightarrow CH_{3ads} \ldots$$

Adsorption equilibria of the first two steps lead to the terms

$$\theta_H = \frac{A\sqrt{p_{H_2}}}{(1 + B\sqrt{p_{H_2}} + C\sqrt{p_{CO}} + \ldots)}$$

$$\theta_C = \frac{A\sqrt{p_{CO}}}{(1 + B\sqrt{p_{H_2}} + C\sqrt{p_{CO}} + \ldots)}$$

The dots represent the terms related to the other products of the reaction which may potentially be present, or the term related to the assumed steady state of some parallel reaction steps, or some interference terms. Again, this has all been analyzed for methanation, but if in the FTS one of the above reactions is the r.d.s., the kinetics would be the same as for the methanation. If the rate determining step is the reaction C + H_{ads} then in the usual approximation $r = k_{tot}\sqrt{p_{H_2}} \cdot \sqrt{p_{CO}}$ when adsorption of all components is weak. When CO adsorption is much stronger than the adsorption of other species $r = k_{tot}\sqrt{p_{H_2}} / \sqrt{p_{CO}}$. The latter equation gives a better agreement with the experimental data than the first one, but in many cases the value α of the power rate law is found to be $\alpha > ½$. To make the agreement better, people make one of two assumptions: (a) the rate determining step is a three-body collision, C + 2H [20]; or (b) the r.d.s. involves CH_x, instead of C and thus, with each equilibrium (steady state) step prior to the r.d.s. hydrogenation step, the rate expression receives one more $\sqrt{p_{H_2}}$ term [21]. For example, in case (b):

$$\theta_{CO} = K\theta_C\,\theta_H \approx \frac{const\sqrt{p_{H_2}}}{\sqrt{p_{CO}}}$$

For CH + H_{ads} as the r.d.s. $r = k_{tot}\,p_{H_2}\,p_{CO}^{-1}$.

An example of kinetic data for methanation, showing the agreement achieved with the theory, is shown in Fig. 5.2.

Another way of getting rid of the problem of too high an α value is to assume an r.d.s. involving an oxygen-containing complex. However, neither of the three suggestions really provides an attractive explanation. With typical FTS and methanation catalysts, oxygen-containing intermediates are not probable, nor is the three-body collision reaction [21]. It is also known that hydrogenation of C is

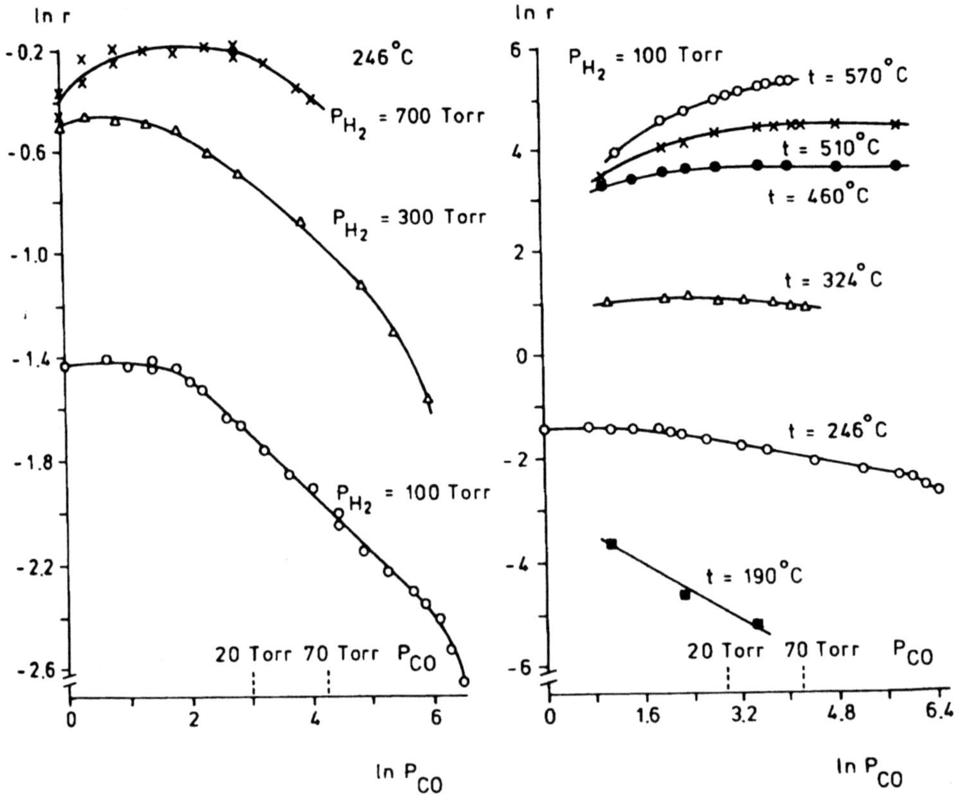

Fig. 5.2. Methanation rate as a function of the ln p_{CO}. *Left*: 519 K, three H_2 pressures, as indicated. *Right*: *T*, as indicated, p_{H_2} = 100 Torr. (From van Meerten et al. [21]).

more difficult than that of CH (see hydrocarbon exchange reactions), so that it does not seem reasonable to shift the r.d.s. further in the sequence of the hydrogenation steps. What else, then?

Most likely, one has to consider the following information [22]. A working FTS or methanation catalyst has (at least) three kinds of carbon on its surface: a reactive type, an almost unreactive (C_s) (perhaps carbidic) and coke (C_{graph}, very unreactive). The working (in the FTS) surface is proportional to $(1–\theta_s – \theta_{graph})$ and the rate of C + H_{ads} is then: $r = k_{tot} (1 – \theta_s – \theta_{graph}) \theta_C\theta_H$.

The θ_s carbon does not contribute noticeably to the FTS but the amount present is a function of p_{H_2}. Hydrogen, when adsorbed, prevents its formation and removes this carbon by a slow reaction up to a certain steady-state concentration, θ_s. This means that the order of α for hydrogen is greater than ½. The situation just described is far from being hypothetical. A group of authors [22a] analyzed the presence of θ_c, θ_s and θ_{graph} on Ni catalysts, a phase diagram has been established and evidence obtained that θ_s and θ_{graph} are functions of temperature and hydrogen pressure [23].

The distribution of products of a FTS of hydrocarbons can be derived in the following way [12–14]. Let us call a fraction of products with i carbon atoms ϕ_i. Now, if a product with nC appears in the gas phase, its fraction, Φ_n is a measure of termination of the anchored chain growth (by addition of CH_x units) at the nC-containing hydrocarbon. At the same time, all products with $i > n$ taken together, are the measure of propagation of the growth. If the probability of growth is α_n and that of termination is $(1 - \alpha_n)$.

$$\frac{\alpha_n}{1-\alpha_n} = \frac{\sum_{n+1}^{\infty} \Phi_i}{\Phi_n}$$

$$\alpha_n = \sum_{n+1}^{\infty} \Phi_i \Big/ \sum_{n}^{\infty} \Phi_i$$

It has been found that $\Phi_{n+1} / \Phi_n \approx$ const., independent of n, and it can easily be verified that const. $\approx \alpha = (\alpha_n)$. By repeated substitution $\Phi_{n+1} = \alpha_n \Phi_1$ or $\Phi_n = \alpha_n (\Phi_1 / \alpha)$. By plotting log Φ_n vs n one can easily determine α.

The Periodic Table, and the periods with metals therein, offer the following pattern of activities:

(a) the block of metals between Sc–Cr–La–W contains those metals which dissociate CO at temperatures around and under 300 K. The oxides arising by that process are not easily reduced at temperatures at which the Fischer–Tropsch synthesis can produce sufficiently high hydrocarbons, so that these metals are not good catalysts for FTS.

(b) Mn, Tc, Re probably behave very similarly to the metals under (a).

(c) Fe, Co, Ni, Ru, Rh dissociate CO at slightly elevated temperatures (Fe 300 K, Rh 470 K) and the oxides of these metals can be reduced by syngas at temperatures around 470 K. Thus, these metals are suitable for the FTS.

(d) Pd, Ir, Pt do not dissociate CO easily enough to be good catalysts for the FTS. However, the surface of these metals, with many defects in the structure, can be active.

(e) Cu, Zn, Ag, Cd, Au, Hg do not adsorb CO strongly enough to be a catalyst for FTS of hydrocarbons. Copper is a component (with ZnO) of the most active catalysts for methanol synthesis.

If one compares the behaviour of the most active FTS catalysts — Fe, Co, Ni and Ru — several things emerge. Unpromoted nickel is probably the most stable catalyst but it produces mainly methane; the other metals, when active in the polymerization of CH_x units, are also susceptible to self-poisoning by inactive carbon deposition.

5.2.3 *Function of Promoters in the Hydrocarbon Synthesis*

It is an old practice [24] to add to metallic catalysts a chemical compound which itself is inactive but which improves the activity/selectivity and/or stability of the metallic catalyst. This is also a common situation in FTS: iron is used in the 'double-promoted' form (Al_2O_3, alkali carbonate) Co supported, and promoted by alkalis and, for example, ThO_2, etc. The additives — promoters — have been shown to have the following functions [25]:

(1) promoters enhance the CO dissociation and the overall rate of synthesis;
(2) promoters shift the selectivity in the FTS to higher hydrocarbons;
(3) promoters favour the formation of unsaturated products;
(4) promoters determine to which extent the FTS of hydrocarbons is accompanied by other reactions (see the next section).

Effect (2) above is most likely a consequence of effect (1). Effect (3) is most likely related to the suppression of the hydrogen adsorption, since a simultaneous hydrogenation of olefins is suppressed when the hydrocarbon-producing catalyst system is promoted by alkalis [48]. Point (4) will be discussed in more detail in the next section. The effect of vanadium promoter (on the Rh/SiO_2 catalyst) on the rate of methane formation from CO preadsorbed at 293K and then subjected to a programmed heating in a flow of hydrogen is shown in Fig. 5.3 [26].

5.2.4 *Synthesis of Higher Oxygenates*

It seems that one conclusion can easily be drawn: alcohols are only produced when the catalyst is properly promoted [27]. Actually, all metals of Group VIII

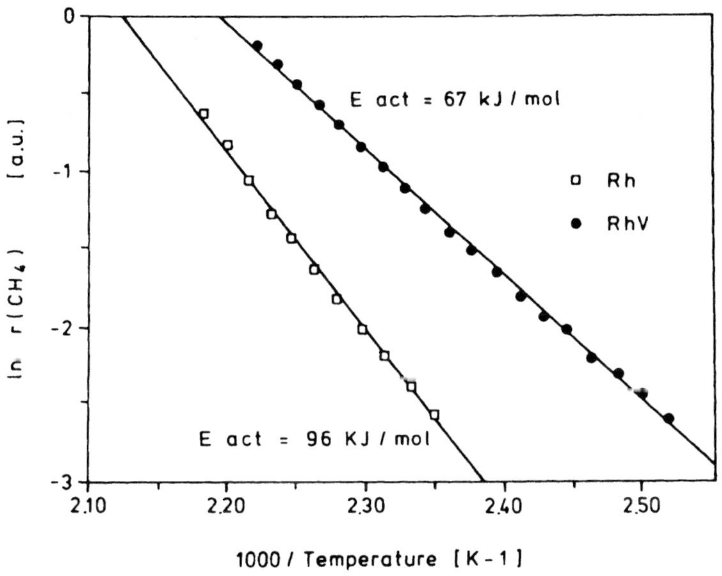

Fig. 5.3. Arrhenius plot for temperature programmed surface reaction: ln(rate) as a function of $1/T$, for the reaction of a pre-adsorbed CO in the flow of H_2. Promoted and unpromoted Rh/SiO_2 catalysts [26].

(8–10 according to the IUPAC terminology) can be made active in the synthesis of higher alcohols. The most suitable one is Rh (right in the middle) and, while in the right lower corner of Group VIII the metals have to be promoted to stimulate CO dissociation and CH_x formation, in the left (upper) corner the dissociation has to be suppressed. Since the promoter is inactive as such and the promoter free metal surface is necessary to produce CH_x units and adsorbed hydrogen, the rates, yields and sometimes also selectivities when plotted as a function of the amount of promoter added, often form a curve with a maximum [28], reflecting that the rate r is indeed described by (see Fig. 5.4):

$$r = k(T)\,\theta_{\text{prom}} \cdot (1 - \theta_{\text{prom}}) \cdot f(\pi(H_2), \pi(CO))$$

Such an expression is expected to be valid when the promoting action is limited to the perimeter of the promoter patches on the metal surface. Three steps can in principle be influenced by promoters (transition metal oxides):

(i) Adsorption of CO. It is known that the C–O bond strength is weakened as a consequence of the presence of the promoter, since the latter enhances the population of the antibonding 2πCO orbitals (see the foregoing section).

(ii) Promoters cause the species containing unsaturated C=O bond (potential and most likely intermediates) are hydrogenated to alcohols [26,29,30].

(iii) A question still being discussed is whether the promoter also stimulates the formation of the aldehydic intermediates. The latter ones can, in principle, also be formed on an unpromoted surfaces [26,27].

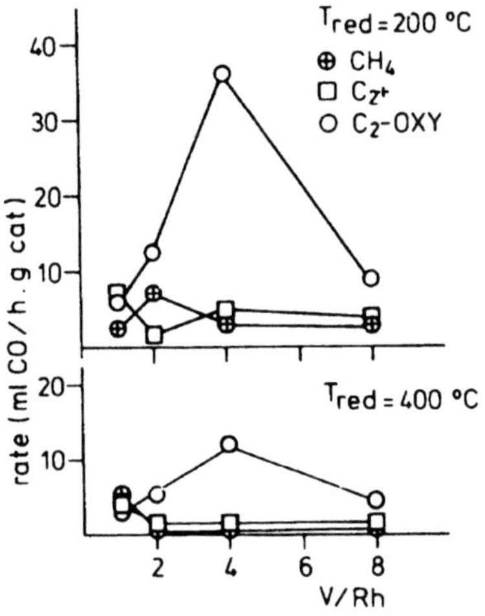

Fig. 5.4. Rate of various product formation, under standard conditions, as a function of the (atomic) promoter/metal ratio, V/Rh catalyst–Rh/SiO_2; promoter introduced as $VOCl_2$ compound [33].

The production of C_2 alcohols is higher on the metals left from the diagonal in Group VIII (Fe, Co, Ru) since these metals tend to accumulate CH_x groups on the whole metallic surface. Rhodium is most suitable for producing ethanol: its CO dissociation activity is weak and thus the CH_x units are preferentially formed on the metal surface round the patches of the promoting oxide.

It is perhaps worthwhile noting that the promoter patches on the metal can be either created during the wet steps of the catalyst preparation, or when a transition metal oxide is used as a support, they can be created by migration of the support material on the metal upon high temperature reduction. The metal surface can be kept almost completely covered *in vacuo* (SMSI effect) [31], but in the presence of CO or of the reaction mixture, the layer of oxide recrystallizes and the metal surface becomes accessible again from the gas phase [28].

5.2.5 *Synthesis of Methanol*

This is an important industrial reaction, alone or in combination with others. The CH_3OH production is often coupled to oxidation to formaldehyde, methanol to gasoline (Mobil) process, methanol to olefins process, carbonylation, etc. Due to this, a large volume of information already exists on catalyst preparation, kinetics, reactors and all other aspects of the related chemical technology [32]. However, let us concentrate our attention here on just one selected problem: the role of the promoter and the nature of the active site on the 'metal on oxides' catalysts. Let us mention in passing that pure metals (promoter free) most likely do not catalyze the synthesis.

Ideas suggested in the literature concerning active sites can be divided into three main streams:

(i) The active metal (Cu, Pd, Pt, Ir) is the locus of the reaction. A promoter is possibly necessary to stabilize a certain size, or shape, of metallic particles [33].

(ii) The surface of the carrier bears the essential intermediates of the synthesis, whereby either the cations (Zn, Cr) are considered to be the active sites, or the OH groups of the support [34–36]. In this picture, the metal is just supplying atomic hydrogen by a spill-over.

(iii) Active sites bearing the essential intermediates are ions of an 'active' metal (Cu, Pd, Co, Cr, Fe) and the zero-valent metallic centres (particles) produce adsorbed hydrogen. The active ions are stabilized by the promoter patches on the metal surface or are localized in the support round the metallic particles. With an oxidizable metal like Cu, additional Cu^{n+} centres can be created by the reaction itself [33].

For a long period of time, the mechanism under (i) was dominant in the literature. One of the main pieces of the support for this mechanism was the finding that with $CO/CO_2/H_2$ mixtures the activity in the CH_3OH synthesis is a linear function of the Cu^0 (metallic) surface area [33]. However, it has been overlooked that, in the presence of CO_2, a large fraction of a Cu^0 surface is covered during a running reaction by oxygen (or oxygen-containing species). This fraction is constant for all catalysts, independent of the Cu^0 surface area [33]. Thus, it could have been also the Cu^+ (or Cu^{2+}) sites which were responsible for the activity. In the last decade the following well established facts seem to support just that last alternative.

(1) With Pd catalysts a linear correlation has been found between Pd^{n+} content and synthesis activity [37].

(2) With Cu a slight preoxidation shortens the induction period which otherwise accompanies the reaction with CO/H_2 mixtures on well reduced catalysts [38].

(3) Anything that stabilizes the presence of Cu^+ indiscriminately enhances the activity of the catalyst; (i) promoters [39]; (ii) CO_2 admixture, up to a certain concentration of CO; (iii) supports (silica treated and impregnated at pH > 7, upon which procedure formation of Cu silicates is stimulated) [40].

Similar observations (but with CO_2) have also been made with Pd, Rh and Pt catalysts [38]. Some relevant information on this point is shown in Fig. 5.5 [37,39]. The hypothesis suggesting that Pd^{n+} or Cu^+ are active sites seems to find strong support in these data (see Fig. 5.5).

The discussion on the active sites and the mechanism is far from closed. The immediate future may bring changes in the picture which is offered above as the most likely one.

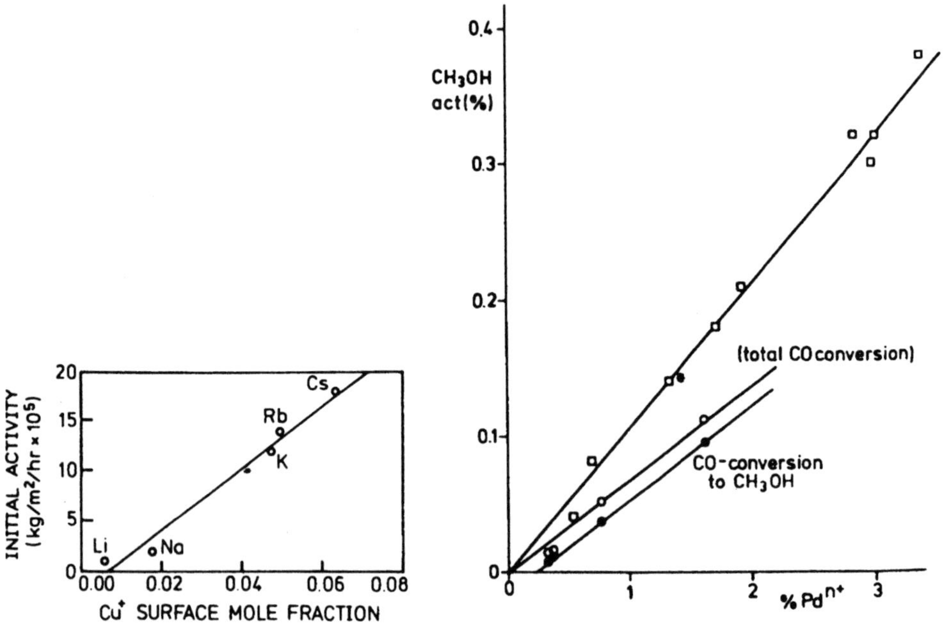

Fig. 5.5. (a) *Left*: Rate of methanol formation as a function of the Cu^+ concentration (detected by XPS). Catalyst unsupported Cu, promoted by various alkalis [61]. (b) *Right*: Rate of methanol formation as a function of the Pd^{n+} concentration (detected by chemical extraction and a.a.s.). Catalyst Pd/SiO_2 promoted by $MgO/MgCl_2$ (various Mg/Pd ratios lead to various Pn^{n+} contents) [59].

5.3 AUTOMOTIVE EXHAUST CATALYSIS

5.3.1 *Air Pollution and Regulations*

Transportation is one of the most important sources of air pollution, particularly in urban environments. The total number of cars, trucks, buses and motorcycles in the world in 1990 was estimated at about 650 million [41]! All these cars produce exhaust emissions.

Typical concentrations of car exhaust gas constituents present in addition to CO_2 and H_2O are given in Table 5.1. The presence of CO, the most toxic component of exhausts, is due to incomplete oxidation of hydrocarbons. Hydrogen is always present at approximately 1/3 the concentration of CO, and originates from the cracking of hydrocarbons. Nitric oxide (almost exclusively NO) is formed during combustion of fuel at high temperatures.

In Section 2.5 the lambda parameter has been described, i.e., the dimensionless air-to-fuel ratio parameter that controls exhaust catalysis chemistry to a significant extent.

TABLE 5.1

Typical concentrations of exhaust gas constituents[a]

Component	Concentration
hydrocarbons[b]	750 ppm
NO_x	1050 ppm
CO	0.68 vol%
H_2	0.23 vol%
O_2	0.51 vol%

[a]From Taylor [42].
[b]Based on C_3.

Gasoline contains additives such as benzene, toluene and branched hydrocarbons to achieve the necessary octane ratings. Direct emission of these volatile compounds, e.g. at gas stations, forms a significant source of air pollution as well. Unburned gasoline and cracked hydrocarbons such as ethylene and propylene also form a substantial constituent of exhaust.

Leaded fuels, containing antiknock additions such as tetra ethyl lead, have been abandoned nowadays because lead is a poison for human beings and for the three-way catalyst, especially for the removal of NO by rhodium.

In the USA the Clean Air Act of 1970 established air-quality standards for six major pollutants: particulate matter, sulphur oxides, carbon monoxide, nitrogen oxides, hydrocarbons, and photochemical oxidants. It also set standards for automobile emissions, the major source of carbon monoxide, hydrocarbons, and nitrogen oxides, see Table 5.2. These emission levels have been achieved owing to the application of catalysis.

TABLE 5.2

US Federal Emission Standard in gram per mile. European standards are comparable to the 1991 values

	1991	2004
Hydrocarbons	0.41	0.125
CO	3.4	1.7
NO_x	1.0	0.2

5.3.2 *The Three-way Catalyst*

The three-way catalyst, consisting of Pt and Rh particles supported on a ceramic monolith, controls emissions of CO, NO and hydrocarbons from automotive exhausts and represents a remarkably successful piece of catalytic technology [42]. The three-way catalyst enables the removal of the three pollutants CO, NO and hydrocarbons, by the following overall reactions:

$$CO + O_2 \rightarrow CO_2$$

$$C_xH_y + O_2 \rightarrow CO_2$$

$$NO + CO \rightarrow N_2 + CO_2$$

Additionally, NO is reduced by H_2 and by hydrocarbons. In order to let the three reactions proceed simultaneously, the composition of the exhaust gas needs to be properly adjusted to an air-to-fuel ratio of 14.7, see Fig. 5.6. At higher oxygen content, the CO oxidation reaction consumes too much CO and hence NO conversion fails. If, however, the oxygen content is too low, all NO will be converted, but the removal of hydrocarbons and CO will not be complete. Oxygen sensors are used to maintain the proper balance of fuel and air.

Catalytic treatment of motor vehicle exhaust has been applied in all passenger cars in the USA since the 1975 models. The first cars with electronic feedback systems and three-way catalysts were 1979 Volvos, sold in California. Today all gasoline cars sold in the Western World are equipped with catalytic converters.

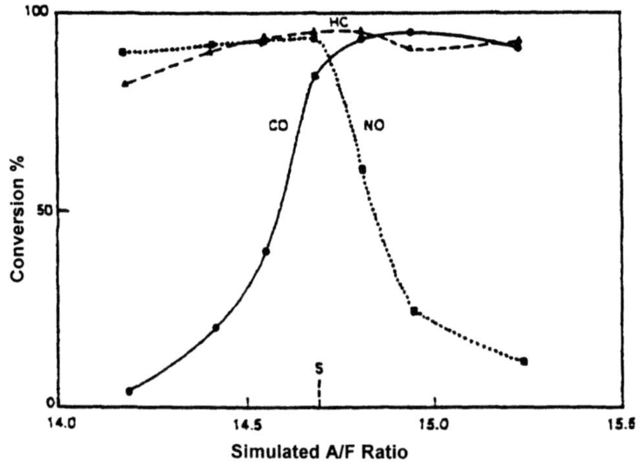

Fig. 5.6. Three-way emission control requires careful adjustment of exhaust gas compositions with air-to-fuel ratios of about 14.7 (from Taylor [42])

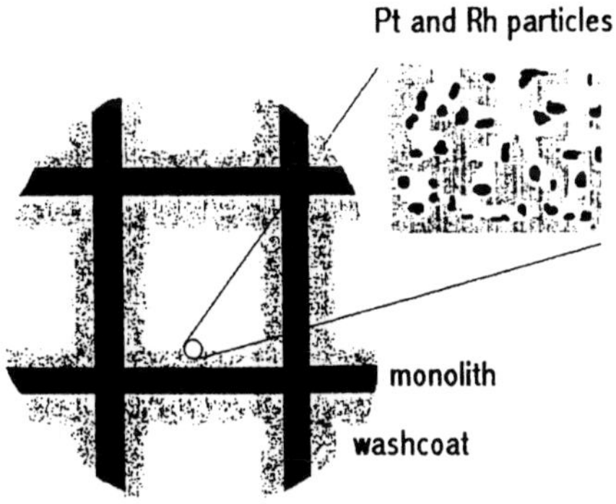

Fig. 5.7. The three-way catalyst consists of platinum and rhodium (or palladium) metal particles on a porous oxidic washcoat, applied on a ceramic monolith.

5.3.3 *The Catalytic Converter*

Figure 5.7 shows schematically how the catalyst is built up. The major supporting structure is a monolith, which is covered by a 30–50 mm layer of porous 'washcoat'. The latter is the actual support and consists largely of γ-Al_2O_3 (70–85%) and other oxides such as cerium oxide (10–30%), lanthanum oxide or alkaline earth oxides (BaO). Some formulations use NiO as H_2S getter. Denser oxides such as α-Al_2O_3 and ZrO_2 are sometimes used as support, to prevent incorporation of rhodium in the support at high temperatures. Only 1–2% of the washcoat weight corresponds to noble metals (Pt, Pd, Rh). Some manufactures use all three, but most converters contain Rh together with Pt. Recently, all-palladium converters have been introduced as well.

The ideal operating temperatures for the three-way catalyst lie between 350 and 650°C. After a cold start it takes at least a minute to reach this temperature, implying that most of the emission of CO and hydrocarbons takes place directly after the start. Temperatures above 800°C should be avoided in order to prevent sintering of the noble metals and dissolution of rhodium in the support.

5.3.4 *Function of the Catalyst Components*

A large number of permutations in composition exists. Usually the precise composition, particularly that of the washcoat, is secret. Here, we describe the function of the different catalyst ingredients [42,43].

- Alumina, present in the gamma modification, is the most suitable high surface area support for noble metals. However, at high temperatures it transforms into the alpha phase, and stabilization is desirable. Another point of concern is diffusion of rhodium into alumina, which calls for the application of diffusion barriers such as lanthanum oxide.

- Ceria is a partially reducible oxide (reduction is promoted by the noble metals). When the air/fuel oscillation swings to the lean side, it takes up its maximum capacity of oxygen; this oxygen is available for CO_2 formation when the composition swings to the rich side. Ceria thus counteracts the effect of oscillating feed gas compositions. It further stabilizes the high surface area of the γ-Al_2O_3 by inhibiting its phase change to the α-phase and it impedes the agglomeration of the noble metals and the loss of noble metal surface area by acting as a diffusion barrier. Oxygen vacancies, conceivably also in regions where CeO_x is in contact with noble metal, are active sites for CO oxidation by O_2, or by H_2O through the water-gas shift reaction:

$$CO + H_2O \rightarrow CO_2 + H_2$$

- Lanthanum oxide is valênce invariant, and does not exhibit any oxygen storage capacity, but it is very effective in stabilizing γ-Al_2O_3. It spreads over the alumina surface and provides a barrier against dissolution of rhodium in the support.
- Platinum serves as the catalyst for oxidation of CO and hydrocarbons. It is relatively insensitive to contamination by lead or sulphur. At high temperatures it is not known to dissolve in the washcoat, but sintering into larger particles is a problem.
- Rhodium is the crucial ingredient of the three-way catalyst. The metal is a by-product of Pt, and is mined in South Africa (±2/3) and Russia (±1/3). However, the average mining ratio equals Pt:Rh = 17:1, and this is much lower than the ratio of Pt:Rh in the catalyst. In 1991, about 87% of the Rh world demand went into catalysts (Pt: 37%). Although expensive, rhodium is difficult to replace owing to its unique properties with respect to NO surface chemistry. Loss of rhodium activity is due to particle growth under reducing conditions (>900 K), and to diffusion into the alumina support under oxidizing conditions (>900 K).
- Palladium can be present in addition to Rh and Pt but may also replace them. Palladium is as good for oxidation as platinum (even for the oxidation of saturated hydrocarbons) but it is somewhat less active for NO reduction. Hence, noble metal loadings of a Pd-only catalyst are 5–10 times higher than for a Pt–Rh catalyst. Palladium is less resistant to residual lead in gasoline than Pt, however, gasoline in the US is essentially Pb-free. Palladium catalysts also require higher ceria loadings, to help prevent deactivation at high temperatures.

5.3.5 Catalyst Deactivation

In the USA, three-way catalysts are required to maintain high activity and meet the emission standards after 50,000 miles or five years. Because catalysts do deactivate with use, fresh catalysts are designed such that they perform well below the emission standards. The extent to which a three-way catalyst deactivates, depends on many factors. The wide range of vehicle operating conditions due to differences in style of driving is an important one.

The major causes of degradation are thermal damage, poisoning by contaminants and mechanical damage of the monolith [42]. High temperatures lead to sintering of the noble metal particles and may also induce reactions between the metals and the support. High temperatures can, for example, be caused by fast driving or by repeated misfiring of the engine, resulting in the (exothermic) oxidation of large amounts of unburned fuel over the catalyst. Excessively high temperatures can damage the support, by promoting the transition to α-alumina and loss of surface area. Shock and high temperatures may lead to formation of channels for the exhaust to pass through the system without contacting the catalyst.

Lead and phosphorus are poisons for the catalyst. Lead is still present at very low levels in unleaded gasoline, but does not present a problem. Misfueling, i.e. using leaded fuel, however, is a serious reason for deactivation, affecting severely and irreversibly the NO reduction activity of the three-way catalyst. The oxidation activity is temporarily lower after misfueling but recovers usually to values within the emission standards. Phosphorus is present in engine oils. It binds strongly to the alumina support and may eventually also block the noble metal surfaces. Sulphur, although a potential poison for all metals, interacts relatively weakly with platinum and rhodium, but becomes a source of concern when palladium catalysts are used. Other contaminants which may be present in gasoline, such as organo-silicon compounds or additives based on manganese, have been found to negatively affect the performance of a three-way catalyst. Note that also the sensitivity of the oxygen sensor to contaminants is a point that needs consideration.

5.3.6 *Catalytic Reactions in the Three-Way Catalyst: Mechanism and Kinetics*

CO oxidation [44–46] and the reaction between CO+NO [47–49] have extensively been studied. Much less is known about hydrocarbon oxidation, however. Also the role of hydrocarbons in reducing NO is only beginning to be explored.

5.3.6.1 The CO Oxidation Reaction

The oxidation of CO is adequately described by the following simplified set of reactions:

$$CO + * = CO_{ads}$$

$$O_2 + 2* = 2\ O_{ads}$$

$$CO_{ads} + O_{ads} \rightarrow CO_2 + 2*$$

in which the adsorption steps are considered to be in equilibrium. The corresponding rate equation is of the form

$$V\frac{d[CO_2]}{dt} = Nk\theta_{CO}\theta_O = \frac{NkK_{CO}\sqrt{K_{O_2}}\,p_{CO}\sqrt{P_{O_2}}}{\left(1+\sqrt{K_{O_2}P_{O_2}}+K_{CO}p_{CO}\right)^2}$$

where V is the reacting volume, N the number of catalytically active sites, k the rate constant of the surface reaction between CO and an O-atom, θ the surface coverage of the indicated molecule, K the equilibrium constant of adsorption, and p the partial pressure of the indicated gas.

The energetics of the CO oxidation reaction is illustrated in Fig. 5.8. The activation energy of the homogeneous gas phase reaction between CO and O_2 would be largely determined by the energy needed to break the O–O bond of O_2, some 500 kJ/mol. The catalyst easily dissociates the O_2 molecule, and the rate-determining step has been shifted to the reaction between CO_{ads} and O_{ads}, which is only about 100 kJ/mol on palladium. The CO oxidation thus illustrates nicely that the essential action of the catalyst lies in the dissociation of a bond. Once this has been accomplished, the subsequent reactions follow, provided the intermediates are not held too strongly by the catalyst as expressed in Sabatier's Principle (see e.g. Ref. [50]).

It is interesting to consider the temperature dependence of the reaction rate, as it reveals a very general phenomenon in catalysis. At low temperatures, the surface is predominantly covered by CO, and the denominator of the rate expression is dominated by the term $K_{CO}[CO]$, giving rise to a negative order in CO. At temperatures above the desorption temperature of CO, oxygen tends to build up on the surface giving rise to a rate expression in which the CO term in the denominator becomes insignificant and the order in CO becomes positive. The term $(K_{O2}[O_2])^{1/2}$ is not negligible, though.

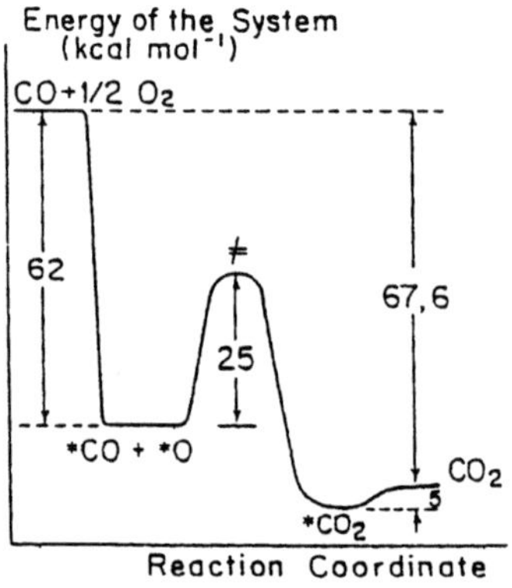

Fig. 5.8. Energetics of the CO oxidation reaction on palladium (from Engel and Ertl [44]).

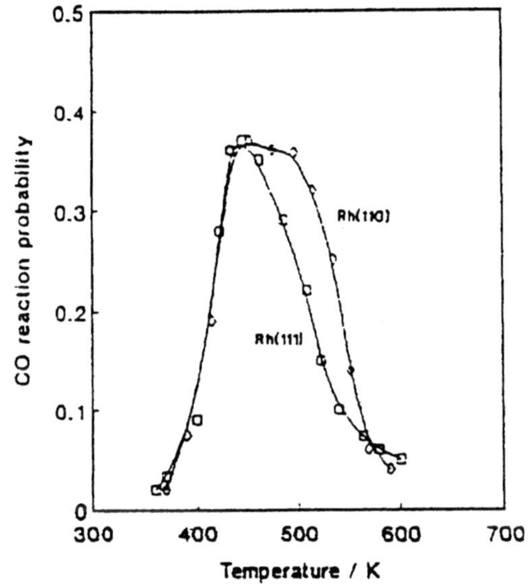

Fig. 5.9. Temperature dependence of the CO oxidation rate over rhodium surfaces (from Bowker et al. [46]).

Figure 5.9 shows the temperature dependence of the CO oxidation rate on a rhodium surface, as reported by Bowker et al [46]. This figure is essentially a Volcano Curve. It expresses that the rate of reaction maximizes when both reactants, adsorbed CO and O, are present in comparable quantities at a temperature where the activation barrier of the reaction can be overcome. At low temperatures the reaction is negatively affected by the lack of oxygen on the surface, while at higher temperatures the adsorption/desorption equilibrium of CO shifts towards the gas phase side, resulting in low coverages of CO. This type of non-Arrhenius like behaviour with temperature is very general for catalytic reactions.

5.3.6.2 The CO + NO Reaction

The precise reaction mechanism according to which the CO + NO reaction proceeds depends very much on the surface coverages of the reacting species. For low total coverage, i.e. high temperature or low pressure, the temperature-programmed experiment of Fig. 5.10 gives a good impression of the individual reaction steps: Approximately 0.25 ml of CO and NO each coadsorbed on a Rh(111) surface are heated at a rate of 5 K/s and give rise to the evolution of CO, CO_2, and N_2, whereas desorption of NO is not observed at these coverages. As we discuss later, NO dissociates at room temperature on rhodium (111). Hence, the TPRS experiment of Fig. 5.10 suggests the following scheme for the CO + NO reaction (at low coverages):

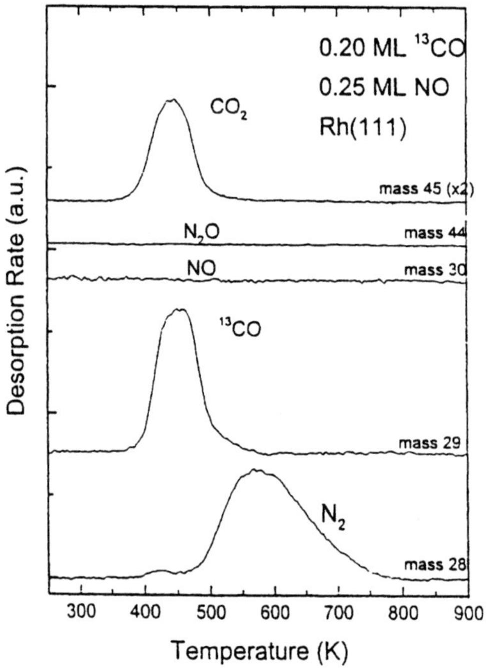

Fig. 5.10. Temperature programmed reaction spectroscopy reveals the products of the CO+NO reaction on Rh(111) (from Hopstaken et al. [51]).

$$NO + * = NO_{ads}$$

$$CO + * = CO_{ads}$$

$$NO_{ads} + * \rightarrow N_{ads} + O_{ads}$$

$$CO_{ads} + O_{ads} \rightarrow CO_2 + 2*$$

$$N_{ads} + N_{ads} \rightarrow N_2 + 2*$$

The fact that CO desorbs along with CO_2 indicates that CO adsorption is probably at equilibrium under typical reaction conditions. For NO, however, the situation is more complex.

5.3.6.3 NO Dissociation on Rh(111)

Breaking of the N–O bond by the surface of rhodium is the most essential step in the catalytic reduction of NO. Although rhodium is sufficiently reactive to achieve this (even without promoters), dissociation can nevertheless be severely impeded if the surface coverage is too high. Experiments using static secondary ion mass spectrometry have clearly demonstrated this [52].

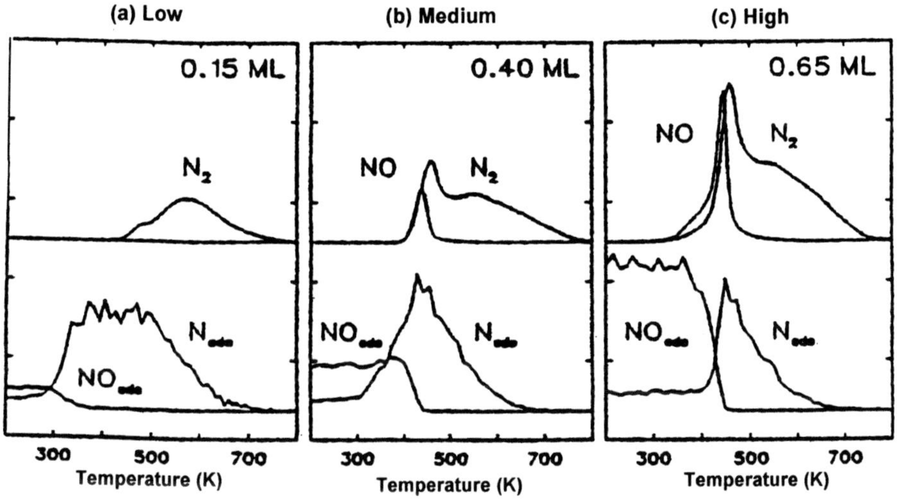

Fig. 5.11. Temperature programmed SIMS and desorption measurements of NO adsorbed on rhodium shows that NO dissociates completely around 300 K at low coverages (left), whereas dissociation at higher coverages (right) is retarded to temperatures where NO desorbs and creates empty sites on the surface (adapted from Borg et al. [52]).

Figure 5.11 shows that NO, preadsorbed at low temperature and at a low coverage of about 15%, dissociates completely at temperatures between 250 and 350 K. This is seen in the static SIMS experiment from the disappearance of the $RhNO^+$ signal, reflecting molecular NO, and the growth of the Rh_2N^+ signal, characteristic of adsorbed nitrogen atoms. Desorption of molecular N_2 starts around 440 K. The results indicate that all NO dissociates at low coverages. If, however, the surface is saturated with NO, dissociation does not set in until a significant fraction of the molecular NO has desorbed, as the SSIMS and TPD measurements of Fig. 5.11c show. This retards the dissociation of NO to temperatures around 400 K, where recombination of N-atoms and desorption of N_2 follow almost instantaneously. Hence, if the initial coverage of NO is high, desorption of a part of the NO molecules has to occur in order to create the free sites that are necessary for its dissociation.

In analogy with the results of theoretical calculations on the dissociation path of CO on rhodium by De Koster and Van Santen [53], we visualize the rupture of the N–O bond as sketched in Fig. 5.12. Starting from a threefold position, the adsorbed NO molecule bends across a rhodium atom to the next threefold site. By stretching over the central rhodium atom, the antibonding NO orbitals have a strong interaction with the Rh *d*-orbitals, and the N–O bond is efficiently weakened. The picture implies that NO requires an ensemble of at least five atoms on the (111) surface of an fcc transition metal in order to dissociate. This is in fair agreement with kinetic modelling, which indicates that three to four NO adsorption sites must be invoked to obtain realistic kinetic parameters [52].

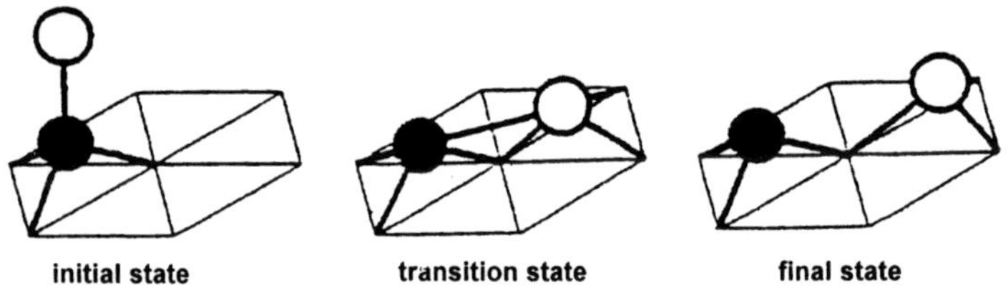

Fig. 5.12. Schematical representation of NO dissociation on an ensemble of seven atoms on the Rh(111) surface, illustrating how NO in the transition state for dissociation is thought to bend over the central rhodium atom. The NO molecule can only dissociate if an empty neighbouring site is available (adapted from De Koster and Van Santen [53]).

TABLE 5.3

Kinetic parameters of elementary reactions on Rh(111)

Reaction	E_{act} [kJ/mol]	ν [s^{-1}]	Ref.
$NO_{ads} + * \rightarrow N_{ads} + O_{ads}$	65±6	$10^{11\pm1}$	52
$NO_{ads} \rightarrow NO_g + *$	113±10	$10^{13.5\pm1}$	52
$CO_{ads} + \rightarrow CO_g + *$	159±5	$10^{15\pm1}$	51
$CO_{ads} + O_{ads} \rightarrow CO_2 + 2*$	70±5	$10^{7\pm1}$	51
$N_{ads} + N_{ads} \rightarrow N_2 + 2*$	118	$10^{10\pm1}$	52

Returning to the mechanism of the CO + NO reaction, we can now list the kinetic parameters for several of the elementary steps, see Table 5.3. As the mechanism of any catalytic reaction is inevitably a sequence of several steps, the surface science approach for studying the kinetics of elementary steps is vitally important, because parameters such as those listed in Table 5.3 form the highly desirable input for the modeling of more complex reaction mechanisms.

A second reason why kinetic parameters of elementary reaction steps are important is that, mainly through spectroscopy and computational chemistry, they provide a link between the intramolecular properties of (adsorbed) reactants and their reactivity. Statistical thermodynamics furnishes the theoretical framework for describing how equilibrium constants and reaction rate constants depend on the partition functions of vibration and rotation [50]. Thus, performing spectroscopy on adsorbed reactants and intermediates gives the input for computing equilibrium constants, while calculations on transition states of reaction pathways starting from structurally, electronically and vibrationally well characterized ground states enable the prediction of kinetic parameters.

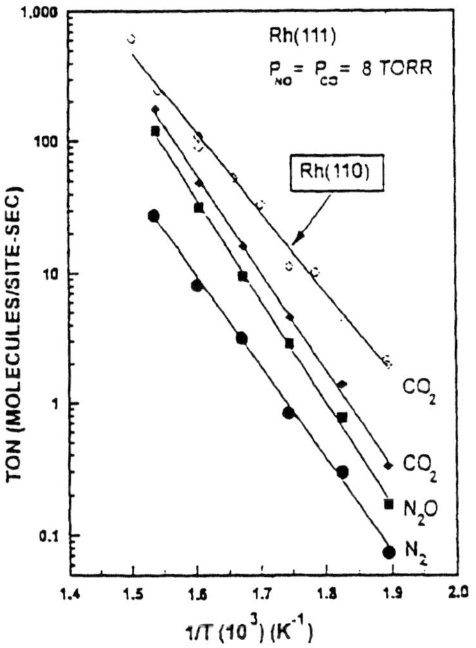

Fig. 5.13. Turnover numbers for the NO+CO reaction at realistic partial pressures on Rh(111) plotted as an Arrhenius diagram. Note that N_2O is a product as well, and the overall reaction is structure sensitive (from Peden et al. [48]).

Note, however, that kinetic descriptions as discussed here represent a serious simplification with respect to ignoring the possibility of lateral interactions. Ordering of adsorbates, even to the extent that reactants organize themselves in islands of substantial dimensions, has a profound influence on the kinetics. Exploration of these effects through Monte Carlo simulations is a field of growing importance [54,55].

5.3.6.4 The CO + NO Reaction at Higher Pressures

The NO+CO reaction is only partially described by the reactions in Table 5.3, as there should also be steps accounting for the formation of N_2O, particularly at lower reaction temperatures. Figure 5.13 shows rates of CO_2, N_2O and N_2 formation on the (111) surface of rhodium, given in the form of Arrhenius plots [48]. Comparison with similar measurements on the more open Rh(110) surface confirms that the reaction is structure sensitive. As N_2O is undesirable, it is important to know under what conditions its formation is minimized. First, the selectivity to N_2O, expressed as the ratio

$$S(N_2O) = \frac{[N_2O]}{[N_2O] + [N_2]} * 100\%$$

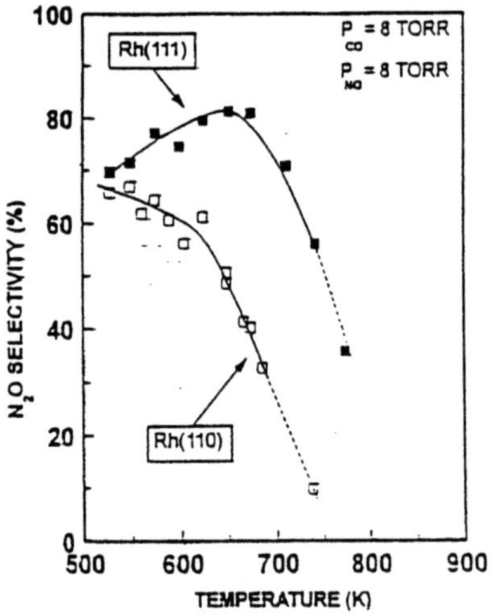

Fig. 5.14. Effect of temperature on N_2O selectivity in the NO+CO reaction on two rhodium surfaces, for NO conversions below 30% (from Peden et al. [48]).

decreases drastically at higher temperatures, see Fig. 5.14. The common operation temperatures of the three-way catalyst are 350–650°C (625–925 K). Secondly, the measurements of Figs. 5.13 and 5.14 were made at relatively low NO conversion levels. The selectivity towards N_2O is seen to decrease at higher conversion, particularly in the case of the more open Rh(110) surface. Third, the real three-way catalyst contains rhodium particles in the presence of CeO_x promoters, and these appear to suppress N_2O formation [56]. Finally, N_2O undergoes further reaction with CO to N_2 and CO_2 which is also catalyzed by rhodium.

The reaction mechanism of the NO + CO reaction is thus a very complicated one. In addition to all the reaction steps considered above, one also has to take into account that intermediates on the surface may organize themselves in islands, or in periodically ordered structures. Monte Carlo techniques are needed in order to account for these effects. As a result, we are still far away from a complete kinetic description of the CO+NO reaction. For an interesting review of the mechanism and kinetics of this reaction we refer to Zhdanov and Kasemo [49].

5.3.6.5 Reactions Involving Hydrocarbons

Hydrocarbons in the exhaust react with oxygen and with NO. Although these reactions have had much less attention than oxidation of CO and reduction of NO by CO, reactions of hydrocarbons play an important role in the overall reaction mechanism of the three-way converter, particularly because the converter is by no means a homogeneously mixed reactor, see Fig. 5.15 [57]. Hence, zones exist where e.g. ethylene and nitrogen atoms are coadsorbed on the noble

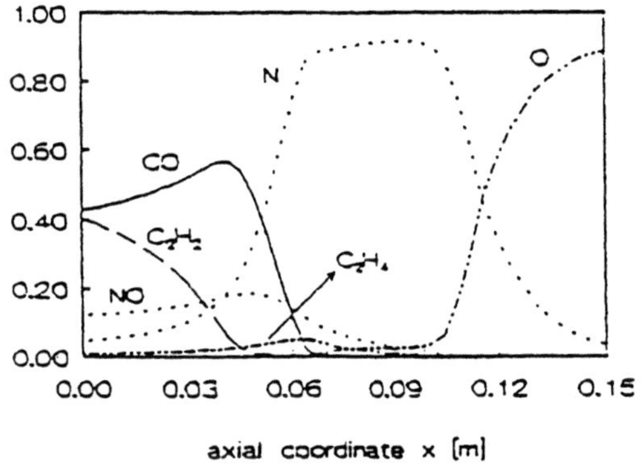

Fig. 5.15. Coverage of the noble metals in the three-way catalyst by various surface species as a function of axial position in the monolith at steady-state conditions and a gas inlet temperature of 500 K. In the middle part, where acetylene and CO have largely been converted, NO dissociation results in a high coverage of N-atoms. O-atoms become dominant when the pollutant conversion is complete (from Marin and Hoebink [57]).

metal surface of the catalyst, which might in principle lead to undesirable byproducts such as HCN.

A kinetic description of these reactions is difficult to give, due to the complicated decomposition pathways of the hydrocarbons on noble metal surfaces. The temperature programmed reaction between adsorbed ethylene and NO on rhodium in Fig. 5.16 illustrates some of the many reactions that may occur [58]. As seen before, the NO molecule starts to dissociate around room temperature. Ethylene decomposes in several steps at different temperatures as evidenced by the release of H_2 and H_2O. The formation of CO and some CO_2 between 500 and 600 K is well above the respective desorption temperatures of these gases, and suggests that the C–C bond of the hydrocarbon breaks in this temperature range and limits the rate of the oxidation on rhodium surfaces. Formation of HCN is observed as well. Note that a large reservoir of surface CN species forms at temperatures of 500 K and remains on the surface until 700–800 K, where it decomposes and is followed by the instantaneous desorption of N_2.

Gases such as HCN and CO are of course totally undesirable in principle, but will, if formed under actual three-way operation conditions, be converted to harmless products during secondary reactions on down-stream locations of the catalyst. Reduction of NO by purposely added hydrocarbons is currently investigated as a possible route to reduce NO_x emissions from the exhaust of so-called lean-burning engines (operating at air-to-fuel ratios above 14.7).

5.3.7 *Concluding Remarks*

The three-way catalyst represents remarkably successful catalytic technology. It is unique in the sense that it has to operate under a wide spectrum of conditions,

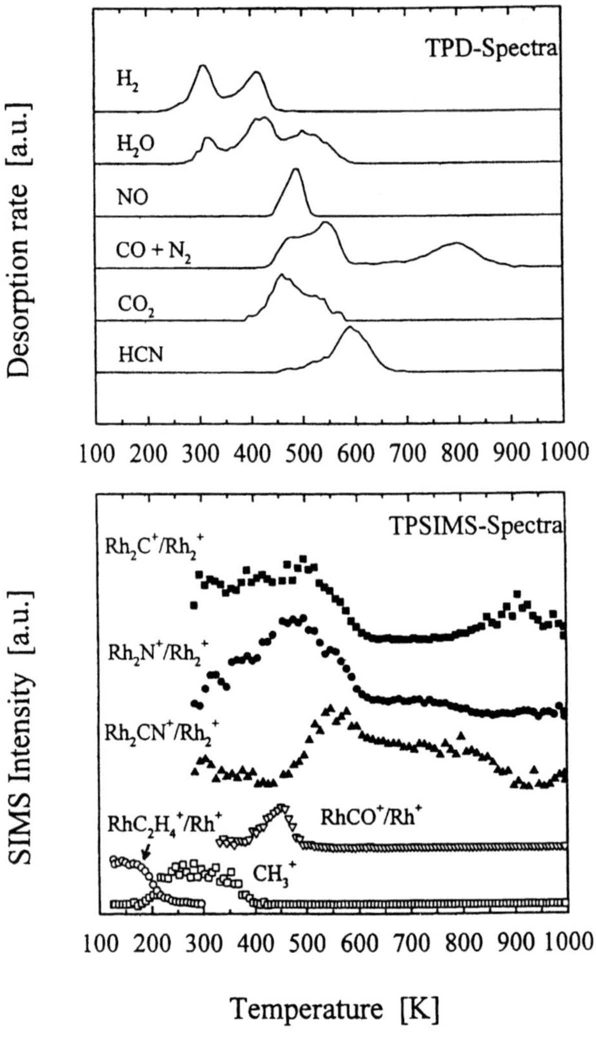

Fig. 5.16. Temperature programmed reaction spectroscopy of ethylene and NO coadsorbed on a rhodium surface reveals the many reactions that are possible. Note the formation of CO and CN species on the surface as visible in SIMS, and the formation of HCN as a gas phase product. The stability of the CN species is the reason that desorption of N_2 occurs at very high temperatures only (from Van Hardeveld et al. [58]).

depending on type of use, personal driving style, climatological conditions, etc. This in contrast to the usual situation in industry, where conditions are selected such that the catalytic process runs optimally.

The three-way catalyst is 'overdesigned' in order to meet specifications after several years of usage. If these specifications become stricter, the TWC system can be further improved to meet these requirements. For example, the highly demanding Californian standards for ultra-low emission vehicles (hc: 0.04 g/mile; CO: 1.7 g/mile; NOx: 0.2 g/mile) have been amply achieved by a TWC that has a provision for preheating before start-up [59]. Further reduction of emissions will not be easy without major innovations in the catalyst, see Heck and Farrauto [60] for an overview of new developments.

Although TWC technology is now standard in the developed countries, effective implementation on a global scale is another story. World-wide annual production is more than 50 million cars, which, ideally should all be equipped with catalytic converters (which is not yet the case). This places great demands on the availability of noble metals, where rhodium is the critical and hence most expensive one. Regeneration of used catalysts is certainly possible but again needs to be implemented on a world-wide scale.

Note also that three-way catalysts can only be applied on gasoline-fuelled engines, and not on diesels. The major problems in controlling the exhaust of diesel engines are how to oxidize soot particles and how to reduce NO in oxidising exhaust streams. Control systems based on filters and catalysts have been developed, but are applicable to new diesel engines only, implying that a large fleet of long-lived trucks and buses will continue to emit soot and/or NO_x in the atmosphere. It will take quite some time and a lot of effort before pollution by automotive exhaust can eventually be considered as a problem of the past.

5.4 SELECTIVE CATALYTIC REDUCTION OF NO BY NH_3

5.4.1 Introduction

In this section, the selective catalytic reduction of NO_x with NH_3 is emphasized. This reaction is the basis of the selective catalytic reduction (SCR) process that removes NO_x from oxygen-rich emissions that occur in power plants, waste incinerators, and gas turbines.

An extensive review on NO_x removal has been published by Bosch and Janssen [61], which may serve as a quick introductory guide and which gives a comprehensive picture of SCR up to 1986.

Bond and Flamerz Tahir [62] reviewed the literature on the preparation, structure and catalytic properties of vanadium oxide monolayer catalysts. This catalyst plays an essential role in the selective catalytic reduction of NO_x.

Japan was the first country to do this, starting with SCR for the removal of NO_x from gas, coal and oil-fired power plants [63]. Utility companies in Japan introduced the catalytic SCR process as early as 1972. At present, many countries utilize SCR with Germany and Japan being the leading countries. Although priority is currently given to combustion modifications for NO_x control in European countries, extended operation of catalytic SCR is expected in the near future.

5.4.2 SCR catalysts

Both titania and titania/silica supported vanadia, molybdena, tungsta and chromia have been applied as SCR catalysts. Low-temperature and high temperature catalysts have been developed. The vanadia on titania catalysts have received most attention.

A variety of reducing agents such as CH_4, CO, H_2, and NH_3 has been used. Ammonia is widely used as reductant in the selective catalytic reduction.

Flue gas produced by a coal-fired burner is passed to the SCR reactor together with ammonia. The catalytic SCR reactor is usually placed between the economiser and the air preheater (see Fig. 5.17).

Downstream the air preheater fly ash is collected by means of the electrostatic precipitators. Via a heat exchanger and flue gas desulphurization, the flue gas is passed to the stack. The SCR catalyst used in this way is in the so-called high dust mode. This means that the resistance of the catalyst against attrition should be very high. That is one of the reasons for using titania as a support.

The overall reactions over highly selective supported vanadia catalysts may be described by the stoichiometry

$$4NO + 4NH_3 + O_2 = 4N_2 + 6H_2O$$

$$6NO_2 + 8NH_3 = 7N_2 + 12H_2O$$

The first overall reaction predominates if the flue gas contains more NO. The high activity of NH_3 combined with the enhanced reaction rate in the presence of oxygen, makes NH_3 the preferred reducing agent. Apart from this main reaction, other reactions are feasible. The variety of reactions occurring is given in Table 5.4.

One of the simplest ways of preparing SCR catalysts is wet impregnation of the carrier with an acidified aqueous solution of ammonium metavanadate, followed by drying and calcination.

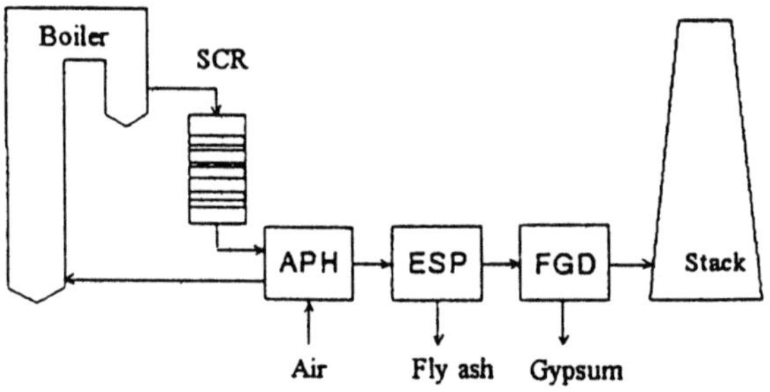

Fig. 5.17. A scheme of a power plant equipped with both selective catalytic reduction (SCR) and flue gas desulfurization (FGD). APH: air preheater; ESP: electrostatic precipitator.

TABLE 5.4

Reaction heats of various reactions (673 K) [64]

No.	Reaction		$-\Delta H$(kJ)
1	$4NO + 4NH_3 + O_2$	$= 4N_2 + 6H_2O$	1625
2	$4NO + 4NH_3 + 3O_2$	$= 4N_2O + 6H_2O$	1304
3	$6NO + 4NH_3$	$= 5N_2 + 6H_2O$	1806
4	$4NH_3 + 3O_2$	$= 2N_2 + 6H_2O$	1263
5	$2NH_3 + 2O_2$	$= N_2O + 6H_2O$	552
6	$4NH_3 + 5O_2$	$= 4NO + 6H_2O$	907

It is well known that the activity of SCR catalysts depends on the amount of vanadia present on the support [61]. Dispersion of vanadia is necessary in order to increase the number of catalytically active species. For instance, four layers of vanadia on titania exhibit an increase of two orders of magnitude in reaction rate with respect to a monolayer. Titania shows a strong interaction with vanadia. In order to decrease the influence of titania, silica is added to the support. It was found that vanadia on silica/titania catalysts are far more active catalysts than vanadia on silica and less active than vanadia on titania materials (see Table 5.5).

In the following paragraphs issues such as the mechanism and kinetics of the reaction of NO, NH_3 and O_2 over vanadium-based catalysts are discussed.

5.4.3 *Species at the Catalyst Surface*

Spectroscopic techniques are used for the detection of all sorts of species present at the surface. However, so far no analysis has been done on the working catalyst.

The infrared spectrum of 4 wt.-% vanadia on titania shows a band at 940 cm^{-1}, which may be assigned to the V=O stretching [66]. Going from 4 wt.-% to 15 wt.-% vanadia in the catalyst, this peak shifts to 1020 cm^{-1}. This peak was assigned to segregated vanadia crystallites [66]. FTIR was used to study the type and distribution of acid Lewis and Brønsted sites of TiO_2/SiO_2 and TiO_2 supports with pyridine as a probe molecule. TiO_2/SiO_2 supports exhibit both Lewis and Brønsted acidity [67] whereas TiO_2 contains mainly Lewis acid sites. These are formed by condensation of adjacent hydroxyl groups. Raman features were found at 1010, 990, and 852 cm^{-1} for vanadia on titania which were attributed to octahedrally co-ordinated surface species. Dehydrated samples accommodate isolated and surface vanadate and polymeric metavanadate species (Fig. 5.18).

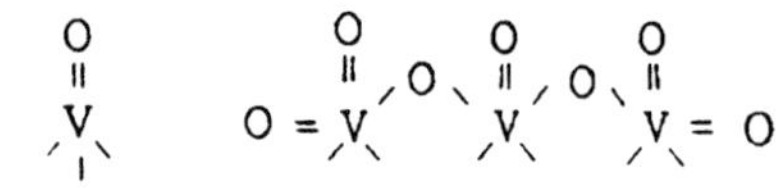

Fig. 5.18. Monomeric vanadyl (left) and polymeric vanadates (right). Adapted from [65].

It has been shown that titania-supported vanadia materials comprise a distribution of monomeric vanadyl, polymeric vanadates and crystalline vanadia, the amount of which is dependent on the vanadia content.

Evidence for polymeric species of the metavanadate type in submonolayer vanadia on titania catalysts is obtained by the appearance of a $\nu_{V=O}$ stretching mode at 870 cm^{-1} [68]. The stretching mode of V=O for isolated vanadyls is found at 1030 cm^{-1}.

Lietti and Forzatti [69] have shown by means of transient techniques such as TPD, TPSR, TPR and SSR (steady-state reaction experiments) that isolated vanadyls and polymeric metavanadate species are present on the surface of vanadia on titania catalysts with V_2O_5 loadings of up to 3.56 wt.-%. Polyvanadate species are more reactive than isolated vanadyls due to the presence of more weakly bonded oxygen atoms.

Went et al. [70] used *in situ* laser Raman spectroscopy and temperature-programmed reduction and temperature-programmed oxidation to establish the presence of monomeric vanadyls, polymeric vanadates and crystallites of V_2O_5 in titania supported catalysts. The distribution of vanadia species as a function of V_2O_5 loading was determined (Fig. 5.19). At low V_2O_5 concentrations a monomeric vanadyl species is present at 1030 cm^{-1} assigned to the V=O stretching mode. At increasing V_2O_5 content in the catalyst the peak at 1030 cm^{-1} broadens significantly and the intensity of the bands located in the region 700–1000 cm^{-1} assigned to polymeric vanadates increases. The band at 840 cm^{-1} is assigned to

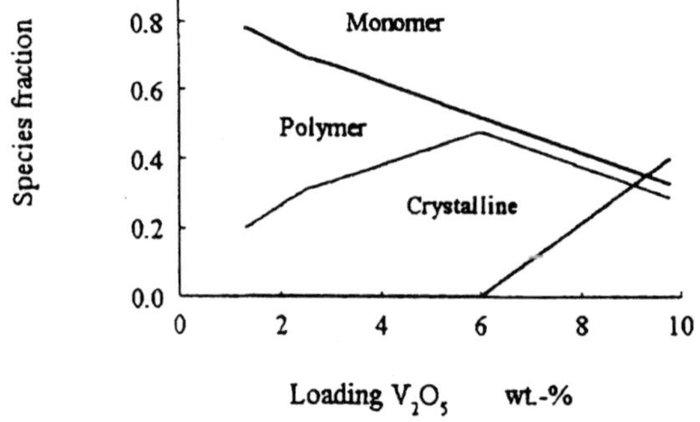

Fig. 5.19. The distribution of monomeric, polymeric, and crystalline vanadia species as a function of the weight loading of vanadia. Adapted from [70].

V–O–V bending vibrations. The average polymer size varies from two to three monomer units depending on the vanadia load. For vanadia amounts of 6 wt.-% and higher, a narrow band at 960 cm^{-1} appears [70].

The Raman feature located at 1030 cm^{-1} was assigned to V=O in V_2O_5 on the TiO_2/SiO_2 catalyst. For lower loadings of vanadia an intensity in the range 920–950 cm^{-1} appears, which is attributed to a tetrahedral metavanadate structure. This observation was supported by means of 51_V NMR measurements. A six-fold co-ordinated species is found at higher vanadium loadings. Small amounts of crystallites of V_2O_5 were present for loadings of 4 wt.-% and above, which appear from the sharp feature at 991–996 cm^{-1} [70].

5.4.4 Kinetics

Supported vanadia catalysts and physical mixtures of V_2O_5 and TiO_2 are described, regarding mechanisms and kinetics. Kinetic data for a variety of supported vanadia catalysts are given in Table 5.5 [71]. Catalysts with TiO_2 as support have the highest activity.

V_2O_5 catalysts are structure-sensitive for various reactions. The V=O group located in the (010) plane plays an essential role in many reactions. Another aspect of these catalysts is the nature of the active site that leads to the SCR reaction or the oxidation of ammonia when ammonia is used as reductant.

Amiridis et al. [99] studied the relationship of the turnover frequency and vanadia loading of the catalyst. Curves 1 and 2 in Fig. 5.20 go through a maximum at full monolayer coverage. Adding SO_2 to the reactants resulted in a

TABLE 5.5

Physicochemical properties of supported vanadia catalysts [71]

Support	S_{BET} ($m^2\ g^{-1}$)	Loading (mmol V m^{-2})	Monolayer fraction	E_{act}[1] (kJ mol^{-1})	ln(TOF)[2]
TiO_2	50	4.14	0.28	65.5	970
SiO_2/TiO_2	108	3	0.20	65.5	500
ZrO_2/TiO_2	42	2.17	0.10	61.7	420
ZrO_2	37	5.27	0.35	61.5	190
Al_2O_3	81	1	0.07	32.9	63
SiO_2	187	3.65	0.24	–	16
Al_2O_3/TiO_2	58	3.59	0.24	65.1	4

[1]Apparent activation energy.
[2]Turnover frequencies in mmol(NO) $mol(V)^{-1}\ ks^{-1}$ at 473 K.

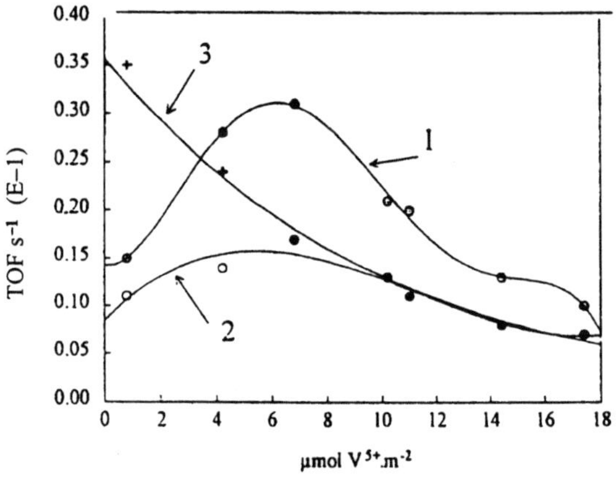

Fig. 5.20. Effect of surface vanadia coverage on SCR activity of V_2O_5/TiO_2 catalysts. [NO] = [NH_3] = 400 ppm; [O_2] = 4 vol.-%; Curve 2 and 3: [H_2O] = 8 vol.-%; curve 3: [SO_2] = 800 ppm. Adapted from [99].

sharp decrease of the activity of the catalyst. The effect of water on the activity was attributed to competitive adsorption of water and ammonia. Upon addition of SO_2 sulphate species were assigned to be responsible for lowering the reaction rate at high vanadia loadings. They concluded from this study that a Brønsted site was the active site.

Ozkan et al. [72] studied the morphological aspects of unsupported vanadia catalysts. Two types of vanadia were prepared. The preparation procedures described resulted in thick particles (V_2O_5(a)) and thin, sheet-like particles (decomposition of ammonium metavanadate V_2O_5(d)). V_2O_5(a) contains relatively more (010) planes than V_2O_5(d). The selectivity to N_2 for V_2O_5(d) is higher than that for V_2O_5(a) whereas over the latter more N_2O is formed. V=O in the (010) planes promote the ammonia oxidation into N_2O more than the reduction of NO into N_2 [72].

The role of ammonia oxidation was studied more comprehensively [73]. From the experimental results it was suggested that ammonia adsorption takes place at two types of sites located on the (010) and (100) planes. V–O–V groups were found to be responsible for the SCR reaction whereas the V=O sites promote the direct oxidation of ammonia to NO and the formation of N_2O from the reaction of NO and ammonia. However, it is stressed here that these studies were carried out with unsupported materials.

Janssen [74] carried out TPR experiments of MoO_3 and V_2O_5 by ammonia and hydrogen. It was concluded that ammonia reduces the oxides forming almost exclusively water and nitrogen. Both catalysts are reduced below 873 K. Above this temperature ammonia decomposes into N_2 and H_2.

The reaction rate of the reaction of NH_3, NO and O_2 over vanadia on titania catalysts at 365 K was 1.9 10^{-18} (mol g^{-1} s^{-1}) and the activation energy 37 kJ mol^{-1}

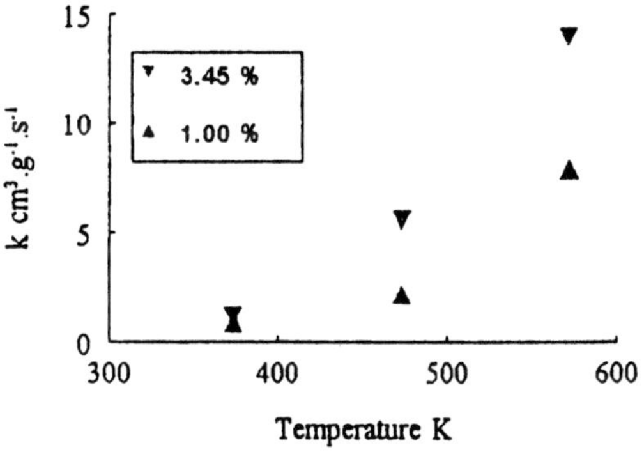

Fig. 5.21. First-order rate constants, k_{NO} cm^3 g^{-1} s^{-1}, as a function of temperature for two vanadia on titania catalysts. $[NO]_i = [NH_3]_i = 1000$ ppm, $[O_2] = 2\%$, balance N_2, GHSV = 15 000. Adapted from [76].

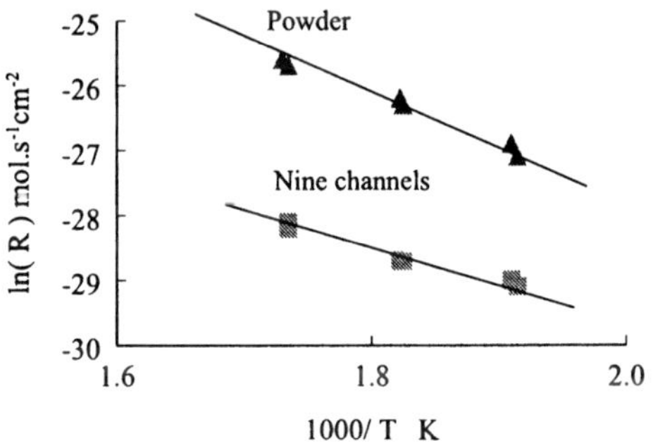

Fig. 5.22. Measured (points) and predicted (lines) NO conversion rates over a vanadia on titania catalyst. $[NO]_i = [NH_3]_i = 1000$ ppm; $[O_2] = 4\%$.-vol. balance nitrogen. Adapted from [77].

[75]. These data were obtained by using ^{13}NO, a positron emitter, at very low concentrations (5 10^{-9} ppm). Moreover, it was confirmed that the nitrogen atom of ammonia combines with the nitrogen atom of nitric oxide.

Buzanowski and Yang [76] studied the kinetics over both unpoisoned V_2O_5 on TiO_2 and alkali poison-doped catalysts. The results for two unpoisoned vanadia on titania catalysts are given in Fig. 5.21. It was observed that the higher the vanadia load the higher the conversion.

Beeckman and Hegedus [77] found a preexponential factor k_0 of 8.64×10^3 cm s^{-1} and an activation energy of 80 kJ mol^{-1}. Arrhenius plots of NO conversion rates on powdered material and nine-channel monolith are shown in Fig. 5.22. Figure 5.22 clearly shows the strong effect of pore diffusion on the observed rates for the nine-channel monolith. The technique of preparing the support influences the activity of the catalyst. Over vanadia on laser-synthesized titania catalysts, large amounts of N_2O were produced [78]. The amount of N_2O increased for vanadia loadings higher than 4 wt.-%. The overall reaction was

$$4NO + 4NH_3 + 3O_2 = 4N_2O + 6H_2O$$

Kotter et al. [79] prepared V and Ti-containing catalysts by dissolving appropriate amounts of ammonium vanadate in a titanium oxychloride solution followed by hydrolysis and evaporation. The 20 wt.-% V_2O_5 catalyst thus formed showed a decrease in BET specific surface area from 85 to 20 $m^2\ g^{-1}$ after heating the sample at 713 K for 100 h.

It was suggested that crystalline V_2O_5 was formed during the heat treatment. Performing the SCR reaction over this pre-treated catalyst, large amounts of N_2O were formed which, indeed, indicates the presence of crystalline V_2O_5 [64]. The catalytic activity of the catalyst could be restored by washing the catalyst with ammonia [79].

Lintz and Turek [80] determined the intrinsic rates of the following three linearly independent overall reactions over 20 wt.-% V_2O_5 on TiO_2 in a recirculation system for temperatures ranging from 423 K to 523 K:

The catalyst consisted of an alumina plate coated with V_2O_5/TiO_2 mixtures on both sides. The mass specific reaction rate decreased with increasing thickness of the catalyst layer. Intrinsic kinetics was obtained for catalyst layers with a thickness lower than 20 mm. A power law dependence of the main reaction rate on the concentration of NO was found in the absence of water in the feed. Water reduces the activity slightly at temperatures lower than 663 K; however, it increases the selectivity with respect to nitrogen.

The authors presented the rate expression given as:

$$r_{m,1} = k_{m,1}[NO]^b \left(\frac{[NH]_3}{1 + a[NH_3]} \right) \left(\frac{1}{1 + c[H_2O]} \right)$$

The main reaction rate is valid for NH_3 and NO concentrations ranging from 10^{-6} to 10^{-4} mol l^{-1} and for water concentrations ranging from zero to 10 vol.-% [80]. The values for a, b, c and $k_{m,1}$ are given in Table 5.6 [80]. The influence of water on the reactivity of vanadia on titania is hardly described.

Turco et al. [81] studied the influence of water on the kinetics of the SCR reaction over a vanadia on titania catalyst in more detail and found that water inhibits the reaction. The influence of water on the SCR reaction is largest at low temperatures (523–573 K) and low at 623 K. They considered a power rate law

$$r = k[NO]^a[NH_3]^b[H_2O]^c$$

The reaction orders a, b and c are for three temperatures summarized in Table 5.7. It can be seen that the reaction order of ammonia varies with temperature.

From the results it was concluded that the sites for ammonia are able to irreversibly adsorb water. The irreversible adsorption of water follows a Temkin isotherm [81].

TABLE 5.6

Kinetic parameters for the SCR reaction on V_2O_5 on TiO_2 on alumina plate [80]

Temperature (K)	$k_{m,1}$ (l^{1-b} mol^{-b} g^{-1} s^{-1})	a (l mol^{-1})	b (–)	c (l mol^{-1})
623	11.7 10^3	1.93 10^5	0.7	271
523	7.03 10^3	21.7 10^5	0.59	n.d

TABLE 5.7

Reaction orders of NO: *a*; NH_3: *b*; and, H_2O: *c*. Reaction conditions are: [NO] = 100–1000 ppm; [NH_3] = 100–3000 ppm; [H_2O] = 200–3000 ppm; [O_2] = 2.7 vol.-% [81]

Reaction order	523 K	573 K	623 K
a	0.82	0.75	0.80
b	0.10	0.12	0.21
c	–0.14	–0.14	–0.10

Topsøe et al. [82] found that water adsorbs on the surface more weakly than ammonia.

Water hydroxylates the surface producing more Brønsted acid sites. Conclusively, water slightly inhibits the catalytic activity at temperatures lower than 663 K and significantly blocks the formation of N_2O at the surface of the catalyst [82]. These results are comparable with those of Odenbrand et al. [83]. The presence of 1 vol.-% of water suppressed the formation of N_2O below 775 K, whereas the maximum of the NO conversion shifted to higher temperatures.

5.4.5 Mechanisms

Three types of vanadium-containing species are present at the surface of the vanadia on titania catalysts. Monomeric vanadyl, polymeric vanadates, and crystalline vanadia depending on the vanadia loading (see Fig. 5.19). Moreover, Brønsted acid sites and Lewis sites are present at the surface of vanadia on titania catalysts. All species are needed to explain the mechanism of the SCR reaction over this type of catalyst. The active sites are related to previously mentioned vanadia/vanadium species. A variety of active sites were proposed such as: two adjacent V^{5+}=O groups or co-ordinated vanadyl centres [81,84–86]; V^{5+}=O and

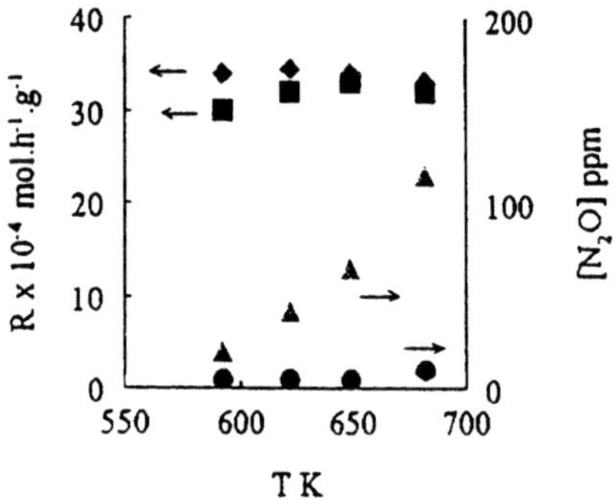

Fig. 5.23. The rate of NO reduction and the amount of N_2O formed as a function of temperature. ◆, and ▲, and ■ and ● correspond to experiments without water and with water, respectively. Adapted from [82].

surface hydroxyl groups [87]; Brønsted sites such as vanadium hydroxyl groups [88]; an oxygen vacancy and hydroxyl groups [89].

Several mechanisms or reaction schemes were proposed by various researchers based on their experimental results. In this paragraph a summary of these mechanisms is given.

The elucidation of the reaction mechanism of the SCR reaction has been carried out using a variety of techniques. Transient studies with isotopically (oxygen-18 and nitrogen-15) labeled molecules have been performed [86,90]. Spectroscopic studies of the working catalysts were performed by Went et al. [90] and Topsøe [91] using laser Raman spectroscopy and FTIR, respectively.

Two types of mechanisms are proposed in the literature: the Langmuir–Hinshelwood and the Eley–Rideal mechanisms. Takagi et al. [92] presented a study of the SCR reaction over V_2O_5 on γ-Al_2O_3. It was suggested that the reaction proceeds via the adsorbed species of NO_2 and NH_4^+ through a Langmuir–Hinshelwood mechanism. It has been shown by IR that ammonia was adsorbed as both NH_4^+ (1410 cm^{-1}) and physisorbed ammonia on alumina (1610, 1275 cm^{-1}). Using FTIR it was shown that NO adsorbed as NO_2.

However, several research groups proposed the Eley–Rideal mechanism for describing the mechanism of the SCR reaction [86–88,90].

Miyamoto et al. [90] proposed a dual-site Eley–Rideal mechanism which is a better description of the mechanism. In this mechanism ammonia is strongly adsorbed as NH_4^+. NO from the gas phase reacts with the adsorbed ammonia species to form N_2, H_2O, and V–OH. The latter is oxidized by oxygen from the gas phase or lattice oxygen to V^{5+}=O. The overall reaction equations are

$$NO + NH_3 + V = O \rightarrow N_2 + H_2O + V\text{–}OH$$

$$2V\text{–}OH + O \rightarrow 2V = O + H_2O$$

The mechanism of the selective reduction of nitric oxide with ammonia in the presence of both labeled oxygen and ammonia over a series of catalysts consisting of unsupported V_2O_5, V_2O_5 on TiO_2, V_2O_5 on SiO_2/Al_2O_3, and V_2O_5 on Al_2O_3 has been reported by Janssen et al. [86].

The experimental results confirm that lattice oxygen participates in the reaction. Water is formed at two sites; the first part comes from the reaction of gaseous nitric oxide and adsorbed ammonia via an Eley–Rideal mechanism; the other part comes from a surface dehydration process. Two adjacent V^{5+}=O groups were supposed to be the active site. From the work of Lietti and Forzatti [69] it appeared that the oxygen atoms of monomeric species are less reactive than those of polymeric metavanadate species.

NO oxidizes the prereduced catalyst to the same extent as oxygen, but at a much lower rate. During the reaction, most of the N_2 and N_2O are formed from one nitrogen atom from nitric oxide and from one nitrogen atom from the ammonia, irrespective of whether oxygen is present or absent. No experimental evidence for isotopic scrambling of gaseous and lattice oxygen at temperatures ranging from 575 to 675 K was observed. Molecular oxygen oxidizes V–OH groups. Janssen et al. [86] have suggested that three combinations of the groups V=O and V–OH are present on the surface of the catalyst at the start of the reaction. These species are in equilibrium due to the mobility of hydrogen atoms of the V–OH groups.

Figure 5.24 summarizes the reaction scheme as suggested by Janssen et al. [86]. Amide-like species NH_2 have also been shown present at the surface [84,86,93]. The formation of N_2O is also described in detail [83,86,87].

It has been suggested by Kantcheva et al. [87] that two molecules of NO adsorb on a doublet oxygen vacancy with simultaneous release of N_2 and N_2O into the gas phase. Again, the oxygen vacancies are reoxidized more rapidly by O_2 than by NO. This explains the enhancing effect of oxygen on the rate of the overall reaction of NO, NH_3 and O_2.

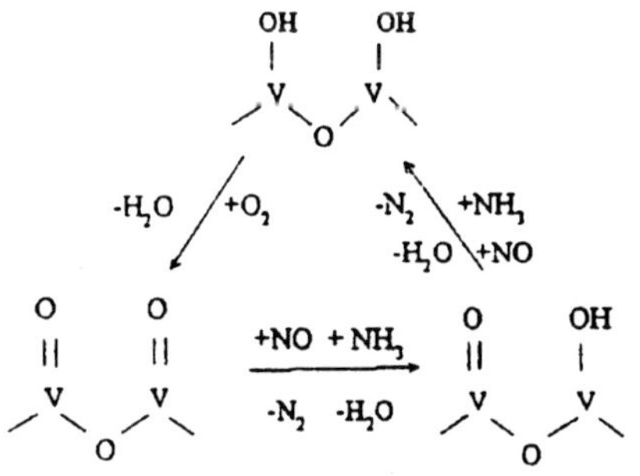

Fig. 5.24. Summarized reaction mechanism of the selective reduction of NO with NH_3 in the presence of oxygen. Adapted from [86].

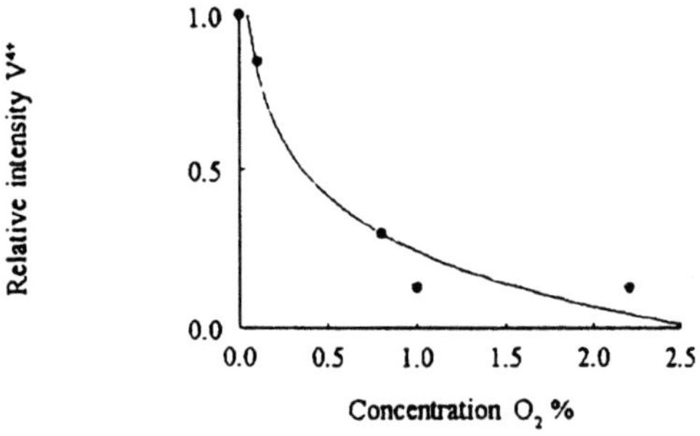

Fig. 5.25. The relative ESR signal intensity of V^{4+} in vanadia on titania catalyst as a function of the gas phase oxygen concentration. Adapted from [93].

During the SCR reaction V^{4+} species are present on the surface of the catalyst. Biffar et al. [93] showed that the concentration of V^{4+} in the catalyst is a function of the oxygen concentration in the gas phase (Fig. 5.25).

Bjorklund et al. [94] found that vanadia loadings up to 10–15 wt.-% did not influence the electrical conductance in TiO_2/SiO_2 supported vanadia catalysts. Above 15 wt.-% the electrical conductivity increases rapidly which may be ascribed to the presence of V^{4+}. Admitting NO or NH_3 or both to the catalyst, the conductivity increased by a factor of 3–4. Increasing the oxygen concentration from 0 to 1.5 vol.-% resulted in a decrease in the conductivity, which is in agreement with the results of Biffar et al. [93].

Gasior et al. [95] stated that the amount of N_2 formed is not a measure of the concentration of V=O groups at the surface. In the absence of oxygen the following reactions occur:

$$V^{5+}\text{–}OH + NH_3 \rightarrow V^{5+}\text{–}ONH_4$$

$$2V^{5+}\text{–}ONH_4 + V^{5+}\text{–}O\text{–}V^{5+} + 2NO + 2e^- \rightarrow 2N_2 + 3H_2O + 2V^{5+}\text{–}OH + V^{4+} \square V^{4+}$$

In the presence of oxygen the vacancy becomes oxidized according to:

$$V^{4+} \text{–}\square\text{–} V^{4+} + \tfrac{1}{2}O_2 \rightarrow V^{5+}\text{–}O\text{–}V^{5+} + 2e^-$$

Odenbrand et al. [83] extended Gasior's model for the formation of N_2O. The active site was suggested to be an oxygen vacancy with one vanadyl group.

$$4V{=}O + 2V^{4+} \text{–}\square\text{–} V^{4+} + 4NH_3 \rightarrow 4V\text{–}NH_2{}^+ \; 4V^{5+}\text{–}OH$$

$$4V\text{–}NH_2 + 4NO \rightarrow 4N_2O + 4H_2O + 2V^{4+} \square V^{4+}$$

Again, the oxygen vacancy is oxidized. Further oxidation of the amido group at higher temperatures leads to

$$4V\text{–}NH_2 + 3O_2 \rightarrow 2N_2O + 4H_2O + 2V^{4+} \square V^{4+}$$

However, Bjorklund et al. [94] were not able to distinguish between the two mechanisms proposed by Janssen et al. [86] and Gasior et al. [95] by using electrical conductance measurements. If ammonia adsorbs on vanadia at room temperature two bands at 1610 cm^{-1} and at 1420 cm^{-1} occurred in the infrared spectrum. These are assigned to coordinatively held ammonia and NH_4^+, respectively. Carrying out this experiment at 623 K only the band at 1610 cm^{-1} was left indicating that ammonia is coordinatively bound to the surface as proposed by Janssen et al. [86].

Chen and Yang [88] concluded from their studies that the active sites for SCR reaction over sulphated titania are Brønsted sites and the mechanism is Eley–Rideal. 52% of the chemisorbed ammonia was active in the SCR reaction while 48% of the chemisorbed ammonia was inactive.

Ciambelli et al. [75] proposed a model for the vanadium oxide monolayer on titania. At loadings of vanadia below 6 wt.-%, a V^{4+} species prevails. At higher loadings the number of V^{5+} species increases, and thus the number of Brønsted sites grows.

Recently, Schneider et al. [96] characterized the active surface species of vanadia on titania by means of diffuse reflectance FTIR. Their results supported the mechanism proposed by Janssen et al. [86]. From the signal intensities at 1435 and 1660 cm^{-1} it was concluded that ammonia adsorbs on Brønsted sites [96]. The 1435 cm^{-1} band was assigned to the n_4(F) bending mode of ammonia adsorbed on Brønsted sites. The corresponding n_2(E) feature was observed as a shoulder at 1660 cm^{-1}.

Ammonia adsorption on Lewis sites is stronger than that on Brønsted sites [97]. *In situ* infrared spectroscopy has been used to monitor surface coverages by various species under reaction conditions. Temperature programmed desorption shows that no NO decomposition occurs in the temperature range 100–600 K. By means of *in situ* FTIR spectroscopy it was observed that the fractional surface coverages by ammonia on the Brønsted and Lewis acid sites were 0.26 and 0.39, respectively, at 573 K. No adsorption of NO was found. Moreover, it was stated that water does not block the sites for ammonia adsorption.

Odenbrand et al. [83] observed that the selectivity toward nitrogen increases upon addition of water, additionally the formation of nitrous oxide is suppressed at temperatures below 775 K.

A kinetic model was proposed by Turco et al. [81] with a nitrosamide as an intermediate compound and vanadyl groups were supposed to be the active sites.

Went et al. [70] used TPD and laser Raman spectroscopy to determine the structure of the catalyst and of the adsorbed species. It was found that the specific activity of polymeric vanadate species was 10 times greater than that of monomeric vanadyls at 500 K. Monomeric species produce N_2 both in the presence and in the absence of oxygen whereas polymeric species produce both N_2 and N_2O. The selectivity towards N_2O increases with increasing O_2 concentration in the feed.

Duffy et al. [89] carried out the reaction of ^{15}NO and $^{14}NH_3$ over V_2O_5 on TiO_2 and over a-Cr_2O_3. The major product over both catalysts is $^{14}N^{15}N$. A simple mechanism was proposed

$$2V\square + V\text{–}OH + {}^{15}NO \rightarrow V{=}{}^{15}NH + 2V{=}O$$

The adsorbed NH group could be hydrolyzed into $^{15}NH_3$ and a V=O group. By-products were found: over vanadia on titania $^{14}N^{15}NO$, and over chromia $^{14}N_2$, $^{15}N_2$ and $^{15}N_2O$.

In situ FTIR has been used to study the reaction steps and intermediates in vanadia on titania catalysts [91]. The changes in concentration of surface sites and adsorbed species of the working catalyst were also obtained. A relationship between the concentration of the Brønsted acid sites and the NO_x conversion was found. Adsorption of ammonia results in two ammonia species. Bands at 3020, 2810, 1670, and 1420 cm^{-1} are assigned to adsorbed NH_4^+ species, whereas intensities at 3364, 3334, 3256, 3170 and 1600 cm^{-1} are responsible for coordinated NH3. A mechanism was suggested in which initially, ammonia is adsorbed on V^{5+}=O forming an activated species. Then NO reacts with this species forming V^{4+}–OH, N_2 and H_2O. Subsequently the vanadia species is reoxidized by NO or O_2 in order to obtain V^{5+}=O.

Ramis et al. concluded from their studies that Brønsted activity is not needed for SCR [98]. They identified NH_3 co-ordinates over Lewis acid sites followed by hydrogen abstraction resulting in adsorbed NH_2 species. At 523 K only NH_3 on Lewis sites were detected because protonated ammonia is thermally less stable than co-ordinated species.

According to Ramis et al. [98] ammonia adsorbs as ammonium ions by protonation over Brønsted sites and co-ordinated ammonia over Lewis acid sites.

So far, no exclusive reaction mechanism is found for the SCR reaction over vanadia on titania catalysts which can explain all experimental observations. The Eley–Rideal mechanism, however, is favoured by many investigators. Ammonia is adsorbed on the surface on Brønsted or Lewis sites, as NH_4^+ ion or as an NH_2 species, respectively. Then NO from the gas phase reacts with the ammonia species forming nitrogen and water leaving oxygen vacancies behind. Thus, lattice oxygen is involved in the reaction mechanism. These vacancies are then occupied by oxygen from the gas phase. More information about the vanadium-containing species at the surface of vanadia on titania catalyst. obtained in the last decade. has helped to explain the reaction mechanism. Isolated vanadyl and polymeric vanadyl species were detected by means of spectroscopic techniques. Polymeric vanadyls are more reactive than isolated vanadyls. The composition of the products of the SCR reaction probably depends on the amount of surface species present at the surface. This issue has still to be clarified.

5.5 SELECTIVE OXIDATION

5.5.1 *Propene Oxidation to Acrolein*

5.5.1.1 *Introduction*

Acrolein is the simplest unsaturated aldehyde and is highly reactive due to conjugation of the carbonyl group with the vinyl group. It is produced as refined acrolein (approximately 120 ktonne per year) and as crude acrolein (approximately 1,400 ktonnes per year). Crude acrolein forms the feedstock for about 85% of worldwide acrylic acid production via further catalytic oxidation [104].

Propene oxidation to acrolein is carried out commercially over a range of bismuth molybdate catalysts to which are added 3–4 additional metal oxides to boost the activity. The final catalysts are mixtures of binary and ternary oxides and some solid solutions. One feature is the ability of lattice oxygen to transfer readily at the reaction temperature between the multiple phases that make up this catalyst and to the reacting propene. Another key feature is that the initial point of activation of the propene is one of the methyl C–H bonds with the production of a surface allyl intermediate, hence the term allylic oxidation.

This section describes

- the composition of multicomponent bismuth molybdate catalysts,
- the surface allyl species,
- source of oxygen inserted into propene to form acrolein, and
- the overall reaction mechanism.

5.5.1.2 *The Composition of Multicomponent Bismuth Molybdate Catalysts*

The allylic oxidation of propene to acrolein over a Cu_2O catalyst was first reported in 1948 by workers from the Shell Development Company [108].

$$CH_2{=}CH{-}CH_3 + O_2 \rightarrow CH_2{=}CH{-}CHO + H_2O$$

Ten years later the bismuth molybdate catalysts were revealed by the SOHIO company, which formed the basis of the first commercial operation [109]. It is

TABLE 5.8

Binary Oxides of Bismuth and Molydenum

$Bi_2Mo_3O_{12}$	α-phase distorted Scheelite structure
$Bi_2Mo_2O_9$	β-phase
Bi_2MoO_6	γ-phase Koechlinite structure

now recognized that there are three binary oxides of bismuth and molybdenum, with distinguishing Bi:Mo atomic ratios that are important in propene oxidation [114]. These are presented in Table 5.8.

Subsequent research has led to the development of the so-called multicomponent bismuth molybdate catalysts. Knapsak found that when half of the bismuth in $Bi_9P_1Mo_{12}O_{52}$ was replaced with Fe to give $Fe_{4.5}Bi_{4.5}P_1Mo_{12}O_x$ an extensive improvement in activity resulted [111]. Since then a large number of multicomponent catalysts have been developed based on each of the three basic structure types shown below, but in which the bismuth component is largely replaced by a mixture of divalent and trivalent cations.

The advantages of the multicomponent catalysts include superior activity (sometimes exhibiting 20 times greater reaction rates per unit mass than the binary oxides), selectivity and mechanical properties. Typically multicomponent catalysts operating in the temperature range 290–350°C gave yields of 90–95% acrolein whereas the simple binary oxides operating in the range 410–460°C gave yields of 40% at identical contact times [114].

A very wide variety of bismuth molybdate catalysts have been prepared for propene oxidation. In addition to the binary oxides listed in Table 5.9, some ternary oxides and solid solutions, such as $Co_{11/12}Fe_{1/12}MoO_x$ have also been observed in these catalysts.

The effect of the additional components has been illustrated by Sleight et al. [117], who prepared a series of Scheelite type (derived from the mineral $CaWO_4$) bismuth molybdate phases to which Pb was added to give a series of solid solutions of composition $Pb_{1-3x}Bi_{2x}\phi_x(MoO_4)$. For every two bismuth ions in the structure a cation vacancy (ϕ) was generated. When this series of catalysts was tested a relationship was observed between vacancy concentration and the rate of propene oxidation, as shown in Fig. 5.26.

X-ray photoelectron spectroscopy has been used to examine the composition of the surface layers of these catalysts. Figure 5.27 presents a typical set of data for the surface composition for a range of catalysts with varying Bi contents namely, $Bi_y Co_8Fe_3Mo_{12}O_x$ (for values of y between 0 and 1.0). These materials always show a high concentration of molybdenum in the surface layers, more than

TABLE 5.9

Compositional range for multicomponent bismuth molybdate catalysts

Component	Range /Atomic %
Mo^{6+}	50–55
Bi^{3+}	3–7
Me^{2+} (Co, Ni, Fe, Mg, Mn)	30–35
Me^{3+} (Fe, Cr, Al)	8–15
	Examples
Elemental Composition	Phase Composition by XRD
$Mo_{12}Bi_1Co_{11}O_x$	β-$CoMoO_4$, α-$Bi_2(MoO_4)_3$
$Mo_{12}Bi_1Co_8Fe_3O_x$	β-$CoMoO_4$, α-$Bi_2(MoO_4)_3$, $Fe_2(MoO_4)_3$, $FeMoO_4$
$Mo_{12}Bi_1Mg_8Fe_3O_x$	β-$CoMoO_4$, α-$Bi_2(MoO_4)_3$, $Fe_2(MoO_4)_3$, $FeMoO_4$
$Mo_{12}Bi_1Co_8Al_3O_x$	β-$CoMoO_4$, α-$Bi_2(MoO_4)_3$, $Al_3(MoO_4)_3$

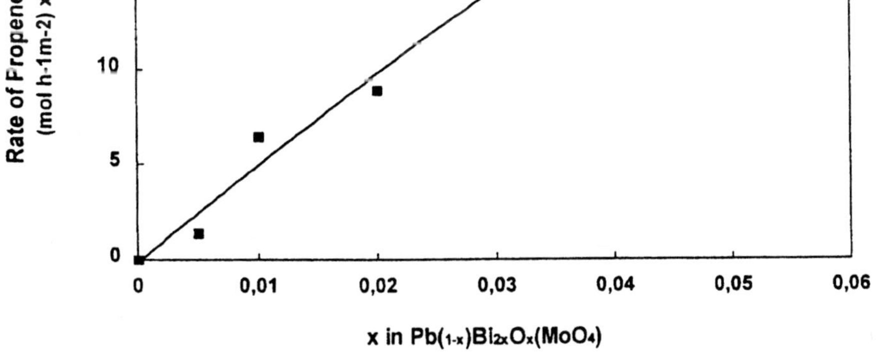

Fig. 5.26. Plot of propene oxidation at 400°C versus cation vacancy concentration for the solid solution $Pb_{(1-x)}Bi_{2x}\phi_x(MoO_4)$ [117].

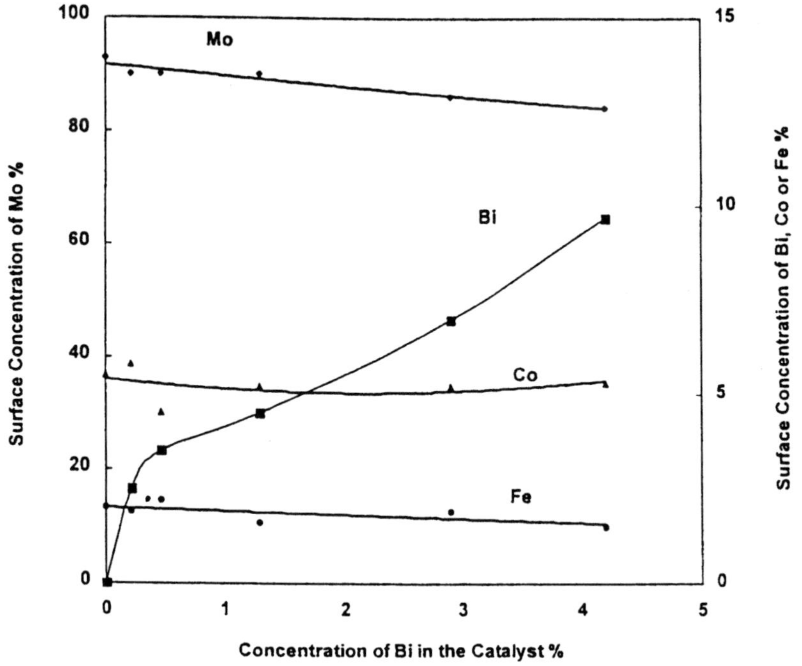

Fig. 5.27. The surface concentration of each metal element in $Mo_{12}Bi_{0-1}Co_8Fe_3O_x$ [114].

would be expected simply on the basis of the elemental composition and when bismuth is added it also appears preferentially in the surface layers [114].

Wolfs and Batist [121] have proposed a structure for these oxide mixtures comprising a core of $Me^{2+}MoO_4$ and $Me^{3+}{}_2(MoO)_4$ encapsulated inside a thin shell of bismuth molybdate. This model has been supported by recent transmission electron microscopy analysis in which a cross-section of a multicomponent bismuth molybdate catalyst was shown to comprise a surface layer of $Bi_2Mo_3O_{12}$ supported on and encapsulating a core of $Co_{11/12}Fe_{1/12}MoO_x$ [107].

5.5.1.3 *The Surface Allyl Species*

A number of classical studies were performed in the 1960s in which isotopically labelled propenes were fed to Cu_2O or bismuth molybdate catalysts. One series of studies used $CH_2{=}CH{-}^{13}CH_3$, $^{13}CH_2{=}CH{-}CH_3$, $^{14}CH_2{=}CH{-}CH_3$ and $CH_2{=}CH{-}^{14}CH_3$ [116,120]. The two key features to emerge were that the isotopic label is never found in the central carbon atom of the acrolein product, but the label is equally distributed between the terminal carbons as follows:

50%	50%
$^{13}CH_2{=}CH{-}CHO$	$CH_2{=}CH^{13}CHO$

The first finding means that a cyclopropane type intermediate is not involved in the reaction mechanism and the second finding points to a linear symmetrical intermediate. The only plausible linear symmetrical intermediate is a π-allyl species:

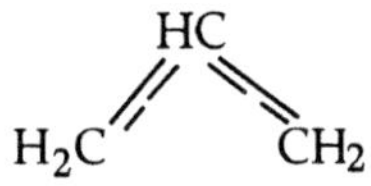

which forms on these surfaces when propene adsorbs.

A second series of experiments involved feeding deuterated propenes, including CH_2=CH–CH_2D [100], CHD=CH–CH_3 [101] and CH_2=CD–CD_3 [112] over a series of bismuth molybdate catalysts. In these experiments a Kinetic Isotope Effect (KIE) was observed provided that one of the C–H bonds in the CH_3 group was substituted by a C–D bond (see Table 5.10). These experiments point to rupture of a methyl C–H bond, namely the weakest C–H bond in the propene molecule, as the slow step in the formation of the π-allyl species and in the overall reaction kinetics.

The expected isotope distribution in the product acrolein can also be calculated on the basis of these results and the results are also consistent with abstraction of an a-hydrogen being the slow step. An example of this type of calculation is shown in Scheme 5.1 for CH_2=CH–CH_2D where the rate constants should be regarded as probabilities. The first hydrogen abstraction can only occur on the CH_2D group and the predicted proportions of d0 and d1 products are shown in the scheme using a k_H/k_D ratio of 2.12. An implicit assumption in this calculation is that isotropic discrimination may occur at each abstraction and will be independent of the kinetically slow step.

Some additional experiments point also to the formation of a second allyl species, namely an oxygen–σ-allyl species (–O–CH_2–CH CH_2) prior to the extraction of the second hydrogen from the evolving acrolein species. Based on the relative stability of allyl (CH_2–CH–CH_2) and carbene (CH_2=CH–CH) species, it is probable that sequential removal of hydrogens would require more energy for the removal of the second hydrogen species [103]. This is not consistent with the experimental results, which clearly show that extraction of the first hydrogen is rate determining. Burrington et al. [102] fed allyl alcohol to bismuth molybdate catalyst and observed rapid exchange of the type:

$$CD_2=CH\text{–}CH_2OH \rightleftharpoons CH_2=CH\text{–}CD_2OH$$

They postulated that either end of the π-allyl species, which forms initially, has an equal probability of collapsing on a lattice oxygen, and the exchange observed with allyl alcohol indicates that this process is rapid and reversible. The same isotopic product distribution is expected as for the sequential hydrogen removal mechanism shown in Scheme 5.1 and the presence of a σ C–O bond in –O–CH_2–CH–CH_2 could reduce the activation energy for removal of the second hydrogen.

TABLE 5.10

Theoretical and experimental kinetic isotope effects k_H/k_D for the oxidation of propene over bismuth molybdate catalysts at 450°C

C–H Bond energies ($kJ\ mol^{-1}$)	Isotope	Experimental (k_H/k_D)	Theoretical (k_H/k_D)	Theoretical (k_H/k_D)	Theoretical (k_H/k_D)	Ref.
			CH_2=	=CH–	–CH_3	
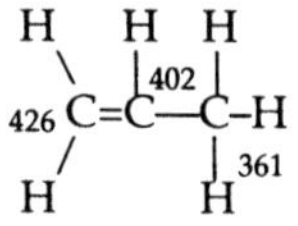	CH_2=CH-CH_2D	2.04	1.00	1.00	1.82	100
	CHD=CH-CH_3	1.02	1.82	1.00	1.00	101
	CH_2=CD-CD_3	1.78	1.00	1.75	1.75	112

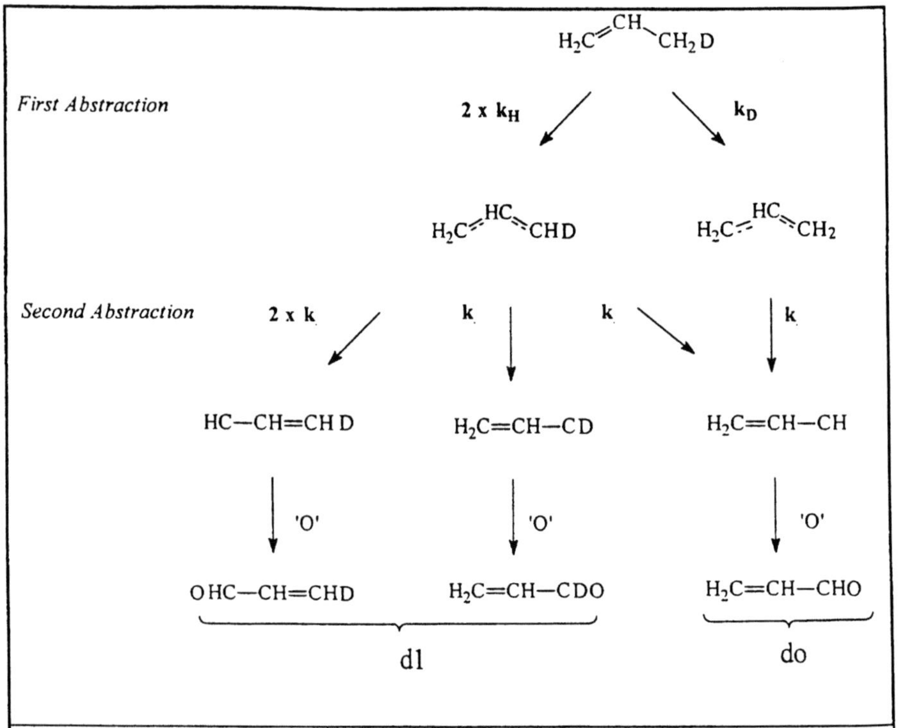

$$d_0 = \{(k_D / (3k_H + k_D) \bullet 2k_H / (2k_H + k_D)\} + \{ k_D / (2k_H + k_D)\} = 0.30 \text{ (expt 0.29)}$$

$$d_1 = \{ 3k_H / (3k_H + k_D) \bullet 2k_H / (2k_H + k_D)\} = 0.70 \text{ (expt 0.71)}$$

Scheme 5.1. Isotopic distribution in acrolein from CH_2=CH–CH_2D [100].

These arguments point to a mechanism in which the rate-determining step is abstraction of the first a hydrogen to form a π-allyl surface species. The next step is the reversible formation of a σ bond between a catalyst oxygen atom and either of the terminal carbon allyl atoms. The second hydrogen abstraction, which is facilitated by the presence of the C–O bond, follows with formation of the carbonyl bond and desorption of the product acrolein. The next question to be addressed is the source of oxygen, which appears in the product acrolein.

5.5.1.4 Source of Oxygen Inserted into Propene to Form Acrolein

A number of experiments were carried out in which ^{18}O was used as the source of gas phase oxygen ($^{18}O_2$), or the source of lattice oxygen ($^{18}O_L$). These experiments were designed to probe the nature of the interaction between propene and the two sources of oxygen available in the reactor, namely gas phase and lattice oxygen. Preliminary experiments [122] confirmed that at 475°C the rate of exchange of gas phase oxygen for lattice oxygen was slow

$$^{18}O_2 + {}^{16}O_L \rightleftharpoons {}^{18}O^{16}O + {}^{18}O_L$$

as was the rate of exchange between the product acrolein and lattice oxygen.

$$C_3H_4{}^{16}O + {}^{18}O_L \rightleftharpoons C_3H_4{}^{18}O + {}^{16}O_L$$

These findings eliminate reactions which could seriously complicate the interpretation of the results.

Wragg et al. [122] found that when propene and $^{18}O_2$ were reacted over a bismuth molybdate catalyst initially comprising only $^{16}O_L$ that the fraction of $C_3H_4{}^{18}O$ in the reaction product initially was very small but increased as the reaction proceeded, consistent with the gradual replacement of $^{16}O_L$ by $^{18}O_L$ derived from $^{18}O_2$.

When the reaction was repeated with propene and $^{16}O_2$ in the gas phase over a bismuth molybdate catalyst comprising only $^{18}O_L$ the amount of $C_3H_4{}^{18}O$ in the reaction product was very high initially but decreased as the reaction proceeded, consistent with the gradual replacement of $^{18}O_L$ by $^{16}O_L$ derived from $^{16}O_2$.

This work has been repeated by many workers using a variety of batch and flow reactors and with binary and multicomponent catalysts, and in all circumstances the essential result has been confirmed, namely the oxygen which appears in the acrolein product is inserted via the lattice and not directly from the gas phase.

An example of the more recent work with multicomponent catalysts is shown in Fig. 5.28.

Involvement of lattice oxygen from deep within bismuth molybdate catalysts is confirmed from the results of Raman spectroscopy analysis of catalysts which

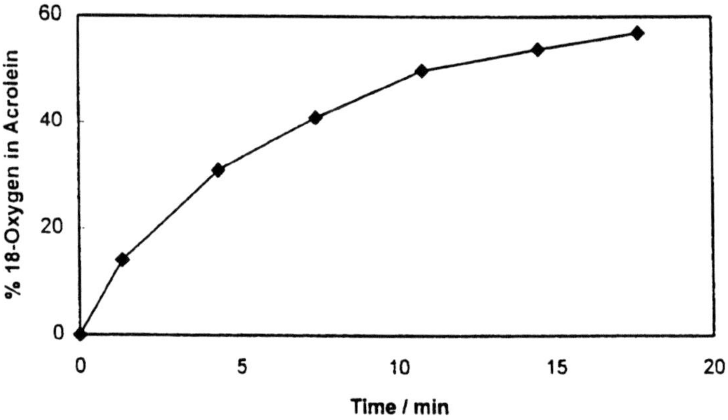

Fig. 5.28. Percentage of ^{18}O in acrolein formed as a function of reaction time from propene and $^{18}O_2$ over $Bi_{0.85}V_{0.55}Mo_{0.45}{}^{16}O_4$ [119].

had been reduced for a short time with propene and reoxidized with $^{18}O_2$ in the gas phase [105]. When γ-$Bi_2Mo_{16}O_6$ was subjected to a number of cycles of reduction by propene and reoxidation by $^{18}O_2$ the positions of the Mo–^{16}O stretching frequencies at 884, 803 and 725 cm^{-1} shifted to lower values as outlined in Table 5.11 This shift is consistent with the gradual replacement of $^{16}O_L$ by $^{18}O_L$.

The surprisingly high level of participation of lattice oxygen in the oxidation of propene is made possible by the very low activation energy measured for the replenishment of lattice oxygen provided that the extent of reduction is not extensive. Typically, activation energies for reoxidation are in the range 5–10 kJ mol^{-1} when the degree of reduction does not exceed 0.1%, but increases sharply thereafter [118].

These experiments essentially indicate that the oxidation of propene over bismuth molybdate catalyst should be viewed as a two step reaction. In the first step propene reacts with lattice oxygen to yield the reaction products and a reduced catalyst:

TABLE 5.11

Raman band shifts for γ-Bi_2MoO_6 reduced in propene and reoxidized in $^{18}O_2$ [105]

Mo–^{16}O Vibrational frequencies (cm^{-1})	Size of Shift (cm^{-1})	
	After 1 cycle	After 6 cycles
	–9	–14
	–11	–18
	–10	–17

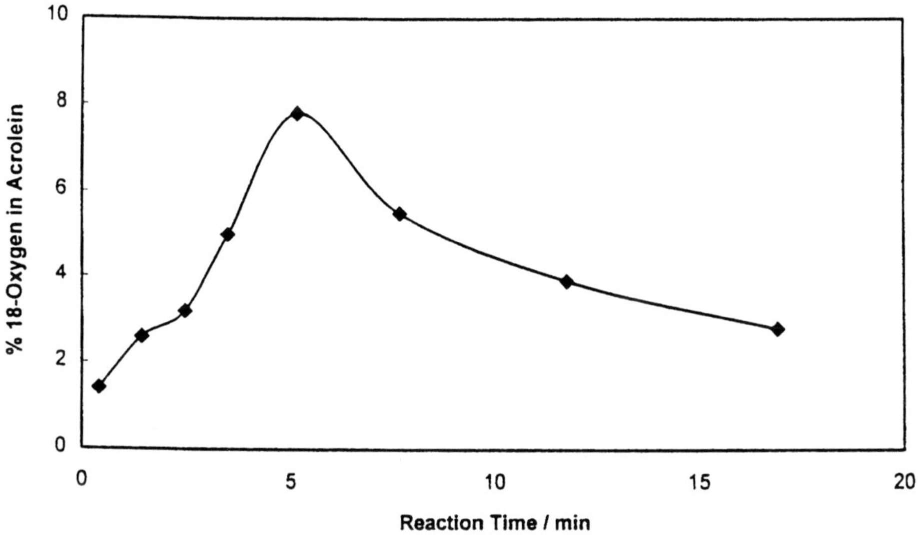

Fig. 5.29. Evolution of 18-Oxygen concentration of acrolein with reaction time during the oxidation of propene over $Bi_2(Mo^{16}O_4)_3$ supported on $\beta Co_{11/12}Fe_{1/12}Mo^{18}O_4$ [114].

$$C_3H_6 + 2O_L \rightarrow C_3H_4O + H_2O + 2\,\square\square_{O_2}$$

In the second step lattice oxygen is replenished by gas phase oxygen:

$$2\,\square\square_{O_L} + O_{2(g)} \rightarrow 2O_L$$

Figure 5.29 presents the results of an experiment in which propene was oxidized to acrolein over a catalyst composed of $Bi_2(Mo_{16}O_4)_3$ supported on $\beta Co_{11/12}Fe_{1/12}Mo^{18}O_4$ [114]. The source of gas phase oxygen used in this experiment was $^{16}O_2$. As seen in this figure the percentage of ^{18}O in the acrolein product increased with reaction time and passed through a maximum, consistent with transfer of ^{18}O from the $\beta Co_{11/12}Fe_{1/12}Mo^{18}O_4$ phase to the $Bi_2(Mo^{16}O_4)_3$ and hence to the acrolein.

There is now a considerable body of evidence to indicate that propene oxidation to acrolein occurs via a Mars and van Krevelen [123] mechanism whereby the reacting hydrocarbon, or a species derived from it, extracts lattice oxygen from the encapsulating surface layer of bismuth molybdate. In a separate step this lattice oxygen is replenished at least in part by lattice oxygen transfer from the encapsulated phases. These phases in turn are reoxidized by gas phase oxygen. This process is shown in Scheme 5.2.

5.5.1.5 *The Overall Reaction Mechanism*

Clearly from the preceding discussion there are three separate steps involved in acrolein formation from propene, namely, abstraction of two hydrogen atoms

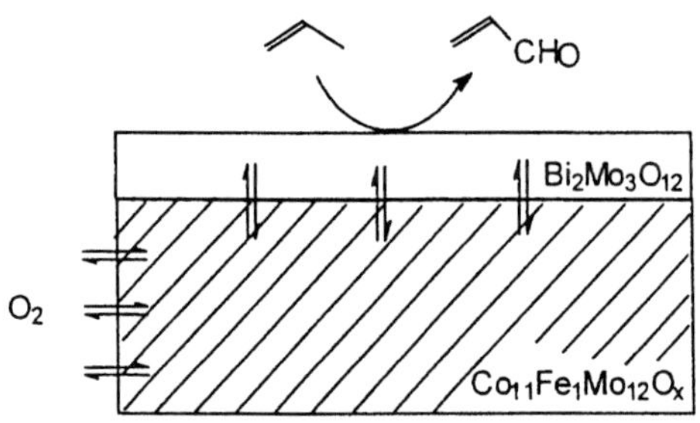

Scheme 5.2. Illustration of Mars and van Krevelen Mechanism with incorporation of gas phase oxygen into one phase and transfer to a separate surface phase for reaction with propene [114].

and insertion of an oxygen atom into the evolving acrolein product. Researchers have attempted, in so far as it is possible to do so, to associate the separate bismuth or molybdenum components of these catalysts with one or other of these steps.

Gryzbowska et al. [106] compared the reaction products formed when pulses of allyl iodide or propene were passed over bismuth oxide or molybdenum oxide. A clear limitation of these experiments is that even the simplest bismuth molybdate catalysts contain neither bismuth oxide nor molybdenum oxide, but instead are made up of a binary oxide of bismuth and molybdenum, whose structure is different to that of bismuth oxide and molybdenum oxide. Gryzbowska et al. selected allyl iodide because of the very low bond dissociation enthalpy associated with the C–I bond, implying that a surface allyl species would readily form from this starting material. In addition, a lower reaction temperature was required for the reaction of allyl iodide than for propene reflecting the greater inherent reactivity of the former.

Over bismuth oxide no acrolein was formed from propene and just small amounts from allyl iodide. Instead the major reaction product was the dimer hexadiene, formed from the dimerization of two allyl species. Over molybdenum oxide no reaction occurred with propene but copious amounts of acrolein formed from allyl iodide. These results are summarized in Table 5.12.

These results led to the postulate that the bismuth component of bismuth molybdate catalysts is responsible for the abstraction of the first hydrogen from propene, but that this oxide is not capable of inserting oxygen into the reaction intermediate which forms. Instead any allyl species which form over this oxide from allyl iodide and less easily from propene, dimerize to hexadiene in the absence of a suitable source of oxygen. By contrast, allyl species do not form from propene over molybdenum oxide. However, when this oxide is exposed to a facile source of allyl species, namely allyl iodide, acrolein forms readily, consistent with the oxygen insertion and second hydrogen abstraction steps occurring over this component of the bismuth molybdate catalysts.

TABLE 5.12

Product distribution from the reactions of propene and allyl iodide with Bi_2O_3 and MoO_3

Catalyst	Reactant	Temperature (°C)	% Acrolein formed	% Hexadiene formed
Bi_2O_3	CH_2=CH–CH_2I	320	5	70
	CH_2=CH–CH_3	480	0	8.6
MoO_3	CH_2=CH–CH_2I	320	98	0
	CH_2=CH–CH_3	480	0	0

TABLE 5.13

Radical species and stable products in the oxidation of propylene [113]

	Radical amount (nmol)		Stable products (nmol)		
Catalyst	Allyl	Allyl peroxy	1,5-Hexa-diene	Acrolein	3-Methyl pentane
Bi_2O_3	31.72	5.60	33.2	7.0	36.1
γ-Bismuth Molybdate	3.72	0.65	0.0	2920	31.0
α-Bismuth Molybdate	0.82	0.15	0.0	311	24.3
MoO_3	0.12	0.02	0.0	25	38.7
Homogeneous Pyrex	0.84	0.18	0.0	3.5	51.6
Bi_2O_3/MoO_3[a]	1.66	0.40	0.0	37	34.6

T = 450°C, Flow rates: Ar: 0.064 $cm^3\,s^{-1}$; C_3H_6 = 0.019 $cm^3\,s^{-1}$: O_2 = 3.76 × 10^{-4} $cm^3\,s^{-1}$; Total pressure – 1 torr.

[a] The Bi_2O_3 catalyst was placed upstream from the MoO_3 catalyst.

A similar approach has been used by Lunsford [113], except that he monitored the presence of gas phase radical species at the exit of a reactor used for propene oxidation. The exit gases from their reactor were cooled and radical species stabilized using the matrix isolation technique. Table 5.13 presents their results. Allyl and peroxy radicals were detected in varying amounts depending on the catalyst that was used. Those catalysts which showed a propensity to form 1,5-hexadiene released more radicals into the gas phase than those over which

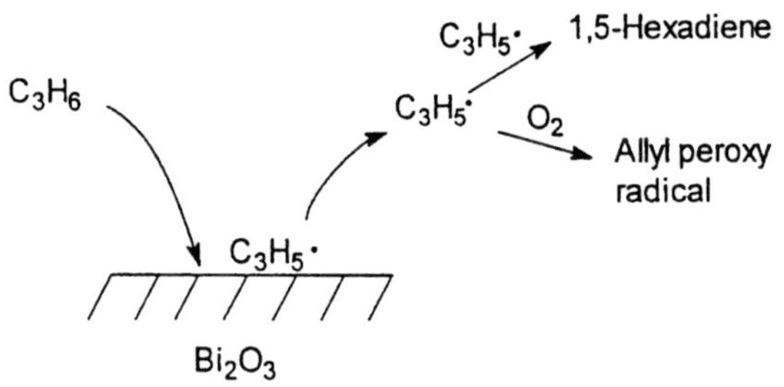

Scheme 5.3. Schematic for the reactions of allyl radicals species over Bi_2O_3 [115].

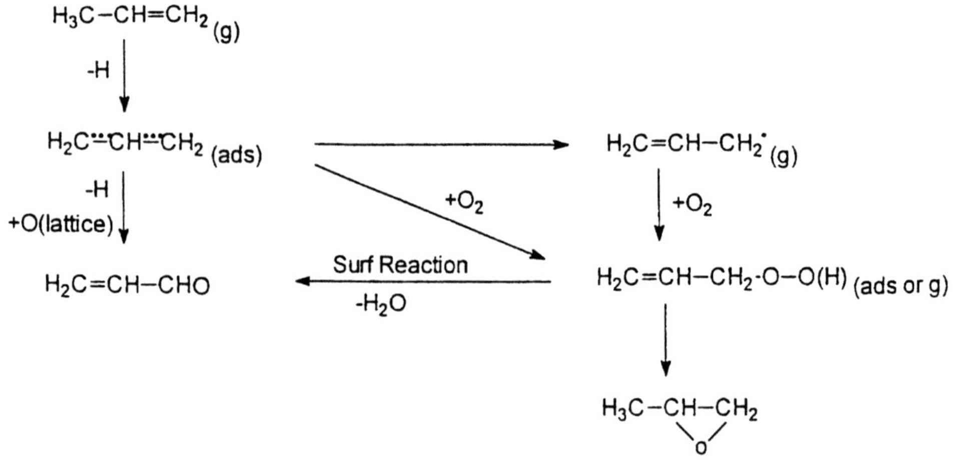

Scheme 5.4. Reaction network for the reactions of radical allyl species over bismuth molybdate catalysts [110].

acrolein was formed. When Bi_2O_3 was placed upstream of MoO_3 in the reactor there was a marked reduction in the amount of radicals detected at the exit of the reactor and an increase in acrolein production. Their data are consistent with the work of Gryzbowska et al. [106] whereby radical species do not react with lattice oxygen in Bi_2O_3, hence they are more readily released into the gas phase with the formation there of 1–5-hexadiene. The proposed network is outlined in Scheme 5.3.

In the presence of the MoO_3 component of the catalyst the radical species formed can react with lattice oxygen leading to acrolein formation as shown in Scheme 5.4.

Compelling evidence that the radial allyl species which forms is important in the catalysis and not just a spectator species comes from measurement of the

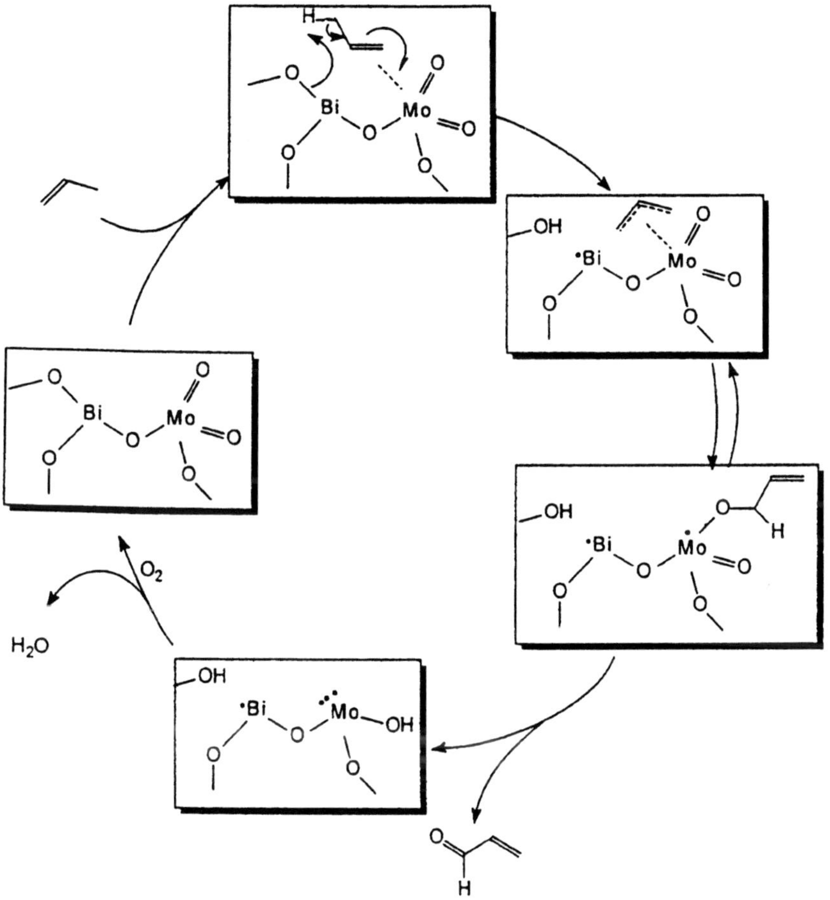

Scheme 5.5. Schematic of the active site for acrolein formation from propene over bismuth molybdate catalysts [105].

kinetic isotope effect for allyl radical formation over Bi_2O_3 catalysts. The data are presented in Table 5.14. The results are almost identical to those reported in Table 5.10 and are consistent with the involvement of a radical allyl species in the overall mechanism.

The most coherent overall picture of the mechanism of propene oxidation has been presented by Grasselli [105] (see Scheme 5.5). The mechanism involved abstraction of a CH_3 hydrogen by the bismuth part of the active site, with the formation of a π-allyl species interacting with the Mo centre. Oxygen insertion occurs next, and the σ-allyl species is more activated for removal of the second hydrogen species than the carbene which would form without oxygen insertion. The final steps involve product desorption and reoxidation of the active site via oxygen transfer from the underlying solid solution.

TABLE 5.14

Kinetic Isotope Effect (kH/kD) during radical formation from propene over Bi_2O_3 [115]

Temperature (°C)	Experimental	Calculated
365	2.3 ± 0.4	2.6
385	2.2 ± 0.4	2.5
425	2.2 ± 0.4	2.2
450	2.1 ± 0.4	2.1
476	1.8 ± 0.4	2.0

5.5.2 *Epoxidation of Ethene*

Ethylene epoxide (EO) is an important intermediate in the chemical industry and the mechanism of its formation has been studied in detail [124–128]. For the industrial aspects see Chapter 2. EO is produced by the selective oxidation of ethylene with oxygen:

$$H_2C{=}CH_2 + \tfrac{1}{2}O_2 \rightarrow H_2C\overset{\overset{O}{/\ \backslash}}{-}CH_2$$

This reaction is accompanied by complete combustion into water and carbon dioxide. The only selective catalyst known is based on silver. This catalyst was known as early as the 1930s and has been continuously improved since then in a rather empirical way. It has been discovered that the catalyst may be promoted by the addition of alkali metal ions. Moreover, the presence of chlorine has a beneficial effect (cf. Fig. 5.30) [124]. Chlorine has to be added continuously because it disappears from the surface by reacting to give chlorinated ethane. It is sufficient to mix 10–40 ppm chlorine with the feed. The feed consists of a mixture of ethylene (24%), oxygen (8%) and the balance of inert gases. The reaction rate was found to be first order in oxygen and zero order in ethylene in the Shell process.

Usually, the selectivity of the reaction is analyzed on the basis of the simplified scheme in Fig. 5.31.

The initial selectivity is determined by the ratio k_1/k_2. At higher conversions k_3 also plays a role.

As catalysts, a low surface area alumina support is used on which large silver particles are present. It is crucial to minimize the consecutive reaction which is enhanced by acid sites of the support. These sites isomerize the epoxide to acetaldehyde which rapidly combusts in a reaction catalyzed by silver.

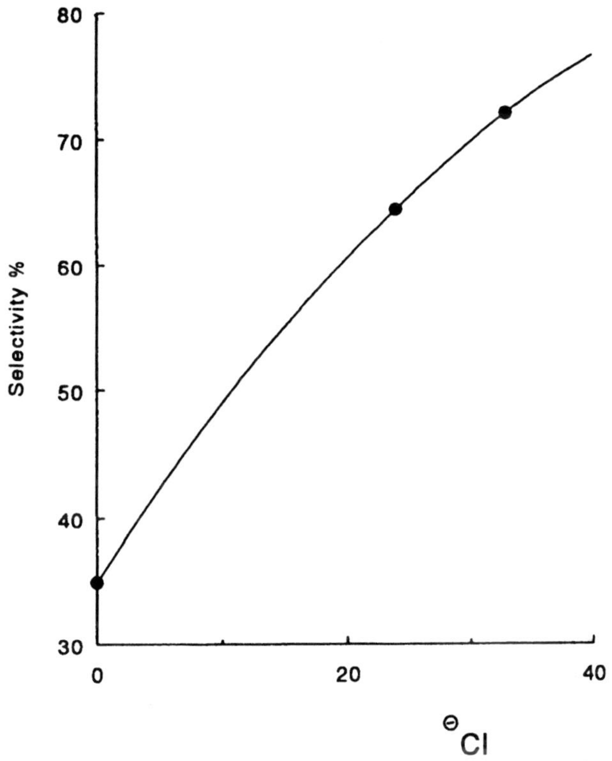

Fig. 5.30. Epoxidation of ethene. Influence of chlorine on selectivity [124].

It is interesting that, until recently, a mechanism was advocated which predicts that the selectivity has a theoretical limit of 86%. This would imply that it does not make sense trying to improve the catalyst much further, because, in practice, plants are already operating at selectivities close to this value. This theory was based on the assumption that the catalyst contains two types of oxygen [125]: molecular O_2, and atomic oxygen. The productive epoxidation reaction was envisaged as shown in Fig. 5.32.

According to this mechanism molecular oxygen is adsorbed at the catalyst surface and ethylene reacts with this species to give epoxide, while one oxygen atom remains at the surface. The oxygen atom remaining at the surface has to be removed in order to restore catalytic activity. In view of this theory, this is only possible by the non-selective complete combustion reaction (Fig. 5.33).

The implication of this mechanism is that the most ideal catalyst under the most optimum conditions would transform six ethylene molecules into EO and one ethylene molecule into water and carbon dioxide. Hence, according to this theory, the selectivity can never exceed 6/7. This theory also explains the favourable influence of chlorine. It was known that chlorine selectively poisons sites on Ag which can adsorb O_2 dissociatively. As atomic oxygen was considered to be

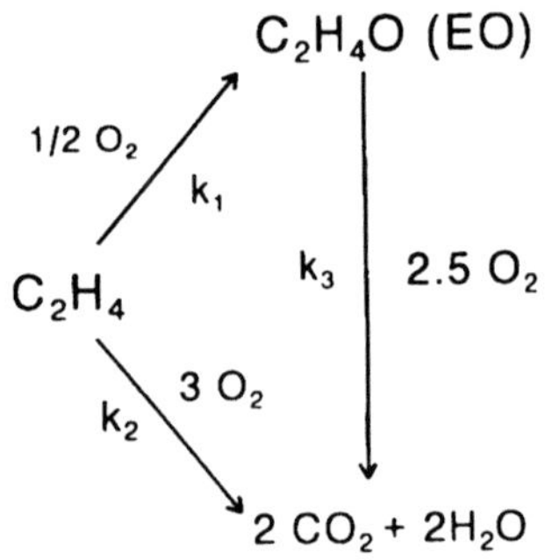

Fig. 5.31. Kinetic scheme of the reactions [124].

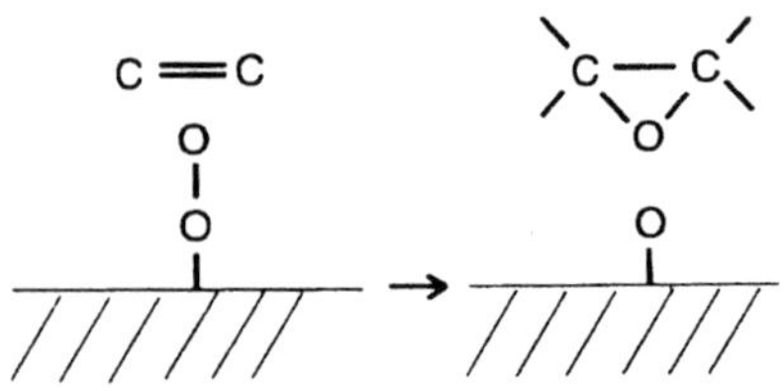

Fig. 5.32. Proposed scheme for selective oxidation of ethylene by preadsorbed molecular oxygen.

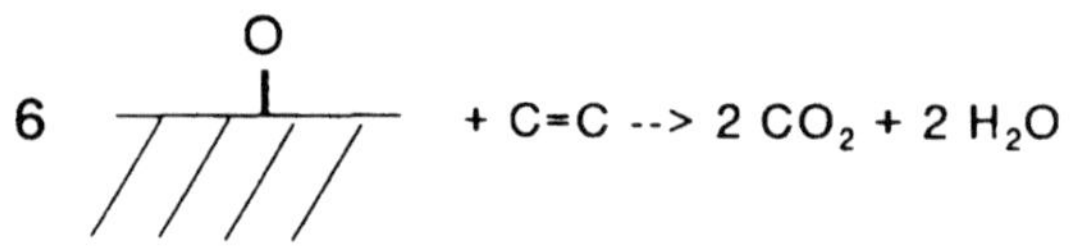

Fig. 5.33. Proposed scheme for complete oxidation of ethylene.

the source of complete combustion, its presence should be minimized and the beneficial action of chlorine was explained.

In 1975 another model was proposed [126], in which atomic oxygen was thought to be the source of selective epoxidation. This was later confirmed experimentally [127]. In the following we will outline the basis of this mechanism [128].

There is no doubt that the initial selectivity is governed by the characteristics of the oxygen at the surface. The subsequent reactions also influence the overall selectivity: support acid sites catalyze the isomerization of EO giving acetaldehyde, which is very reactive towards complete combustion. The beneficial influence of alkali promoters is due to the poisoning of the acid sites at the catalyst support.

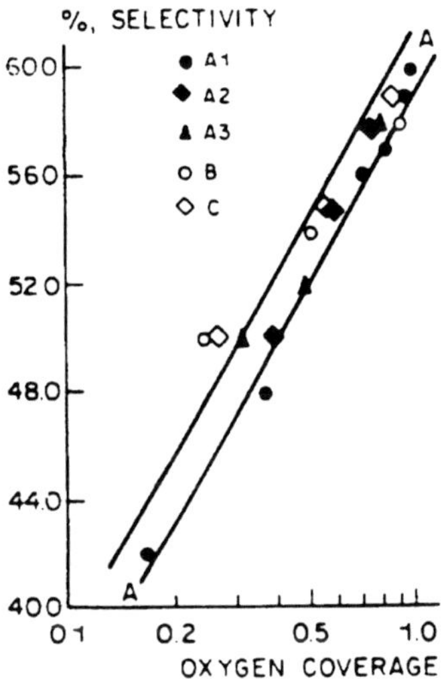

Fig. 5.34. Selectivity versus oxygen coverage of various supported Ag catalysts [124].

The selectivity depends strongly on the oxygen coverage. At low oxygen coverages the selectivity is much lower than at high coverages (cf. Fig. 5.34).

According to the current view of the mechanism this is explained as follows. At low coverages the oxygen is very strongly adsorbed, whereas at higher coverages the adsorption is much weaker. From TPD data [129] it was found that, below half monolayer coverage, the binding strength is 50–60 kJ/mol. This contrasts with the data at higher coverages which give only 0–13 kJ/mol. When we consider the reaction enthalpy for the formation of the epoxide from atomic oxygen, the reason why selective epoxidation at low oxygen coverage is slow becomes clear: the reaction enthalpy is only 50–70 kJ/mol. This value barely exceeds the bond strength of oxygen at low coverages. Thus, in this case total oxidation — a reaction that is much more exothermic — is thermodynamically favoured, whereas epoxidation is hardly possible.

The influence of the promoters and modifiers can be understood from a comparison of the desired reaction, leading to EO, and the undesired reactions, leading to complete combustion. In the desired reaction an electrophilic attack has to take place (Fig. 5.35).

In the undesired reaction nucleophilic attack takes place at a C–H bond (Fig. 5.36).

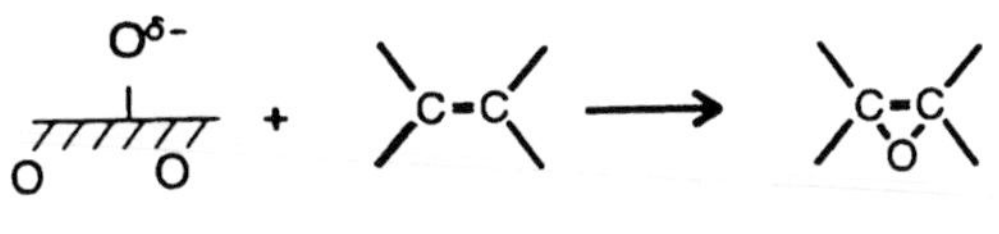

Fig. 5.35. Electrophilic attack of oxygen [124].

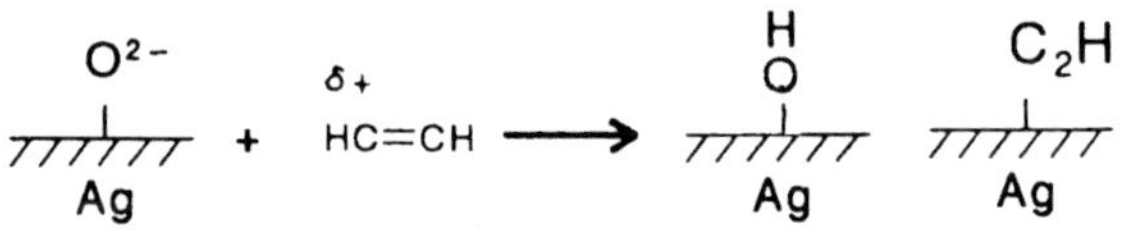

Fig. 5.36. Nucleophilic attack of oxygen [124].

At increasing oxygen coverages the charge on the oxygen atoms will decrease and, as a consequence, the EO production rate increases, whereas the rate of the undesired complete combustion reactions decreases. The influence of Cl is probably to decrease the charge of oxygen. Chlorine is a promoter because, under reaction conditions, it performs this function better than subsurface oxygen atoms. Under reaction conditions the steady-state coverage of oxygen is rather low, whereas chlorine will chlorinate the surface to a higher degree. The influence of the alkali metal might lie in the further stabilization of chlorine. Moreover, alkali metal atoms will poison acidic sites of the weakly acidic alumina support (upon which the silver particles are dispersed) which catalyze the undesired following reactions, increasing the combustion rates.

A summary of the mechanism described above according to Ref. [124] is presented in Fig. 5.37. Of course, one can think of alternative detailed pictures. For example, silver is a catalyst with subsurface oxygen exposed to the gas phase as a positively charged centre and can interact with π-electrons of ethene. Obviously, this practically very important reaction still has some detailed points which require elucidation.

It is striking that only silver has been found to be a good catalyst. Why is silver unique? The mechanism described here may provide an answer. The key factor might be that silver can dissociate O_2, but the oxygen atoms are so weakly chemisorbed that epoxidation is possible. Moreover, silver only weakly activates the C–H bonds in ethylene. Definitive answers on the potential and the explanation of promoters and modifiers are waiting for more information.

Up till now it has not been possible to carry out the analogous reaction with propene. Numerous researchers have attempted to develop a process for the direct oxidation of propene into propene epoxide (PO). Only indirect routes have, up to now, been applied in successful selective processes (see Section 5.5.4). Those indirect processes involve the use of hydrogen peroxide, organic peroxides and peracids, hypochlorides, etc. (see e.g. SMPO, Chapter 2). The reason that it is difficult to epoxidize propene is the facile formation of an allylic intermediate because the C–H groups in the methyl group become activated.

Because of their success in ethylene epoxidation, it is not surprising that specially modified Ag catalysts have received intense attention. Although promising developments have been reported, there is still no commercial process for the direct oxidation of propene into PO. Here is a real challenge for the scientific community.

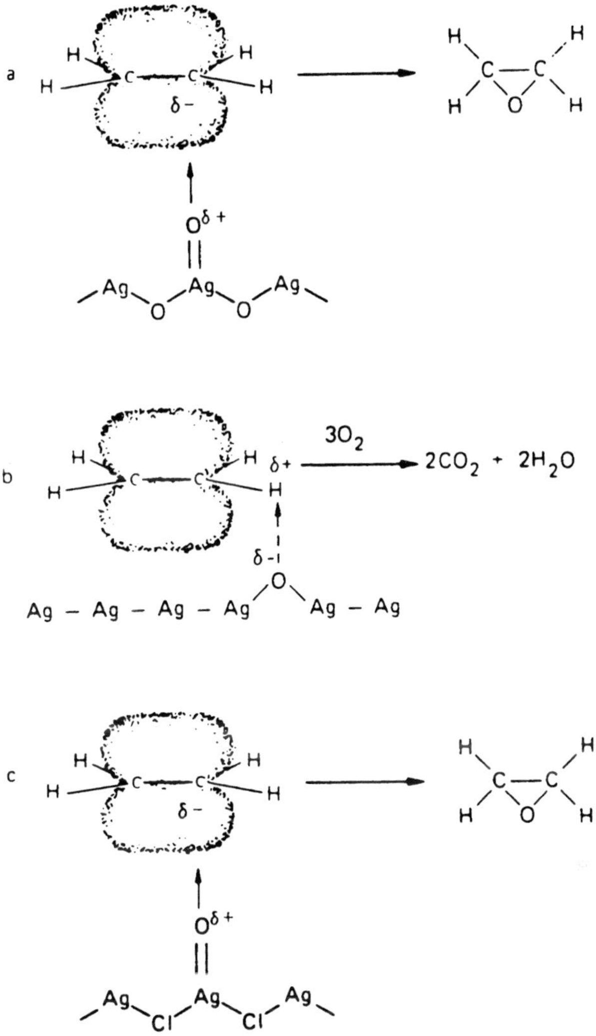

Fig. 5.37. Schematic representation of the proposed transition state leading to epoxidation [124].

5.5.3 The Wacker Reaction; Vinylacetate Production

In the silver catalyzed epoxidation reaction insertion of oxygen into the ethylene π-bond is accompanied by reduction of silver ions formally of valency 3+, to silver ions of 1+ valency. The original state of the silver surface is re-established by reoxidation with oxygen.

Oxidation of ethylene to the aldehyde catalyized by homogeneous Pd^{2+} complexes occurs in an analogous fashion. In $PdCl_4^{2-}$, Cl^- is replaced by ethylene and water and the metal atom is reduced:

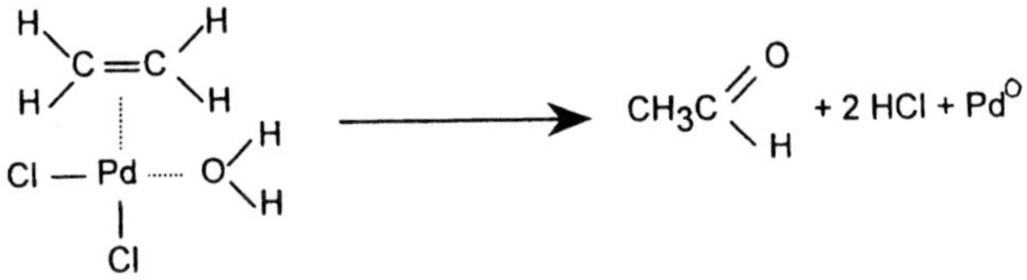

Scheme 5.6.

Pd° is reoxidized in a consecutive step with oxygen using the Cu^{+1}/Cu^{+2} redox couple as a catalyst.

An interesting question arises whether this reaction can also occur on the Pd surface. Co-adsorption of alkenes and oxygen on Pd- or Rh surfaces indeed has been found to yield some aldehyde or keto formation.

Moiseev has proposed that the analogous reaction between ethylene and acetic acid to vinylacetate takes place on the Pd surface [130].

The stoichiometric reaction is

$$C_2H_4 + CH_3COOH + \frac{1}{2}O_2 \rightarrow H_2O + C_2H_3OOCH_3$$

The Moiseev proposal is that reaction occurs on the Pd surface via an adsorbed vinyl species:

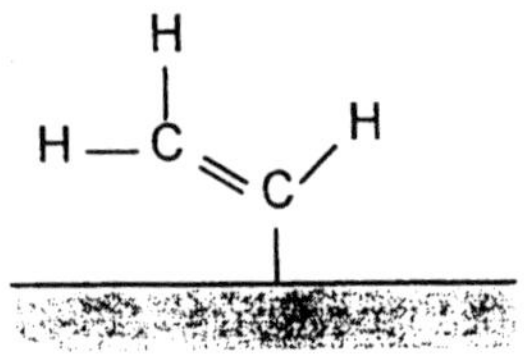

Scheme 5.7.

Surface science experiments have shown that such intermediates can be formed rather easily [131]. It has also been demonstrated that acetic acid can readily absorb to a Pd surface by reacting with preadsorbed oxygen [132] (Scheme 5.8).

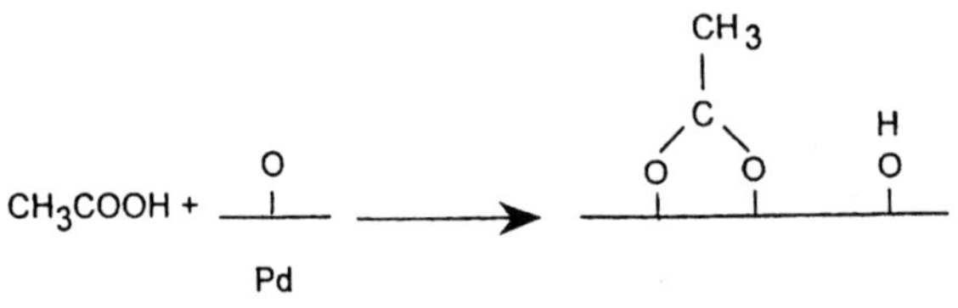

Scheme 5.8

Vinyl acetate is then formed by associative desorption by adsorbed vinyl and acetate. Current evidence strongly favours the Moiseev proposal.

When catalyzed by the homogeneous Pd^{2+} complex acetate can be inserted into the ethylene bond analogously to H_2O. Vinylacetate then is formed by the βCH cleavage reaction with formation of PdH.

5.5.4 *Epoxidation using Hydro- or Hydrogenperoxide*

Whereas so far oxidation reactions with O_2 have been discussed, an important heterogeneous catalytic oxidation reaction is the epoxidation of propylene to propylene epoxide. Using hydroperoxide the reaction can be done over Ti dispersed on silica [133]. Ti becomes 4 coordinated as Si in SiO_2. With hydrogen-peroxide the reaction has been found to occur for Ti incorporated in the lattice of MFI zeolite [134]. The unique property of these systems is that the electrophilic nature of oxygen intermediates does not cause reaction with the allylic CH bond. It is proposed that the adsorbed intermediate is:

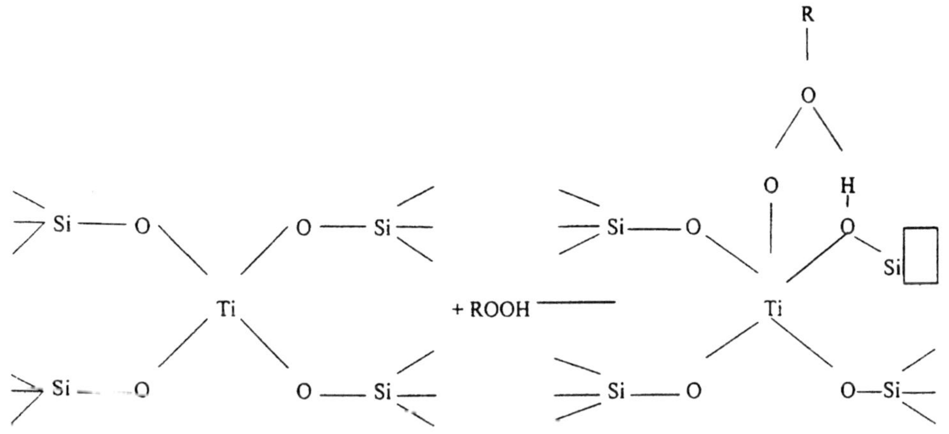

Scheme 5.9

Upon contact with peroxide one Ti–O is ruptured with formation of a silanol group. The OH bond of peroxide cleaves in this reaction. One of the oxygen atoms of the peroxide radical is then used for selective oxidation. Water or alcohol is formed as a coproduct. A significant issue in the solid state chemistry of the catalyst is prevention of leaking of Ti from the catalyst. This could happen if subsequent bondcleavage of Ti–O occurs. Enough strain has to be present around the catalytic Ti site so that after reaction the Ti–O bond is repaired.

5.6 ELECTROCATALYSIS

5.6.1 Introduction

Electrochemistry (electrodics) is concerned with chemical reactions that involve the transfer of electric charge across a solid/electrolyte interface. Charge-transfer reactions are of two types: one is electron transfer, the other is ion transfer together with neutralization of the ion at the surface of the solid.

Electrochemical conversions can be classified as anodic or cathodic: oxidations take place, by definition, at an anode (electron transfer from species in solution to the electrode), while reductions take place at a cathode (electron transfer from the electrode to species in solution). In the majority of cases it is not sufficient to stick any good conductor into solution to get your electrochemical conversion going — electrode reactions usually have to be accelerated by employing active catalysts. Electrocatalysis is a term applied to describe the processes occurring at electrode surfaces. In both heterogeneous catalysis and electrocatalysis there is a strong interaction between reactants and the catalyst / electrocatalyst surface in the reaction transition state; that is, adsorbed species are of prime importance in both areas. On the other hand, in electrocatalysis the reactant and the reaction product(s) differ in electric charge, the direction and rate of the reaction can be controlled by variation of the electrode potential, and the reaction takes place in the presence of an electrical double-layer (see below). Nevertheless, it has become apparent over the last few decades, that the surface processes in electrocatalysis and heterogeneous catalysis have much in common, and indeed this commonality will be evident from the examples discussed in this section.

Equilibrium at a solid–electrolyte interface is established by an exchange of charged species. We are dealing, therefore, with an electrified interface [135]. Assuming a sufficiently high conductivity, the excess charge in the solid (e.g. a metal) can be treated as a surface charge. The counter-charge of opposite sign in the electrolyte, however, remains at a certain distance, viz. the outer Helmholtz plane (OHP, Fig. 5.38). The electrified interface, then, is a double layer, viz. a layer of excess charge on the metal and, in concentrated solutions*, a layer of counter-charge at the OHP. In the presence of strongly adsorbed anions in the inner Helmholtz plane (IHP, Fig. 5.38b) we have a triple layer. The electrical properties of such a double (or triple) layer can be understood in terms of a (series of) parallel plate condenser(s). The potential distribution in the double layer is also given in Fig. 5.38.

---

* Here we consider only aqueous solutions. In dilute solutions the counter-charge is not confined to the OHP, but extends further into the solution, forming the so-called Gouy–Chapman layer.

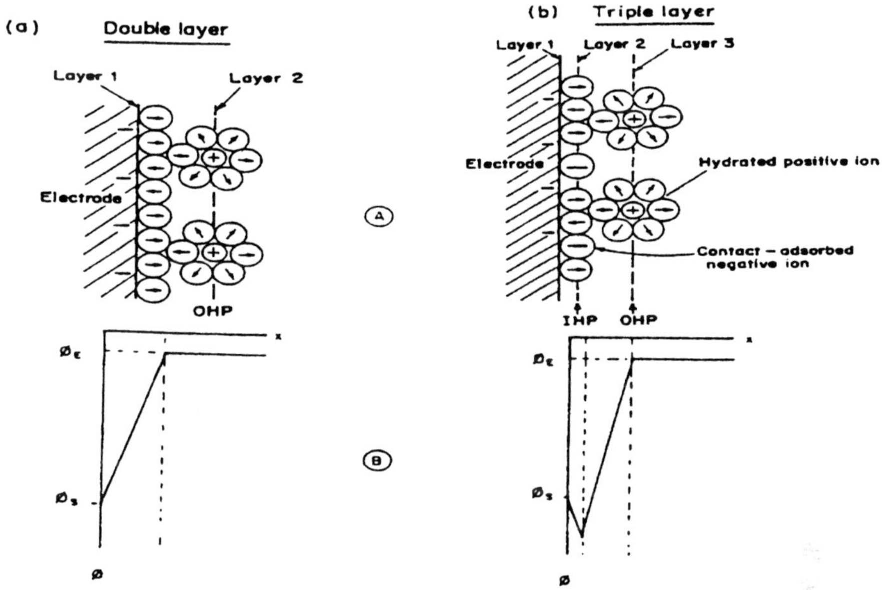

Fig. 5.38. A. (a) An electrical double layer and (b) an electrical triple layer. B. Potential distribution at the interface. OHP = Outer Helmholtz Plane, IHP = Inner HP, ϕ = Galvani potential.

One may wonder whether the presence of such a double layer could affect the structure of adsorbed intermediates. It is found, e.g., that whereas formic acid adsorbs on Pt from the gas phase as formate + H, it binds with its C-atom to the metal in the electrochemical situation:

$$HCOOH + 2Pt \rightarrow Pt - COOH + Pt - H \quad (5.1)$$

$$Pt - H \rightarrow Pt + H^+ + e^- \quad (5.2)$$

Exactly why this should be so has not yet been elucidated, however. It is possible that it is simply due to the tendency of the O and OH moieties to be solvated by water molecules, thus orienting the molecule with its C end towards the Pt surface and therefore increasing the probability of the reaction proceeding there.

At the equilibrium potential E_o of a given electrochemical reaction, e.g.

$$Ox + ne^- \rightarrow Red \quad (5.3)$$

no nett current flows; electrons are still exchanged but the forward and back reactions exactly cancel. The rate of charge transfer, expressed as the current density i (which equals nF * (reaction rate), with F the Faraday constant), can, in the absence of mass transfer limitations, be empirically related to the over-potential η, which is the potential difference between the equilibrium potential and the actual electrode potential ($\eta = E - E_o$), via the so-called Butler–Volmer equation [136]:

$$i = i_o\,(\exp[\alpha_a F\eta/RT] - \exp[-\alpha_c F\eta/RT] \tag{5.4}$$

where i_o is the exchange current density (i.e. the current that flows back and forth, so to speak, at E_o), and α_a and α_c the anodic and cathodic charge-transfer coefficients respectively, which determine the relative response of the respective partial current densities to a change in overpotential η. If the potential is far from E_o one of the partial currents densities becomes negligible: e.g. at $E >> E_o$ only the anodic part remains:

$$i = i_o \exp[\alpha_a F\eta/RT] \tag{5.5}$$

and this can be simplified to

$$\eta = a + b \log i \tag{5.6}$$

the ubiquitous Tafel equation, in which b ($=2{\cdot}3RT/\alpha F$) represents the Tafel slope (a plot of E vs $\log i$ is accordingly called a Tafel plot).

For simple one-electron redox reactions, in which *Ox* and *Red* both remain in solution, the charge transfer coefficients α are often found to be approximately 0.5; for more complex reactions they vary considerably. Thus, an electrocatalytic effect can not only express itself in a change in i_o, but also by affecting α (i.e. in general by changing the reaction pathway). This is schematically indicated in Fig. 5.39. A generalized expression for the exchange current density is

$$i_o = nFk_a C_A^a C_B^b \exp(\alpha_a F\Delta\Phi_e / RT) \tag{5.7}$$

where the chemical rate constant k_a incorporates the chemical activation energy, $\Delta\Phi_e$ = the potential difference across the solid–solution interface at equilibrium, and a,b, ... are the electrochemical reaction orders (to be determined at constant potential!). In the following, we will briefly discuss some examples of electro-catalytic reactions mainly taken from the fields of water electrolysis (the evolution of H_2 and O_2, reactions also frequently encountered as counterreaction in electro-(in) organic processes) and of low-temperature fuel cells [137]. A few words will also be said about the link between electrochemistry and liquid-phase heterogeneous catalysis. A more extensive treatment of the subject of electro-catalysis can be found in a recent book [138] and a review paper [139].

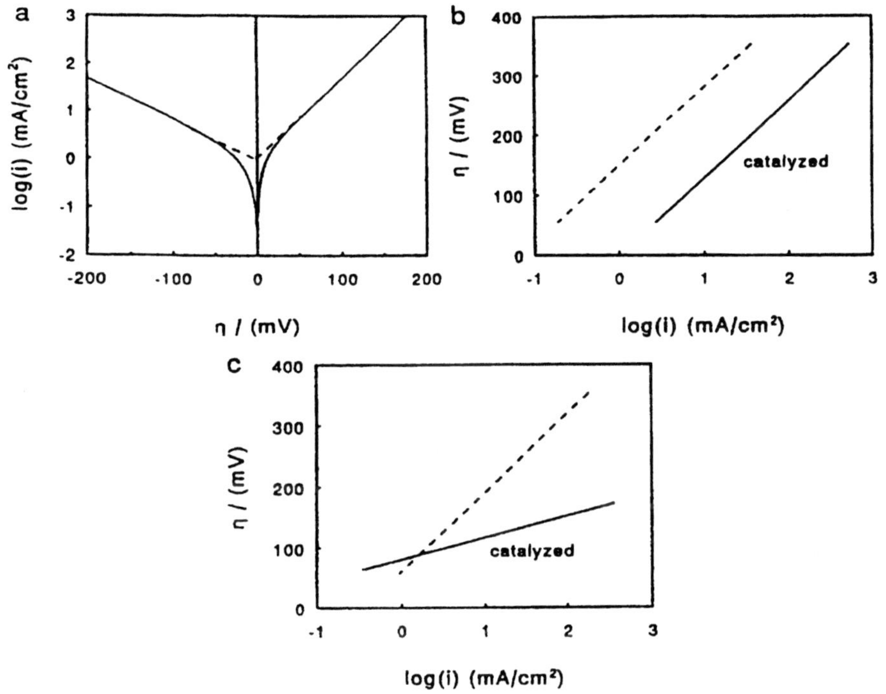

Fig. 5.39. (a) Extrapolation of semilogarithmic current voltage curves to the equilibrium potential, E_0, for determining the exchange current density i_0. (b) Increase of the exchange current density due to electrocatalysis. (c) Access of alternative mechanisms with decreased overpotential at high current densities due to electrocatalysis with the consequence of reduced slope of the semilogarithmic current voltage curve (from Ref. [139]).

5.6.2 Electrochemical Evolution of Hydrogen

The most frequently investigated electrochemical reaction is undoubtedly the hydrogen electrode, which has the overall reaction (assuming an acidic electrolyte)

$$H_2 \leftrightarrow 2H^+ + 2e^- \tag{5.8}$$

All the basic laws and concepts of electrode kinetics were developed and verified with the hydrogen electrode. Unfortunately, the H_2 electrode must be considered to be extremely complicated. This may well have been the reason for the relatively slow development of electrode kinetics [140].

The anodic reaction, H_2 oxidation, will be discussed in the next section. Here we will devote a few words to the cathodic reaction, H_2 evolution, the main aim being to show that the effectiveness of different metals in promoting this reaction can be understood in terms of a classical volcano relationship (see Chapter 3). Generally speaking, the elementary steps in the H_2-evolution reaction are the following: the first step is the neutralization of a proton to form an adsorbed H atom, the so-called Volmer reaction

$$H^+ + e^- \rightarrow H_{ad} \tag{5.9}$$

This is followed by either the Tafel reaction,

$$2H_{ad} \rightarrow H_2 \tag{5.10}$$

or the Heyrovsky reaction

$$H_{ad} + H^+ + e^- \rightarrow H_2 \tag{5.11}$$

(corresponding to a Langmuir–Hinshelwood and an Eley–Rideal mechanism, respectively). The complications encountered in the study of this reaction derive from the fact that the coverage with H_{ad} varies greatly from one material to another, that it depends on the applied potential, and that it, moreover, is sensitive to the electrolyte employed (specific adsorption of anions), and to electrode pre-treatment (e.g. whether or not there are still some oxygen ions in the surface layers of the metal).

With careful experimentation one can get around these problems, however, and independent of whether the Tafel or the Heyrovsky equation obtains in a particular case, one may expect the volcano relationship to apply when we plot the exchange current density vs. the M–H bond strength for a variety of metals. This is indeed what is observed (see Fig. 5.40): at intermediate values of the M–H bond strength, the evolution of H_2 is clearly the most effectively catalyzed.

5.6.3 Electro-oxidation of Hydrogen

A lot of attention has been paid to the electrochemical oxidation of H_2 in the context of fuel-cell research [141]. Obviously, materials that adsorb H_2 dissociatively should be the better catalysts, and this is borne out in practice. The best electrocatalyst is Pt (in acid), as indeed for the reverse reaction (Fig. 5.40), the elementary steps being simply [142]:

$$H_2 \rightarrow 2H_{ad} \tag{5.12}$$

$$H_{ad} \rightarrow H^+ + e^- \tag{5.13}$$

This is one of the few electrode reactions that does not follow the Butler–Volmer equation. The reason is that the dual-site dissociation of the H_2 molecule is the rate-controlling step. But although this is a non-electrochemical step, the reaction rate is still a function of the potential, because the hydrogen oxidation is self-

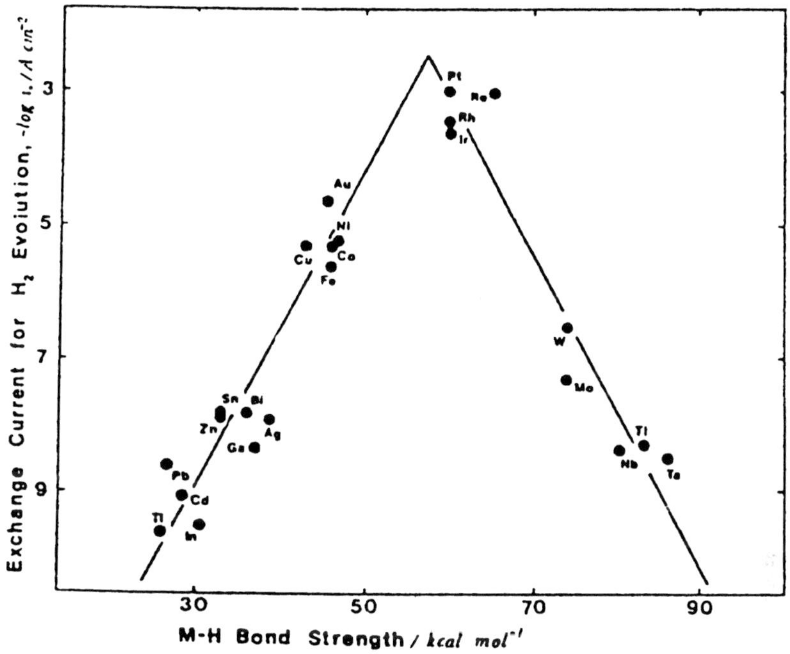

Fig. 5.40. Exchange currents for electrolytic hydrogen evolution vs. strengh of intermediate metal-hydrogen bond formed during electrochemical reaction itself. *E*(M–H) from Krishtalik.

poisoned by adsorbed hydrogen atoms, H_{ad}, and the hydrogen atom adsorption isotherm is a function of the polarization. At the equilibrium potential ($E_o = 0$ V), the Pt surface is nearly completely covered with H_{ad}, and the coverage decreases approximately linearly with the polarization to reach zero at about 0.3 V, where therefore the oxidation current is at its maximum.

Thus, the current density of H_2 oxidation on Pt is given by (remember that the Pt/H_2 electrode is reversible):

$$i = 2Fk_aC_{H_2}(1-\Theta)^2 - 2Fk_c\Theta^2 \tag{5.14}$$

At equilibrium, we have

$$i_o = 2Fk_aC_{H_2}(1-\Theta_0)^2 = 2Fk_c\Theta_0^2 \tag{5.15}$$

i_o on Pt has the high value of 10^{-2} A/cm^2, which makes H_2 by far the most reactive fuel for a fuel cell. The limiting current density (for $E > 0.3$ V) can be calculated to be

$$i_{lim} = i_o/(1-\Theta_0)^2 \tag{5.16}$$

with Θ_o = H_{ad} surface coverage at equilibrium (about 0.95).

We note in passing, that there is no particle size effect on the H_2-oxidation rate over Pt, although the Volmer reaction, the non-limiting step, is faster on the larger crystallites.

A problem in the context of low-temperature fuel cells is the extreme sensitivity of H_2 oxidation on Pt to traces of CO (ppm level; reformed fuel always contains some of it). The reason is that CO adsorbs much more strongly on Pt than H_2 and is only removed electro-oxidatively at rather high potentials ($CO_{ad} + H_2O \rightarrow CO_2 + 2H^+ + 2e^-$, $E > 0.5$ V), thus blocking the surface for the desired H_2-oxidation reaction. Adding some Sn, Ru, or perhaps Mo, to the Pt alleviates the problem to a certain extent, by promoting partial removal of CO at lower potentials, presumably through 'activation' of water molecules, but does not solve it. Anode catalysts exist that are insensitive to CO, an example being WC, but their intrinsic activity for the main reaction is unfortunately much too low.

5.6.4 Electrochemical Evolution of Oxygen

The interest in the electrochemical evolution of O_2 is connected mainly with its role in water electrolysis and in secondary batteries (e.g. Zn/O_2). In both cases an alkaline electrolyte is applied, almost without exception, and the discussion here will be limited to this milieu. The overall reaction is

$$4OH^- \rightarrow O_2 + 2H_2O + 4e^- \qquad (5.17)$$

The potential at which this reaction can be carried out is so high ($E > 1.3$ V) that even noble metals are covered with an oxidic layer; in fact, conducting oxides (e.g. Co-based spinels and perovskites) are often used in this process. Because of the forcing conditions, anode stability looms large, but that is another story. Many, very many mechanisms have been proposed for the reaction. Fortunately, they can be reduced to essentially two types, the same two types we have encountered in the case of H_2 evolution (*vide supra*), viz. the recombination of two adsorbate atoms and the reaction of an adsorbate with a solution species under the formation of an O–O bond ('peroxide' path). The prototype of the former mechanism is that of Krasil'shikov:

$$S + OH^- \rightarrow S - OH + e^- \qquad (5.18)$$

$$S - OH + OH^- \rightarrow S - O^- + H_2O \qquad (5.19)$$

$$S - O^- \rightarrow S - O + e^- \qquad (5.20)$$

$$2S - O \rightarrow O_2 + 2S \qquad (5.21)$$

where S denotes a metal ion at the electrode surface. The peroxide mechanism can be summarized as follows: after (5.18)–(5.20) the reaction continues with

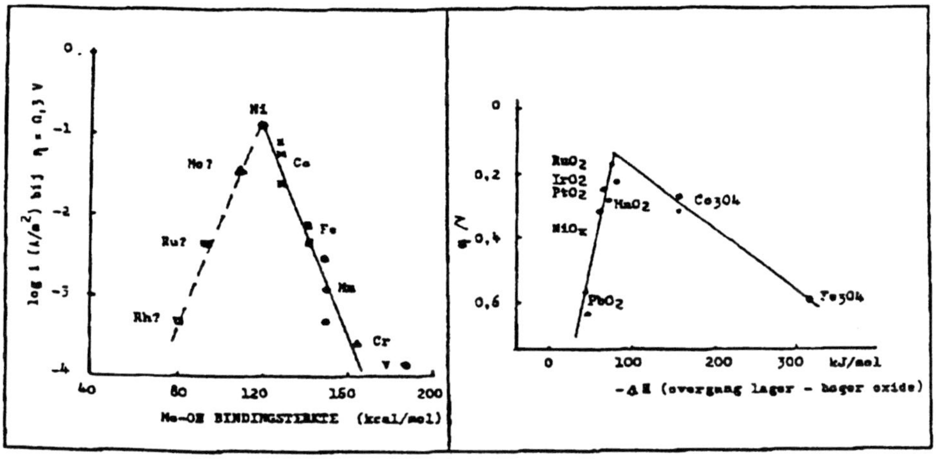

Fig. 5.41. A (left panel): Specific current density for O_2 evolution as a function of Me–OH bond strength. B (right panel): O_2-overpotential as a function of the enthalpy change of the transition from a lower to a higher oxide.

$$S - O + OH^- \rightarrow S - OOH + e^- \quad (5.22)$$

$$S - OOH + OH^- \rightarrow S + O_2 + H_2O + e^- \quad (5.23)$$

Interestingly, theoretical calculations [143] have indicated that both mechanisms can occur on one and the same electrocatalyst, in casu $SrFeO_3$, viz. as a function of the precise position (co-ordination sphere) of the metal ion at the surface (edge vs corner positions of Fe^{4+} in $SrFeO_3$). Since the proposed mechanisms are so similar to those pertaining to the electrochemical evolution of I I_2, it is obvious to look for a volcano-type relationship between the O_2-evolution reaction rate and, e.g., the metal ion–OH bond strength. Two attempts in this direction are summarized in Fig. 5.41 [144]: in Fig. 5.41a the activity is plotted against a calculated Me–OH bond strength, and in Fig. 5.41b against the enthalpy change of the transition from a lower to a higher oxide. The idea behind the latter parameter is that the active site S (a metal ion) shuttles between S and SO, so that the ease with which this cycle occurs will be a function of the energy difference between the two states. The observed correlations are not bad, but only present nevertheless a rather global view of the relation between the catalytic performance and the chemical properties of a given oxidic material.

We note in passing that ruthenium dioxide is not only an effective catalyst for O_2 evolution, but also for Cl_2 evolution. The latter reaction follows the path $Cl^- \rightarrow Cl_{ad} + e^-$, similar to (5.18) above, followed by $2Cl_{ad} \rightarrow Cl_2$, similar to (5.21) above. It turns out that at high current densities RuO_2 catalyzes the evolution of Cl_2 much more effectively than that of O_2, wherefore it is extensively used as an electrode material in commercial chlorine production. The origin of this subtle catalytic effect does not appear to be well understood.

5.6.5 *Electroreduction of Oxygen*

The interest in the electrochemical reduction of O_2 again stems from its pivotal role in low-temperature fuel cells [137,141]. And again, by far the best catalyst for this reaction, in acid electrolytes, is platinum.

The first two steps in the electroreduction of oxygen over Pt are:

$$O_2 + H^+ + e^- + Pt \rightarrow Pt - O_2H \tag{5.24}$$

$$Pt - O_2H + Pt \rightarrow Pt - O + Pt - OH \tag{5.25}$$

after which the adsorbed O and OH species are further reduced to H_2O (note that O_2 is thought to adsorb molecularly, the O–O bond breaking only occurring in a subsequent step).

The molecular oxygen reduction reaction is very slow, unfortunately, so that appreciable currents can be drawn only at relatively high overpotentials (i_0 is only of the order of 10^{-10} A/cm^2, and the Tafel slope varies from 60 to 120 mV/dec). One has also to guard against strongly adsorbing impurities in the electrolyte, as they impede reaction (5.25) and, thus, lead to the formation of H_2O_2 rather than H_2O, with a concomitant loss of efficiency ($2e^-$ vs $4e^-$ reduction), not to mention its corrosiveness. Improvement of the activity can sometimes be achieved through alloying the platinum with other metals. An interesting rationalization of the effect of alloying is due to Jalan and Taylor [145]: they propose that the activity increases with increasing 'fit' between the oxygen molecule and two Pt nearest neighbours (lower energy of activation for the bond-breaking reaction). And in fact, there is a reasonable correlation between specific activity and Pt–Pt nearest neighbour distance (Fig. 5.42). Of course, this does not prove a direct cause–effect relationship, but it is a pleasing chemical model, which moreover has recently been corroborated in a study of the effect of Pt microstructure on the O_2-reduction performance [146]. Nevertheless, an explanation in terms of changes in electronic properties may well turn out to be more appropriate in the end, but such an explanation does not exist yet.

Again, alternatives for Pt (-alloys) are not available (in acid), although there has been an active search for them. For a time, transition metal porphyrins and phthalocyanins supported on carbon were thought to be promising candidates, where their use was motivated by nature's use of metalloporphyrins for oxygen activation. The catalytic action of these chelates can be described in terms of the following mechanism:

$$Me^{3+} + e^- \leftrightarrow Me^{2+} \tag{5.26}$$

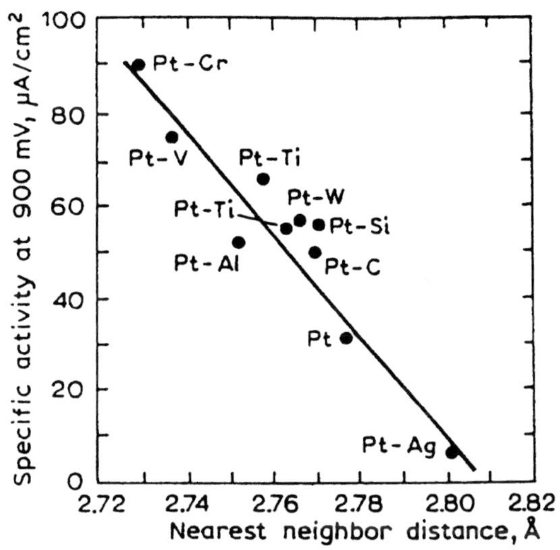

Fig. 5.42. Specific activity for oxygen reduction vs. electrocatalyst nearest neighbour distance; 100 % H_3PO_4, 200°C [145].

$$Me^{2+} + O_2 + H^+ \rightarrow Me^{3+}\,(O_2H) \tag{5.27}$$

$$Me^{3+}\,(O_2H) + H^+ + e^- \rightarrow Me^{2+} + H_2O_2 \tag{5.28}$$

where Me denotes the central metal ion in the large porphyrin ligand. The major drawback of these catalysts is immediately apparent from this mechanism, viz. they tend to reduce O_2 to the undesirable H_2O_2, rather than to H_2O. The point is that the reduction has to take place on a single site, which impedes O–O bond breaking — for such a thing to happen the Me^{4+} state should be accessible

$$Me^{3+}\,(O_2H) + H^+ + e^- \rightarrow Me^{4+}\,(O) + H_2O \tag{5.29}$$

and this is rarely the case, and certainly not for the most active systems [147]. Furthermore, in this case a volcano-type relationship can also be observed, viz. between the electrocatalytic activity and the Me^{2+}/Me^{3+} redox potential, but its interpretation is more involved than in the cases discussed above, and the interested reader is referred to the pertinent literature [148,149].

5.6.6 Electrochemical Oxidation of Alcohols

In automotive applications of fuel cells, where methanol is often considered to be a very attractive fuel, it would be advantageous to be able to oxidize methanol directly at the anode, thus avoiding the necessity of reforming it first. The product of the reaction being CO_2, a CO_2- rejecting electrolyte is needed, i.e. in general an acid.

Once again, the best catalyst turns out to be (promoted) Pt. Roughly speaking, three reaction steps can be distinguished in the electrochemical oxidation of MeOH over Pt:

$$CH_3OH \rightarrow \equiv C\text{–}OH + 3H^+ + 3e^- \tag{5.30}$$

$$H_2O \rightarrow \text{–}OH + H^+ + e^- \tag{5.31}$$

$$\equiv C\text{–}OH + OH \rightarrow CO_2 + 2H^+ + 2e^- \tag{5.32}$$

The first (fast) step is essentially a dehydrogenation reaction involving the methyl group. Interestingly, in UHV studies of the MeOH/Pt system the adsorption reaction has been found to form initially a methoxy species, i.e. the O–H bond is the first to break. In agreement with this, an isotope effect in MeOH adsorption kinetics is observed with CD_3OH/CH_3OH in the electrochemical case, and with CH_3OD/CH_3OH in the UHV case. As to why methanol should adsorb differently in the two cases, we refer to the HCOOH discussion above (Section 5.6.1).

The general view is that it is the activation of water, step (5.31), coupled with the strong adsorption of COH_{ad} (or its product, CO_{ad}), that is the difficulty here (as in the case of O_2 reduction, i_0 is only of the order of 10^{-10} A/cm^2). Actually, the status of CO_{ad}, i.e. whether it is a true reaction intermediate or merely a poison (as it is in the case of H_2 oxidation, cf. Section 5.6.3), is quite controversial What is certain is that CO_{ad} is there on the surface during the reaction — as shown by, e.g., *in situ* IR spectroscopy — and also that CO_{ad} obtained via adsorption of CO shows different oxidation kinetics from CO_{ad} derived form MeOH. What causes this latter phenomenon is not completely understood yet, but adsorption geometry (e.g. island formation) is surmised to be the main factor. Also, while CO can easily replace H_{ad}, MeOH cannot, and so its adsorption can only start in earnest when the Pt surface is substantially free from H_{ad} (i.e. at $E \geq 0.25$V, whereas $E_0 = 0.01$V), thus constituting another reason why MeOH cannot react at low potentials.

Many elements and compounds have been evaluated as possible promoters of the activity of Pt [150], the most promising system being PtRu supported on carbon (or, perhaps, on WO_3). The action of Ru is generally discussed in terms of accounting for reaction (5.31) — the bifunctional mechanism — but there are also indications that it modifies the bonding between Pt and the methanolic residue (ligand effect) [151]. It is this possibility of electronically influencing the catalytic action of Pt that keeps the hope of a sufficiently active anode material alive. In any case, the surface coverage of Pt by Ru should be relatively low, because the adsorption of methanol requires quite a large ensemble of Pt atoms, which would not be available were the Ru coverage to increase much beyond 0.1 monolayer.

In this context, it should be mentioned that once the Pt surface really starts being oxidized, i.e. when the oxygen coverage becomes noticeable, the methanol oxidation rate goes down as well. This is the origin of a particle-size effect for this reaction [151]: when the particle diameter falls below ~4 nm, the increased strength of adsorption of O(H), and by the way, also CO(H) [155], causes the specific MeOH electroxidation activity to decrease with decreasing Pt particle size.

Much has been published on the selective electrochemical oxidation of a large variety of organic compounds, among which the higher alcohols (e.g. ethanol to acetic acid, propanol-2 to acetone). The interesting point in the present context is that some of these conversions have also been studied purely catalytically in the liquid phase, employing catalysts such as Pt/C, with O_2 or air as the oxidant. It has been remarked [152] that such systems should also be considered from an electrochemical point of view. Indeed, it stands to reason that the overall oxidation reaction is essentially the sum of the two constitutive electrochemical half-reactions [153]. In the case of alcohol oxidation we would then have

$$R\text{–}CH_2\text{–}OH + 5OH^- \rightarrow R\text{–}COO^- + 4H_2O + 4e^- \tag{5.33}$$

$$O_2 + 2H_2O + 4e^- \rightarrow 4OH^- \tag{5.34}$$

where the reactions involve OH^- rather than H^+ because such conversions are usually carried out in an alkaline environment (e.g. at pH = 8). Showing that this view is correct is another matter, but it has been done [154] — the complete argument is however too long to be reproduced here. Essentially, the proof consists of showing (i) that the (diffusion-limited) O_2 reduction current is independent of the irreversible adsorbates deriving from the substrate alcohol, (ii) that the rate of alcohol oxidation at a given potential is the same in the electrochemical (no O_2 present, potential imposed) and in the catalytic (both O_2 and alcohol present, potential resultant) situation, (iii) that the amount and type of adsorbed intermediates is the same in both these situations, and (iv) that at the resulting catalyst potential, the anodic and catholic half-reactions are exactly equal (since no net current flows).

A summary picture is presented in Fig. 5.43. The catalytic process can be carried out in the kinetic and in the diffusion-limited regime. In the figure the former case is represented by the 'smooth Pt' data. The catalyst potential stabilizes at a rather high value (~1 V) — at lower values the alcohol oxidation would not be able to keep up with the O_2-reduction rate. At this potential the Pt surface is covered to an appreciable extent with O_{ad} species (that are not involved in the

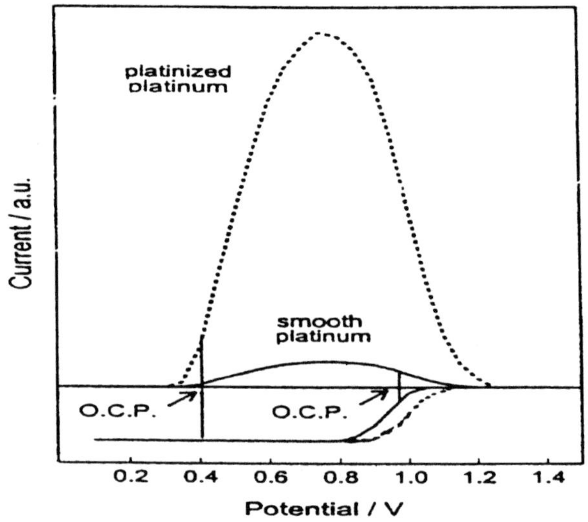

Fig. 5.43. Current-potential curves for ethanol oxidation and oxygen reduction on smooth and platinized platinum. OCP = Open Circuit potential (catalyst potential at which no net current flows). Upper part: alcohol electro-oxidation curves; bottom part: oxygen electro-reduction curves).

catalytic reaction, but simply block the surface — this indeed is why the alcohol oxidation current decreases above a certain potential), and it can be shown that deactivation occurring in this regime is connected with changes in the properties of this O_{ad} layer. In the diffusion-limited case, represented in Fig. 5.42 by the 'platinized Pt' data (a platinized Pt electrode has a very rough surface, thus allowing much higher current densities than smooth Pt — as long as the diffusion can keep up, which is not the case for O_2 reduction), the potential stabilizes at a much lower value, i.e. where the diffusion-limited O_2 reduction rate is matched by the alcohol oxidation rate. The electrode is now covered by alcohol adsorbates. For complex molecules like methyl-α-D-glucopyranoside (MPG) — where the idea is to oxidize it to the corresponding acid, an intermediate in the synthesis of vitamin C — deactivation is observed in this regime as well, and this can be shown to be due to the accumulation of an irreversible carbonaceous residue. While the initial chemisorption reaction is simply

$$C_7H_{14}O \rightarrow 6\text{–}CO + \text{–}CH_x + (14-x)H^+ + (14-x)e^- \qquad (5.35)$$

the CH_x species are surmised to transform over time (to graphite-like structures?) while blocking an increasing part of the surface. In consequence, the number of sites where the oxidation reaction can take place decreases over time, and so the activity per site needs to increase to keep up with the O_2-reduction rate, and this leads to an increase in the potential (Butler–Volmer again). Since the deactivation mechanisms just described have been elucidated using electrochemical methods, it is hoped that it is clear that the electrochemical view has much to commend itself here. We close by remarking that most liquid-phase heterogeneous reactions, whether oxidations or reductions (e.g. the hydrogenation of NO, or nitrate, to hydroxyl amine), will be amenable to the same approach.

REFERENCES

1 P. Sabatier and J.B. Senderenas, *C.R. Hebd. Seances, Acad. Sci.*, 134 (1902) 514.
2 E.J. Orlov, *Zhur. Ross. Khim. Ob-a*, 40 (1908) 1588.
3 F. Fischer and H. Tropsch, *Brennst. Chem.*, 4 (1923) 276.
F. Fischer and H. Tropsch, *Ber. Dtsch. Chem. Ges.*, 59 (1926) 830.
F. Fischer and H.Tropsch, *Brennst. Chem.*, 7 (1926) 97.
4 M. Ritschel and W. Vielstich, *Chem. Eng. Tech.*, 52 (1980) 327;
D. Kitzelmann and W. Vielstich, *Z. Phys. Chem.*, 112 (1978) 215.
5 S.R. Craxford and E.K. Rideal, *J. Chem. Soc.*, 1604 (1939).
R.B. Anderson in P.H. Emmett (Editor), *Catalysis*, Reinhold, N.Y., Vol. 4, 1956.
6 H.H. Storch, N. Columbic and R.B.Anderson in *The F.T. and Related Synthesis*, Wiley, N.Y., 1951.
H. Pichler, *Adv. Catal.*, 4 (1952) 271.
G. Blyholder and P.H. Emmett, *J. Phys. Chem.*, 64 (1960) 470 and refs. therein.
7 H. Pichler and H. Schulz, *Chem. Ing. Tech.*, 42 (1970) 1162.
8 (a) P.R. Wentrczek, B.J. Wood and H. Wise, *J. Catal.*, 41 (1976) 363.
(b) M. Araki and V. Ponec, *J. Catal.*, 44 (1976) 439;
J.W.A. Sachtler, J.M. Kool and V. Ponec, *J. Catal.*, 56 (1979) 284.
(c) J.A. Rabo, A.P. Risch and M.L. Poutsma, *J. Catal.*, 53 (1978) 295.
(d) P. Biloen, J.N. Helle and W.M.H. Sachtler, *J. Catal.*, 58 (1979) 95.
9 V. Ponec and W.A. van Barneveld, *Ind. Eng. Chem. Prod. Res. Dev.*, 18 (1979) 268.
10 N.A. van Barneveld and V. Ponec, *J. Catal.*, 88 (1984) 382.
11 R.C. Braddy III and R. Pettit, *J. Am. Chem. Soc.*, 103 (1981) 1287.
12 P.J. Flory, *J. Am. Chem. Soc.*, 58 (1936) 1877.
G.V. Schultz, *Z. Phys. Chem.*, 29 (1935) 299.
13 E.F.G. Harrington, *Chem. Industry (London)*, (1946) 346.
14 R.B. Anderson, R.A. Friedel and H.H. Storch, *J. Chem. Phys.*, 19 (1951) 313.
15 O.A. Hougen and K.M. Watson, *Ind. Eng. Chem.*, 35 (1943) 529.
K.H. Yang and O.A. Hougen, *Chem. Eng. Progr.*, 46 (1950) 146.
16 J. Ross, *J. Catal.*, 71 (1981) 205.
17 S. Van Ho and P. Hariott, *J. Catal.*, 64 (1980) 272.
R.L. Palmer and D.A. Vroom, *J. Catal.*, 50 (1977) 244.
18 C.P. Huang and J.T. Richardson, *J. Catal.*, 51 (1978) 1.
V.M. Vlasenko, L.A. Kukhar, M.T. Rusov and N.P. Sachenko, *Kinet. Katal.*, 37 (1972) 762.
19 M.A. Vannice, *J. Catal.*, 37 (1975) 462.
D.I. Ollis and M.A. Vannice, *J. Catal.*, 38 (1975) 514.
G.C. Bond and B.D. Turnham, *J. Catal.*, 45 (1976) 128.
20 A.O.I. Routavuouma and H.S. van der Baan, *Appl. Catal.*, 1 (1981) 243.
21 R.Z.C. van Meerten, J.G. Vollenbroek, M.H. de Croon, P.F. van Nisselroy and J.W.A. Coenen, *Appl. Catal.*, 3 (1982) 29.

22 (a) D.W. Goodman, R.D. Kelley, T.E. Madey and J.M. White, *J. Catal.*, 64 (1980) 479.
(b) N.M. Gupta, V.S. Kemble, K.A. Rao and R.M. Jver, *J. Catal.*, 60 (1979) 57.
(c) T.M. Duncan, P. Winslow and A.T. Bell, *J. Catal.*, 93 (1985) 1.
23 D.W. Goodman, R.D. Kelley, T.E. Madey and J.T. Yates, *J. Catal.*, 63 (1986) 226.
24 R.B. Underwood and A.T. Bell, *Appl. Catal.*, 21 (1986) 157. J.S. Rieck and A.T. Bell., *J. Catal.*, 96 (1985) 88; 99 (1986) 262. D.W. Goodman, *Appl. Catal.*, 19 (1984) 1. G. Broden, G. Haufer and H.P. Bonzel., *Surf. Sci.*, 84 (1979) 295. T. Okuhara, T. Enomoto, H. Tamaru and M. Misono, *Chem. Lett.*, (1984) 1491. T. Mori, A. Miyamoto, N. Takahishi, M. Fukugaya, M. Miizuma, T. Hattori and Y. Murakami, *Chem. Soc. Chem. Commun.*, (1984) 678.
25 D.J. Dwyer and J.H. Aardenbergh, *Appl. Surf. Sci.*, 19 (1984) 14.
26 T. Koerts, W. Welters, R.A. van Santen, L.E.Y. Nonneman and V. Ponec in *Natural Gas Conversion*, Stud. in Surf. Sci., 61 (1990) 235.
27 L.E.Y. Nonneman, A.G.T.M. Bastein, V. Ponec and R. Burch, *Appl. Catal.*, 68 (1990) L23.
28 A.G.T.M. Bastein, W.J. van der Boogert, G. van der Lee, H. Luo, B. Schuller and V. Ponec, *Appl. Catal.*, 29 (1987) 243. G. van der Lee and V. Ponec, *Catal. Rev. Sci. Eng.*, 29 (1987) 183.
29 H.Y. Luo, A.G.T.M. Bastein, A.A.J.P. Mulder and V. Ponec, *Appl. Catal.*, 38 (1988) 241.
30 A.A. Kiennemann, R. Breault, J.P. Hindermann and M. Laurin, *J. Chem. Soc. Faraday Trans I*, 83 (1987) 2119.
31 S.J. Tauster, *Acc. Chem. Res.*, 20 (1987) 389.
32 M.W. Twigg (Editor) *Catalyst Handbook*, Wolfe, London, 1989. J. Bart and R.P.A. Sneeden, *Catal. Today*, 2 (1987) 1.
33 G.C. Chinchin, M. Spencer, K.C. Waugh and D.A. Whan, *J. Chem. Soc. Faraday Trans I*, 83 (1987) 2193. G. Chinchin, K.C. Waugh and D.A. Whan, *Appl. Catal.*, 25 (1986) 101.
34 H.P. Nguyen, P. Meriaudeau and C. Naccache, *Appl. Catal.*, 21 (1986) 337.
35 Y. Kikuzono, S. Kagami, S. Naito, T. Onishi and K. Tamaru. *J. Chem. Soc. Faraday Disc.*, 72 (1981) 137.
36 V. Pitchon, H. Praliaud and G.A. Martin in *Natural Gas Conversion*, Stud. Surf. Sci. Catal., 61 (1990) 265.
37 J.M. Driessen, E.K. Poels, J.P. Hindermann and V. Ponec, *J. Catal.*, 82 (1983) 20.
38 V. Ponec, *J. Chem. Soc. Faraday Trans I*, 84 (1987) 2244. V. Ponec and L.E.Y. Nonneman in *Natural Gas Conversion*, Stud. Surf. Sci. Catal., 61 (1990) 225.
39 G.R. Sheffer and T.S. King, *J. Catal.*, 115 (1989) 376; 116 (1989) 488.
40 M. Amaru, M. Bettahar and D. Olivier, *Appl. Catal.*, 51 (1989) 141.
41 M.P. Walsh, *Platinum Metals Rev.*, 36 (1992) 126.
42 K.C. Taylor, *Catal. Rev.-Sci. Eng.*, 35 (1993) 457.
43 M. Shelef and G.W. Graham, *Catal. Rev.-Sci. Eng.*, 36 (1994) 433.
44 T. Engel and G. Ertl, *J. Chem. Phys.*, 69 (1978) 1267.
45 D.W. Goodman, and C.H.F. Peden, *J. Phys. Chem.*, 90 (1986) 4839.
46 M. Bowker, Q. Guo, and R.W. Joyner, *Catal. Lett.*, 18 (1993) 119.
47 S.H. Oh, G.B. Fisher, J.E. Carpenter, and D.W. Goodman, *J. Catal.*, 100 (1986) 360.
48 C.H.F. Peden, D.N. Belton, and S.J. Schmieg, *J. Catal.*, 155 (1995) 204.
49 V.P. Zhdanov and B. Kasemo, *Surf. Sci. Rep.*, 29 (1997) 31.
50 R.A. van Santen and J.W. Niemantsverdriet, *Chemical Kinetics and Catalysis*, Plenum, New York, 1995.
51 M.J.P. Hopstaken, W.J.H. van Gennip, and J.W. Niemantsverdriet, *Surface Sci.*, to be published.
52 H.J. Borg, J.F.C.J.M. Reijerse, R.A. van Santen and J.W. Niemantsverdriet, *J. Chem. Phys.*, 101 (1994) 10052.

53 A. de Koster and R.A. van Santen, *Surface Sci.*, 233 (1990) 366.
54 H.C. Kang and W.H. Weinberg, *Acc. Chem. Res.*, 25 (1992) 253.
55 R.J. Gelten, A.P.J. Jansen, R.A. van Santen, J.J. Lukkien,. J.P.L. Segers, and P.A.J. Hilbers, *J. Chem. Phys.*, 108 (1998) 5921.
56 S.H. Oh, *J. Catal.*, 124 (1990) 477.
57 G.B. Marin and J.H.B.J. Hoebink, *CaTTech*, 1 (1997)137.
58 R.M. van Hardeveld, A.J.G.W. Schmidt and J.W. Niemantsverdriet, *Catal. Lett.*, 41 (1996) 125.
59 B.J. Cooper, *Platinum Metals Rev.*, 38 (1994) 2.
60 R.M. Heck and R.J. Farrauto, *CaTTech*, 1 (1997) 117.
61 H. Bosch, F. Janssen, *Catal. Today*, 2 (1987) 369-532.
62 G.C. Bond, S. Flamerz Tahir, *Appl. Catal.* 1 (1991) 1–31.
63 T. Inui, A. Miyamoto, K. Aika, *Catal. Today*, 10 (1991) 1–118.
64 F. Janssen, *Thesis*, University Twente, the Netherlands 1987.
65 G.T. Went, L. Leu, A.T. Bell, *J. Catal.*, 134 (1992) 479–491.
66 P. Ciambelli, G. Bagnasco, L. Lisi, M. Turco, G. Chiarello, M. Musci, M. Notaro, D. Robba, P. Ghetti, *Appl. Catal.*, 1 (1992) 61–77.
67 J.P. Solar, P. Basu, M.P. Shatlock, *Catal. Today*, 14 (1992) 211–224.
68 L. Lietti, P. Forzatti, G. Ramis, G. Busca, F. Bregani, *Appl. Catal.*, 3 (1993) 13-35.
69 L. Lietti, P. Forzatti, *J. Catal.*, 147 (1994) 241–249.
70 G.T. Went, L. Leu, A.T. Bell, *J. Catal.*, 134 (1992) 492–505.
71 J. Nickl, D. Dutoit, A. Baiker, U. Scharf, A. Wokaun, Ber. Bunsen Ges. *Phys. Chem.*, 97 (1993) 217–228.
72 U.S. Ozkan, Y. Cai, M.W. Kumthekar, *Appl. Catal.*, 96 (1993) 365–381.
73 U.S. Ozkan, Y. Cai, M.W. Kumthekar, L. Zhang, *J. Catal.*, 142 (1993) 182–197.
74 F. Janssen, *J. Thermal Anal.*, 37 (1991) 1281–1288.
75 U. Baltensperger, M. Ammann, U.K. Bochert, B. Eichler, H.W. Gaggeler, *J. Phys. Chem.*, 97 (1993) 12325–12330.
76 M.A. Buzanowski, R.T. Yang, *Ind. Eng. Chem. Res.*, 29 (1990) 2074–2078.
77 J.W. Beeckman, L. Hegedus, *Ind. Eng. Chem. Res.*, 30 (1991) 969–978.
78 P. Ciambelli, G. Bagnasco, L. Lisi, M. Turco, G. Chiarello, M. Musci, M. Notaro, D. Robba, P. Ghetti, *Appl. Catal.*, 1 (1992) 61–77.
79 M. Kotter, H.-G. Lintz, T. Turek, D.L. Trimm, *Appl. Catal.*, 52 (1989) 225–235.
80 H.-G. Lintz, T. Turek, *Appl. Catal.*, 85 (1992) 13–25.
81 M. Turco, L. Lisi, R. Pirone, *Appl. Catal.*, 3 (1994) 133–149.
82 N.Y. Topsøe, T. Slabiak, B.S. Clausen, T.Z. Srnak, J.A. Dumesic, *J. Catal.*, 134 (1992) 742–746.
83 C.U. Odenbrand, P.L.T. Gabrielsson, J.G.M. Brandin, L.A.H. Andersson, *Appl. Catal.*, 78 (1991) 109–123.
84 G. Ramis, G. Busca, F. Bregani, *Catal. Lett.*, 18 (1993) 299–303.
85 M. Inomata, A. Miyamoto, Y. Murakami, *J. Catal.*, 62 (1980) 140–148.
86 F. Janssen, F. van den Kerkhof, H. Bosch, J.R.H. Ross, *J. Phys. Chem.*, 91 (1987) 5921-5927; *J. Phys. Chem.*, 91 (1987) 6633–6638.
87 M. Kantcheva, V. Bushev, D. Klissurski, *J. Catal.*, 145 (1994) 96–106.
88 J.P. Chen, R.T. Yang, *J. Catal.*, 139 (1993) 277–288.
89 B.L. Duffy, H.E. Curry-Hyde, N.W. Cant, P.F. Nelson, *J. Phys. Chem.*, 97 (1993) 1729–1732.
90 A. Miyamoto, K. Kobayashi, M. Inomata, Y. Murakami, *J. Phys. Chem.*, 86 (1982) 2945–2950.

91 N.Y. Topsøe, *Sci.*, 265 (1994) 1217–1219.
92 M. Takagi, T. Kawai, M. Soma, T. Onishi, K. Tamaru, *J. Phys. Chem.*, 4 (1976) 430–431.
93 W. Biffar, R. Drews, K. Hess, R. Lehnert, W.D. Mroß, O. Scheidsteger, *Brenstoffen, Wärme, Kraft*, 38 (1986) 211–216.
94 R.B. Bjorklund, C.U.I. Odenbrand, J.G.M. Brandin, H.L.H. Andersson, B. Liedberg, *J. Catal.* 119 (1989) 187–200.
95 M. Gasior, J. Haber, T. Machej, T.J. Czeppe, *J. Mol. Catal.*, 43 (1988) 359–369.
96 H. Schneider, S. Tschudin, M. Schneider, A. Wokaun, A. Baiker, *J. Catal.* 147 (1994) 5–14.
97 T.Z. Srnak, J.A. Dumesic, B.S. Clausen, E. Törnqvist, N.-Y. Topsøe, *J. Catal.*, 135 (1992) 246262.
98 G. Ramis, Y. Li, G. Busca, *Catal. Today*, 28 (1996) 373–380.
99 M.D. Amiridis, I.E. Wachs, G. Deo, J-M J, D.S. Kim, *J. Catal.*, 161 (1996) 247–253.
100 C.R. Adams and T.J. Jennings, *J. Catal.*, 2 (1963) 63–68.
101 C.R. Adams and T.J. Jennings, *J. Catal.*, 3 (1964) 549–558.
102 J.D. Burrington, C.T. Kartisek and R.K. Grasselli, *J. Catal.*, 63 (1980) 235–254.
103 J.L. Callaghan, R.K. Grasselli, E.C. Milberger and H.A. Streicker, *Ind. Eng. Chem., Prod. Res. Develop.*, 9(2), 134–142.
104 W.G. Etzkoen, J.J. Kurland and W.D. Neilsen, *Kirk-Othmer Encyclopedia of Chemical Technology*, 1 (1991) 232–251.
105 R.K. Grasselli, *Appl. Catal.*, 15 (1985) 127–139.
106 B. Gryzbowska, J. Haber and J. Janas, *J. Catal.*, 49 (1977) 150–163.
107 D.-H. H. He, W. Ueda and Y. Moro-oka, *Catal. Lett.*, 12 (1992) 35–44.
108 G. Hearne and M. Adams, 1948, US Patent 2,451,485, Shell Development Co.
109 J.D. Idol, 1958, US Patent, 2,904,580, Standard Oil Co, Ohio.
110 G.W. Keulks, L.D. Krenzke and T.M. Notermann, *Adv. Catal.*, 27 (1978) 183–225.
111 A.G. Knapsak, 1962, Dutch patent, 1,125,901.
112 Krenzke L.D. and G.W. Keulks, *J Catal.*, 64 (1980) 295–302.
113 J.H. Lunsford, *Langmuir*, 5 (1989) 12–16.
114 Y. Moro-oka W. Ueda, *Adv. Catal.*, 40 (1994) 233–273.
115 D.J. O'Driscoll, K.D. Campbell and J.H. Lunsford, *Adv. Catal.*, 35 (1987) 139–186.
116 W.M.H. Sachtler, *Rec. Trav. Chim.*, 82 (1963) 243–245.
117 A.W. Sleight, K. Aykan and D.B. Rogers, *J. Solid State Chem.*, 12 (1975) 231–236.
118 T.P. Snyder and C.G. Hill Jr., 31 (1989) 43–95.
119 W. Ueda, K. Asakawa, Cs.-L. Chen, Y. Moro-oka and T. Ikawa, *J. Catal.*, 101 (1986) 360–368.
120 H.H. Voge, C.D. Wagner and D.P. Stevenson, *J. Catal.*, 2 (1963) 58–62.
121 M.W.J. Wolfs, and Ph.A. Batist, *J. Catal.*, 32 (1974) 25–36.
122 R.D. Wragg, P.G. Ashmore and J.A. Hockey, *J. Catal.*, 28 (1973) 337–337.
123 P. Mars, D.W. van Krevelen, *Chem. Eng. Sci.*, 3 (1954) 41.
124 R.A. van Santen and H.P.C.E. Kuipers, The mechanism of ethylene epoxidation, *Adv. Catal.*, 35 (1987) 265–321.
125 P.A. Kilty and W.M.H. Sachtler, *Catal. Rev. Sci. Eng.*, 10 (1974) 1.
126 E.L. Force and A.T. Bell, *J. Catal.*, 44 (1976) 175.
127 R.A. van Santen and C.P.M. de Groot, *J. Catal.*, 98 (1986) 530.
128 R.A. van Santen, Theoretical Heterogeneous Catalysis, in *World Scientific Lecture and Course Notes in Chemistry*, Vol. 5, World Scientific, Singapore, 1991.
129 A.W. Czanderna, *J. Vac. Sci. Technol.*, 14 (1977) 408.
130 I.I. Moiseev, Catalytic oxidation, (R.A. Sheldon, R.A. van Santen, eds.), in *NIOK course*, World Scientific, Singapore, 1995, pp. 203–239.

131 H.J. Borg, R.M. van Hardeveld and J.W. Niemantsverdriet, *J. Chem. Soc. Faraday Trans.*, 91 (1995) 3679–3684.
132 J.L. Davis and M.A. Barteau, *Surf. Sci.*, 256 (1991) 50.
133 R.A. Sheldon, *J. Mol. Catal.*, 7 (1980) 107.
134 M.G. Clerici, G. Belussi and U Romano, *J. Catal.*, 129 (1991) 159. M.G. Clerici and P. Ingallina, *J. Catal.*, 140 (1993) 71.
135 J. O'M. Bockris and A.K.N. Reddy, *Modern Electrochemistry*, Vol. 2, Plenum Press, New York, 1970.
136 W.J. Albery, *Electrode Kinetics*, Clarendon Press, Oxford 1975.
137 J.A.R. van Veen, in R. A. Sheldon and R. A. van Santen (Eds.), *Catalytic Oxidation — Principles and Applications*, World Scientific, Singapore, 1995, pp. 137–150.
138 J. Lipkowski and Philip N. Ross (Eds.), *Electrocatalysis*, Wiley-VCH, 1998.
139 H. Wendt, S. Rausch and T. Borucinski, Advances in Applied Electrocatalysis, *Adv. Catal*, 40 (1994) 87.
140 K.J. Vetter, *Electrochemical Kinetics*, Academic Press, New York, 1967.
141 L.J.M.J. Blomen and M.N. Mugerwa (Eds.), *Fuel Cell Systems*, Plenum Press, New York, 1993.
142 P. Stonehart and P.N. Ross, *Catal, Rev. - Sci. Eng.*, 12 (1975) 1.
143 S.P. Mehandru and A. B. Anderson, *J. Electrochem. Soc.*, 136 (1989) 158.
144 S. Trasatti, in H. Wendt (Ed.), *Electrochemical Hydrogen Technologies*, Elsevier, Amsterdam, 1990, p. 104.
145 V. Jalan and E.J. Taylor, *J. Electrochem. Soc.* , 130 (1983) 2299.
146 J.A. Poirier and G. E. Stoner, *ibid.*, 141 (1994) 425.
147 A.L. Bouwkamp-Wijnoltz, W.H.M. Visscher and J.A.R. van Veen, *Electrochim. Acta*, 43 (1998) 3141.
148 J.A.R. van Veen and J.F. van Baar , *Rev. Inorg. Chem.*, 4(1982) 293.
149 J. Zagal, *Coord. Chem. Rev.*, 119 (1992) 89.
150 M.M.P. Janssen and J. Moolhuysen, *J. Catal.*, 46 (1977) 289; *Electrochim. Acta*, 21 (1976) 861-869.
151 T. Frelink, W. Visscher and J.A.R. van Veen, *Surf. Sci.*, 335 (1995) 353; *Langmuir*, 12 (1996) 3702.
152 S. Trasatti and R. Parsons, *Pure Appl. Chem.*, 58 (1986) 437.
153 C. Wagner and W. Traud, *Z. Elektrochem.*, 44 (1938) 391.
154 J.F.E. Gootzen, A. H. Wonders, A.P. Cox, W. Visscher and J.A.R. van Veen, *J. Mol. Catal. A.*, 127 (1997) 113.
155 S. Mukerjee and J. McBreen, *J. Electroanal. Chem.*, 448 (1998) 163.

Chapter 6

Homogeneous catalysis with transition metal complexes

6.1 INTRODUCTION

Three topics in homogeneous catalysis with organo-transition metal complexes will be dealt with in this chapter:
- rhodium catalyzed hydroformylation (6.2);
- zirconium catalyzed polymerization of alkenes (6.3); and
- enantioselective hydrogenation (6.4).

Homogeneous catalysis involves a much broader area than will be presented here, but a selection was necessary in order to remain within the scope of this book. The aim of this chapter is to give an impression of the current status of the field of homogeneous catalysis, expectations concerning the future developments of the crucial concepts and techniques, and all this in relation to present and potential industrial applications.

Formally speaking, homogeneous catalysis implies that all reacting partners are present in one phase, usually the liquid phase. A great variety of dissolved homogeneous catalysts are known: Brønsted and Lewis acids and bases, metal complexes, metal ions, organometallic complexes, organic molecules, enzymes, artificial enzymes, etc. Today, 'homogeneous catalysis' is often used in a restricted sense i.e. to refer to catalysis with organometallic and coordination complexes. This is certainly the area that has led to the largest number of industrial applications in the last three decades. From a scientific and application viewpoint the other types of catalysts are equally interesting. In this chapter we will focus on (organo)transition metal complexes as catalysts.

Homogeneous transition metal catalysts are used in such industrial processes as:

- toluene oxidation to benzoic acid;
- xylene oxidation to phthalic acid;
- ester condensation to polyesters;
- carbonylation of methanol to acetic acid (Rh, Ir);
- carbonylation of methyl acetate to acetic anhydride;
- polymerization of dienes to unsaturated polymers;
- hydroformylation of alkenes to aldehydes (solvents, detergents alcohols, plasticizer alcohols);
- oligomerization of ethene and propene;
- hydrocyanation of 1,3-butadiene to adiponitrile;
- cyclodi(tri)merization of 1,3-butadiene;
- asymmetric hydrogenation (e.g. L-DOPA, Metolachlor, Cilazapril));
- asymmetric isomerization (menthol);
- asymmetric epoxidation (disparlure)
- codimerization of 1,3-butadiene and ethene;
- oxidation of ethene to ethanal;
- ring opening metathesis polymerization of dicyclopentadiene and norbornene derivatives;
- hydrosilylation of alkenes;
- alternating polymerization of ethene and carbon monoxide.

For further details about these reactions the reader is referred to other works on this topic [1,2].

In this chapter the reader will be introduced to some of the concepts that are used in the study of homogeneous catalysts and the basic techniques applied to *in-situ* measurements. The study of ligand effects is one of the main themes in homogeneous catalysis [3]; the rate and selectivity of a given process can be optimized to the desired level through proper control of the ligand environment. As will be seen, this initially requires an empirical approach, but once certain trends are known, the variation of ligands can be carried out with some degree of sophistication. It is often clear what factors will influence the optimization process, but the sign of the effect is not always easy to predict. New conversions are most often found by accident; the role of design is still rather modest here.

The three topics described in the following sections have been chosen to be complementary to one another in many senses:

- bulk chemicals vs fine chemicals;
- early-transition metal catalysis vs noble metal catalysis;
- carbonylation vs hydrogenation vs alkene insertions;
- low molecular weight products vs polymers;
- commercialized examples, including very recent ones, vs a discovery that is just in its infancy.

The three processes have in common that a considerable body of fundamental and detailed knowledge about intermediates and the reaction pathways is already available, a typical feature of homogeneous catalysts.

6.2 RHODIUM CATALYZED HYDROFORMYLATION

6.2.1 Introduction

Functionalization of hydrocarbons from petroleum sources is mainly concerned with the introduction of oxygen into the hydrocarbon molecule. In general, two ways are open to achieve oxygen functionalization: oxidation and carbonylation. Oxidation is commonly encountered in the synthesis of aromatic acids, acrolein, maleic anhydride, ethene oxide, propene oxide, and acetaldehyde. Hydroformylation (CO/H_2) (older literature and the technical literature refer to the 'oxo' reaction) is employed for the large-scale preparation of butanol, 2-ethylhexanol, and detergent alcohols. The main use of 2-ethylhexanol is in phthalate esters which are softeners in PVC. The catalysts applied are based on cobalt and rhodium. (For a general review see Ref. [2]).

The hydroformylation of alkenes was accidentally discovered by Roelen [4] while he was studying the Fischer–Tropsch reaction (conversion of syn-gas to liquid fuels with a heterogeneous cobalt catalyst) in the late thirties. In what was probably designed as a mechanistic experiment Roelen examined whether alkenes were intermediates in the 'Aufbau' process for converting syn-gas (from coal, Germany 1938) to fuel. It took more than a decade before the reaction was taken further, but now it was the conversion of petrochemical hydrocarbons into oxygenates that was the driving force. It was discovered that the reaction was not catalyzed by the supported cobalt, but in fact by $HCo(CO)_4$ which was formed in the liquid state.

A key issue in the hydroformylation reaction is the ratio of linear to branched product produced. Figure 6.1 explains this colloquial expression. The linear product is the desired product, since the value of butanal is higher; also because this is the product which can be converted to 2-ethylhexanol via base-catalyzed aldol condensation and hydrogenation. Sometimes the aldol condensation reaction is carried out simultaneously with hydroformylation. When butanal is the desired product the condensation reaction is suppressed. The detergent alcohols

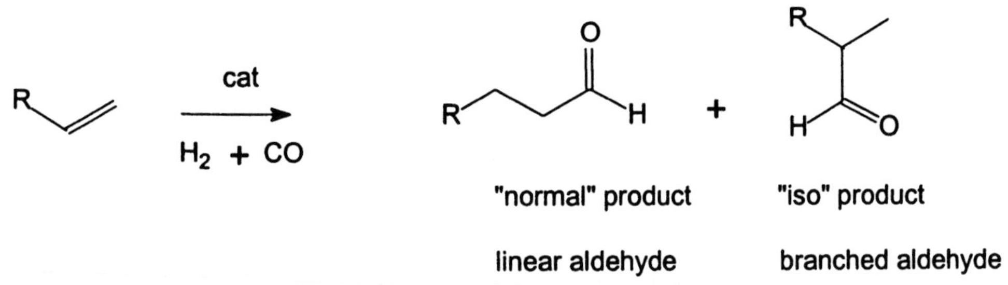

Fig. 6.1. Hydroformylation of terminal alkenes to linear and branched aldehydes.

should preferably be linear because their biodegradability has been reported to be better than that of the branched product. The linearity of the aldehyde product is therefore of great industrial importance. It is an interesting scientific question to see how the linearity can be influenced and maximized.

In the sixties it was recognized that ligand substitution on the cobalt carbonyl might influence the performance of the catalyst. It has been found that aryl phosphines or phosphites have little influence; in fact they may not even coordinate to cobalt under such high CO pressures. Tertiary alkyl phosphines, however, have a profound influence [5]; the reaction is much slower, the selectivity to linear products increases, the carbonyl complex formed, $HCoL(CO)_3$, is much more stable, and the catalyst acquires activity for hydrogenation. This process has been commercialized by Shell. As a result of the higher stability of the cobalt complex, the Shell process can be operated at lower pressures and higher temperatures (50–100 bar vs 200–300 bar for $HCo(CO)_4$, 170°C vs 140°C).

6.2.2 Rhodium-based Hydroformylation

With all the characteristics of the cobalt process in mind, it can be seen that there was a sizeable incentive to improve on this performance. Fundamental work by Evans, Osborn and Wilkinson [6] demonstrated that rhodium triphenylphosphine catalysts allowed the operation of the hydroformylation reaction at much lower pressure (1 bar) thus offering the prospect of much lower capital and operating costs. The selectivity was also reported to be considerably higher; virtually no hydrogenation was observed and the linearity was in some instances as high as 90%. The rhodium catalysts were reported to be three orders of magnitude faster in rate. The resulting milder reaction conditions would give much less aldol condensation product, if any. In 1971 Union Carbide, Johnson and Matthey, and Davy Powergas joined forces to develop a process based on this new discovery. For a comparison between the hydroformylation catalysts, see Table 6.1.

A convenient catalyst precursor is $RhH(CO)(PPh_3)_3$. Under ambient conditions this will slowly convert 1-alkenes into the expected aldehydes, while internal alkenes hardly react. At higher temperatures pressures of 10 bar or more are required. Unless a large excess of ligand is present the catalyst will also have some isomerization activity for 1-alkenes. The internal alkenes thus formed, however, will not be hydroformylated. Accordingly, the 2-alkene concentration will increase while the 1-alkene concentration will decrease; this will slow down the rate of hydroformylation. This makes the rhodium triphenylphosphine catalyst less suited for the conversion of alkenes other than propene, for which isomerization is immaterial. To date, industrial hydroformylation of higher alkenes is still carried out with cobalt catalysts.

TABLE 6.1

Summary of hydroformylation catalysts

Catalyst	Co	Co/phosphine	Rh/phosphine
Pressure, bar	200	70	30
Temperature, °C	140	170	120
Substrate alkene	C_3	C_3	C_3
	internal C_{10+}	internal C_{10+}	terminal
Product	aldehyde	alcohol	aldehyde
Linearity, %	60–70	70–90	60–95
Alkane by-product, %	2	10–15	0
Corrosion	++	++	–
Metal deposition	++	++	–
Heavy ends	++	++	–
Catalyst/Co costs	1	2	5000

Propene hydroformylation can lead to products with a linearity ranging from 60 to 95%, depending on the phosphine concentration. At very high phosphine concentration the rate is low, but the linearity achieves its maximum value. Low ligand concentrations, with concomitant low linearities, will give turnover frequencies in the order of 10,000 mol mol^{-1} h^{-1} at 10 bar and 90°C. In the presence of carbon monoxide this rhodium catalyst has no activity for hydrogenation and the selectivity, based on starting material, is virtually 100%. The butanal produced contains no alcohol and can be converted both to butanol and to other products, as desired.

In 1984 a new process has come on stream, the Ruhrchemie/Rhone-Poulenc process, employing a two-phase system with rhodium in a water phase and the substrate and the product in an organic phase. Two phase catalysis has become an important tool in homogeneous catalysis [7,8]. From the point of view of atom economy and energy utilization, the RC/RP process is highly attractive [7,9]. The catalyst used is a rhodium complex of meta-trisulphonated triphenylphosphine (sodium salt), which is highly water-soluble. The ligand, tppts, forms complexes with rhodium which are most likely very similar to the well-known triphenylphosphine complexes (i.e. $RhH(CO)(PPh_3)_3$). Ligands giving faster and even more selective catalysts have been reported [9]. The process using tppts is also suited for butenes. It is not applicable to the hydroformylation of higher 1-alkenes, because some isomerization to unreactive 2-alkenes takes place. The isomerization reaction is not fast enough to equilibrate alkene isomers, as cobalt catalysts would do. The selectivity to terminal products from such a mixture is low when phosphine based catalysts are used. Furthermore the solubility of higher alkenes in water is very low.

6.2.3 Ligand Effects

In the last two decades, the key to the successful development of homogeneous catalysts has undoubtedly been the exploitation of the effects that ligands exert on the properties of metal complexes: by tuning the electronic and steric properties of a catalytically active complex, selectivities and rates can be dramatically altered. There are many examples for each metal in the periodic table which show how one and the same metal can catalyze totally different reactions, depending on the surrounding ligands. Ligands that have been utilized to induce changes (by variation of their structure) in the outcome of catalytic reactions include phosphines, amines and imines, alkoxides, and cyclopentadienyl anions. In the following section one group of ligands, the phosphines, will be introduced in somewhat greater detail; similar approaches have been attempted for the other ligands but our current knowledge in the area of phosphine ligands is most advanced.

6.2.4 Phosphine Ligands

Phosphorus-based ligands have found widespread application not only in organometallic chemistry but also in industrial applications of homogeneous catalysis. Alkyl phosphines are strong bases and as expected they are good σ-donor ligands. Organophosphites are very weak bases but nevertheless they form very stable complexes with (electron-rich) transition metals. The metal-to-phosphorus bonding resembles that described for metal-to-ethene and metal-to-carbon monoxide (see Chapter 4). The σ-donation with the lone pair needs no further explanation. The strong binding of phosphites, especially to low-valent metals, suggests that π-back-bonding should also play a role. This is indeed the case. The current view is that the antibonding σ-orbitals of phosphorus to carbon (in the case of a phosphine) or to oxygen (in the case of a phosphite) play the role of the π-acceptor orbital on phosphorus. The σ-basicity and π-acidity of phosphorus ligands has been studied [2] in an elegant manner by looking at the stretching frequencies of the coordinated carbon monoxide ligands in complexes such as $NiL(CO)_3$ or $CrL(CO)_5$ where L is the phosphorus ligand. Strong σ-donor ligands give a high electron density on the metal and hence a substantial back-donation to the CO ligands and lowered IR frequencies result. Strong π-acceptor ligands will compete with CO for the electron back-donation and the CO stretch frequencies will remain high. The IR frequencies represent a reliable measure of the electronic properties of a series of phosphine ligands toward a particular metal. The latter restriction is important because it is clear that a stability series of, e.g., the proton basicity scale is not the same as the metal phosphorus ligand stability series. The electronic parameters of phosphorus ligands may differ somewhat from one metal to another. The electronic parameter for phosphorus

TABLE 6.2

χ-Value of phosphorus ligands

Ligand PR_3, R=:	χ-value:
t-Bu	0
n-Bu	4
Ph	13
CH_3O	20
PhO	29
Cl	41
F	55
CF_3	59

centred ligands, as determined for a certain metal becomes more reliable when the catalysis of interest utilizes the same metal complex. Thus, in a study of ligand effects for nickel (zero)-catalyzed reactions the Tolman parameters (see below) represent a good measure, but in a study of ligand effects on rhodium-catalyzed hydroformylation the electronic parameters should preferably be taken from a series of related rhodium complexes. Often the latter data are not available.

Tolman has defined an electronic parameter for phosphorus ligands based upon the vibrational spectra of $NiL(CO)_3$; with L=P(*t*-Bu)$_3$ as the reference, the electronic parameter χ (chi) for the other ligands is simply defined as the difference in the IR frequencies of the symmetric CO stretches of the two complexes. The variability of the phosphorous ligands is nicely reflected in the IR frequencies, which can be measured with sufficient accuracy to give different values for the relevant substituents. Carbon substituents give χ-values ranging from 0 to 20, phosphites are found in the range from 20 to 40, and halogen substituted compounds are found up to 59 (the unit should be cm^{-1}). For comparison, the ligands at the top of the range, such as PF_3, are as good electron acceptors as carbon monoxide, or even better. Some typical values have been collected in Table 6.2.

Studies have been published aiming at the use of χ-values in a quantitative manner in linear free-energy relationships in the same way as Hammett, Taft, and Kabachnik constants. Since the χ-values stem from IR frequencies rather than stability or chemical kinetic data, it is obvious that such a direct translation is not permissible.

Thus, the electronic parameter for ligands with the same donor atom can be fairly well measured and applied in a qualitative sense. Many attempts have been undertaken to define a reliable *steric* parameter complementary to the electronic parameter. Tolman's parameter θ (theta) is most often used. Tolman

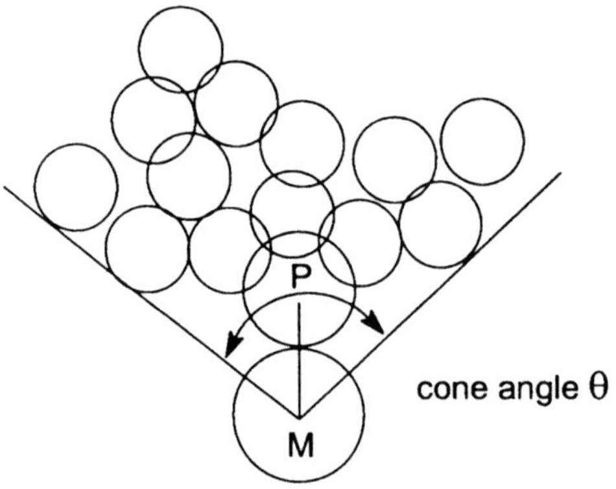

Fig. 6.2. Cone angle as defined by Tolman.

proposed measuring the steric bulk of a phosphine ligand from CPK models in the following way. From the metal centre, located at a distance of 2.28 Å from the phosphorus atom in the appropriate direction, a cone is constructed which embraces all the atoms of the substituents on the phosphorus atom (see Fig. 6.2). The cone angle is measured, and these cone angles θ (simply in degrees) are the desired steric parameters. Crystal structure determinations have shown that in practice the angles realized in the structures are smaller than the θ-values would suggest. For example, two *cis* triphenylphosphines may have a P–M–P angle as small as 95° whereas the θ-values would predict 145°. In reality intermeshing of the R substituents leads to smaller effective cone angles.

A complete separation between steric and electronic parameters is not possible; bending the ideal angles may be energetically more favourable for the overall energy of the complex than the 'inflexible' CPK approach, but changing the angles will also change the electronic properties of the phosphine ligand. Bending the alkyl groups away from their ideal angles will destabilize their bonding σ-orbitals and lower the antibonding π^* orbitals, thus enhancing π-back donation. Both the χ- and θ-values should therefore be used with some reservation. Some typical values for the steric parameter are presented in Table 6.3.

6.2.5 Ligand Effects in Rhodium Catalyzed Hydroformylation

Soon after the initial discovery of the hydroformylation activity of $RhH(CO)(PPh_3)_3$ it was found that ligands have a profound influence on the activity and selectivity of the rhodium catalysts. The majority of the numerous published patents and publications are concerned with the effect of ligands and it is therefore impossible to review this matter here. Instead we have made a choice that includes a study of the electronic and steric effects of phosphines and phosphites and a few examples of bidentate ligands. A few trends will be briefly summarized.

TABLE 6.3

θ-Values of phosphorus ligands, PR_3

R=:	θ-value:
H	87
CH_3O	107
PhO	128
n-Bu	132
Ph	145
i-Pr	160
C_6H_{11}	170
2-*t*-BuC_6H_4O	175
t-Bu	182
2,6-$Me_2C_6H_3O$	190
2-MeC_6H_4	194

6.2.5.1 *Catalyst preparation*

The 'catalyst' for hydroformylation is a rhodium(I) hydride species, which is clearly distinct from the species that are active for hydrogenation. The hydrogenation catalysts are cationic $Rh(I)^+$ or neutral Rh(I)Cl species. Carbonylation of alcohols also requires an ionic Rh(I) species, e.g. $[Rh(CO)_2I_2]^-$. Often rhodium(I) salts are used as the precursor for hydroformylation catalysts. Under the reaction conditions (H_2, CO, ligands, temperature >50°C) these salts are converted to a rhodium hydride complex, although there are several papers that seem to invoke cationic rhodium species as the catalysts. Chlorides have a particularly deleterious effect on the activity (i.e. they are not converted into hydrides under mild conditions) and it has been reported that the addition of bases such as amines has a strong 'promoting' effect on such systems:

$$L_3RhCl + H_2 + CO + R_3N \rightarrow L_3Rh(CO)H + R_3NH^+Cl^-$$

Rhodium salts of weaker acids are smoothly converted into rhodium hydride without the addition of base:

$$L_3RhA + H_2 + CO \rightarrow L_3Rh(CO)H + HA$$

A = acetate, 2,4-pentanedionate, hydrocarbyl sulphide etc.

6.2.5.2 *Phosphites: electronic effects*

Aryl phosphites were among the first ligands to be studied extensively. The results obtained with triphenylphosphine and triphenyl phosphite are strikingly

TABLE 6.4

Hydroformylation with rhodium phosphite and phosphine catalysts

Ligand PR_3 R=	χ-Value	θ-Value	Linearity of product[a]
n-Bu	4	132	71
n-BuO	20	109	81
Ph	13	145	82
PhO	29	128	86
2,6-$Me_2C_6H_3O$	28	190	47
4-ClC_6H_4O	33	128	93
CF_3CH_2O	39	115	96
$(CF_3)_2HCO$	51	135	55

[a](conditions: 90°C, 7 bar $CO/H_2 = 1:1$, substrate 1-heptene, ligand/rhodium varies from 3:1 to 20:1). (From Ref.[10]9; last two entries from Ref. [11]).

similar at low ligand concentrations. At higher ligand:metal ratios phosphites may retard the reaction. The results of the early work by Pruett and Smith [10] together with two other data [11], are summarized in Table 6.4. We have added the χ- and θ-values in this overview, which were not known at the time the catalytic work was reported.

One general trend that emerges from the table is that ligands with high χ-values give a higher selectivity toward linear products. In addition more isomerization is obtained; unfortunately this is often neglected in the reports. Preferably one would like to have also values for the absolute rates in addition to the selectivities and hence a detailed explanation of the results is not always possible. The trend towards higher linearities breaks down at two instances; one involves a rather bulky ligand with a θ-value of 190°, the other one involves hexafluoroisopropyl phosphite having a very high χ-value. Both give rise to unstable rhodium carbonyl complexes either for steric reasons (see below) or for electronic reasons. The χ-value of the hexafluoroisopropyl phosphite ligand is very high indeed and it is thought that this value of 51 means that electronically it is very similar to CO: i.e. a strong electron acceptor. Hence, the propensity of its mixed carbonyl hydride rhodium complexes to lose CO is likely to be similar to the rather unstable $HRh(CO)_4$.

6.2.5.3 Phosphites: steric effects

Very bulky phosphite ligands (see Fig. 6.3) [12,13,30,36] also yield unstable rhodium complexes (sterically hindered phosphites are commercially available as antioxidants for polyalkenes). The cone angle of the ligand shown is as high as

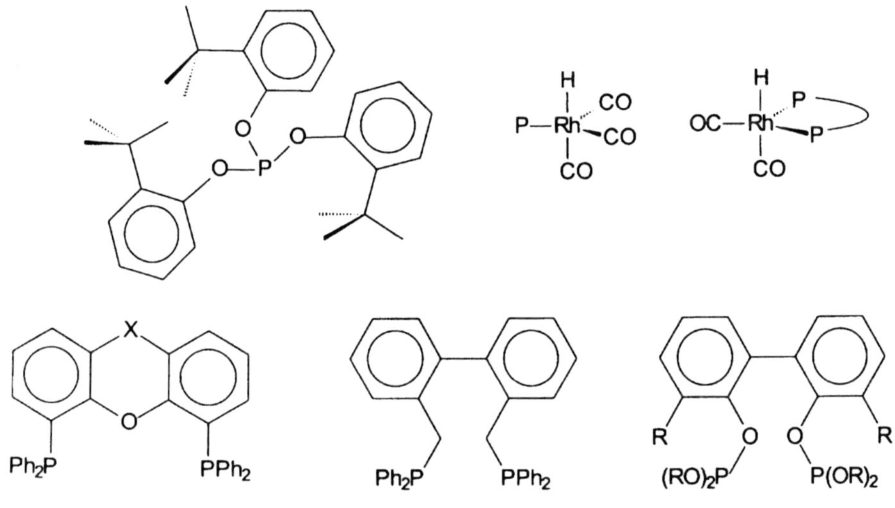

Xantphos BISBI diphosphite

Fig. 6.3. 2-*t*-Butylphenyl phosphite, a convenient bulky phosphite [36] and new bidentate ligands [21–25].

175° and the complex formed with rhodium has the formula $HRh(CO)_3L$; apparently there is not enough space for the coordination of a second phosphite. (In a trigonal bipyramidal complex there is room for two bulky ligands at the two axial positions but this would leave an equatorial position for a σ-bonded hydride which is energetically unfavourable.) The monoligand complex cannot be isolated and has only been observed under a pressure of CO and H_2 [12]. This complex easily loses CO which enables coordination of a molecule of alkene. As a result, complexes with bulky phosphite ligands are very reactive towards otherwise unreactive substrates, such as internal or 2,2-dialkyl-1-alkenes. The rate of reaction reaches the same values as those found with the triphenylphosphine catalysts and monosubstituted 1-alkenes, i.e. up to 15,000 mol of product per mol of rhodium complex per hour at 90°C and 10–30 bar. When 1-alkenes are subjected to hydroformylation with these monodentate bulky phosphite catalysts an extremely rapid hydroformylation takes place with turnover frequencies up to 160,000 mole of product per mole of rhodium per hour [13]. A moderate linearity of 65% can be achieved. Due to the very rapid consumption of CO the mass transport of CO can become rate determining, which results in highly unsaturated rhodium complexes giving a rapid isomerization of terminal to internal alkenes. In the extreme situation this means that it makes no difference whether we start with terminal or internal alkenes. Since hydroformylation of internal alkenes is also relatively fast with this catalyst the overall linearity obtained may become rather low (20–30%). This explains the low linearity quoted in Table 6.4 for 2,6-dimethylphenylphosphite.

6.2.5.4 Alkyl phosphines

Phosphorus ligands with low χ-values such as trialkyl phosphines, form very stable complexes in combination with CO and the activity for hydroformylation decreases. The strong binding due to a combination of donor and acceptor ligands inhibits ligand dissociation and no alkene complexation can occur. Mixed alkyl–aryl phosphines give active rhodium hydroformylation catalysts, although higher temperatures are required [14]. A novel, bimetallic rhodium catalyst has been reported by Stanley and coworkers [15]. This catalyst involves alkyl tetraphosphine ligands and positively charged dimeric rhodium(II) species. The positive charge reduces back-donation to carbon monoxide and active catalysts are obtained.

6.2.5.5 Diphosphines: electronic effects

Electronic effects of diphosphine ligands have been reported by Unruh, Casey and Van Leeuwen [16–18]. Unruh studied the hydroformylation of 1-hexene using substituted ferrocene derived diphosphines $Fe(C_5H_4PR_2)_2$ (Fig. 6.4).

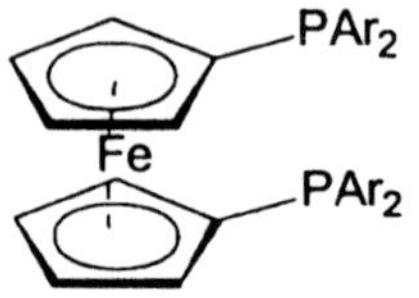

Fig. 6.4. Ferrocene based diphosphine ligands. (Ref. 16; for Ar see Table 6.5).

The electronic influence of the substituted aryl groups, R, follows a similar trend as that observed for phosphites; higher χ-values lead to higher linearities (Table 6.5). Interestingly, the authors also reported on the rates and selectivities

TABLE 6.5

Hydroformylation with substituted ferrocene based aryl diphosphines (Fig. 6.4)

Ar=	3.χ_i-Value (Ar)	Linearity %	Relative rate	Isomerization % 2-hexene
Ph	4.3	81	7.2	4
p-Cl-C_6H_4	5.6	87	9.3	5
m-F-C_6H_4	6.0	89	13.7	5
p-CF_3-C_6H_4	6.8	92	13.8	6

(Conditions 110°C, 8 bar CO/H_2 = 1:1, 1-hexene, Ref. [16], χ_i is defined as the χ-parameter for one substituent in the equation $\chi = \Sigma\chi_i$ for $(R_i)_3P$).

to the isomerization side reaction to 2-hexene. It turns out that under the conditions chosen only a slight increase to isomerized alkene occurs. The rate of hydroformylation increases with increasing π-acceptor capability of the ligand. This study clearly shows that the selectivity to normal products truly increases with increasing χ-values.

The effect can be explained along two lines and, in the absence of complete kinetic data, discrimination between the two is difficult. Firstly, there may be a direct electronic preference for the formation of a higher proportion of the linear alkyl intermediate when the π-back-donation to the phosphine ligand increases. The effectiveness also depends on the kinetics of the catalytic process. Alternatively, but with similar kinetic constraints, the effect of the χ-value may be indirect. It is known from the work with triphenyl phosphine that the linearity of the aldehyde product increases at increasing phosphine concentration. A steric explanation has been accepted [16,19] for this phenomenon; a more crowded complex will favour a linear alkyl intermediate (see also Figs. 6.7 a–f). Complexes of aryl phosphine ligands with higher χ-values are more stable, leading to the formation of more crowded complexes. If steric effects were dominant one would expect at least a slight decrease of the reaction rate with increasing χ-value and linearity. This certainly holds for the triphenyl phosphine modified catalysts. Since this is not the case here, it is tempting to accept a 'direct' electronic effect on the linearity, be it kinetic in origin (alkene insertion) or thermodynamic (*n*-alkyl-metal vs *i*-alkyl-metal stability).

6.2.5.6 *Effect of large bite angles*

An important development of the last decade has been the design, synthesis, and use of bidentate phosphorus ligands containing large bite angles [20–27]. The majority of bidentate ligands known hitherto coordinate to a metal in a *cis* fashion, i.e. the "natural" bite angle of the ligand (Casey) is close to 90°. In a trigonal bipyramidal compound such a ligand will assume an equatorial-apical position (see 6.2.7). The newly developed ligands have bite angles in the range 110–120°. These coordinate in a bis-equatorial fashion. Examples are shown in Fig. 6.3. Many of these ligands give rise to high linear-branched ratios in rhodium catalyzed hydroformylation. It is thought that the steric bulk created in the equatorial plane hinders the formation of branched alkyl species. The rates for the phosphine ligands can be enhanced by using electron-withdrawing substituents, as outlined in the previous paragraph. The effects of the electronic properties on the selectivity are not yet fully understood.

Both diphosphite and diphosphine based catalysts have been reported that will even hydroformylate internal alkenes to mainly linear products [26,27]. The ligands are based on bisphenol and Xantphos backbones shown in Fig. 6.3.

6.2.6 Kinetic Studies

In spite of its importance, only very few data have been reported on the kinetics of the hydroformylation reaction. The generally adopted reaction scheme is rather complex and an analytical expression for the rate, even for a system with only one ligand, would be very complicated.

The reaction sequence as proposed by Heck and Breslow [28] is shown in Scheme 6.1.

$$\underset{\mathbf{1}}{RhH(CO)_2L_2} \rightleftharpoons \underset{\mathbf{2}}{RhH(CO)L_2} + CO \qquad (1)$$

$$\underset{\mathbf{2}}{RhH(CO)L_2} + ol \rightleftharpoons \underset{\mathbf{3}}{RhH(CO)L_2(ol)} \qquad (2)$$

$$\underset{\mathbf{3}}{RhH(CO)L_2(ol)} \rightleftharpoons \underset{\mathbf{4}}{RhR(CO)L_2} \qquad (3)$$

$$\underset{\mathbf{4}}{RhR(CO)L_2} + CO \rightleftharpoons \underset{\mathbf{5}}{RhR(CO)_2L_2} \qquad (4)$$

$$\underset{\mathbf{5}}{RhR(CO)_2L_2} \rightleftharpoons \underset{\mathbf{6}}{RhC(O)R(CO)L_2} \qquad (5)$$

$$\underset{\mathbf{6}}{RhC(O)R(CO)L_2} + H_2 \rightleftharpoons \underset{\mathbf{2}}{RhH(CO)L_2} + RC(O)H \qquad (6)$$

$$\underset{\mathbf{6}}{RhC(O)R(CO)L_2} + CO \rightleftharpoons \underset{\mathbf{7}}{RhC(O)R(CO)_2L_2} \qquad (7)$$

Scheme 6.1.

In Fig. 6.5 the catalytic cycle is shown in a form that was introduced into this field by Parshall [1].

The scheme reduces to its most simple form when carbon monoxide is the only ligand present in the system, because equilibria of mixed ligand/carbon monoxide complexes do not occur. The kinetics of the hydroformylation reaction using hydrido rhodium carbonyl as the catalyst was studied by Markó [29]. For 1-pentene the rate expression found is:

$$\text{rate} = k[Rh][H_2][CO]^{-1} \qquad (6.1)$$

(conditions 75°C, CO pressure 40–170 bar, H_2 pressure 33–126 bar).

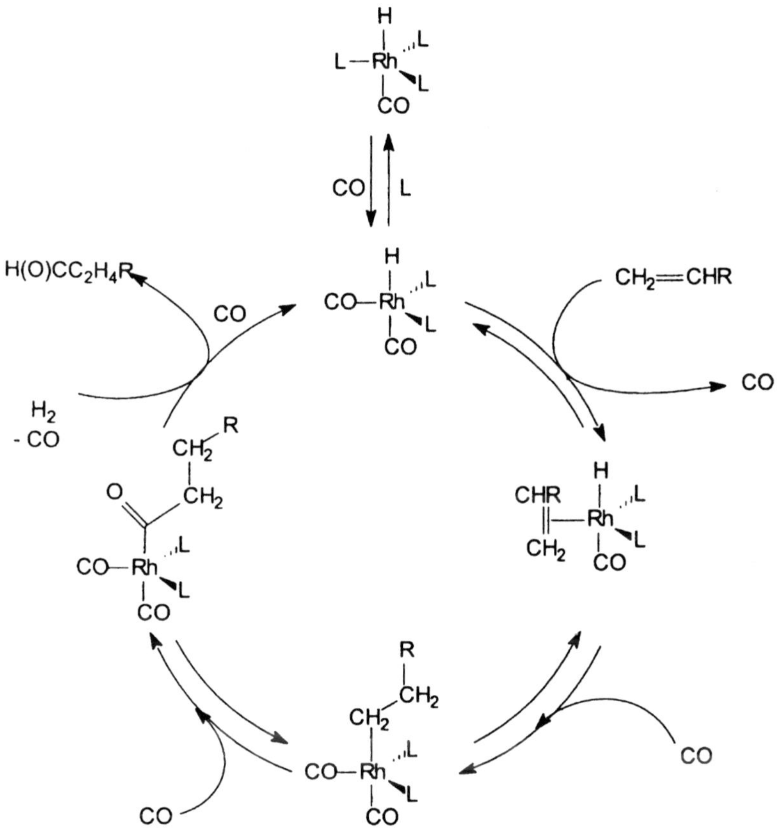

Fig. 6.5. Hydroformylation of propene by $HRh(CO)L_3$.

The reaction is first order in rhodium catalyst concentration, first order in dihydrogen pressure and has an order of minus one in carbon monoxide pressure. In our Scheme 6.1 this would be in accord with a rate-determining step at the end of the reaction sequence, e.g. reaction 6. Since the reaction order in substrate is zero, the rhodium catalyst under the reaction conditions predominates as the alkyl or acyl species; any appreciable amount of rhodium hydride occurring under fast pre-equilibria conditions would give rise to a positive dependence of the rate of product formation on the alkene concentration. The minus one order in CO suggests that the acyl species rather than the alkyl species is dominant under the reaction conditions. The negative order in CO is explained [29,30] by equilibrium 7. The saturated complex loses CO, and subsequently the unsaturated 16-electron species reacts with H_2 to give aldehyde and rhodium hydride (reaction 6).

The hydroformylation of 1-hexene catalyzed by rhodium carbonyl has been studied by Lazzaroni and coworkers [31]. They were particularly interested in the influence of reaction parameters on the regioselectivity and the chemoselectivity (to aldehyde and 2-hexene). To understand their results we have to extend Scheme 6.1 by taking account of the formation of linear and branched aldehydes, as well as of isomerization. This is shown in Scheme 6.2.

$$\underset{\mathbf{3}}{RhH(CO)L_2(ol)} \rightleftharpoons \underset{\mathbf{4i}}{Rh(i\text{-}R)(CO)L_2} \tag{3i}$$

$$\underset{\mathbf{4i}}{Rh(i\text{-}R)(CO)L_2} \rightleftharpoons \underset{\mathbf{2}}{RhH(CO)L_2} + i\text{-}ol \tag{8i}$$

$$\underset{\mathbf{4i}}{Rh(i\text{-}R)(CO)L_2} + CO \rightleftharpoons \underset{\mathbf{5i}}{Rh(i\text{-}R)(CO)_2L_2} \tag{4i}$$

$$\underset{\mathbf{5i}}{Rh(i\text{-}R)(CO)_2L_2} \rightleftharpoons \underset{\mathbf{6i}}{RhC(O)i\text{-}R(CO)L_2} \tag{5i}$$

$$\underset{\mathbf{6i}}{RhC(O)i\text{-}R(CO)L_2} + H_2 \rightleftharpoons \underset{\mathbf{2i}}{RhH(CO)L_2} + i\text{-}RC(O)H \tag{6i}$$

$$\underset{\mathbf{6i}}{RhC(O)i\text{-}R(CO)L_2} + CO \rightleftharpoons \underset{\mathbf{7i}}{RhC(O)i\text{-}R(CO)_2L_2} \tag{7i}$$

Scheme 6.2.

The rate constants and R groups with the subscript *i* represent the pathways and products of the *iso* (or branched) species; the linear products and their rate constants are without subscript, as in Scheme 6.1. Reaction 8i has been added to the scheme to represent the β-hydride elimination from the secondary or iso-alkyl intermediate to give 2-alkene, the isomerization product. They varied the temperature from 15 to 110°C, and the pressure from 40 to 200 bar [31]. Unfortunately the ratio of CO and H_2 was kept constant at 1:1 throughout this study. Chemoselectivity to aldehydes is complete at room temperature, the selectivity to linear aldehyde being about 50%. At higher temperatures and lower pressures the selectivity to aldehyde decreases as a consequence of isomerization to 2-hexenes; the 'escape' step (8i) in Scheme 6.2. As long as 1-hexene is present 2-hexene is not converted. At a later stage, when all 1-hexene is consumed 2-hexene is also converted to branched aldehydes (with formyl groups at the 2- and 3-positions), albeit at a much lower rate. When isomerization occurs the apparent initial linearity of the aldehyde product is much higher than 50%, as is observed in the absence of isomerization. After total conversion of the 1- and 2-alkenes the average linearity is again 50% (rounded figures). Hence the authors conclude that the branched alkyl intermediate Rh(*i*-R)(CO)L_2 (in Scheme 6.2) has a tendency to form the isomerized alkene at higher temperatures and lower pressures (particularly CO, reaction 4i, must be responsible) instead of reacting further to the acyl species and product. Many studies concerning product linearity have been in error due to the neglect of this phenomenon.

A system kinetically very similar to the phosphine-free rhodium carbonyl catalyst is obtained with bulky phosphites (Fig 6.3) [30]. At temperatures from 50 to 80°C, and CO and H_2 partial pressures ranging from 10 to 70 bar, the rate of aldehyde formation is first order in H_2 and approximately minus one order in CO. The reaction rate is independent of the concentration of 1-octene at conversions below 30%. The reaction was found to be first order in rhodium concentration and insensitive to the phosphite/rhodium ratio, provided that the absolute concentration was sufficiently high to generate a hydride complex from the pentanedionate precursor (reaction 9).

$$Rh(CO)_2(acac) + L + H_2 + CO \rightarrow RhH(CO)_3L + Hacac \qquad (9)$$

In these experiments the isomerization was also recorded and was found to be rather low (6%). The low selectivity to isomerized products has consequences with regard to the mechanism. Let us return to the kinetic expression (eqn. 1) which states that the reaction is first order in rhodium and H_2, zero order in alkene, and minus one order in CO. In the extreme case of rapid pre-equilibration up to reaction 6 one would expect the system to go back and forth very fast between species **1** and **5**. As we have learned from the work by Lazzaroni this would implicate that the isoalkyl species **3i** would regenerate alkene complex **2** now containing the isomerized alkene. The isomerized alkene, however, is not observed, or only in very minor quantities. This means that backward reactions 3 and 8i do not occur. The first reactions of the cycle determining the regio- and chemoselectivity are therefore irreversible while reaction 6 is still rate-determining.

Two mechanisms have been proposed for the last step of the catalytic cycle, reaction 6 (Scheme 6.1), a direct reaction of complex **7** with H_2 and a reaction of the hydride complex **1** and the acyl complex **6** or **7** to give aldehyde plus a rhodium dimer:

$$RC(O)RhL_n + H_2 \rightarrow RC(O)H + HRhL_n \qquad (6a)$$

$$RC(O)RhL_n + H_2 \rightarrow RC(O)H_2RhL_n \qquad (6b)$$

$$RC(O)H_2RhL_n \rightarrow RC(O)H + HRhL_n \qquad (6c)$$

$$RC(O)RhL_n + HRhL_n \rightarrow RC(O)H + Rh_2L_{2n} \qquad (6d)$$

Reaction 6a may proceed directly without intermediates as an hydrogenolysis reaction as has been found experimentally for ruthenium analogues [32] and as has been found to be a likely pathway for cobalt acyl complexes on the basis of *ab initio* calculations [33]. Alternatively, reaction 6a may proceed via an oxidative addition mechanism followed by reductive elimination. The product of the oxidative addition reaction has been observed in the stepwise hydroformylation cycle of the iridium analogue of this rhodium catalyst [34]. When 6c is fast with respect to 6b the two mechanisms 6a and 6b,c cannot be distinguished experimentally. There are also several experiments in support of mechanism 6d being the last step of the hydroformylation cycle [19,35 and references therein]. If 6d were operative and rate-determining the rate equation would of course show the bimolecular nature of step 6d; this type of dependency has not been reported for rhodium catalysts. All of the experiments concerning reactions 6a–d, however, refer to model reactions rather than to active catalytic systems, with one exception. During the hydroformylation of cyclohexene or cyclooctene with $HRh(CO)_3L$ (L=bulky phosphite) a very rapid exchange of the rhodium bound hydride with the cycloalkene 'sink' occurs, which is an order of magnitude faster than hydroformylation. Hence, when deuteroformylation (i.e. a mixture of D_2 and CO is used) is carried out, a fast exchange of $DRh(CO)_3L$ with the solvent hydrogen sink takes place converting the rhodium complex into the hydride, as was confirmed by *in situ* IR [12]. The two equations 6a and 6d now read as follows:

Fast exchange with solvent: $\quad DRhLn \rightarrow HRhLn$

$$RC(O)RhL_n + D_2 \rightarrow RC(O)D + DRhL_n \qquad (6a_d)$$

$$RC(O)RhL_n + HRhL_n \rightarrow RC(O)H + Rh_2L_{2n} \qquad (6d_d)$$

In this system we can now distinguish between the two pathways. A direct reaction with D_2 will give a deuterium label in the aldehyde product. Due to the fast exchange with excess hydrogen in the alkene solvent, rhodium hydride will carry hydrogen, initially at least, instead of deuterium. Consequently reaction $6d_d$ will give an undeuterated product aldehyde. In the actual experiment, 60% of the product consisted of the deuterated aldehyde, which proves that reaction 6a is the mechanistic pathway in this active catalytic system. That the degree of deuteration is not quite 100% was ascribed to the production of HD and H_2 during the experiment, kinetic isotope effects, and mass transfer limitations. When a mixture of D_2 and H_2 was used, only hydridoaldehyde was observed. At least for one rhodium catalyst it is now clear that a direct reaction of dihydrogen with the acyl rhodium species concludes the catalytic cycle.

6.2.6.1 Kinetics of the triphenylphosphine catalyst

The most detailed published study on the kinetics of the hydroformylation using triphenylphosphine containing rhodium catalysts is by Cavalieri d'Oro et al. [38]. As the overall kinetic equation they found:

$$\text{rate} = k[Rh][PPh_3]^{-0.7}[\text{alkene}]^{0.6}[H_2]^{0.05}[CO]^{-0.1}$$

(conditions: 90–110°C, 1–25 bar CO, 1–45 bar H_2, PPh_3/Rh ratio 300:1–7:1).

This rate equation tells us that the reaction rate is almost independent of the syn-gas pressure; only a very slight suppression of the rate is observed while raising the CO partial pressure. Triphenyl phosphine retards the reaction within the limits studied (reaction order in PPh_3 –0.7). A positive order is found in alkene substrate concentration, although the order is not quite 1.0 but only 0.6. These results point to a rate-determining step that occurs early in the catalytic cycle. Step (1) as the rate-determining step would lead to a minus one order in CO (or PPh_3 when this is the dissociating ligand) and zero order in alkene. Complexation of alkene to the unsaturated complex (step 2) would give a plus one order in alkene and, depending on the correct kinetics, roughly a minus one order in CO pressure (or PPh_3 concentration). Fast insertion of alkene (step 3), followed by a rate-determining reaction with CO (or PPh_3)(step 4) and/or rate-determining insertion of CO (step 5) would lead to approximately the rate expression found experimentally.

The linearity of the aldehyde product increases with the concentration of triphenyl phosphine. This is being exploited in the Union Carbide process for the hydroformylation of propene in which linearities >90% are obtained. The rate, however, drops to lower values and the most likely explanation for the higher linearities in this system would seem to be the steric congestion around the rhodium atom at high phosphine concentrations, which kinetically and thermodynamically favours the formation of linear alkyl rhodium complexes relative to branched alkyl rhodium complexes. Product linearity decreases with the number of triphenyl phosphines present in the series of precursor complexes:

$$HRh(PPh_3)_3(CO) > HRh(PPh_3)_2(CO)_2 > HRh(PPh_3)(CO)_3$$

In this system higher CO pressures lead to lower linearities simply by shifting the complex equilibria to the species containing less phosphine. When isomerization plays a role under the reaction conditions applied, higher CO pressures will also give lower initial linearities by suppression of reaction (8i), the isomerization reaction as outlined above for the carbonyl or phosphite rhodium catalysts. These two effects can be clearly distinguished by monitoring the isomerization reaction.

While the dependence on phosphine concentration and carbon monoxide strongly depends on the reaction parameters, it is important to note most catalyst systems will show a first-order dependence in alkene and rhodium concentration. A recent study involving fluoroalkylphosphine ligands illustrates this once again [37]. Thus, with very few exceptions, complexation of alkene to the metal centre is the rate-determining step. One exception was discussed above, the systems containing no ligand or containing bulky monophosphite ligands. Substrates (or their impurities) that are able to give π-allylic intermediates lead to different kinetics as the insertion of carbon monoxide in these stabilized species is relatively slow.

6.2.7 The Characterization of Intermediates

The intermediates which play a role in a cycle of a homogeneous catalyst can be characterized by various spectroscopic techniques such as NMR, IR, Raman spectroscopy, and UV-vis spectroscopy. Also, intermediates may crystallize from a reaction mixture and the structure can then be solved with a single-crystal X-ray determination. Only on rare occasions do intermediates crystallize from the reacting systems since their concentrations are low. Often one turns to model compounds of the actual catalyst by changing the ligand or the metal. For example, iridium complexes show the same catalytic behaviour as the rhodium complexes. Since they are often much slower as catalysts the intermediates can be intercepted (see below). Another common approach is the synthesis of a ligand that simultaneously contains the substrate of the catalytic reaction; this may also lead to the isolation of likely intermediates.

Of the spectroscopic techniques NMR is a particularly useful tool because many of the nuclei of interest in homogeneous catalysts are amenable to NMR studies: ^{1}H, ^{13}C, ^{15}N, ^{31}P, ^{103}Rh, and ^{107}Pt. Two general drawbacks should be mentioned. First the concentration region that can be studied by NMR (10–100 mmol l^{-1}) is well above the region for active catalysts (<1 mmol l^{-1}). Metal–ligand equilibria may shift considerably with concentration, which makes a direct comparison of NMR data and catalytic runs difficult. The reaction may become very fast at the high concentration used for NMR: too fast for an *in-situ* study. Secondly, when gases are consumed during a reaction it is impossible to run *in-situ* spectra; high-pressure NMR tubes allow the study of complexes under pressure, but when gas is consumed the pressure will drop quickly (high catalyst concentration) and mass transfer may also be a limiting factor in the absence of stirring.

In-situ IR measurements overcome the problems mentioned for *in situ* NMR spectroscopy. The information that we obtain from vibrational spectroscopy is far less detailed, however, than that from NMR. The concentration of the catalyst may be equal to the one used in practical catalytic systems. Secondly, autoclaves have been equipped with IR cells, either as flow cells or via real *in-situ* monitoring in the 'Moser' cell (see below), which allows one to work with gaseous reactants. In the following we will mention a (very) few examples of complexes that may be intermediates in the hydroformylation reaction observed with these two techniques.

6.2.7.1 NMR measurements

NMR measurements of ^{1}H, ^{13}C, and ^{31}P may be very informative about the structure of the complex. The hydride precursor of the catalyst has a trigonal bipyramidal structure in which the two phosphorus ligands present can occupy either two equatorial sites or one axial and one equatorial site (Fig. 6.6).

When ligand exchange is slow on the NMR timescale the structure can indeed easily be deduced from the spectra. A bidentate ligand such as $PPh_2NMeCHMe$ $CHPhOPPh_2$ [39] (we neglect here the asymmetric nature of the ligand), which has two very distinct absorptions for the amino phosphorus and the phosphinite phosphorus nuclei, gives ^{1}H and ^{31}P spectra that can readily be interpreted. The hydride gives a doublet around –8.7 δ with a large coupling constant (116 Hz) with the phosphorus atom *trans* to it and smaller couplings with rhodium (10.1 Hz) and the phosphinite phosphorus nucleus (9.9 Hz). In the phosphorus spectra the expected couplings are with rhodium, the mutual phosphorus coupling; and in the proton coupled spectra the proton coupling for the phosphorus *trans* to it.

Many catalyst precursors have been identified by NMR recently [16,21,23,24, 39,40]. Especially those containing bidentate ligands can be easily analyzed. The coupling constant of hydrogen to phosphorus is very small when they are cis. Values of 10 Hz are often indicative of fast exchange between isomers containing equatorial and apical ligands. Coupling constants of phosphorus-to-phosphorus are very large for two ligands in equatorial positions, but small for ligands in a cis orientation.

Brown and Kent [19] studied the species involved in the hydroformylation of styrene and 1-decene with the catalyst precursor $RhH(PPh_3)_3(CO)$ at ambient pressure and temperatures of –80° to 25°C. Under an atmosphere of syn-gas $RhH(PPh_3)_3(CO)$ (a) (Fig. 6.7) transforms into $RhH(PPh_3)_2(CO)_2$ (b) and (c) and free PPh_3. Typical concentrations are 0.02 mol/l, which is considerably higher than the concentrations used in typical catalytic experiments (<0.5 mmol/l). At room temperature a rapid exchange with free phosphine and CO is observed. Two conformations for $RhH(PPh_3)_2(CO)_2$ are observed: isomer (b) has two equatorial phosphines and the other isomer (c) has one equatorial and one axial

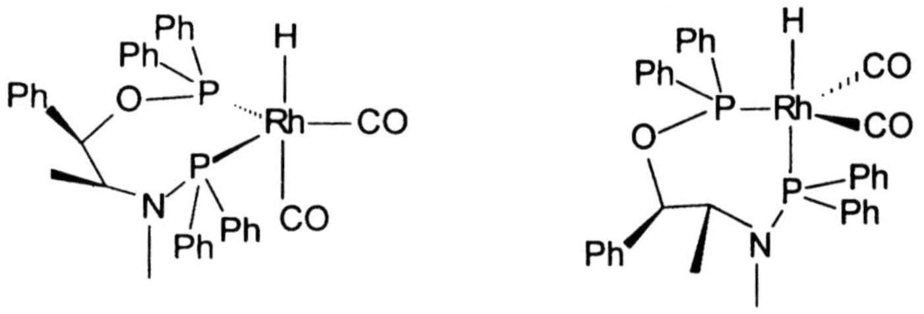

Fig. 6.6. Isomerism in rhodium hydrido carbonyls. The dominant species is the right-hand one (Ref. [39]).

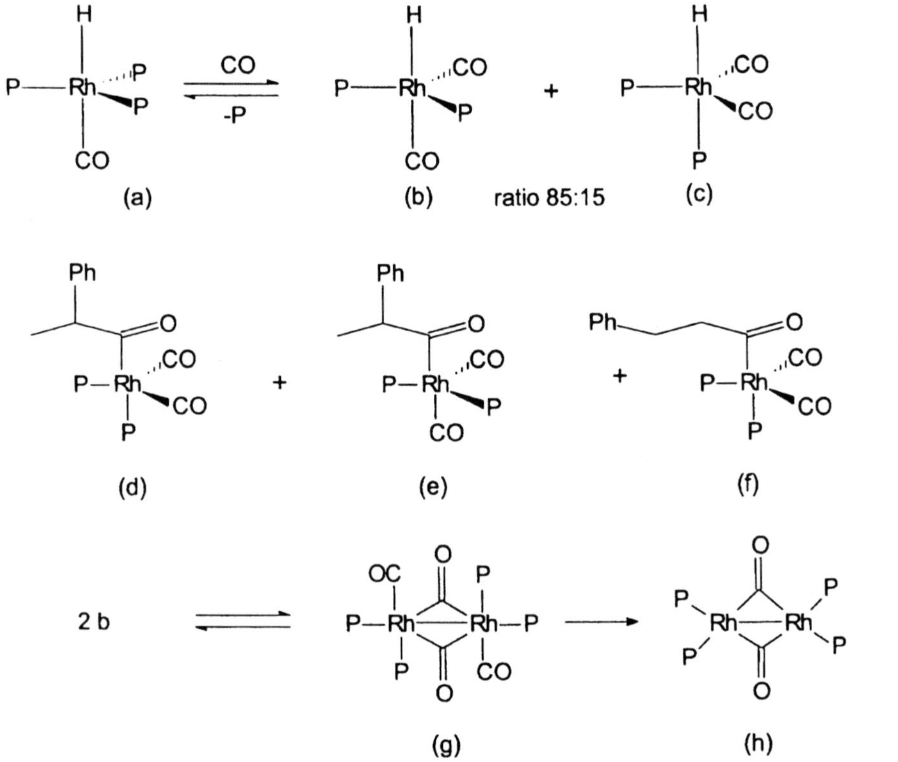

Fig. 6.7. Rhodium complexes found during hydroformylation of styrene (P=PPh_3) (Ref. [19]).

phosphine. Isomer (b) is the dominant species. When $RhH(PPh_3)_2(CO)_2$ is made to react with styrene the branched acyl complexes (d) and (e) are formed. Now the isomer (e) is the major species. This is presumably due to steric factors. The branched isomers isomerize slowly to the linear acyl complex (f). This shows that, at low temperature, the kinetic product in the first reaction with styrene is the branched styryl intermediate. With 1-decene only the linear acyl complex is observed. Interestingly, a fast intramolecular exchange of the two distinct phosphines in the complex can be observed. At higher temperatures a rapid exchange with free CO and PPh_3 is also found. Formation of dimeric species (g) and (h) is a general feature of these studies.

The reaction with dihydrogen could not be tracked down in this study. Iridium forms much more stable complexes than rhodium and therefore it has been studied to obtain more details about the complexes and the mechanism of rhodium hydroformylation. Using an iridium complex, Eisenberg and Deutsch [34] succeeded in the identification by NMR of all species involved in the catalytic cycle: the ethyl complex (i) (Fig. 6.8) the propionyl complex (j) (including X-ray structure determination), and, moreover, the dihydride adduct (k). The latter

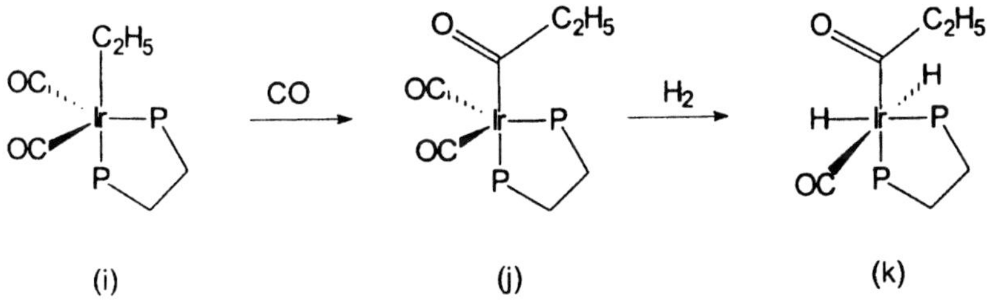

Fig. 6.8. Iridium complexes showing stepwise hydroformylation (Ref. [34]).

gives the aldehyde product upon decomposition. This series of complexes and their reactions suggests that in the case of iridium the last step in the hydroformylation involves an oxidative addition of dihydrogen. This is the first direct proof of such a mechanism. It is not necessarily true, however, for the rhodium triphenylphosphine modified catalyst, let alone all rhodium catalysts. For instance, when the acyl complexes mentioned above were heated in the Brown and Kent experiments, aldehyde was formed even in the absence of dihydrogen, which would support a bimetallic step for the product formation. The lesson to be learned is that several processes may be very close in activation parameters in homogeneous catalysis and one has to be aware of the several possibilities rather than one single mechanism.

6.2.7.2 *IR studies of ligand-free rhodium carbonyl catalysts*

Recently, thorough studies providing great detail on the rhodium carbonyl catalysts have appeared from Pino and Garland [41]. We will refer here to one aspect of their study only; the reaction of 3,3-dimethyl-1-butene, CO, H_2, with $Rh_4(CO)_{12}$ at 4–14°C. This substrate was chosen because it lacks the possibility to isomerize, while still giving an acceptable rate of hydroformylation. The cluster carbonyl reacts slowly with the syn-gas and alkene to give the monomeric acyl complex $RC(O)Rh(CO)_4$. The characteristic IR frequencies in the CO stretch region of the latter compound are: 2111, 2065, 2039, 2020, and 1698 cm^{-1}. The latter complex builds up as the dominant intermediate while the catalytic hydroformylation starts. Clearly, the reaction of the acyl complex with dihydrogen is the rate-determining step in this catalytic system. The rate equation for the reaction with hydrogen under the conditions (5–30 bar of H_2 and 10–40 bar of CO) is:

$$\text{rate} = k\,[\text{RCORh(CO)}_4]^1[\text{CO}]^{-1.1}[\text{H}_2]^1[\text{alkene}]^{0.1}$$

The turnover frequency under these conditions is rather low (<20 mol mol^{-1} h^{-1}) but the kinetic results, in return, are highly accurate. The minus one order in CO and the plus one order in H_2 agree very well with the results from Markó [29].

The rate of hydroformylation was proportional to the concentration of the acyl complex. The apparent activation parameters were $\Delta H^{\ddagger}$ = 49.3 kJ/mol and $\Delta S^{\ddagger}$ = 121 J/mol K. The activation parameters and the reaction order are consistent with reaction 6 and 7 as being rate determining. A low order of 0.1 in alkene was found, which indicates that the 'rate-determining step' is not purely reaction 7 and that either a pre-equilibrium contributes as well or that one of the earlier steps in the cycle is also somewhat slower.

6.2.7.3 IR studies of ligand-modified rhodium carbonyl catalysts

The triphenylphosphine modified rhodium hydroformylation catalyst has been studied with IR under a variety of conditions. As far as high pressure spectra are concerned they are usually measured in cells that are connected via a pump with the autoclave in which the reaction is taking place. The disadvantage is that mass transport limitations of the reacting gas and temperature variations may strongly influence the results. To circumvent these problems Moser introduced a new type of cell which utilizes cylindrical internal reflectance (CIR) through a transmitting crystal that is mounted in the autoclave, submersed in the liquid phase where the reactions are taking place [42]. There is no doubt that one now has the possibility of looking at the catalytic system in operation. The signal-to-noise ratio of the CIR-FTIR cell is less than that of the common transmission cell. Deposition of solid material onto the crystal may distort the measurements. The difference from the NMR measurements lies in the fact that in the non-reacting, concentrated solution in the NMR tube, the system will have equilibrated to a new composition whereas the *in situ* IR measurements give the distribution of the reacting species in the steady state of the catalytic cycle.

Moser studied the $RhH(PPh_3)_2(CO)_2$ catalyst at 70°C and pressures of 11–34 bar. First of all it was found that under a pressure of syn-gas the species present also have the formula $RhH(PPh_3)_2(CO)_2$ also when an excess of phosphine is present. In the presence of 1-hexene the main species observed is still the hydride $RhH(PPh_3)_2(CO)_2$. The absorptions of a minor species are assigned to the corresponding alkyl rhodium complex $Rh(C_6H_{13})(PPh_3)_2(CO)_2$. This observation is consistent with a slow dissociation of a ligand from the starting hydride, followed by a reaction with the alkene. Such a mechanism leads to an inverse dependence on CO or PPh_3 (or both) and a proportional relation of the rate with the alkene concentration.

A transmission cell was used in the studies reported by van Leeuwen [18,21, 30,43]. The autoclave and the transmission cell are only a few millimetres apart and the liquid is pumped with a centrifugal pump from the autoclave to the cell. The time delay between autoclave and cell can be as low as a few milliseconds. The advantage is a higher signal-to-noise ratio. The *in-situ* measurements on Xantphos type ligand systems showed that the only species that can be observed

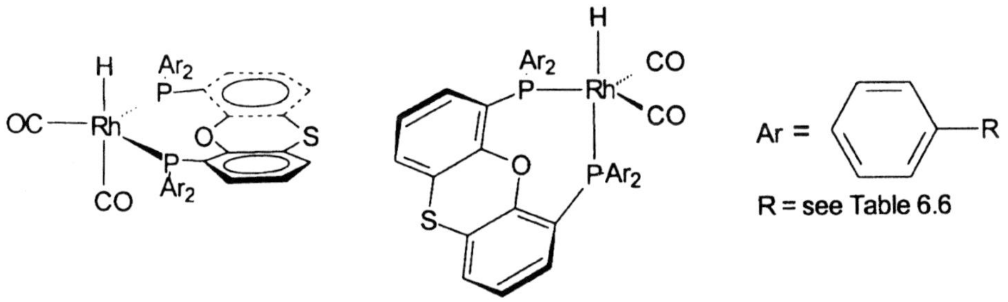

Fig. 6.9. Structure of the complexes [19].

during the hydroformylation are of the type RhH(Xantphos)$(CO)_2$. Interestingly a mixture of isomers was observed containing bis-equatorial and equatorial-apical diphosphines (Fig. 6.9). NMR spectroscopy showed time averaged spectra of these species. Stronger electron donors give a higher proportion of the isomer containing an apical phosphine. Table 6.6. shows the frequencies for both isomers and the expected electronic effect on the position of the bands; electron donating phosphines lead to more back donation to carbon monoxide and hence lower CO stretch frequencies. In parentheses the ratios of the two isomers are given for two ligands. Rhodium hydride absorptions are usually very weak in the dicarbonyl hydride species.

Another point worth mentioning is the observation of the dimers (g) (Fig. 6.7) under hydroformylation conditions, especially when the rhodium concentration is high, and hydrogen pressure and temperature are low [43]. The presence of the orange dimers under reaction conditions is important, because it explains why in many cases the rate of reaction increases when the hydrogen pressure is raised; the equilibrium between (g) and (2b) shifts to the left, giving more of the active rhodium hydride species and the rate will increase. Without *in situ* studies this dependence of the reaction rate on the hydrogen pressure might be interpreted as an indication that the hydrogenolysis step (6) (Scheme 6.1) is rate-determining, which it is not.

TABLE 6.6

IR frequencies of complexes RhH(diphosphine)$(CO)_2$ [18]

Substituent R	3.χ_i-Value	CO eq-ap (cm^{-1})	CO eq-eq (cm^{-1})
$N(CH_3)_2$		2027, 1960 (50%)	1983, 1935 (50%)
OCH_3	3.4	2034, 1966	1990, 1942
H	4.3	2037, 1972	1994, 1946
F	5.0	2041, 1974	1997, 1950
Cl	5.6	2042, 1977	1999, 1952
CF_3	6.4	2046, 1982 (90%)	2004, 1957 (10%)

6.3 ZIRCONIUM CATALYZED POLYMERIZATION OF ALKENES

6.3.1 Introduction

In this section we shall discuss some of the recent advances in the area of alkene polymerizations catalyzed by homogeneous complexes of zirconium and titanium. One of the main research themes for these catalysts has been the polymerization of propene. In less than a decade a completely new family of catalysts has been developed which enables one to synthesize regioselective and stereoselective polymers of a wide variety of monomers. Several new catalysts have found application on an industrial scale (syndiotactic polystyrene, ethene-norbornene copolymers, long-chain branched polyethylene), but soon more applications will be taken into production. Before outlining the main concepts of this relatively new field, a few words will be said about the present heterogeneous titanium catalysts used for the production of polypropylene. For a more extensive review on this matter the reader is referred to the specialist reports [44]. The titanium catalyst for the stereoselective polymerization of propene to isotactic polymer was discovered by Ziegler and Natta in the mid fifties. Soon after its discovery it was taken into commercial exploitation and today isotactic polypropene is still one of the most important commodity polymers. As mentioned in Section 2.4, the catalyst and the concomitant technology have undergone drastic changes over the years. Initially the polymer was made in a solution or slurry with moderate selectivity to the isotactic polymer, the atactic material being removed by extraction. Catalyst residues were removed by washing with alcohol. Catalysts with higher activities and selectivities have made separation from catalyst residues and from atactic material superfluous. Two processes are operated today — one in liquid propene and one in the gas phase — which involve only very few process steps compared to the older processes.

6.3.2 Supported Titanium Catalysts

The modern catalyst is a $MgCl_2$-supported $TiCl_3$ complex which replaces the $TiCl_3$ crystallites used in the older processes. $TiCl_3$ is bonded to the $MgCl_2$ support via bridging chloride ions and the titanium site is modified by complexing Lewis bases. Catalyst preparation is a complicated synthetic procedure, although most of the steps are fairly well understood. We will mention briefly here the types of chemical and physicochemical operations involved, without claiming any authenticity for this particular 'recipe'! Small particles of $MgCl_2$ are prepared by ball milling (other techniques have been reported). The surface of the $MgCl_2$ is modified by ligands such as alcohols or esters before they are treated with $TiCl_4$. Alternatively the support is treated with donor–acceptor complexes of $TiCl_4$ and esters, ethers or alkoxy silanes. These donor molecules are a prerequisite for obtaining a high isotactic index. Titanium chloride is bonded to the surface via bridging chloro anions. Soluble titanium complexes are removed by washing with hydrocarbons, and excess donor compound may be removed by treatment of the solid particles with liquid $TiCl_4$. The titanium is activated by reaction with triethyl aluminium or the Lewis adduct of the latter with the donor compounds is applied.

6.3.3 Isotactic Polypropylene

As has already been mentioned in Chapter 2 Section 2.4, the synthesis of the commercial polymer involves regioselective 1,2-insertions of propene in a stereoregular fashion, leading to isotactic polymer, see Figs. 6.10 and 6.11.

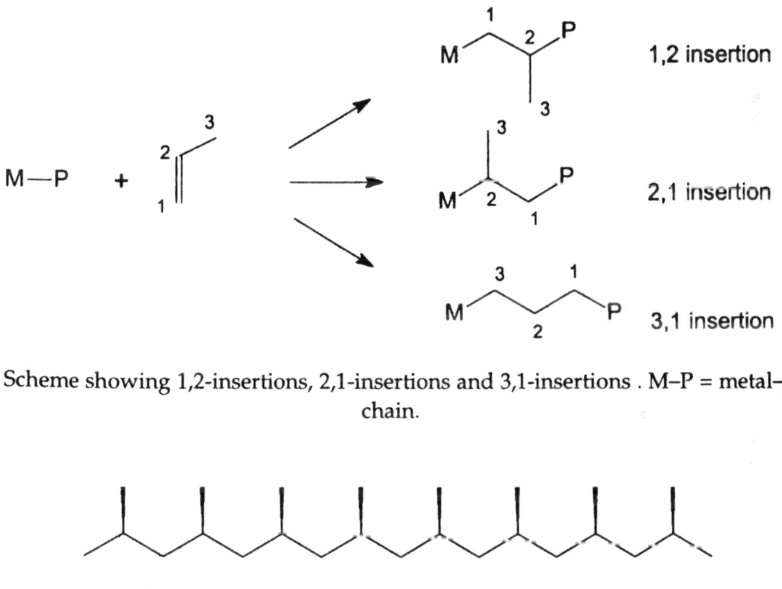

Fig. 6.10. Scheme showing 1,2-insertions, 2,1-insertions and 3,1-insertions . M–P = metal–polymer chain.

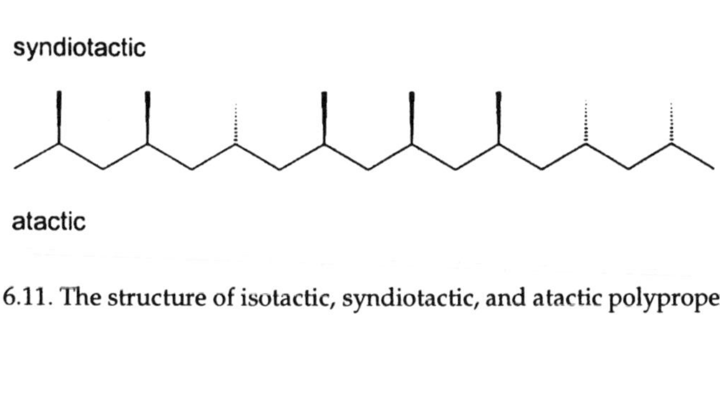

Fig. 6.11. The structure of isotactic, syndiotactic, and atactic polypropene.

Three stereoregularities are distinguished, but, in practice more products may be expected, differing in sequences of regularities and mistakes in the chains. Making a polymer is much more complicated than making butanal (see Section 6.2) because the properties of a polymeric material depend on several more molecular properties than those for a low molecular weight product such as butanal. In the catalytic preparation the following properties of the polymer have to be tuned:

- regioselectivity;
- stereoselectivity;
- molecular weight, M_w: weight average molecular weight, M_n: number average molecular weight;
- molecular weight distribution, or dispersion ($=M_w/M_n$);
- morphology of the polymer particles.

Physical properties that have to be controlled by these molecular factors include:

- melting point;
- glass transition temperature;
- strength;
- modulus;
- crystallinity; and
- viscosity of the molten polymer.

6.3.4 The Cossee–Arlman Mechanism

The mechanism proposed for the solid titanium chloride catalysts is essentially the same for all catalysts and it is usually referred to as the Cossee–Arlman mechanism [46]. Titanium is hexacoordinated in the $TiCl_3$ or supported catalysts by four bridging chlorides and one terminal chloride that is replaced by an alkyl (P, for polymer chain) from the alkylating agent (Et_2AlCl or Et_3Al), and a vacancy that is available for propene coordination (see Fig. 6.12). The front and back of the complex shown are not equivalent; the "blocked" chlorine at the front causes more steric hindrance than the "exposed" one at the back [46].

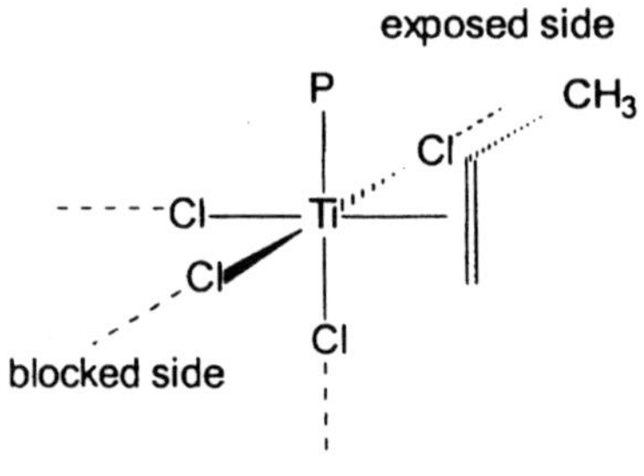

Fig. 6.12. Cossee active site on asymmetric α-$TiCl_3$ surface (surface not shown).

The asymmetry of the site regulates the mode of coordination of the propene molecule; in other words it steers the direction the methyl group will point to. We will not go into the detail of the mechanism of the heterogeneous reaction as there is, relative to the homogeneous catalysts, a considerable amount of speculation involved. A slightly different mechanism from that of Cossee was proposed by Allegra [47]. In the following we will discuss the homogeneous catalysts and the stereochemistry of the polymerization reaction; for the heterogeneous catalysts and their mechanisms the reader is referred to other texts [44].

6.3.5 Homogeneous versus Heterogeneous Catalysts

From the early days on there has been an interest in homogeneous catalysts for the polymerization of propene and ethene with titanium and chromium similar to the vanadium based Ziegler catalysts, which are homogeneous. It was thought that in a homogeneous system all metal atoms (titanium, chromium) could in principle participate in the catalytic process and, activity and other things being equal, this would make removal of the catalyst unnecessary. Also, the molecular weight distribution would be much narrower and could be controlled at will. Drawbacks were also envisaged, such as the morphology of the powder to be formed and stereocontrol. For stereocontrol the $TiCl_3$ lattice seemed to play an important role and stereocontrol was not considered to be an easy task in a homogeneous catalyst. Polymerization of ethene with homogeneous catalysts has been known for a very long time (1957, [48]), but these catalysts failed to polymerize propene and only dimerization may be observed. In the seventies the first claims appeared concerning the homogeneous stereospecific polymerization [44] but they received relatively little attention as, during the same years, the first highly active heterogeneous titanium catalysts were reported and the industrial interest in the more risky field of homogeneous catalysts diminished. In the following, the general principles of polymerization catalysis will be discussed, with an emphasis on stereocontrol, together with some recent achievements.

6.3.6 Site Control versus Chain-end Control

Over the years two mechanisms have been proposed as being responsible for the stereocontrol of the growing polymer chain; firstly the site-control mechanism and secondly the chain-end control mechanism. In the site-control mechanism the structure of the catalytic site determines the way the molecule of 1-alkene will insert (enantiomorphic site-control). Obviously, the Cossee mechanism belongs to this class. In the chain-end control mechanism the last monomer inserted determines how the next molecule of 1-alkene will insert. Several Italian schools have supported the latter mechanism. For heterogeneous catalysts it would seem that site control was strongly gaining preference [44], at least until the more

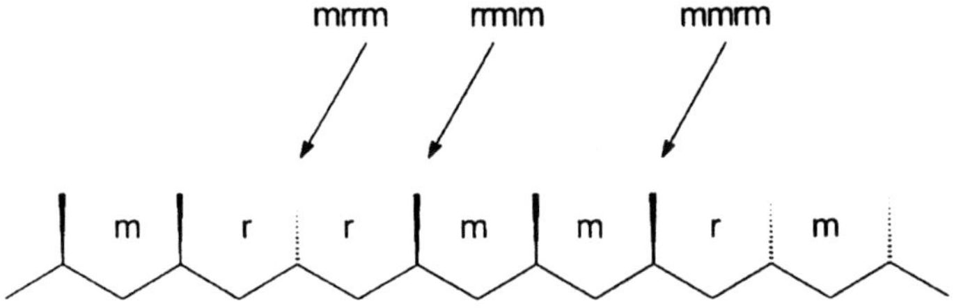

Fig. 6.13. *m* and *r* Relationships in polypropene.

detailed work on homogeneous catalysts became known. The work of Corradini [45] in particular has shown that, at least for polypropene, the two mechanisms cannot be separated and that together they afford such high stereospecificities. By way of introduction we will first discuss the two mechanisms separately.

The analysis of the products using high resolution ^{13}C NMR has greatly contributed to the mechanistic insight and distinction between the various catalysts. Two simple examples will serve as an illustration of this point. The relation between two adjacent methyl groups is indicated by *r* (racemic) or *m* (meso), which means that, with respect to the stretched chain, they are on the same side or on opposite sides (see Fig. 6.13). The local connectivities *m* or *r* are somewhat confusing in view of the definition of isotacticity. In an 'isotactic' polymer the same absolute configuration is reproduced at every insertion according to the original proposal by Natta [49] which is indeed the case in sequences of *m* (although 'meso' does not seem to express this). Hence, a syndiotactic polymer consists of a sequence of *r* configurations, whereas the isotactic polymer consists of purely *m* configurations. With two adjacent *m* stereocentres, *mm*, we say that we are dealing with a triad of methyl groups with an isotactic stereochemistry. The chemical shift in the NMR of the ^{13}C atoms depends on the stereochemistry of the substituted carbons in the vicinity; not only can we tell the stereochemical relation with the two nearest neighbours (the *mm*, *mr*, and *rr* triads), but we can also deduce the relation with the two next ones from the chemical shifts observed. In other words, with ^{13}C NMR we are able to distinguish pentads of methyl groups, i.e. we can distinguish from the NMR spectrum, for example, an *rmmm* sequence from an *mmmm* sequence. We can expect ten different pentads, two of which coincidentally have the same chemical shift and hence, in modern NMR analysis of propene polymers, nine signals are observed. Stereochemical control by the catalytic site and control by the growing chain can be easily distinguished from one another by studying the 'mistakes' as they occur in the ^{13}C NMR spectrum. In a system with site control the mistake ideally leads to a single odd insertion in the chain: the site enforces, for instance, the growth of an isotactic chain with all *m* configurations and after one mistake has occurred it will return to producing the same configurations. In the polymer chain this gives a series of *m* relationships interrupted by two *r* configurations. In the NMR

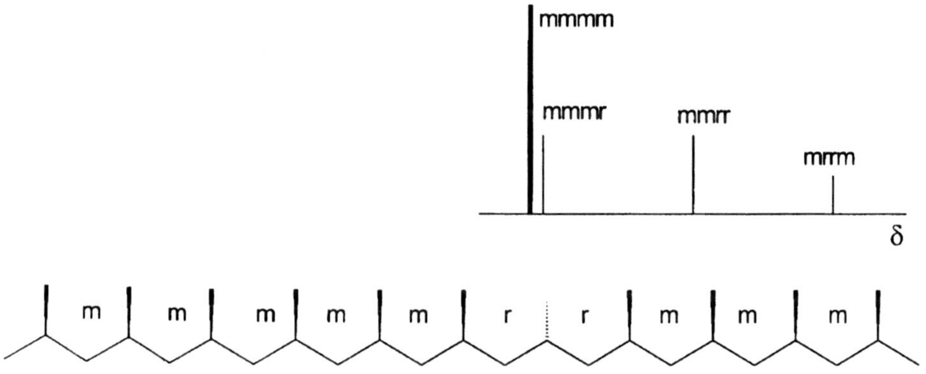

Fig. 6.14. Single mistake during a site-controlled isotactic polymerization and schematic ^{13}C NMR spectrum of the methyl region.

spectrum this gives rise to three new peaks belonging to *mmmr, mmrr,* and *mrrm* in a 2:2:1 ratio (see Fig. 6.14). Since recently nonad sequences can be identified by ^{13}C NMR spectroscopy [50], yielding even more detailed information.

When a chain-end control mechanism is operative the result is different. If one mistake is made in a sequence of isotactic insertions the next propene that comes in will be directed by the previous one and the 'mistake' will propagate in the growing chain. Hence, after a block having all methyl groups pointing to one direction, a block will form having the methyl groups pointing to the other side of the growing chain (Fig. 6.15). An example of polypropene with isotactic stereoblocks was reported in 1984 [51]. Naturally, this achiral catalyst can only give chain-end control as it lacks the necessary chiral centre for site control. In the ^{13}C NMR the stereoblocks can be clearly observed as they lead to the typical 1:1 ratio of *mmmr* and *mmrm* absorptions in addition to the main peak of *mmmm*

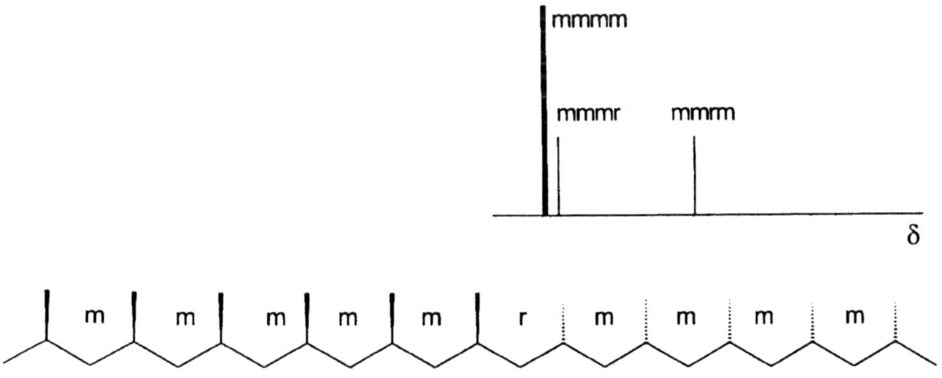

Fig. 6.15. Formation of stereoblocks in a chain-end controlled isoatactic polymerization and schematic ^{13}C NMR spectrum of methyl region.

pentads. These are two simple examples showing how the analysis of the ^{13}C NMR spectra can be used for the determination of the most likely mechanism of control of the stereochemistry. Obviously, further details can be obtained from the statistical analysis of the spectra and very neat examples are known [52].

A general difference between the stereochemistry of the products of the homogeneous catalysts on the one hand, and the heterogeneous catalysts on the other, has been noted in the literature. Measurement of the overall isotactic index of the products of the homogeneous catalysts revealed that polymers with isotactic indices as high as 96% still have a melting point of only 140°C, whereas the products of heterogeneous catalysts with such isotactic indices show melting points of 160–166°C. This can be explained as follows. The commercial heterogeneous catalysts, in spite of their high degree of optimization, contain different sites, one group producing polymer with a high isotactic index (99+%) and another group of sites producing a small amount, say 4%, of atactic polymer. Hence a mixture of two polymers is formed, the melting point of which is determined by the isotactic component. The homogeneous catalyst, on the other hand, ideally contains only one type of site that produces one polymer with an isotactic index of, say, 96%. While the isotactic index may look very promising, the melting point may be disappointing since all the errors made occur in the main chain of the product, giving a lower melting point. Improved homogeneous catalysts have recently been synthesized that also give polypropene melting points above 160°C and isotactic indices of 99.7% [53].

We now turn to the actual polymerization process and we will try to present a series of pictures that clarifies how chain-end control can be used to obtain either syndiotactic or isotactic polymers. Subsequently we will see how a chiral site can influence the production of syndiotactic or isotactic polymers. Finally, after the separate stories of 'chain-end control' and 'site control', the reader will be confused by introducing the following elements: (1) pure chain-end control can truly occur when the catalyst site does not contain chirality; (2) but since we are making chiral chain ends in all instances, pure site control does not exist. In a polymerization governed by site control there will potentially always be the influence of chain-end control. This does not change our story fundamentally; all we want to show is that stereoregular polymers can indeed be made, and which factors play a role; but their relative importance remains hard to predict.

6.3.7 Chain-End Control: Syndiotactic Polymers

Achiral catalysts derived from $VCl_4/Al(C_2H_5)_2$ or $V(acac)_3/Al(C_2H_5)_2$ are the best known examples, giving syndiotactic polymers [54]. Termination experiments have proved that this polymerization involves a 2,1-insertion instead of the expected 1,2-insertion. This is to say that the intermediate vanadium alkyl contains a branched or secondary alkyl group. Two simple assumptions now

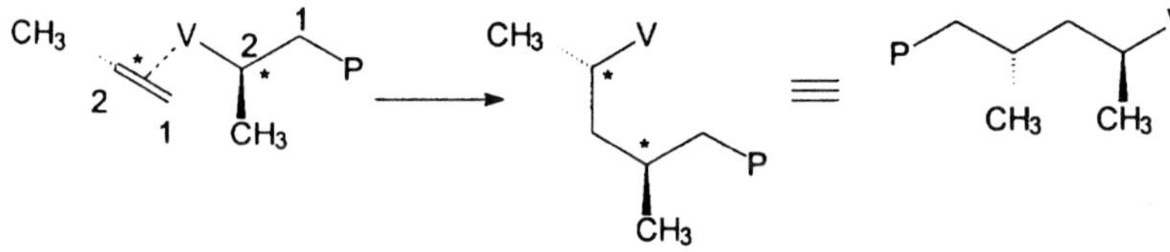

Fig. 6.16. Chain-end control leading to syndiotactic polymer. The polymer chain is arranged in such a way that all carbons of the chain lie in the plane of the figure. The double bond of the propene to be inserted also lies in the plane, with its substituents in a plane perpendicular to the plane of the figure. The methyl groups above or below the plane of the figure are drawn in the usual manner.

explain why a syndiotactic polymer is obtained. Firstly, the growing chain assumes a conformation in which the two largest groups, the polymer chain and the vanadium complex, are *trans* to one another in the chain end formed by the last two carbon atoms (see Fig 6.16). Secondly, the insertion of propene takes place via a mechanism best described as a 2+2 addition. The methyl group of the propene to be inserted and the one of the previously inserted propene molecule will assume positions leading to a minimum of steric interaction (see Fig. 6.16). Since the two largest groups attached to the last inserted propene are now in mutual *syn* positions, the chain must be rotated in order to obtain the more stable *trans* configurations at both linkages, i.e. the chain being stretched as it is usually represented. When this procedure is repeated a couple of times we see that a syndiotactic microstructure is indeed formed. The most peculiar aspect of this family of catalysts is the apparent formation of a secondary alkyl group at the vanadium centre, where one would normally expect a primary alkyl group. Direct observation has not yet been achieved, but the termination experiments unequivocally prove that the reacting species is a secondary alkyl vanadium complex, both during the fast polymerization reaction with a subsequent molecule of propene, but also, more surprisingly, after presumed equilibration of the living polymer before reacting with, e.g., I_2. There still is the possibility of a very fast equilibrium between the primary and secondary alkyl, the latter being far more reactive.

6.3.8 Chain-End Control: Isotactic Polymers

By the same token, 1,2-insertion gives an isotactic microstructure. If the methyl group of the previously inserted molecule is upwards, the methyl group of the next molecule will be pointing downwards (see Fig. 6.17). The interaction may not be quite as strong as in the case of 2,1-insertion discussed above, but there will always be a tendency of the growing chain to arrive at an isotactic stereochemistry when 1,2-insertion occurs. One example of chain-end control leading to isotactic polymer was reported by Ewen [51] using Cp_2TiPh_2/alumoxane as the catalyst. The stereoregularity increased with lower temperatures; at –45°C

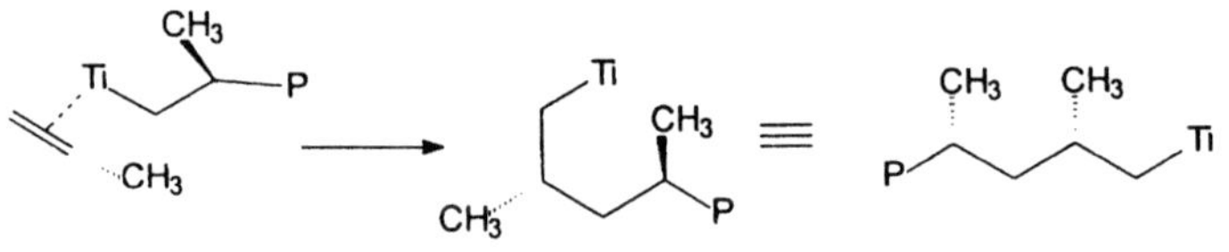

Fig. 6.17. Chain-end control leading to isotactic polymer. The polymer chain is arranged in such a way that it lies in the plane of the figure. The double bond of the propene to be inserted also lies in the plane with its substituents in a plane perpendicular to the plane of the figure. The methyl groups above or below the plane of the figure are drawn in the usual manner.

the isotactic index as measured on pentads amounted to 52%. As mentioned above, the polymer contains stereoblocks of pentads. At 25°C the polymerization gives almost random 1,2-insertion and an atactic polymer is formed.

6.3.9 Site Control: Recent History

Until 1984 no well-defined homogeneous propene polymerization catalysts were known, based on Group 4 metals. The development of the new family of homogeneous catalysts based on biscyclopentadienyl Group 4 metal complexes for the stereoselective polymerization of alkenes is mainly due to Kaminsky, Ewen and Brintzinger. The first breakthrough came from the laboratory of Sinn and Kaminsky, who had been studying the interaction of zirconium and aluminium alkyls for a few years. In 1980 they reported [56] on an extremely fast homogeneous catalyst for the polymerization of ethene, formed from the interaction of $Cp_2Zr(CH_3)_2$ and $(CH_3AlO)_n$ [Cp=C_5H_5=cyclopentadienyl, the aluminium compound, alumoxane or aluminoxane, was obtained from the reaction of trimethylaluminium and a hydrated salt, $CuSO_4{\cdot}5H_2O$, as a slow-release source of water]. At 8 bar of ethene and 70°C an average rate of insertion of ethene was reported amounting to $3{\times}10^7$ mol of ethene per mol of Zr complex per hour! With propene this catalyst led to completely atactic polymer [57].

Ewen was the first to report the synthesis of stereoregular polymers with soluble Group 4 metal complexes and alumoxane as the co-catalyst [51]. He found that Cp_2TiPh_2 with alumoxane and propene gives isotactic polypropene via chain-end control, see above. The second important breakthrough was how to build a chiral centre into the Group 4 metallocene. Substituted chiral metallocene complexes had already been reported in 1982 by Brintzinger [58]. Using an intrinsically chiral titanium compound (*rac* ethylene-bis-indenyl titanium dichloride) [51] he obtained polypropene that was in part isotactic. Subsequently Kaminsky and Brintzinger showed that highly isotactic polypropene can be obtained using the racemic zirconium analogue of the ethylene-bis(indenyl) compound [59]. The initial results gave a material with a relatively low molecular weight and a low melting point, 140°C compared to 165° for the material obtained with commercial heterogeneous catalysts. The molecular weight distribution was very narrow, M_w/M_n=1.8, as is to be expected from a homogeneous catalyst. The molecular weight may be as high as 300,000 [60], strongly decreasing upon raising the temperature. Although the isotactic index is high (>96%) the melting point may be low at 140°C (see above). Modification of the cyclopentadienyl ligands has led to a very rich chemistry and today a great variety of microstructures and combinations thereof can be synthesized as desired, including isotactic polymer with melting points above 160°C, syndiotactic polypropene [61], block polymers, hemiisotactic polymers etc.

Fundamental studies have led to a detailed insight into the mechanism of the polymerization and the control of the microstructure through the substituents on the cyclopentadienyl ligands. For a survey of these studies and the dominant effects playing a role, the reader is referred to contributions by Ewen [52] and Erker [62], and the reviews by Brintzinger and Bochmann [44]. In the present chapter we will summarize a few of the features that play a role in determining the microstructure; the field is still in development and obviously the definite answer to several questions has not yet been given.

6.3.10 Site Control: Isotactic Polymers

The structure of the catalyst has been proved by Jordan, Turner, and others [64] to be a cationic species, $Cp_2M{-}R^+$ (M = Ti, Zr, Hf). The basic structure is depicted in Fig. 6.18. Zirconium is tetrahedrally surrounded by two cyclopentadienyl ligands and two chlorides. The latter two ligands are not essential as they are

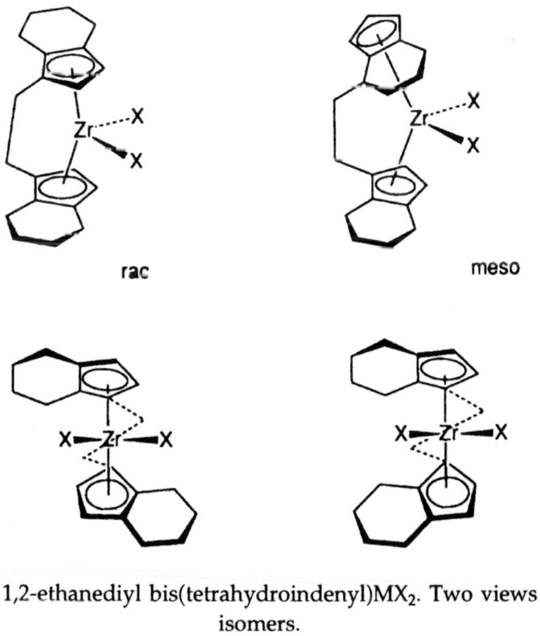

Fig. 6.18. Structure of 1,2-ethanediyl bis(tetrahydroindenyl)MX_2. Two views of the *rac* and *meso* isomers.

replaced before the polymerization can start by an alkyl group (e.g. CH_3 from methylalumoxane) and by a solvent molecule thus generating the required cationic species. The two cyclopentadienyl anions are linked together by a bridge, here an ethane bridge, and extended with an organic moiety which affords the molecule its chirality. A 1,2-ethane bridged bis(1,1'-tetrahydroindenyl) dianion has been drawn in the example in Fig 6.18. Two compounds are obtained when this complex is synthesized, the *rac*-isomer mixture and the *meso*-isomer. They can often be separated by crystallization. The sterically less hindered *rac*-isomers may be formed in excess. The isomers can only interconvert by breaking a cyclopentadienyl–metal bond and reattaching the metal at the other face of the cyclopentadienyl ligand. This process is very slow compared to the rate of insertion reactions with propene.

The *rac*-isomers have a twofold axis and therefore C_2 symmetry. The *meso*-isomer has a mirror plane as the symmetry element and therefore C_s-symmetry. The racemic mixture can be used for polymerization reactions since the two chains produced by the two enantiomers are identical, if begin and end groups are not considered. The asymmetry possibly introduced initially in the crystal during crystal growth from a chiral catalyst will be removed by melting and crystallization. Note: when catalysts of this type are to be used for asymmetric synthesis, e.g. as Lewis acids in Diels–Alder reactions, separation of the enantiomers is a prerequisite.

For polymerization to occur, the X atoms in Fig. 6.18 are replaced by an alkyl chain and a coordinating propene molecule. Coordination of propene introduces a second chiral element and several diastereomers can be envisaged. The step-by-step regulation of the stereochemistry by the site can be most clearly depicted by drawing the molecule as shown in Fig 6.19. First we will discuss an over-simplified scheme adopting a purely site-controlled, hypothetical mechanism.

Here we have taken the plane of the paper identical to the plane determined by Zr and the two X atoms of Fig. 6.19 or, in other words, the plane bisecting the two cyclopentadienyl planes. Propene coordinates in a π-fashion to zirconium with the two alkene carbons situated in the plane. The migratory insertion reaction can be described as a 2+2-addition following 1,2-regioselectivity. For the 1,2 insertion two possible structures remain, one with the methyl group pointing down in the present Fig. 6.19.**a** and one with the methyl group above the plane, Fig. 6.19.**b**. The tetrahydroindenyl group below the plane (drawn as a dotted line) will favour situation 6.19.**b**. If we accept that the transition state resembles the coordination complex, the stereochemistry of the insertion product follows from structure **b** → **c**. The alkyl group has migrated to the other coordination site of the complex. The question is, what will be the relation between the next insertion and the one that we have just described?

The next molecule of propene will coordinate onto the complex with the methyl group pointing downwards (6.19.**d**); migration of the new alkyl anion

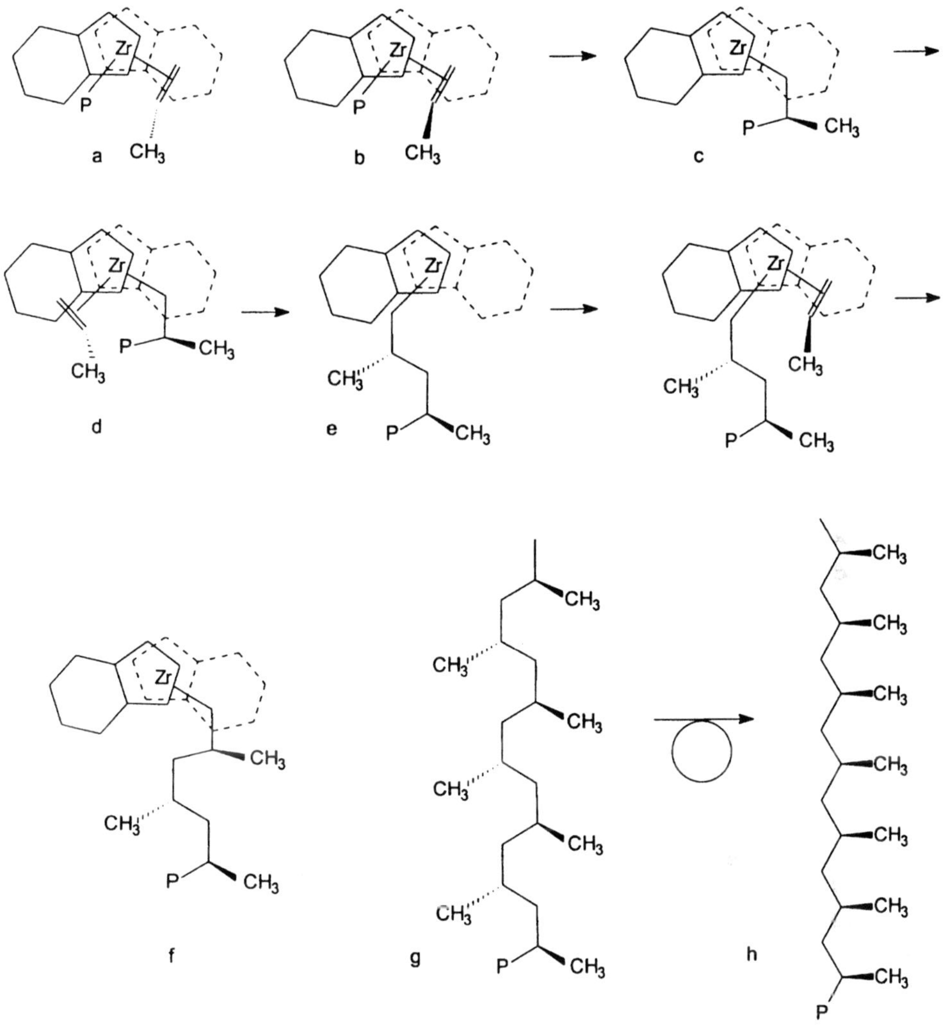

Fig. 6.19. Isotactic polymerization of propene with *rac*-1,2-ethanediyl-*bis*(indenyl)zirconium complexes. The plane of the drawing coincides with the plane bisecting the two planes of the indenyl ligands. Propene coordination takes place with the alkene π-orbitals in the plane of the drawing. The carbon atoms of the growing chain are depicted in the plane of the figure; in between insertions no rearrangements to more stable conformations have been drawn. Indenyl ligand above the plane drawn in full.

gives **e** with the stereochemistry shown. When this process is repeated several times, forming the polymer carbon chain in the plane of the figure, structure **f** is obtained. From this structure we cannot immediately deduce what the micro-structure is, since the polymer chain is not stretched as shown before (Fig. 6.11). By rotating carefully around the bonds indicated we obtain the isotactic stereo-chemistry, **h**, from **g**. Hence, the experimentally found isotactic stereochemistry

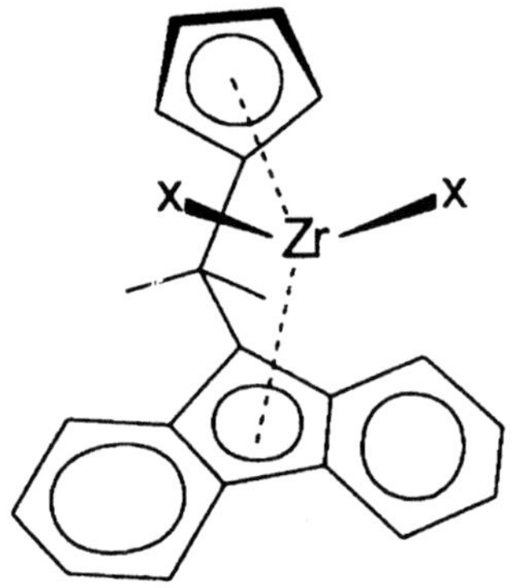

Fig. 6.20. Isopropyl(1-fluorenyl-cyclopentadienyl) zirconium complex for syndiotactic polymerization of propene. The cyclopentadienyl ligand above the plane is drawn in full. The fluorenyl ligand below the plane is drawn with dotted lines.

with *rac-bis*(indenyl) derived complexes can be explained with a very simple set of pictures.

A formal approach leads to the same result more easily; since the complex has C_2 symmetry coordination of propene will lead to the same stereochemistry at both sites of the catalyst. When propene coordinates with its *re* face in structure **b**, it will also do so in structure **d**. The new chiral carbon atoms will all have the same absolute configuration and isotactic polypropene will be obtained.

A series of elegant experiments which support this mechanism involve the use of isopropyl(1-fluorenyl-cyclopentadienyl) ligands [51]. This complex is not chiral (i.e the dichloride precursor of Fig. 6.20), but has a plane of symmetry instead. The catalyst was found to give syndiotactic polymer. Coordination of propene at either site now leads to mirror images. Migration of the alkyl chain will create carbon atoms with the opposite absolute configuration (i.e syndiotactic polypropene). This is an important result, since hitherto it was thought that syndiotactic polymers could only be obtained via 2,1-insertion, controlled by the stereochemistry of the chain end.

Errors can be introduced in several ways, for the indenyl as well as for the fluorenyl complexes:

- 1,2-insertion with the methyl group pointing to the sterically more hindered direction;
- 2,1-insertion (with two orientations), ;see below
- 2,1-insertion followed by isomerization (3,1 insertion);
- epimerization of the alkyl may occur [63];
- alkyl migration without insertion ("back skip") [70]; a catalyst having C_1 symmetry will produce isotactic polymer as proposed in the original Cossee mechanism (!); and
- alkyl exchange between metals.

Impurities in the zirconium precursor complex can also lead to lower tacticities. The activities of complexes may differ by several orders of magnitude and therefore the results can be considerably distorted by impurities. With the present knowledge one can estimate the extent to which impurities will distort the results. For example, it has been found that the presence of the *meso*-isomer in a bis(indenyl)ethane-type catalyst may sometimes produce only little of the atactic isomer. The *meso*-isomer containing one alkyl group and a propene coordinated to the vacant site will have no preference for the methyl group pointing up or down, regardless of the position of the chain. The two sites, however, will show a great difference towards propene coordination, one being very open and the other being strongly sterically hindered. The hindrance at one site may be such that this site is virtually blocked for propene coordination and as a result the rate of propene polymerization of such a catalyst may be very low.

6.3.11 Double Stereoselection: Chain-End Control and Site Control

So far we have looked at chain-end control and site control as if they were independent. As has already been mentioned, a site-control *only* mechanism does not exist; since we are, by definition, making a chiral chain end, site control must be accompanied by chain-end influences. Recent results have shown that this is indeed the case [52,55,62]. The simple explanation given above has to be modified.

Insertion of propene into alkylzirconium model compounds derived from isospecific bis-indenyl catalysts may lead to low enantiospecificity [65]. Thus, site control alone does not lead to high stereoregularities. Molecular mechanics calculations [66] and a thorough analysis of the substituent effect on the statistical distribution of microstructure defects indicate that, firstly, the polymer chain assumes the energetically most favourable position with respect to the (asymmetric) site. According to Corradini, the indenyl ligand or the chlorine atoms of the lattice will direct the polymer chain. The polymer chain occupies the "free space" near the site. In Fig. 6.21 we have schematically drawn this for the $TiCl_3$

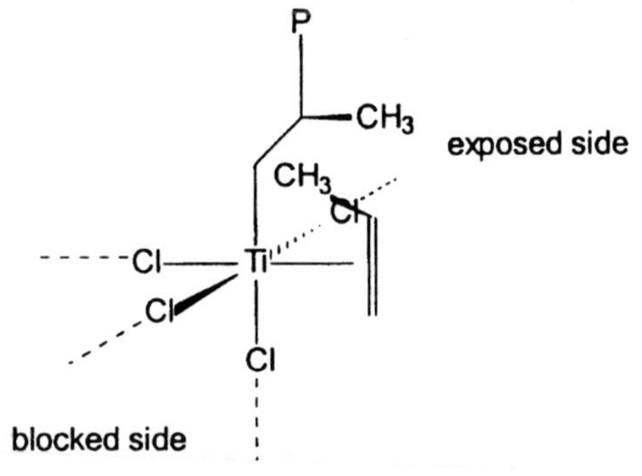

Fig. 6.21. Corradini's active site on an asymmetric α-$TiCl_3$ surface.

catalyst. Subsequently, the last inserted methyl group of the polymer chain will direct the methyl group of the incoming propene into the most favourable position. This *reverses* the propene orientation given in Figs. 6.12 and 6.19. It does not basically change the explanation because *all* propene molecules will orient themselves in the same way. Thus, the previous methyl directs the subsequent insertion and the control is similar to that in Figs. 6.12 and 6.17, but now the position of the polymer chain P is fixed by the catalyst site.

This brings us to double stereoselection and reinforcement of the mechanisms. If the site (a)symmetry were to control the orientation of the chain, and if, then, the orientation of the incoming propene is controlled by both the chain and the site, the highest stereoselection is obtained when the two influences reinforce one another. For 1,2-insertion this can be done most effectively for isotactic polymerization, since chain-end control 'naturally' leads to isotactic polymer and this we can reinforce by site control with ligands of the bis(indenyl)ethane type. The chain-end influence of short chains is smaller than that of longer polymer chain and therefore short chain ends lead to lower selectivities. It may also be inferred that making syndiotactic polymer via a 1,2-insertion mechanism on Ti or Zr complexes is indeed more difficult than making an isotactic polymer, because the two mechanisms now play a counterproductive role.

A further consequence of the directing influence of the polymer chain by the site is the effect it may have on β-hydride elimination. Chain termination via β-hydride elimination requires a coplanar arrangement of the metal atom, the two carbon atoms involved and the hydrogen atom. One can imagine that the site may prohibit this and that therefore in certain sterically more hindered catalysts higher molecular weights can be achieved. An interesting study of the substituent effects on molecular weight, stereoregularity, and rate of reaction using bridged zirconocenes has been published by Spaleck and coworkers [68].

6.3.12 *Effect of hydrogen*

Hydrogen is added during alkene polymerizations in order to regulate the molecular weight. Chain transfer using hydrogen is attractive as it gives saturated chain ends via a σ-bond metathesis reaction. When hydrogen is added during a propene polymerization, one obtains not only the expected lower molecular weight, but also higher reaction rates and higher enantioselectivities (the melting point of the polymer increases). How Chadwick and Busico [67] unravelled this phenomenon is an interesting detective story but for our purposes it is more convenient to present the explanation straight to the point.

In Fig. 6.10 we outlined the two modes of insertion that can occur, 1,2 and 2,1 insertion. The latter often leads to isomerization via β-elimination and re-insertion, which will also lower the stereoregularity. Another result, especially for catalysts that show hardly any to no β-elimination, would be to continue the

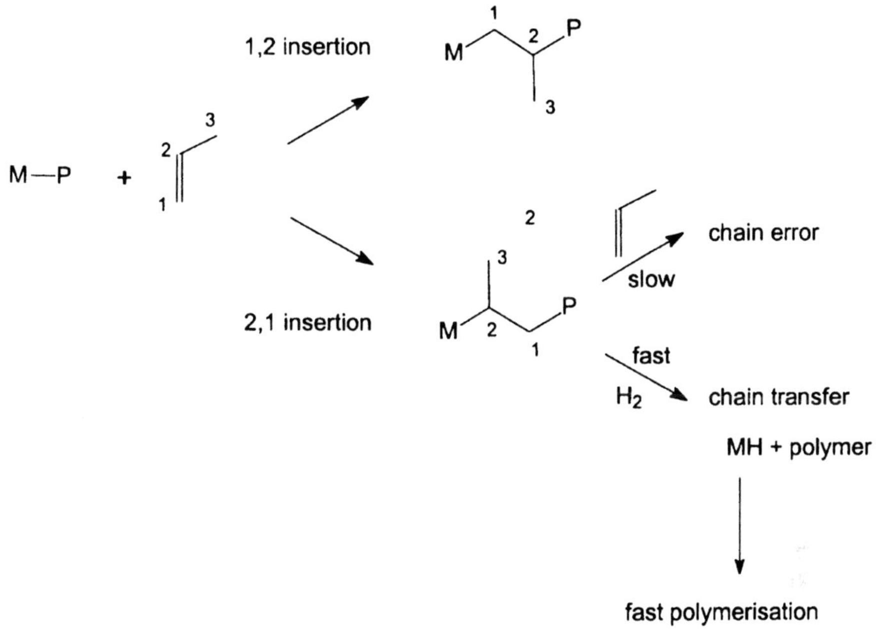

Fig. 6.22. Effect of dihydrogen on rate and stereoselectivity in propene polymerization.

polymerization which also leads to an error in the chain and hence a lower melting point of the polypropene. It turns out that the insertion of the next molecule of propene in the branched alkyl metal complex is much slower than the insertion of propene in a regular chain formed after a 1,2 insertion. In several catalysts studied this leads to a considerable amount of "dormant" catalyst sites, i.e. the metal centre is tied up in unreactive secondary alkyl metal complexes. If eventually an insertion does occur it leads to an error in the chain. By admitting dihydrogen into such a system the unreactive secondary alkyl metal complexes are removed by the reaction with dihydrogen, which is of course less hampered by the steric hindrance of the alkyl group. The metal hydride formed will start a new fast growing polymer chain. Thus,

1. the molecular weight is lowered somewhat,
2. but the overall activity increases,
3. and the number of errors in the chains diminishes (see Fig. 6.22).

6.3.13 Further Work

In addition to propene and ethene many other alkenes (higher alkenes, styrenes, cyclic alkenes) have been polymerized with these new catalysts and a great variety of new polymers and oligomers have been synthesized, including plastic materials with melting points as high as 500°C, novel rubber materials etc. Several new polymers can now be made by catalyst design. Basically, the concepts developed by Cossee and his contemporaries have not changed, but today we have the tools to design the catalysts that may give the desired polymer. Stereoselectivity and molecular weight may not be high enough in some instances, but the achievements are impressive with more to follow:

- A cyclopentadienyl, indenyl-ethane ligand undergoes insertion alternately in an isotactic and an atactic fashion to give a hemiisotactic polymer, as is to be expected.
- One could imagine a complex that, at one site, will insert only one certain type of alkene due to high steric constraints (e.g. ethene) and a second alkene at the other site, thus giving an alternating copolymer that may even contain stereoregularity.
- Block polymers can be made in which the tacticity of the blocks alternates via an inversion reaction of the site [69].
- Block copolymers could be made via sequential polymerization of two monomers at a 'living' catalyst, although the catalyst would produce only one polymer molecule per metal site.

Here we have discussed only very few of the many ligands that have been made. Many variations of substituents at the cyclopentadienyl ligand have been studied, and there are more to come, as well as variation in the structure of the bridge, the anions, the central metal, and replacement of the cyclopentadienyl ligand by other bulky ligands. In summary, the toolbox is quite extensive; prediction of properties of new polymers to be made could guide the catalysis research in the design of new catalysts.

6.4 ASYMMETRIC HYDROGENATION

6.4.1 Introduction

To date many examples are known which show that tailor-made chiral metal complexes, comprising a metal atom bound to chiral organic ligands can discriminate with high precision (energy differences in the range of 10 kJ/mol) between enantiotopic atoms or faces of an achiral molecule. Many of these chiral complexes also have an interesting catalytic activity in the production of a broad range of enantiomerically pure compounds. As the chiral complexes can be synthesized in both enantiomers, in principle, both enantiomers of a product are accessible. (This is a clear distinction from enzymatically catalyzed reactions, which can produce only one hand of a chiral product.)

The demand for enantiomerically pure compounds in fine chemical synthesis is growing rapidly. An obvious reason for this development is that the opposite hand of a chiral pharmaceutical or chemical with a desired biological activity has at best no activity, or worse, causes side effects.

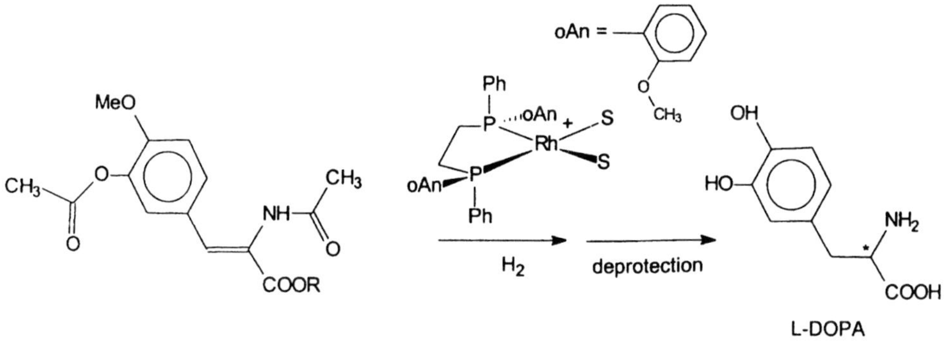

Fig. 6.23. Synthesis of L-DOPA with the cationic Rh-DiPAMP catalyst.

6.4.2 Cinnamic Acid Derivatives

The first industrially interesting production of a chiral pharmaceutical with a chiral metal complex catalyst is the asymmetric hydrogenation of cinnamic acid derivatives to give L-DOPA, see Fig. 6.23.

The synthesis of L-DOPA, a drug for the treatment of Parkinson's disease, has been developed by Knowles at Monsanto [71] and was applied on an industrial scale. This kind of enantioselective catalysis became possible when Horner and Knowles in 1968 discovered methods for the synthesis of chiral triorganophosphines. The L-DOPA synthesis is carried out with a cationic rhodium–bisphosphine complex in which the enantioselectivity is induced by the chiral diphosphine. Surprisingly, the reaction is not very sensitive to the type of chiral diphosphine used, although it must be added that most ligands tested are bis(diphenylphosphino) derivatives and that the enantiomeric excesses obtained do show some variation. On the other hand the reaction is very sensitive to the type of substrate. The presence of polar substituents in these substrates is a prerequisite for a successful asymmetric hydrogenation. This polar substituent, e.g. an acetamido group, is of crucial importance because it functions as a secondary complexation function in addition to the alkene functionality, present in cinnamic acid, that has to be hydrogenated.

In the first step the prochiral alkene entity coordinates to the cationic chiral rhodium centre with either one of the two enantiotopic faces (due to the asymmetry in the diphosphine ligand) which leads to the formation of two possible structures (see Fig. 6.24). Coordination of the auxiliary amide oxygen is of pivotal importance; if it were absent the adduct would have a much less rigid structure (leading to a less-organized transition state during the hydrogenation step) resulting in a lower enantiomeric excess of the product.

Detailed studies of the hydrogenation of cinnamic acid derivatives (Fig. 6.25) by Halpern [72] and Brown [73] have revealed that the most stable intermediate

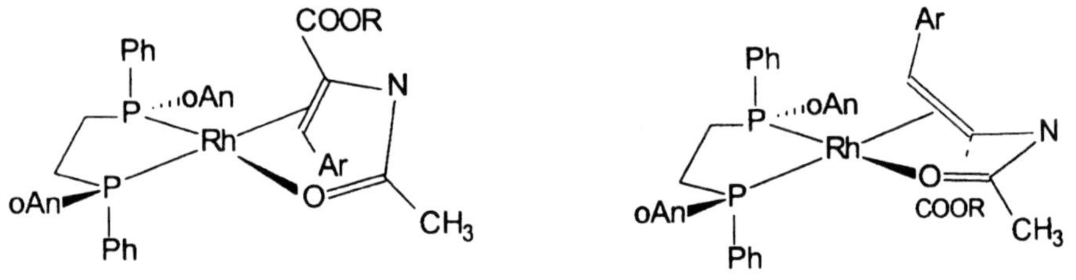

Fig. 6.24. Two diastereomeric intermediate alkene–rhodium complexes.

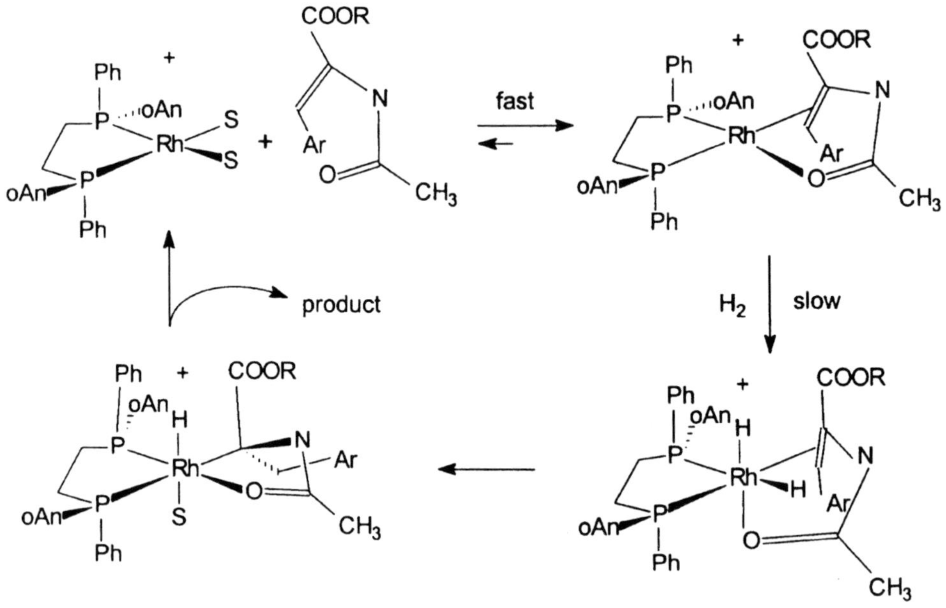

Fig. 6.25. Asymmetric hydrogenation shown for one diastereoisomer.

of the two alkene adducts is not the one that leads to the major observed enantiomeric product, see Fig. 6.26. This means that the least stable intermediate alkene–rhodium complex reacts faster in the subsequent reaction step involving the oxidative addition of dihydrogen to the alkene–rhodium complex (Fig. 6.25.). The oxidative addition of H_2 is irreversible, the two alkene–rhodium enantiomers are formed in this step. If we assume that no dissociation of the alkene occurs, this step determines the enantioselectivity of the alkene hydrogenation.

Migration of the hydride locks the configuration of the enantiomeric alkene–carbon centre; in general this step is also reversible but probably not in this instance. In the examples studied neither the dihydride intermediates nor the alkyl intermediates have been observed and so it seems reasonable to assume

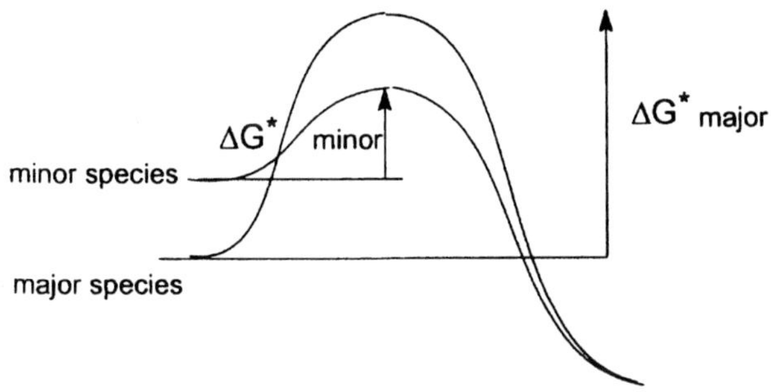

Fig. 6.26. Minor species leading to dominant product route.

that addition of H_2 is also the rate-determining step. Since the latter is a bimolecular reaction and the other ones are monomolecular rearrangement reactions, one cannot in absolute terms say that oxidative addition is rate-determining.

The example illustrates two points:

1. The intermediate alkene adduct observed in the NMR spectra is not the one leading to the major catalytic pathway, see also Fig. 6.25.
2. Enantioselectivity is determined in the first irreversible step after the stereogenic-centre was formed which is not always the rate-determining step.

One might ask the question why a reaction involving such a small dihydrogen molecule can lead to such enormous differences in rate for the diastereomers present. Tentative answers were developed by Brown, Burk and Landis [74]. Their studies included the use of iridium instead of rhodium as the iridium hydride intermediates can be spectroscopically studied. Consider the oxidative addition depicted in Fig. 6.25. The oxidative addition can take place from the top of the molecule (as shown), but it can also take place from the bottom, giving another diastereomeric intermediate that probably does not undergo migration. The two oxidative additions require rotations in opposite directions of the substrate with respect to the rhodium phosphine complex. Likewise, both the major and the minor diastereomeric substrate complex require such a rotation upon oxidative addition. The formation of the two productive diastereomeric dihydrides may involve two different barriers for rotation. Thus, a very stable alkene-rhodium complex resulting from a very good fit of the substrate in the ligand-rhodium cavity, may lead to a more hindered rotation of the alkene upon oxidative addition [75]. The overall result of these processes is hard to predict; molecular mechanics can only show us the energies of the intermediates, while kinetics (free energies of activation) are pivotal in determining the enantiospecificity.

The difference of the catalytic system discussed above from that of the Wilkinson catalyst lies in the sequence of the oxidative addition and the alkene complexation. The hydrogenation of the cinnamic acid derivative involves a cationic catalyst of which the intermediate alkene (enamide) complex has been spectroscopically observed. In subsequent steps oxidative addition of H_2 and insertion of the alkene occur, followed by reductive elimination of the hydrogenation product.

A large series of asymmetric ligands have been developed, most of which have the stereogenic centre(s) in the bridge connecting the phosphorus donor atoms rather than at the phosphorus atoms themselves (cf. DIPAMP in Fig. 6.23). Their synthesis is similar to one developed by Kagan for DIOP [76] (see Fig. 6.27) and starts from asymmetric acids which can be obtained commercially (tartaric acid in the case of DIOP). A third group of very effective asymmetric ligands carry the asymmetry in the substituents at phosphorus, the DuPhos family of ligands, developed by Burk at Du Pont [77] (see Fig. 6.27). An example of a fourth group of ligands containing atropisomerism is BINAP (see Section 6.4.4).

Crucial to the success of the DIOP type ligands is the C_2 'propeller' type symmetry, which divides the space around rhodium (or any metal) into four quadrants; two relatively empty and two filled (see Fig. 6.28). The two phenyl groups at one face of the chelate ring adopt axial and equatorial positions. This explains the similarity of the effect of this large family of ligands on the enantio-

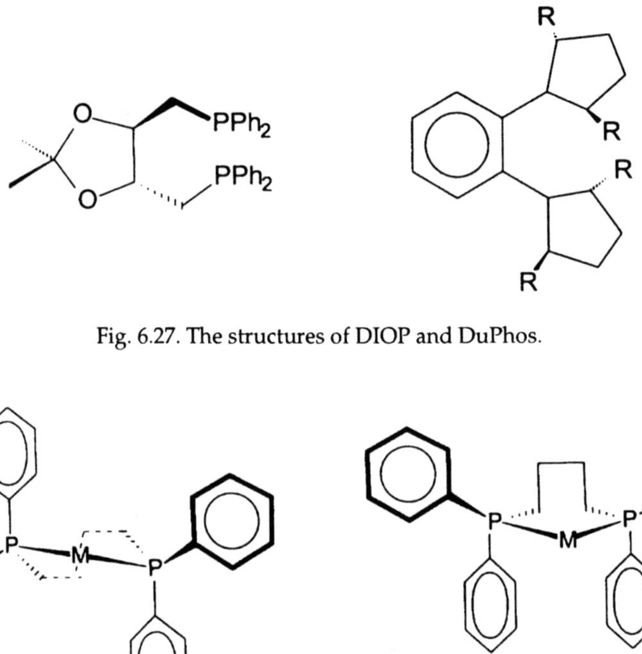

Fig. 6.27. The structures of DIOP and DuPhos.

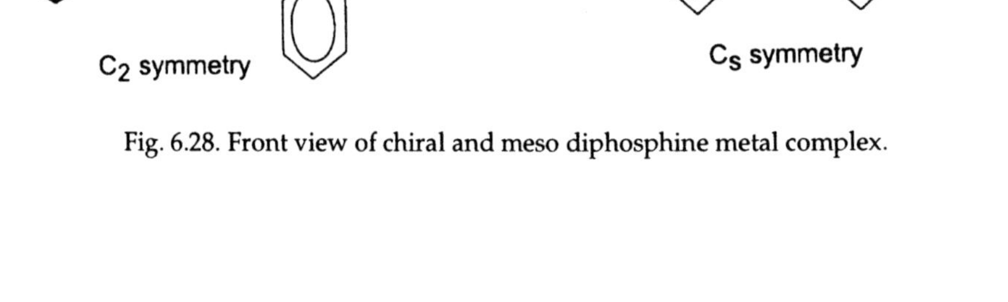

Fig. 6.28. Front view of chiral and meso diphosphine metal complex.

selective reactions. Coordination of enamides (see above) will now take place in such a way that the auxiliary donor atom coordinates to one side, and the phenyl substituted alkene will coordinate to the other side with the face that gives the least interaction of its substituents with the phenyl group of the ligand pointing into the same quadrant. This gives the predominant metal alkene adduct; as we have seen in the above example, this is not to say that the dominant reaction path will arise from this species that is the major component in the equilibrium mixture.

Even in the absence of an asymmetric bridge, a diphosphine complex may have C_2 symmetry, but it will be a racemic mixture. Not all diphosphine ligands adopt C_2 symmetry with respect to the four phenyl groups. For instance, $Ph_2P(CH_2)_4PPh_2$ adopts C_s symmetry (Fig. 6.28). In this configuration two phenyl groups are at the side of the chelating ring and will take (more or less) axial positions, while the other two at the opposite face of the ring will assume equatorial positions. Even though there may be an asymmetric centre in the backbone of a diphosphine, if the complex prefers a C_s 'local' symmetry for the phenyl groups, the asymmetry is not optimally passed on to the substrate coordinating to the diphos ligand–metal complex. Many crystal structures have been determined that support this view and refine it even further, as do molecular graphics (or molecular mechanics) for a broad range of ligands [78].

6.4.3 *BINAP Catalysis*

At present much research is being directed to the development of catalytic routes to a series of asymmetric carboxylic acids that lack the acidamido ligand as additional functionality. Four are listed in Fig. 6.29, which are important as

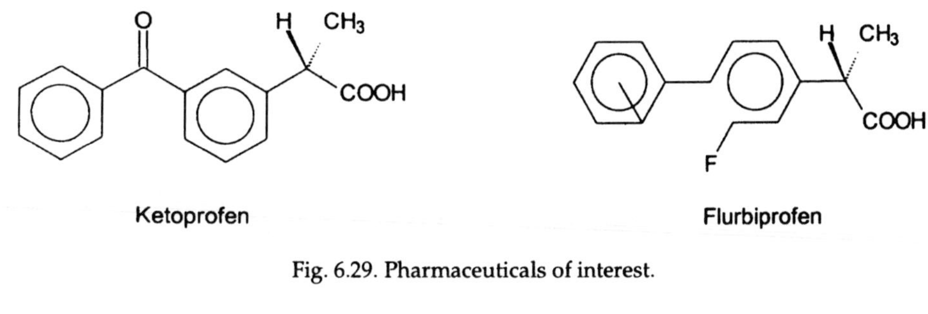

Fig. 6.29. Pharmaceuticals of interest.

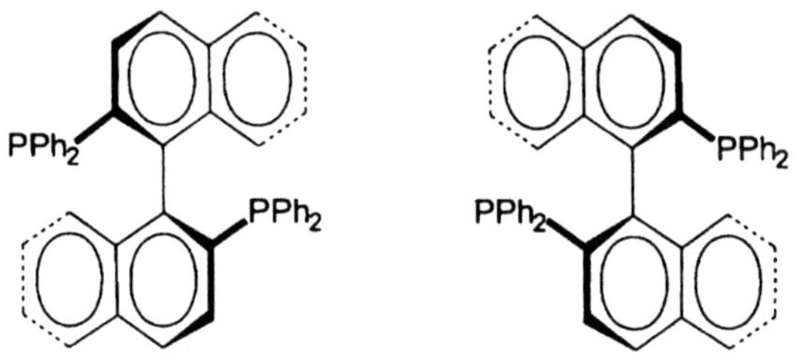

Fig. 6.30. Asymmetric BINAP ligand.

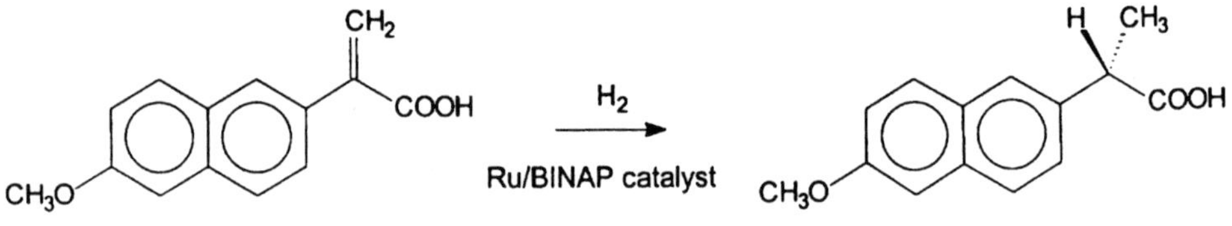

Fig. 6.31. Hydrogenation route to S-Naproxen.

anaesthetics for rheumatic diseases. Their turnover in dollars beats many of the bulk chemicals: the turnover of Naproxen in 1990 amounted to \$700 million.

S-Naproxen is being produced by Syntex by resolution with a chiral auxiliary. The main patents from Syntex expired in 1993, which represents the main drive for an increase in the efforts to find alternative synthetic routes which may lead directly to an asymmetric carbohydroxylation of the styrene precursor. Asymmetric hydrogenation of molecules of this type has been successfully accomplished by Noyori using ruthenium complexes of the ligand BINAP (Fig. 6.30). With this system a high enantioselectivity can be achieved (97%) albeit at rather high pressures (135 bar). Monsanto [79] claims a similar route that operates at a much lower pressure. The reaction scheme is shown in Fig. 6.31.

BINAP was first introduced by Noyori [80]. It has been particularly explored for reduction with ruthenium catalysts. BINAP is an atropisomeric ligand because rotation around the central C–C bond is blocked. Accordingly BINAP exists in two enantiomers. Complexes of Ru(II) with BINAP are extremely powerful catalysts for enantioselective hydrogenations of prochiral α,β- and β,γ-unsaturated carboxylic acids, enamides, allylic and homoallylic alcohols, imines etc. [83]. In many cases, the hydrogenation is quantitative with enantiomeric excesses of over 95%. A wide variety of vitamins, terpenes, β-lactam antibiotics, etc. are accessible by the use of catalysts containing the BINAP stereogenic element. An example for 3-oxo carboxylic esters is shown in reaction (1) of Fig. 6.32.

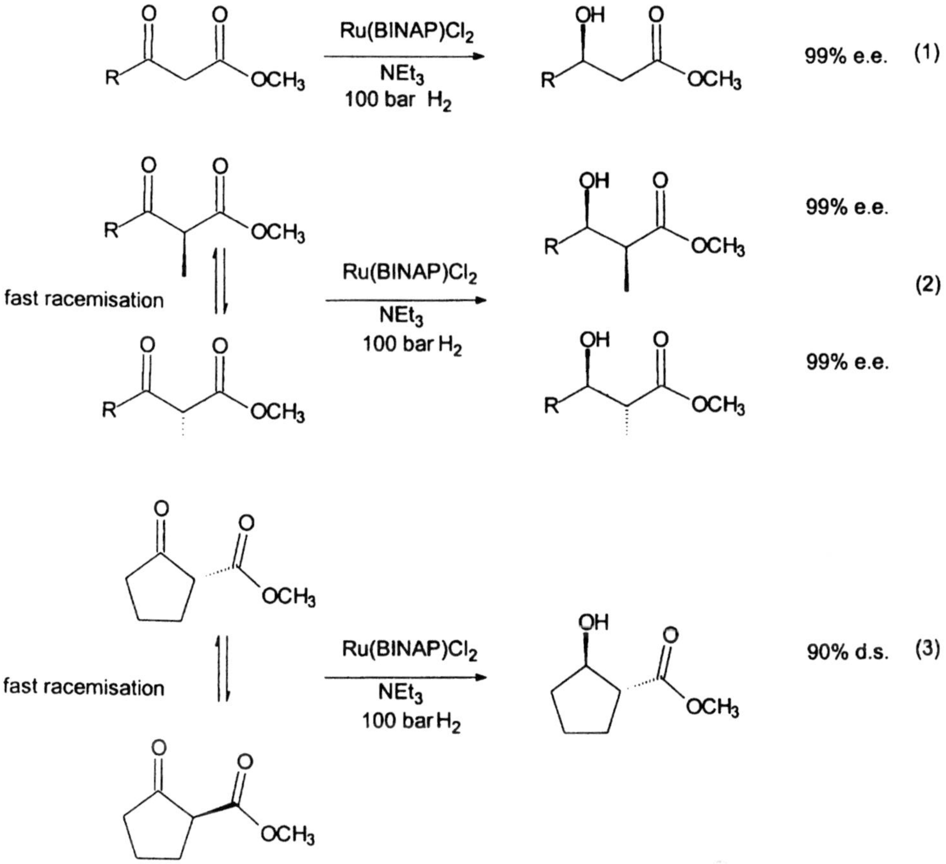

Fig. 6.32. Asymmetric hydrogenation using Ru^{2+}(BINAP) of ketoesters.

An interesting and technically important variation is the hydrogenation of 2-substituted 3–oxo carboxylic esters (reaction (2), Fig. 6.32). The enantioselectivity of the BINAP-Ru catalyzed hydrogenation at C-3 is very high, which means that the chirality of BINAP controls the facial selectivity at the carbonyl function. Thus, independent of the configuration of C-2, only one configuration is formed at C-3, giving two diastereoisomers, each with high e.e.

When so-called kinetic resolution occurs in a system like this, one enantiomer of the racemic substrate reacts much faster than the other yielding preferentially only one diastereoisomer. The chemical yield in those systems is only 50% or less. Racemic compounds with a chirally labile stereogenic centre, however, may yield full conversion to one diastereoisomer via racemization prior to hydrogenation.

The diastereoselectivity is low for most of the substrates tested (reaction 2). It only reaches values above 90% for cyclic compounds (reaction 3). For this

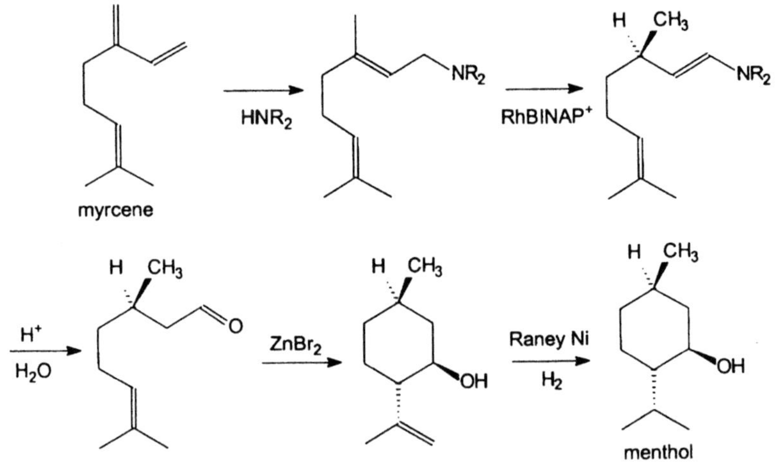

Fig. 6.33. The Takasago process for (–)menthol.

so-called second-order stereoselectivity hydrogenation to be successful, (a) the rate of racemization must be fast compared to the rate of hydrogenation of the slowest enantiomer, (b) the carbonyl face discrimination by the chiral BINAP-Ru(II) catalyst must be efficient, and (c) the stereogenic centre at C-2 must lead to an appreciable difference between the rates of the two enantiomers.

Finally, another important application of BINAP is found in the Takasago process for the commercial production of (–) menthol from myrcene. The catalyst used is a rhodium complex of BINAP. Figure 6.33 gives the reaction scheme [81]. The key reaction is the enantioselective isomerization of the allylamine to the asymmetric enamine. It is proposed that this reaction proceeds via an allylic intermediate.

6.4.4 Chiral Ferrocene Based Ligands

One last type of asymmetric ligand will be described here. The ligands discussed so far all contained C_2 symmetry. An important new class of ligands having C_1 symmetry was introduced by Togni [82] (see Fig. 6.34). They can be easily made

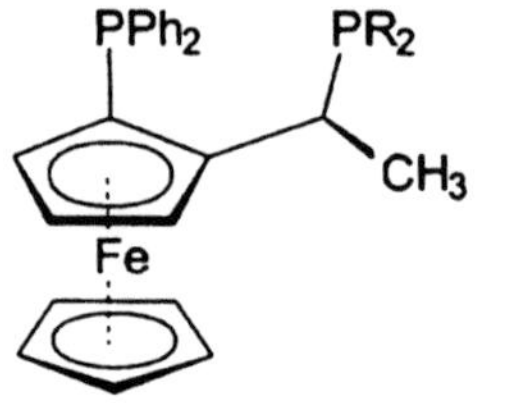

Fig. 6.34. Ferrocene derived C_1 symmetric chiral ligands.

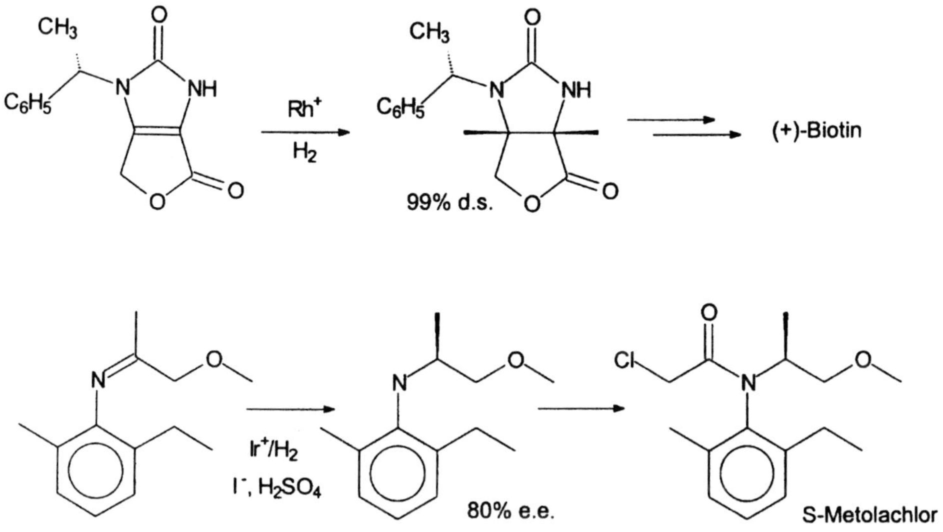

Fig. 6.35. Commercialized asymmetric hydrogenations using ferrocene based ligands.

from an enantiomeric amine as the precursor in a few steps. Different substituents can be introduced at the phosphorus atoms. In addition to the chiral carbon atom the molecule has planar chirality as well.

Recently two new processes have come on stream utilizing these ferrocene based chiral ligands [82]. The first one is the rhodium catalyzed hydrogenation of an already chiral precursor for Biotin. A very high diastereoselectivity is obtained, 99%. The second example involves the asymmetric hydrogenation of an imine using an iridium catalyst. The enantioselectivity for the intermediate used for the production of Metolachlor is only 80%, which is as yet tolerable for an agrochemical (Fig. 6.35). Nothing is known about the mechanisms of these reactions.

Asymmetric catalysis involving metal catalyzed hydrogenations and isomerizations is becoming increasingly important in the production of pharmaceuticals, agrochemicals and flavours and fragrances. More examples of asymmetric homogeneous catalysts used industrially can be found in the literature [1,83].

REFERENCES

1 G.W. Parshall and S.D. Ittel, Homogeneous Catalysis. *The Applications and Chemistry of Catalysis by Soluble Transition Metal Complexes*, Wiley, New York, 2nd Ed., 1992. R. Ugo (Editor), *Aspects of Homogeneous Catalysis*, Vols. 1–7, Kluwer Academic Publishers, Dordrecht, 1990.

2 Cornils, B.; Herrmann, W. A., (Editors), *Applied Homogeneous Catalysis with Organometallic Compounds*, Vols 1–2, VCH, Weinheim, Germany, 1996.

3 C.A. Tolman, *Chem. Rev.*, 77 (1977) 313.
4 O. Roelen, *Ger. Pat.* 849548 (1938).
5 L.H. Slaugh and R.D. Mullineaux, *U.S. Pat.* (to Shell) 3 239 569 and 3 239 570 (1969).
6 D. Evans, J.A. Osborn and G. Wilkinson, *J. Chem. Soc.* (A), (1968) 3133.
7 B. Cornils,; W.A. Herrmann, *Aqueous-Phase Organometallic Catalysis*, Wiley-VCH: Weinheim, Germany, 1998; Vol. 1–2.
8 F. Ungváry, *Coord. Chem. Rev.*, 160 (1997) 123.
9 W.A. Herrmann and C.W. Kohlpainter, *Angew. Chem., Int. Ed. Engl.*, 32 (1993) 1524.
10 R.L. Pruett and J.A. Smith, *J. Org. Chem.*, 34 (1969) 327.
11 P.W.N.M. van Leeuwen and C.F. Roobeek, *Br. Pat. Appl.* (to Shell) 2,068,377 (1980).
12 T. Jongsma, P.W.N.M. van Leeuwen and G. Challa, *J. Organomet. Chem.*, 421 (1991) 121.
13 A. Van Rooy, E.N. Orij, P.C.J. Kamer, F. van den Aardweg and P.W.N.M. van Leeuwen, *J. Chem. Soc. Chem. Commun.*, (1991) 1096.
14 A.A. Oswald, D.E. Hendriksen, R.V. Kastrup and E.J. Mozeleski, in *Homogeneous Transition Metal Catalyzed Reactions*, W.R. Moser and D.W. Slocum (Editors), Adv. in Chem. Series 230, ACS, Washington, DC, 1992, p. 395.
15 M.E. Broussard, B. Juma, S.G. Train, W.-J. Peng, S.A. Laneman, G.G. Stanley, *Science*, 260 (1993) 1784. R.C. Matthews, D.K. Howell, W.-J. Peng, S.G. Train, W.D. Treleaven and G.G. Stanley, *Angew. Chem. Int. Ed. Engl.* 35 (1996) 2253.
16 J.D. Unruh and J.R. Christenson, *J. Mol. Catal.*, 14 (1982) 19.
17 C.P. Casey, E.L. Paulsen, E.W. Beuttenmueller, B.R. Proft, L.M. Petrovich, B.A. Matter, and D.R. Powell, *J. Am. Chem. Soc.*, 119 (1997) 11817.
18 L.A. van der Veen, M.D.K. Boele, F.R. Bregman, P.C.J. Kamer, P.W.N.M. van Leeuwen, K. Goubitz, J. Fraanje, H. Schenk, C. Bo, *J. Am. Chem. Soc.*, 120 (1998) 11616.
19 J.M. Brown and A.G. Kent, *J. Chem. Soc. Perkin Trans. II*, (1987) 1597.
20 E. Billig, A. Abatjoglou and D.R. Bryant, *U.S. Patents* 4,668,651 (1987) and 4,769,498 (1988).
21 A. van Rooy, P.C.J. Kamer, P.W.N.M. van Leeuwen, K. Goubitz, J. Fraanje, N. Veldman, and A.L. Spek, *Organometallics*, 15 (1996) 835.
22 T.J. Devon, G.W. Philips, T.A. Puckette, J.L. Stavinoha, and J.J. Vanderbilt, *PCT Int. Appl.* WO 87 07,600 (1987), *U.S. Patent* 4,879,416 and 4,694,109 (1989), and *Eur. Pat. Appl.* EP 0 326 286 (1989).
23 C.P. Casey, G.T. Whiteker, M.G. Melville, L.M. Petrovich, J.A. Gavney Jr. and D.R. Powell, *J. Am. Chem. Soc.*, 114 (1992) 5535.
24 M. Kranenburg, Y.E.M. van der Burgt, P.C.J. Kamer, and P.W.N.M. van Leeuwen, *Organometallics*, 14 (1995) 3081.
25 K. Yamamoto, S. Momose, M. Funahashiand M. Myazawa, *Synlett.*, 1990, 711. C. .P. Casey and G.T. Whiteker, *J. Org. Chem.*, 55 (1990) 1394.
26 P.M. Burke, J.M. Garner, W. Tam, K.A. Kreutzer, A.J.J.M. Teunissen, C.S. Snijder, C.B. Hansen, *WO* 9733854 (1997) (to Du Pont and DSM).
27 L.A. van der Veen, P.C.J. Kamer, P.W.N.M. van Leeuwen, *Angew. Int.Ed.Engl.*, 38 (1999).
28 R.F. Heck and D.S. Breslow, *J. Am. Chem. Soc.*, 83 (1961) 4023.
29 G. Csontos, B. Heil and L. Markó, *Ann. NY Acad. Sci.*, 239 (1974) 47.
30 A. van Rooy, E.N. Orij, P.C.J. Kamer, P.W.N.M. van Leeuwen, *Organometallics*, 14 (1995) 34.
31 R. Lazzaroni, P. Pertici and G. Fabrizi, *J. Mol. Catal.*, 58 (1990) 75.
32 A.M. Joshi and B.R. James, *Organometallics*, 9 (1990) 199.
33 L. Versluis and T. Ziegler, *Organometallics*, 9 (1990) 2985.
34 P.P. Deutsch and R.Eisenberg, *Organometallics*, 9 (1990) 709.

35 W.D. Jones, J.M. Huggins and R.G. Bergman, *J. Am. Chem. Soc.*, 103 (1981) 4415.
36 P.W.N.M. van Leeuwen and C.F. Roobeek, *J. Organomet. Chem.*, 258 (1983) 343 and *U.S. Patent* 4,467,116 (1984).
37 I.T. Horváth, G. Kiss, R.A. Cook, J.E. Bond, P.A. Stevens, J. Rábai, and E.J. Mozeleski, *J. Am. Chem. Soc.*, 120 (1998) 3133.
38 P. Cavalieri d'Oro, L. Raimondo, G. Pagani, G. Montrasi, G. Gregorio and A. Andreetta, *Chim. Ind. (Milan)*, 62 (1980) 572.
39 Y. Pottier, A. Mortreux and F. Petit, *J. Organomet. Chem.*, 370 (1989) 333.
40 A. van Rooy, P.C.J. Kamer, P.W.N.M. van Leeuwen, *J. Organomet. Chem.*, 494 (1995) C15. G.J.H. Buisman, L.A. van der Veen, P.C.J. Kamer, and P.W.N.M. van Leeuwen, *Organometallics*, 16 (1997) 5681.
41 M. Garland and P. Pino, *Organometallics*, 10 (1991) 1693.
42 W.R. Moser, C.J. Papile, D.A. Brannon and S.J. Weiniger, *J. Mol. Catal.*, 41 (1987) 271.
43 A. Castellanos-Páez, S. Castillón, C. Claver, P.W.N.M. van Leeuwen and W.G.J. de Lange, *Organometallics*, 17 (1998) 2543.
44 B.L. Goodall, "Polypropylene: Catalysts and Polymerization Aspects" in this work, Chapter 1, pp. 1–125 in S. van der Ven, *Polypropylene and Other Polyolefins: Polymerization and Characterization*, Elsevier Science Publishers, New York, 1990. H.H. Brintzinger, D. Fischer, R. Mülhaupt, B. Rieger and R.M. Waymouth, *Angew. Chem. Int. Ed. Engl.* 34 (1995) 1143. M. Bochmann, *J. Chem. Soc. Dalton Trans.*, 1996, 255.
45 P. Corradini, V. Busico, L. Cavallo, G. Guerra, M. Vacatello and V. Venditto, *J. Mol. Catal.*, 74 (1992) 433. M. Toto, L. Cavallo, P. Corradini, G. Moscardi, L. Resconi and G. Guerra, *Macromolecules*, 31 (1998) 3431.
46 E.J. Arlman and P. Cossee, *J. Catal.*, 3 (1964) 99.
47 G. Allegra, *Makromol. Chem.*, 145 (1971) 235.
48 D.S. Breslow and N.R. Newburg, *J. Am. Chem. Soc.*, 82 (1957) 5072.
49 G. Natta, P. Pino, P. Corradini, F. Danusso, E. Mantica, G. Mazzanti and G. Moraglio, *J. Am. Chem. Soc.*, 77 (1955) 1708.
50 V. Busico, R. Cipullo, G. Monaco, M. Vacatello, J. Bella, and A. Segre, *Macromolecules*, 31 (1998) 8713.
51 J.A. Ewen, *J. Am. Chem. Soc.*, 106 (1984) 6355.
52 J.A. Ewen, M.J. Elder, R.L. Jones, L. Haspeslagh, J.L. Atwood, S.G. Bott and K. Robinson, *Makromol. Chem., Makromol. Symp.*, 48/49 (1991) 253.
53 T. Mise, S. Miya and H. Yamazaki, *Chem. Lett.*, (1989) 1853.
54 Y.Takegami and T. Suzuki, *Bull. Chem. Soc. Jpn.*, 42 (1969) 848. A. Zambelli, G. Natta and I. Pasquon, *J. Polym. Sci.*, C4 (1963) 411. Y. Doi, S. Suzuki, F. Nozawa, and K. Soga, in *Catalytic Polymerization of Olefins*, T. Keii and K. Soga (Editors), Kodansha, Tokyo, 1986, p. 257.
55 R. Waymouth and P. Pino, *J. Am. Chem. Soc.*, 112 (1990) 4911.
56 H. Sinn, W. Kaminsky, H.-J. Volmer and R. Woldt, *Angew. Chem., Int. Ed. Engl.*, 19 (1980) 390.
57 H. Sinn and W. Kaminsky, *Adv. Organometal. Chem.*, 18 (1980) 99.
58 F.R.W.P. Wild, L. Zsolnai, G. Huttner and H.H. Brintzinger, *J. Organomet. Chem.*, 232 (1982) 233.
59 W. Kaminsky, K. Külper, F.R.W.P. Wild and H.H. Brintzinger, *Angew. Chem. Int. Ed. Engl.*, 24 (1985) 507.
60 W. Kaminsky, K. Külper and S.D. Niedoba, *Makromol. Chem., Makromol. Symp.*, 3 (1986) 377.
61 J.A. Ewen, *J. Am. Chem. Soc.*, 110 (1988) 6255 and *Scientific American* (May 1997) 60.

62 G. Erker, R. Nolte, R. Aul, S. Wilker, C. Krüger and R. Noe, *J. Am. Chem. Soc.*, 113 (1991) 7594.
63 V. Busico, D. Brita, L. Caporaso, R. Cipullo and M. Vacatello, *Macromolecules*, 30 (1997) 3971. M.K. Leclerc and H.H. Brintzinger, *J. Am. Chem. Soc.*, 118 (1996) 9024.
64 R.F. Jordan, R.E. LaPointe, C.S. Bajgur, S.F. Echols and R. Willett, *J. Am. Chem. Soc.*, 109 (1987) 4111. G.G. Hlatky, H.W. Turner and R.R. Eckman, *J. Am. Chem. Soc.*, 111, (1989) 2728.
65 M.C. Sacchi, E. Barsties, I. Tritto, P. Locatelli, H.H. Brintziger, and U. Stehling, *Macromolecules*, 30 (1997) 3955.
66 P. Corradini, G. Guerra, M. Vacatello and V. Villani, *Gazz. Chim. Ital.*, 118 (1988) 173. L. Cavallo, G. Guerra, M. Vacatello and P. Corradini, *Macromolecules*, 24 (1991) 1784. L. Cavallo, G. Guerra, L. Oliva, M. Vacatello and P. Corradini, *Polym. Commun.*, 30 (1989) 17.
67 V. Busico, R. Cipullo and P. Corradini, *Makrom.Che. Rapid Comm.*, 13 (1992) 15. J.C. Chadwick, A. Miedema and O. Sudmeijer, *Macrom. Chem. Phys.*, 195 (1994) 167. V. Busico, R. Cipullo, J.C. Chadwick, J.F. Modder and O. Sudmeijer, *Macromolecules*, 27 (1994) 7538.
68 W. Spaleck, F. Küber, A. Winter, J. Rohrmana, B. Bachmann, M. Antberg, V. Dolle, and E.F. Paulus, *Organometallics*, 13 (1994) 954. M. Toto, L. Cavallo, P. Corradini, G. Moscardi, L. Resconi and G. Guerra, *Macromolecules*, 31 (1998) 3431.
69 G.W. Coates and R.M. Waymouth, *Science*, 267/5195 (1995) 217. J.L. Maciejewski Petoff, M.D. Bruce, R.M. Waymouth, A. Masood, T.K. Lal, R.W. Quan and S.J. Behrend, *Organometallics*, 16 (1997) 5909.
70 J.A. Ewen, R.L. Jones, A. Razavi and J.D. Ferrara, *J. Am. Chem. Soc.*, 110 (1988) 6255. T.A. Herzog, D.L. Zubris and J.E. Bercaw, *J. Am. Chem. Soc.*, 118 (1996) 11988. V. Busico, R. Cipullo, G. Talarico, A.L. Segre and L. Caporaso, *Macromolecules*, 31 (1998) 8720.
71 W.S. Knowles, M.J. Sabacky and B.D. Vineyard, *Adv. Chem. Ser.*, 132 (1974) 274.
72 J. Halpern et al., *Science*, 127 (1982) 401; A.S.C. Chan, J.J. Pluth and J. Halpern, *J. Am. Chem. Soc.*, 102 (1980) 5922. C.R. Landis and J. Halpern, *J. Am. Chem. Soc.*, 109 (1987) 1746.
73 J.M. Brown and P.A. Chaloner, *J. Am. Chem. Soc.*, 102 (1980) 3040. J.M. Brown, P.A. Chaloner and G.A. Morris, *J. Chem.Soc. Perkins Trans*,. 2, 1987, 1583.
74 S.K. Armstrong, J.M. Brown and M.J. Burk, *Tetrahedron Lett.* 34 (1993) 879. J.S. Giovannetti, C.M. Kelly and C.R. Landis, *J. Am. Chem. Soc.*, 115 (1993) 4040. B.F.M. Kimmich, E. Somsook and C.R. Landis, *J. Am. Chem. Soc.* 120 (1998) 10115. C.R. Landis and T.W. Brauch, *Inorg. Chim. Acta*, 270 (1998) 285.
75 E.P. Kyba, R.E. Davis, P.N. Juri and K.R. Shirley, *Inorg. Chem.* 20 (1981) 3616.
76 H. B. Kagan, T.-P. Dang, *J. Am. Chem. Soc.*, 94 (1972) 6429.
77 M.J. Burk and J.E. Harlow, *J. Am. Chem. Soc.*, 114 (1992) 6266.
78 J.M. Brown and P.L. Evans, *Tetrahedron*, 44 (1988) 4905.
79 A.S.C. Chan, *199th National Meeting*, Boston, Am. Chem. Soc., 1990.
80 A. Miyashita, A. Yasuda, H. Tanaka, K. Toriumi, T. Ito, T. Souchi and R. Noyori, *J. Am. Chem. Soc.*, 102 (1980) 7932.
81 K. Tani, T. Yamagata, Y. Tatsuno, Y. Yamagata, K. Tomita, S. Akutagawa, H. Kumobayashi and S. Otsuka, *Angew. Chem. Int. Ed. Engl.*, 24 (1985) 217.
82 Togni, *Angew. Chem. Int. Ed. Engl.*, 35 (1996) 1475. H.-U. Blaser and F. Spindler, *Top. Catal.*, 4 (1998) 275. A. Togni, C. Breutel, A. Schnyder, F. Spindler, H. Landert and A. Tijani *J. Am. Chem. Soc.*, 116 (1994) 4062.
83 *Catalytic Asymmetric Synthesis*, I. Ojima, Editor, VCH Publishers, New York 1993. *Asymmetric Catalysis in Organic Synthesis*; R. Noyori, Editor, John Wiley and Sons, Inc. New York, 1994.

Chapter 7

Biocatalysis

7.1. INTRODUCTION

Biocatalysis covers a broad range of scientific and technical disciplines, which are geared to develop biocatalysts and biocatalytic processes for practical purposes. The natural pool of biocatalysts is extremely diverse and includes whole cells of microbial, plant or animal origin, as well as cell-free extracts and enzymes derived from these sources. The wide range of catalytic power offered by nature remains, however, largely unexplored. Currently, only a very small fraction of the known biocatalysts are actually being applied on a commercial scale. For example, of the approximately 4,000 known enzymes, about 400 are available commercially, but only about 40 are actually used for industrial applications.

Biocatalysts are generally more efficient than chemical catalysts. Typically, enzymes can exhibit turnover numbers of ≥100,000 s^{-1} vs. the values of 0.01–1 s^{-1} typically observed for hetero- and homogeneous catalysts. Furthermore, enzymes can operate under relatively mild conditions and usually exhibit a very high degree of substrate-, chemo-, regio-, and enantioselectivity. Another important factor, which has contributed significantly to the use of biocatalysts in practical applications, is the enormous technical advances made in the last decade in molecular biology, which permit one to overproduce and redesign enzymes and to redirect metabolic fluxes essentially at will by genetic, protein, and metabolic pathway engineering. As a result, biocatalysts can be increasingly regarded as relatively inexpensive but complex chiral catalysts, which can be tailored to some extent to meet specific requirements. Nevertheless, biocatalysts are still not routinely applied in the chemical industry for a variety of reasons. For example, biocatalysts for all types of desirable conversions do not necessarily exist in nature, and biocatalysts often suffer from instability and substrate/product inhibition. In addition, some enzymes require a cofactor (a small dissociable molecule that is essential for activity and, in the case of enzymes catalyzing oxidation-reduction reactions, provides the reducing or oxidizing equivalents needed for the reaction), which must be recycled to reduce costs to acceptable levels. Although good progress has been made in designing cofactor-regenerating systems *in vitro*, the alternative *in vivo* approach, using whole cells to secure an efficient cofactor-recycling system, is often preferred in applications of biocatalysis, as we shall see.

In general, biocatalytic processes are perceived as being more environmentally friendly than chemical production methods, which renders the former politically and socially more attractive. In fact, however, this assumption is not necessarily true. There are many well-developed chemical processes that have a higher product to waste ratio than their biocatalytic counterparts. The potential benefits of biocatalytic processes need to be examined on a case by case basis, and often it is of particular interest to combine the best of both worlds by designing hybrid processes.

It is beyond the scope of this chapter to cover the whole area of biocatalysis. Instead, the chapter will focus initially on the question of when it is advantageous to use biocatalysts and what decisions have to be made so that they can be optimally utilized. Examples that illustrate the decision-making process are given throughout. Subsequently, selected applications of biocatalysts will be examined, using either isolated enzymes or enzymes that operate on their own in resting (immobilized) or permeabilized cells, and focusing on the individual classes of enzymes as introduced in Chapter 4 (hydrolases, lyases, isomerases, transferases, ligases, and oxidoreductases). For each class, the practical scope and relevant recent advances will be discussed. This approach will allow the student to assess the strengths and limitations of each of these classes of biocatalyst. Topics such as process engineering and reactor design in biocatalysis and methods for stabilizing enzymes through immobilization are covered in appropriate sections of subsequent chapters.

7.2 BIOCATALYSIS VS. CHEMICAL CATALYSIS

Biocatalysts are becoming more and more familiar as shelf reagents in chemistry. Nevertheless, there is still some uneasiness regarding their use in industry, due largely to unfamiliarity with these catalysts provided by Nature. In this section, a strategy is discussed that will allow the prospective user to answer the question 'When and how to put a biocatalyst to work?'. The decision tree shown in Table 7.1 lists the important questions that must be answered to make optimal use of a biocatalyst.

The first question to answer is: Is biocatalysis an attractive alternative to the far more familiar chemical conversions? Aspects to be considered in this initial

TABLE 1

Decision scheme: when and how to use biocatalysts?

Level	Consideration
Reaction	chemical conversion vs. biocatalysis
biocatalyst	isolated enzyme vs. whole cell
form	free vs. immobilized biocatalyst
medium	water vs. organic solvent
bioreactor	standard vs. novel
operation	(fed-)batch vs. continuous
integration	reactions/process steps/overall process

decision making are the availability, purity and costs of reactants and other medium components, the number of reaction steps involved, the selectivity and yield, reaction conditions, the ease of product purification, final product quality, and environmental aspects ('greenness'). Experience has demonstrated that biocatalysis is perceived as a favourable alternative only when most of these factors are positive for biocatalysis.

The second level of decision making concerns the form of the biocatalyst, i.e., a whole cell, a cell organelle, an enzyme complex, a crude enzyme preparation, or an isolated enzyme, all either free or immobilized. Availability, price, cofactor need, etc., are points that must be considered at this level. The costs associated with immobilization obviously should be regained by the possibility of developing a more efficient process.

The third level concerns the reaction medium. Will it be the conventional aqueous environment with the label 'green', or a non-conventional medium such as an organic solvent? The latter can have convincing advantages with respect to solubility, substrate/product inhibition, reaction equilibrium, and product purification.

On the basis of the kinetics of the biocatalysts and the foregoing decisions, the reactor configuration must then be chosen: standard or novel? Here, availability, experience, desired mixing, and mass-transfer properties play a dominant role. The mode of operation of the bioreactor can be (fed-)batch or continuous. In making this decision, the kinetics, stability and form of the biocatalyst, desired substrate conversion and product concentration, and need for process control all play a role.

Finally, throughout the whole decision making process, the level of process integration with respect to reactions (if more than one is involved) and process steps should be taken into account. These aspects are considered in more detail in the following sections.

7.2.1 Chemical Conversion vs. Biocatalysis

When the number of reaction steps can be drastically reduced by means of the introduction of a biocatalyst, the odds in favour of opting for biocatalysis go up dramatically. Even then, however, conventional chemistry remains a strong competitor, not the least because of the fact that an enormous amount of knowledge and experience is available in this field. Other aspects such as the need for fewer and cheaper reagents and other essential compounds, higher yields, faster reaction rates and milder reaction conditions should speak for biocatalysis as well. Enantioselectivity and stereoselectivity can be strong points in favour of biocatalysis, but enormous progress has lately been made in this respect in chemical catalysis. If using the labels 'natural' and 'green' help in selling a product, this factor can also help biocatalysis become the production procedure of choice. Furthermore, a proper fit with existing processes is also desirable. Obviously, this short list of considerations is far from being exhaustive and is not given in order of priority. In addition, each individual application will have its own specific questions to be answered in the decision making process.

An illustrative example of biocatalysis replacing a chemical conversion is the production of 6-aminopenicillanic acid (6-APA). Figure 7.1 shows the chemical

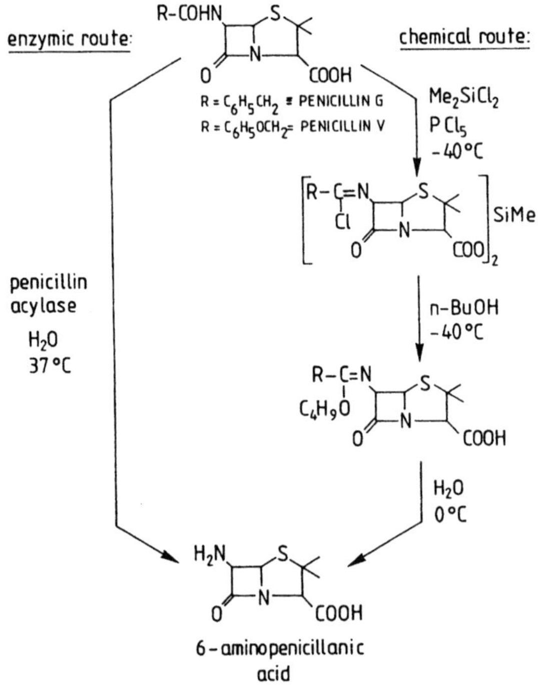

Fig. 7.1. Production of 6-aminopenicillanic acid (6-APA) from penicillin G or V by biocatalysis and chemical synthesis.

and enzymic routes. The chemical synthesis comprises three reaction steps that require harsh reagents, the environmentally unfriendly solvent dichloromethane, and a reaction temperature of –40°. In contrast, the enzymic route involves only one step and uses mild conditions, requiring merely the biocatalyst and the substrate. Given these considerations, it is not surprising that biocatalysis has largely replaced chemical synthesis of 6-APA during the last decade. The enzyme processes use immobilized penicillin V or G acylases derived from fungi such as *Bovista plumbea* and bacteria such as *E. coli*, respectively. Productivities of up to 2000 kg of 6-APA/kg immobilized enzyme are obtained, with operating lifetimes in excess of 10^3 hours [1].

7.2.2 Isolated Enzyme versus Whole Cell

In this text, the term biocatalyst is used to refer to an enzyme, an enzyme complex, a cell organelle or a whole cell. The latter can either be growing or non-growing, and viable or not. The biocatalysts can be of microbial, plant or animal origin. In the latter two cases in particular, enzyme purification is usually unavoidable, as tissues are generally not suitable for use in biocatalytic processes. Plant and animal cells are not cultivated for enzyme production to any considerable extent, nor are they genetically modified with the purpose of developing a whole-cell biocatalyst. This is in sharp contrast to microbial cells, which are genetically engineered both for the purpose of increased enzyme production and for use as biocatalysts. The example discussed below is an illustration of this. Factors to consider in deciding whether an enzyme or whole cells should be used include availability, the number of steps involved, yield, cofactor need, byproduct formation, required product purity, price, etc. If, for instance, a biocatalyst is not commercially available, it may have to be made in house, adding at least one extra step.

A beautiful example of biocatalysis by means of an engineered whole cell is the biosynthesis of indigo [2]: "The green route to a blue dye!" For centuries, indigoid dyes were obtained in minute quantities from mollusc and more readily from fermented leaves of the European woad plant and the Asian indigo plant. Since 1897, indigo has been chemically synthesized from hazardous reagents (aniline, formaldehyde and cyanide) with formation of toxic byproducts. Genencor International is developing a commercially competitive route to indigo using recombinant (genetically altered) *E. coli* cells that can directly synthesize indigo from glucose. Glucose is converted to the amino acid tryptophan, then to indole, and finally to indigo (Fig. 7.2). Several enzymes are involved, making a process utilizing the isolated enzymes unattractive. To make the microbial process competitive, one must remove the bottleneck in the pathway between glucose and tryptophan. *E. coli* normally makes tryptophan from glucose, but only in the relatively small quantities needed for growth. Genencor scientists have removed

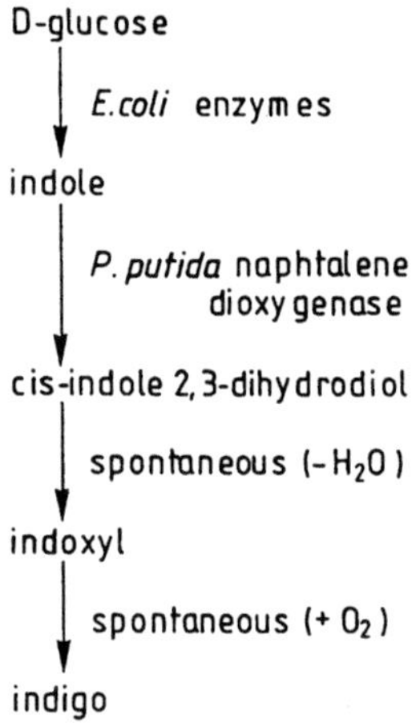

Fig. 7.2. Biosynthesis of indigo by recombinant *E. coli* cells.

the ability of the cell to sense this amino acid, thus preventing the cell from shutting off this pathway; as a result, the tryptophan biosynthesis pathway is always turned on. Practical yields of tryptophan and of indigo are expected to be close to the organism's theoretical maximum. This return to biological production of the dye represents a step toward the 'greening' of the chemical process industry.

7.2.2.1 Cofactor regeneration

The problem of cofactor regeneration is an important limitation in the use of many biocatalysts, and hence merits specific consideration. Many enzymatic reactions require coenzymes, e.g. NAD^+/NADH, to provide the oxidizing/reducing equivalents for oxidations/reductions. An example is the reduction of the steroid cortisone to 20-dihydrocortisone with the enzyme 20-β-hydroxysteroid dehydrogenase. A stoichiometric amount of NADH is consumed by this reaction. Because of the high price of NADH (currently about \$27000/kg), the NAD^+ that is formed has to be regenerated to NADH via a second reaction if such a process is to have any hope of being competitive in price. With a system that is able to carry out n regenerations, the cost of NADH drops to (\$27000/n) × ($moles_{NADH}/moles_{product}$) per kg of product. Thus, a large value of n ($n >>> 1000$) is required. In general, the requirements for a good regeneration system are:

(a) The thermodynamics should be favourable ($\Delta G < 0$)
(b) The kinetics should be favourable (the regeneration should be fast and the regeneration reaction should be selective)
(c) The cofactor stability should be good (e.g., NADH should not decompose during the regeneration).

Regeneration reactions may be: uncatalyzed, chemically catalyzed, electrochemically (catalyzed), enzymatically catalyzed, or "catalyzed" by living cells.

The cost of the substrate for the regeneration reaction must be minimized to make the process cost effective. With living cells cheap co-substrates can often be used, e.g., during the reduction of cortisone to 20-dihydrocortisone, NADH may be regenerated by reduction of NAD^+ to NADH by glucose. However, living cells are sometimes not very selective due to the presence of interfering enzymes. An example of coenzyme regeneration using isolated enzymes is production of L-alanine from pyruvate with an NADH-dependent alanine dehydrogenase (AlaDH):

$$\mathrm{NADH} + \mathrm{pyruvate} + NH_4^+ \rightleftharpoons \mathrm{alanine} + H_2O + NAD^+$$

A regeneration reaction for NADH that may be used is the conversion of pyruvate with a NADH-dependent lactate dehydrogenase (LDH):

$$\mathrm{NAD+} + \mathrm{lactate} \rightleftharpoons H^+ + \mathrm{pyruvate} + \mathrm{NADH}$$

These two reactions have to be carried out simultaneously, so the overall reaction will be

$$\mathrm{lactate} + NH_3 \rightleftharpoons \mathrm{alanine} + H_2O$$

The advantage of this reaction is that lactate is much cheaper than pyruvate, so that the reaction is not only useful for the regeneration of NADH but also for the production of pyruvate. The disadvantage is that lactate dehydrogenases are enantiospecific, so a D-lactate as well as an L-lactate dehydrogenase is required for full conversion of racemic lactate (Fig. 7.3).

The thermodynamics of the alanine synthesis reaction are favourable if a large excess of ammonium is used. In a batch reactor the number of regenerations of

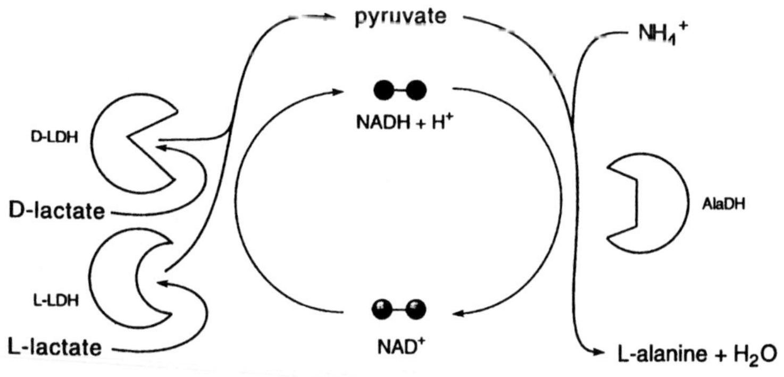

Fig. 7.3. Alanine production with cofactor regeneration. The black and gray dumbbell shapes are the reduced and oxidized forms of the NAD cofactor bound to polyethylene glycol.

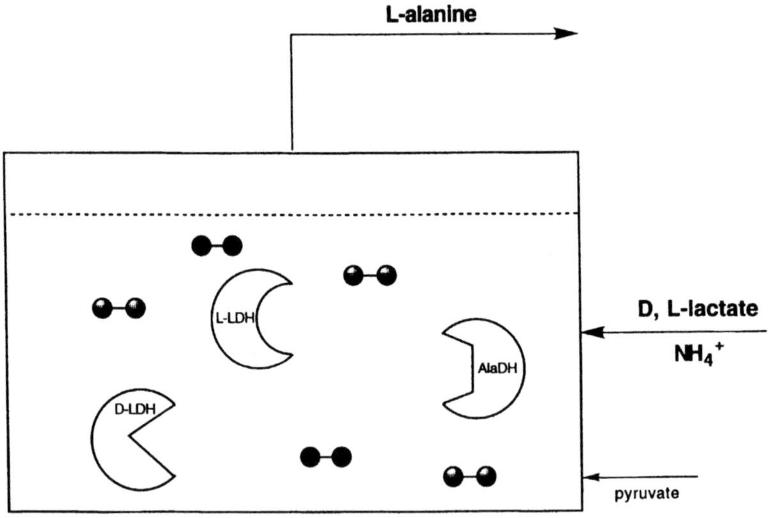

Fig. 7.4 Scheme of enzyme membrane reactor.

NADH may be very high, but then the reaction should be started with a very small amount of NADH and the reaction rate would be very low. A sufficiently high rate is obtained if the concentration of NAD(H) is around its K_m, but at those concentrations it will be necessary to recover it after the reaction. A solution to this problem is the enzyme membrane reactor (Fig. 7.4). This is a kind of CSTR (continuous stirred tank reactor) that retains the enzyme and the cofactor using an ultrafiltration membrane. This membrane has a molecular weight cut-off of about 10,000. Enzymes usually have a molecular mass of 25,000–250,000, but the molecular mass of NAD(H) is much too low for retention. Hence, it is first derivatized with polyethylene glycol (PEG 20,000). The reactivity of NAD(H) is usually not greatly affected by the derivatization with this soluble polymer. As a result, alanine can now be produced continuously by high concentrations of both enzymes and of NAD(H) in this reactor.

In order to obtain a sufficiently large membrane area on an industrial scale, hollow fibre ultrafiltration modules are applied (similar types to those used for protein purification). Since the concentration of NAD(H) does not change in the steady state and there is no NAD(H) in the entrance or exit flow, it becomes clear that the actual concentration of pyruvate in the reactor is equal to the concentration in the feed. Since there is a non-zero concentration of pyruvate in the reactor (and in the exit flow), some pyruvate has to be added to the reactor continuously in order to keep the reaction going. The rate equations of both reactions are required to calculate the precise rate of pyruvate addition.

There are many other examples of cofactor regeneration reactions and/or reactions that may be performed in an enzyme membrane reactor. The only other example to be considered here, however, is the regeneration of NADH by formate dehydrogenase (FDH) using formate. The advantage of this reaction is that it is irreversible, because carbon dioxide is liberated.

$$NAD^+ + \text{formate} \rightleftharpoons H^+ + CO_2 + NADH$$

7.2.3 *Free versus Immobilized Biocatalyst*

Biocatalyst costs are often an important factor in the overall cost of the product. One of the most important reasons to consider the immobilization of a biocatalyst is therefore the possibility of facilitated reuse or continuous utilization. This enables prolonged use of the biocatalyst and can thus significantly reduce the process costs. Furthermore, the stability of biocatalysts, in particular of enzymes and recombinant cells, can be improved dramatically in many cases by immobilization. Immobilization also creates the possibility of using the biocatalyst in a packed-bed or fluidized-bed reactor. Due to easy retention of the immobilized biocatalyst in the reactor, high volumetric activities can be realized and, in case of immobilized growing cells, the operation can be under wash-out conditions with respect to free cells. Obviously, the extra costs associated with the immobilization must be earned back by the possibility of a more efficient use. Many techniques are available for immobilizing biocatalysts [3]; some of the more useful are discussed in Chapter 9.

An example illustrating the benefits of immobilization is nitrification by immobilized cells. Nitrification is the biological oxidation of ammonia to nitrate via nitrite, a process widely used in waste-water treatment. It involves two microbial species, *Nitrosomonas* and *Nitrobacter*, both of which are notoriously slow-growing and sensitive bacteria that are easily washed out of the bioreactor, resulting in loss of nitrifying capacity. By immobilization of the cells, they are much more easily retained in the bioreactor, and high nitrifying capacities can be achieved even under relatively extreme conditions. The cells grow well in gelatinous supports such as alginate and carrageenan, and form a thin layer of biomass just underneath the surface. Growth limitation gradually changes into diffusion limitation as more biomass is formed in the gel beads. Since diffusion is less sensitive to temperature than is growth, a bioreactor with beads grown full of cells at 30°C loses only about 30% of its activity when the temperature is lowered to 10°C; in contrast, in the case of freely suspended cells only 5% of the activity remains at 10°C.

7.2.4 *Water versus Organic Solvent*

The use of water as the reaction medium for biocatalysis has long been advocated as one of the major advantages of the application of biocatalysts. This so-called advantage at the same time severely limits the potential scope of applications. In particular, when poorly water-soluble substrates and/or products are involved, or when the substrate or product is an inhibitor of the biocatalyst, the eventual product concentrations are generally very low, making product recovery laborious and expensive. In such cases, an organic solvent phase or another non-conventional medium can be used as a substrate reservoir and/or product extractant. This can enhance the final product concentration, which facilitates downstream processing [4]. It is also possible to shift equilibria in the direction of synthesis if a solvent can be found that selectively extracts the desired product into the organic phase. Complete replacement of free water by a suitable organic solvent with a concomitant decrease in the water activity can reverse hydrolytic reactions, resulting in, e.g. transesterification instead of fat hydrolysis by lipases. Lower water activities can also greatly enhance the stability of enzymes. Finding a biocompatible solvent is obviously the first requirement in cases where the use of water is more of a limitation than an advantage. Another aspect that has to be dealt with is the more complex system created in cases where only part of the free water is replaced by a water-insoluble organic solvent, yielding a two-liquid-phase bioreactor.

An example of a reaction that has been studied in such a two-liquid-phase bioreactor is the epoxidation of short-chain alkenes. Alkenes are poorly soluble in water, and the product epoxide is toxic to the biocatalyst (in this case, whole cells because of the need for cofactor regeneration); both considerations make the use of a suitable organic solvent attractive. Immobilization appeared to be essential in most solvents to prevent aggregation of cells with loss of activity and to eliminate formation of emulsions. A systematic study of the biocompatibility of solvents [5] eventually yielded the log P concept of Laane et al. [6], on the basis of which solvent toxicity can be predicted. In this case, P is the partition coefficient of a compound (organic solvent) in a standard octanol/water mixture, and can be calculated quite easily and accurately by the 'fragmental-constants' procedure [7]. Even though log P is a rather crude parameter lacking a detailed mechanistic basis, so far no better or easier alternative has been developed. Solvents with a log P below 2 can be expected to be toxic to all viable cells and most enzymes, while those with P values ranging between 2 and 4 are unpredictable in this respect. Solvents with log P above 4, i.e. the more hydrophobic ones, are generally not toxic. Plant cells, however, are slightly more sensitive, and the above values shift about two units up when dealing with plant cells.

7.2.5 *Standard versus Novel Bioreactor*

The stirred-tank reactor has long been and is still the work horse in industry. Operators have, so to say, grown up with it, and there is great industrial reluctance to change to another configuration. However, the characteristics of

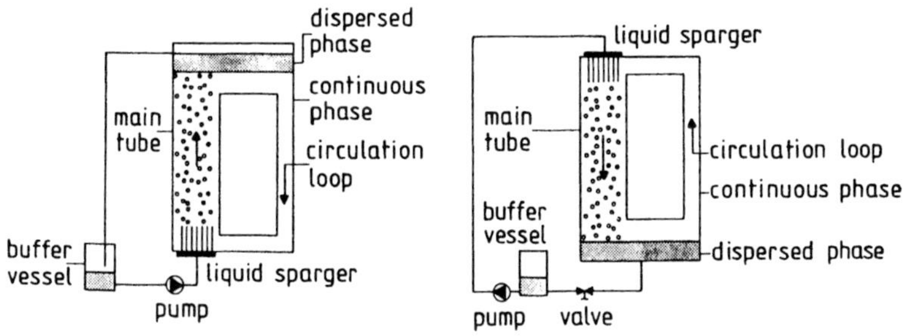

Fig. 7.5. The liquid-impelled loop reactor: organic solvent less dense (left) and denser (right) than water.

biocatalysts sometimes call for other bioreactor types. For instance, Michaelis–Menten kinetics in combination with product inhibition call for a plug-flow reactor such as the packed-bed reactor. For this, immobilization of the biocatalysts is required. This is also true for fluidized-bed reactors, which can be advantageous if, for example, the medium feed contains particulate contaminants. Obviously immobilization only makes sense if the biocatalyst is sufficiently stable or is (apparently) sufficiently stabilized by the immobilization to run a continuous reaction. The liquid-impelled loop reactor is a two-liquid-phase bioreactor specifically designed for biocatalysis in non-conventional media. Figure 7.5 shows two of the sixteen possible configurations [8]. Tonnage is another important factor in the choice of multi-purpose standard vessel versus dedicated novel bioreactor.

When isolated cofactor-requiring enzymes are involved, the membrane reactor can be a good choice. For example, L-tertiary leucine is an unnatural amino acid used as chiral intermediate for the synthesis of a number of products. Chemical asymmetric synthesis of L-tertiary leucine is very difficult, but Degussa carries out a small-scale (kilogram) commercial production in a membrane enzyme reactor. Leucine dehydrogenase is used in combination with formate dehydrogenase for cofactor regeneration. The cofactor is coupled to polyethylene glycol to increase its size, so as to facilitate retention in the membrane reactor. Good reactor productivity (0.64 kg/day), enzyme reuse and a cofactor-recycling efficiency of 130,000 (mol product formed/mol cofactor used) are thus achieved, making the cofactor costs very low [1].

7.2.6 (Fed-)Batch versus Continuous Operation

Many factors play a role at this decision level. Type of kinetics, operational stability of the biocatalyst, form of biocatalyst, type of bioreactor, bioreactor operating costs, necessity of process control, compatibility with down-stream processing, tonnage or scale of operation, existing facilities and experience, feedstock, type of product, GMP, and others, are all involved in an intricate manner. Economy usually turns the scale in the end. In the chemical industry in particular, bulk products with a huge market are manufactured in fully continuous, specially designed processes. In the fermentation industry, on the contrary, the products with the largest market volumes are made in batch and especially fed-batch fermenters. Biological wastewater-treatment plants are continuous.

Cheetham [1] discusses the twelve most successful commercial biotransformations (excluding steroid modification). About half of these use continuous bioreactors with immobilized biocatalysts. These processes yield relatively low value, high volume products, whereby even small improvements in the production process represent large amounts of money, justifying the necessary research and development in process optimization. No unifying criteria exist that form a basis upon which to choose the mode of operation.

One of the very few examples of a successful biotransformation process for a commodity chemical is the conversion of acrylonitrile into acrylamide. Acrylamide is an important industrial chemical that is mainly processed into water-soluble polymers and copolymers, which find applications as flocculants, paper-making aids, thickening agents, surface coatings, and additives for enhanced oil recovery. The chemical manufacture of acrylamide has been established for a long time. It requires, however, a high energy input, involves extensive purification of the product, and generates large quantities of toxic waste as well. Nitto Chemicals (Japan) has developed a biocatalytic process in which acrylonitrile is quantitatively converted into acrylamide. A key advantage of this process appears to be the fact that, because fewer impurities are produced, polyacrylamide of higher molecular weight can be manufactured. Thus, lower in-use concentrations can be employed, creating considerable savings for the end user [1]. In this process, *Pseudomonas chloraphis* cells, containing nitrile hydratase, were first used; more recently, use of a *Rhodococcus* strain has been introduced. Cell-associated nitrile hydratase is more stable than the isolated enzyme, and therefore whole cells are directly immobilized in polyacrylamide particles and used in column reactors operated at temperatures below 10°C. A first column is fed with a 3% acrylonitrile solution (both acrylonitrile and acrylamide inhibit the enzyme), with more added in sequential batches until the acrylamide concentration reaches 20%, after which it is simply decolorized and concentrated. Nitto operates this process on a 6000 ton/year scale using immobilized cells produced in-house. It is unclear whether the process is batch or continuous, but the biocatalytic process was apparently quite easy to integrate into the existing chemical process.

An example of a process where the fed-batch mode has a decisive advantage over a continuous operation is the biotransformation of γ-butyrobetaine into L-carnitine. This fine chemical has pharmacological, nutritional, and animal-feed applications, making product purification an essential part of the production process. Consequently, process optimization of biotransformation and downstream processing should be done in an integral way. Researchers of Lonza examined several options, e.g. chemostat-with-cell-recycling and fed-batch processes, and used Blackman kinetics to define the optimal process with respect to productivity, down-stream processing and costs [9]. The fed-batch process they developed has a low volumetric productivity when compared to the continuous process with cell recycle (30 versus 130 $kg^3 d^{-1}$). However, the yield is much higher, more than 99% in contrast to 90–95%. This strongly facilitates the downstream processing, making the fed-batch process about 40% cheaper than the continuous process. Furthermore, their process integration considerations determined the choice of raw materials for the biotransformation, resulting in an optimized number of unit operations in the downstream processing.

7.2.7 Integration of Reactions/Process Steps/Overall Process

The process discussed above is a specific example where integration played a key role in the development. In general, integration is an important, multi-faceted issue in the development of a biocatalytic process. As always in biotechnology, integration of biological and engineering sciences immediately from the start is essential for optimal results. If more than one reaction step is involved, integration of these steps into a 'one-pot' synthesis should be considered, as in that case intermediate products do not have to be isolated. In addition, considering the possibility of physical integration of process steps is important, e.g., including part of the downstream processing in the actual biocatalysis. Overall process integration goes further; each individual process step is optimized, not on its own, but with respect to the whole process, taking into account all interactions. It is a current topic in both chemical and biochemical engineering, but illustrative industrial examples within the scope of this chapter are still almost non-existent.

An interesting example of the integration of reaction steps is the production of D-(–)-*p*-hydroxyphenylglycine, an important intermediate in the manufacture of semi-synthetic antibiotics. Racemic hydantoins, readily prepared from the corresponding aldehydes, are converted by microbial cells to the D-N-carbamoyl amino acid. Under the pH conditions of the reaction, the unwanted L-enantiomer undergoes rapid racemization, thus allowing a theoretical yield of 100%. The resulting D-N-carbamoyl amino acid can then be converted into product by treatment with nitrous acid. Kanegafuchi has operated this process since 1983, but was rapidly followed by Recordati, who further integrated reaction steps. This company uses *Agrobacterium radiobacter*, which contains in addition to D-hydantoinase also N-carbamoyl-D-amino acid amidohydrolase. Thus, direct formation of product is possible without the nitrous acid treatment required after the D-hydantoinase reaction. The amidohydrolase is especially high yielding, because CO_2 is formed as a byproduct, driving the reaction to completion.

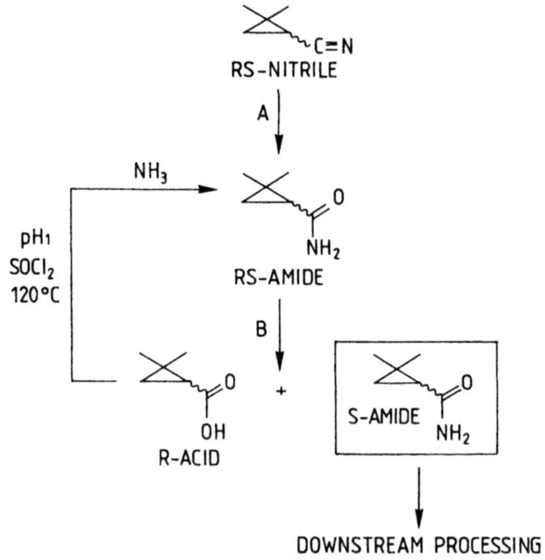

Fig. 7.6. The production of S-2,2-dimethylcyclopropanecarboxamide: A and B 'one-pot' biotransformation steps.

The analogous process to produce D-(–)-phenylglycine has yet to be commercialized, as it is still too complex and expensive compared to the relatively old chemical process, which is well optimized and uses depreciated equipment.

An industrial process for the production of S-2,2-dimethylcyclopropanecarboxamide (an intermediate for the production of Cilastatin, which is used in conjunction with certain antibiotics to prevent their degradation in the kidney) based on an integrated chemical and biocatalytic approach has been developed by Lonza researchers [10] (Fig. 7.6). The two biotransformation steps are executed subsequently by a strain containing a highly active nitrile hydratase and by an *E. coli* strain containing a cloned stereoselective amidase. It is claimed that the biocatalytic route has a higher enantioselectivity than the chemical alternatives (less recycle) and that it is a one-pot synthesis (at least the two biotransformation steps). The process runs efficiently on an industrial scale (15 m^3) and allows cost-effective production of a pure product in high yield.

A multi-step process involving reactions with less compatible conditions is the biological removal of ammonia from waste water. Under aerobic conditions nitrification occurs (see above), while the second step, denitrification, requires an anoxic environment. In conventional installations the steps are separated. By means of a 'magic-bead concept' the steps can be integrated (Fig. 7.7). Physical separation of the two strains is necessary, however, and is accomplished by immobilization in double-layer beads. In the outer layer oxygen is consumed by the obligate aerobic nitrifying cells, thus creating an anoxic centre for the denitrifying cells. The latter usually are facultative anaerobic, heterotrophic, fast-

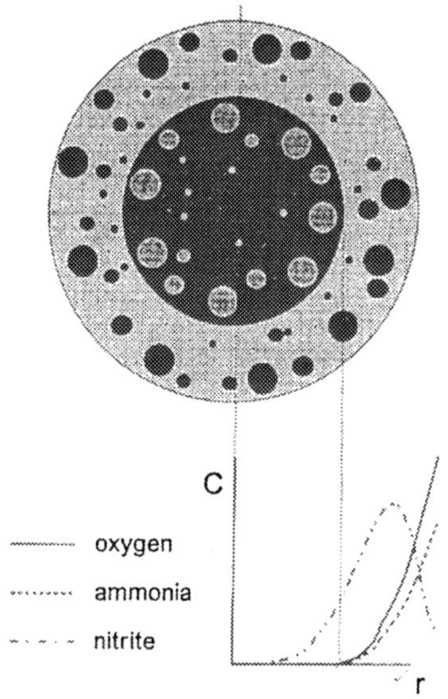

Fig. 7.7. Magic-bead concept for integration of oxidative and reductive biodegradation processes (adapted from Ref. [11]).

growing bacteria that exhibit the highest denitrifying capacity under anaerobic conditions. Therefore, when an organic C-source is present in addition to ammonia, physical confinement of the denitrifiers to the anoxic core is essential to prevent overgrowth of the slow-growing nitrifiers and to establish a stable nitrifying and denitrifying system with a high capacity.

An example of integration of part of the downstream processing in the actual bioreactor is two-liquid-phase biocatalysis (see above), in which the organic-solvent phase is used as extractant for the product. The liquid-impelled loop reactor (Fig. 7.3) is specifically designed for this purpose. It is based on the well-known air-lift principle, but instead of air, a water-immiscible, heavier or lighter organic solvent is injected.

7.3 AREAS OF ENZYME APPLICATIONS

For each class of enzymes introduced in Chapter 4.5, Table 7.2 shows the most important current commercial application areas. In 1995 the world sales of industrial enzymes were estimated to be about $1.2 billion. By far the most important enzymes commercially are (isolated) hydrolases. Over 90% of enzyme sales is generated by hydrolases such as proteases, lipases and (hemi)cellulases, which are typically used in bulk applications such as detergent formulations, food & feed, and in the treatment of pulp & paper. Next are isomerases and lyases, and clearly emerging are the oxidoreductases. At the moment, the world leader is NOVO Nordisk in Denmark, followed by the US-based company

TABLE 7.2

Classification and application areas of enzymes

	Detergents	Pulp & paper	Food & feed	Waste treatment	Specialties	Research
Hydrolases	•	•	•	•	•	•
Lyases			•		•	•
Oxidoreductases			•		•	•
Isomerases			•			•
Transferases						•
Ligases						•

Genencor. Two other important players, Gist-brocades and Quest, are located in the Netherlands and focus mainly on food and feed enzymes.

7.3.1 Hydrolases

By far the best-studied class of enzymes is the hydrolases. They include enzymes that act on ester (lipases, esterases), ether (ether hydrolases), glycosyl (carbohydrases), and peptide or amide bonds (proteases, peptidases) by splitting C–O and C–N bonds in the presence of water. The same enzymes can also be used to form these bonds in media containing little water. In addition, less well-characterized hydrolases exist for C–halogen, P–C, P–N, and S–N bonds. Of the about 640 known hydrolases, approximately 125 are commercially available to some extent.

Some examples of hydrolases are listed in Table 7.3 according to the industrial sector in which they are used. Most of these hydrolases operate in the hydrolytic mode, since only a few processes are currently known that make use of the synthetic power of hydrolases. Increasingly, however, these enzymes are being tailored by genetic tools such as random or site-directed mutagenesis and over-expression to lower the operational costs of the biocatalyst and to obtain enzymes of high purity, as well as to meet specific customer requirements such as stability, substrate specificity, and optimal pH and temperature of operation. A typical example is the protease subtilisin from *Bacillus subtilus*, which was made bleach-resistant by replacing one bleach-sensitive amino acid (cysteine) by the chemically inert alanine [12].

Shown in Figure 7.8 for each market segment are the preferred enzyme characteristics with respect to temperature and pH profile. Screening natural sources for the existence of such enzymes has been and is still a powerful tool that can yield enzymes with the desired characteristics. Further fine tuning can be achieved by the above-mentioned recombinant DNA technologies.

TABLE 7.3

Hydrolases and their actual application in various industries

Industrial sector	Purpose	Type of hydrolase
Detergents	Stain removal, anti-pilling	Protease, lipase, amylase, cellulase
Pulp & paper	Bleaching	xylanase
Food & feed		
— Bakery	Mixing time	Protease
	Proofing time, tolerance	Xylanase
	Bread volume	Xylanase, amylase
	Flavour	Protease, peptidase
	Colour	Amylase, xylanase
— Beverages	Improve brewing process	Protease, glucanase
	Clarification fruit juices	Pectinase, cellulase, amylase
	Coffee, tea and cocoa proc.	Pectinase, tannase
— Dairy	Cheese manufacturing	Chymosin, lipase
	Fat and oil modification	Lipase, esterase
— Meat & fish	Tenderization	Protease (e.g. papain)
— Nutrition	Protein hydrolysates	Protease, peptidase
Waste treatment	Degradation of polymers	Protease, cellulase, hemicellulase, lipase, pectinase, amylase
Specialties		
— Fine chemicals	Aspartame synthesis	Protease (*e.g.* thermolysin)
	Amino acid synthesis	Aminoacylase
	Antibiotics	Penicillin acylase
— Pharmaceuticals	Improve digestion	Protease, lipase, amylase
	Tissue healing	Collagenase
	Anti-leukomenia	Asparaginase

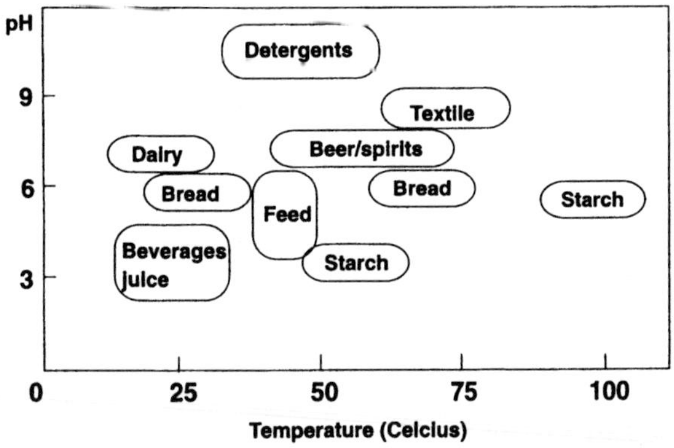

Fig. 7.8. Desired enzyme characteristics per market segment (from Gist-brocades).

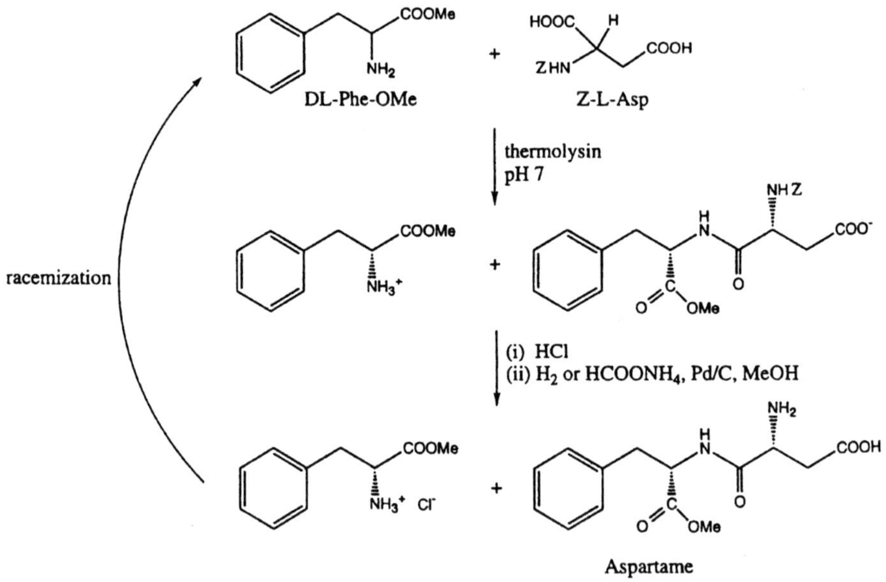

Fig. 7.9. Combined chemo/biosynthetic route to aspartame (from Ref. [15]).

An example of a commercial process that uses a hydrolase in the hydrolytic mode is the production of the antibiotic intermediate 6-APA by Gist-brocades [13], discussed above in Section 7.2.1 (Fig. 7.1). The biocatalyst is a penicillin acylase, which generates 6-APA from the starting material penicillin G in one step at a moderate temperature in water. The atom utilization of the enzymic process is much higher than that of the corresponding chemical process [14].

Another example of a protease-catalyzed commercial process, which in this case uses the enzyme in a synthetic mode, is the completely regio- and stereoselective production of the low-caloric sweetener aspartame developed by DSM-TOSOH [15] (Fig. 7.9). Aspartame is a dipeptide consisting of the amino acids phenylalanine and aspartic acid, which are coupled by the enzyme thermolysin from *Bacillus thermoproteolyticus*. For an efficient coupling, relatively high temperatures are required and the amount of water in the system must be kept low to drive the reaction in the desired direction. Thermolysin, which is a metallo-endoprotease, meets these two requirements. It is thermostable, and it works in an organic solvent, which is required to keep the water activity low. In practice, however, organic solvents were not necessary, since the product aspartame forms an insoluble complex with unreacted D-Phe-OMe, which crystallizes out of the aqueous medium.

7.3.2 *Lyases*

Lyases catalyze either the reversible cleavage of C–C, C–O, C–N and other bonds by elimination to produce double bonds or, in the reverse direction, the addition of groups across double bonds [16]. The type of reaction catalyzed can be decarboxylation (decarboxylases), dehydration (dehydrases), deamination (deaminases), cyanohydrin cleavage (oxynitrilases), aldol cleavage (aldolases), and α, β, or γ-elimination (pyridoxal phosphate-containing lyases). In cases where the reverse reaction is the more important, or the only one to be demonstrated, 'synthetase' may be used in the name.

Over 300 lyase enzymes, catalyzing a wide variety of reactions, have been described [16], of which about 35 are commercially available. In most cases, these enzymes do not require cofactor recycling, exhibit an absolute stereospecificity, and can give a theoretical yield of 100% vs. 50% for typical enantiomeric resolutions. Lyases are therefore attractive biocatalysts for the production of optically active compounds, and have already found a home in several commercial process (Table 7.4). Typically, either growing, immobilized, or permeabilized cells are used in these applications. This approach is preferred to the use of the more expensive isolated enzymes, as long as there are no interfering enzymatic side reactions taking place in the cell. More recently, aldolases are attracting interest for the production of unusual sugars for the pharmaceutical and food industries. These lyases catalyze a reversible aldol condensation by reacting the deprotonated α-carbon atom of an aldehyde or ketone with the carbonyl carbon of another aldehyde or ketone, resulting in a β-hydroxyaldehyde or a β-hydroxyketone. To date, more than 30 aldolases have been described, and some of these are used on a pilot plant scale for the production of unusual sugars (*e.g.* as sweeteners and in pharmaceuticals) and certain amino acids.

Other commercial applications are found in the food industry, where lyases already have a long history of safe use. In these applications, isolated enzymes are preferred for safety reasons. Well-known examples include lyases that degrade polysaccharides, such as pectin (pectin lyases), alginate (alginate

TABLE 7.4

Examples of the use of lyases in commercial processes for the production of fine chemicals

Product	Substrate	Type of lyase	Annual production
L-Aspartate	Fumarate	Aspartase	6000 tons
L-DOPA	Pyruvate, catechol, NH_4^+	Tyrosine-phenol lyase	200 tons
L-Malate	Fumarate	Fumarase	500 tons
L-Carnitine	γ-Butyrobetaine	An enoyl-CoA hydratase	Started
Cyanohydrins	Aldehydes, HCN	(R)-Oxynitrilase	Multi kg scale
Acrylamide	Acrylonitrile	Nitrile hydratase	3000 tons
L-Alanine	L-Aspartate	L-Aspartate-β-decarboxylase	100 tons

lyases), and xanthan gum (gelrite lyase). They are mainly used to reduce the viscosity of food products, or for the production of specific oligomers with interesting chelating and physiological properties.

A remarkable example of a lyase-catalyzed commercial process, developed by the Japanese company Nitto and operative since 1985, is the biocatalytic production of acrylamide using immobilized whole cells of *Rhodococcus rhodochrous* [17]. The enzyme responsible for water addition to the double bond of acrylonitrile is nitrile hydratase:

$$CH_2{=}CHCN + H_2O \xrightarrow{\text{Nitrile hydratase}} CH_2{=}CHCONH_2 \qquad (7.1)$$

The cells are immobilized in a polyacrylamide gel, and during fed-batch processing yields of 400 g acrylamide per litre are achieved. The biocatalytic process is far simpler than the classical chemical process, which requires a copper catalyst and a polymerization inhibitor. Furthermore, the bioroute has a selectivity of >99% at >99% conversion.

7.3.3 Isomerases

Relatively little is known about applications of this class of enzymes. In general, they catalyze geometric or structural changes within one molecule. According to the type of isomerization effected, they may be called racemases, epimerases, (*cis-trans*) isomerases, and tautomerases. Currently, about 126 isomerases are known, of which only about six are available commercially.

The best-known and, to date, the only example of a commercial process based on an isomerase employs glucose isomerase, which is used in the food industry to increase the sweetness of glucose syrups by the conversion of glucose into the sweeter fructose [18]. Another type of isomerase that may find eventually an application in the food sector is the so-called protein disulfide-isomerase [19]. This type of enzyme catalyzes the sulfhydryl-disulfide exchange and should therefore, in principle, be capable of controlling texture/mouth-feel by cross-linking cysteine-containing proteins (*e.g.* glutens). Indirectly, this enzyme may affect the taste of a product as a result of an altered mouth-feel. The other types of isomerases mentioned in Table 7.5 could also be of interest to the flavour industry. As yet, however, no applications have been described for the other industry sectors, although several potential applications can be envisaged.

An interesting possibility would be the use of a double-bond isomerase that is able to convert eugenol into isoeugenol. Eugenol itself is readily available from clover, but cannot be converted directly by lipoxygenase into vanillin. In contrast, isoeugenol is not readily available, but is easily converted to vanillin by lipoxygenase (see Section 7.3.6). That an eugenol/isoeugenol converting isomerase may exist in nature arises from the observation that in certain lemon grass oils both compounds are present.

TABLE 7.5

Some (potential) application areas for isomerases

Type of isomerase	Purpose	Application example
Glucose isomerases	Interconversion of glucose into fructose	Sweetener
Protein disulfide-isomerases	Cross-linking proteins	Texturiser that may affect mouth-feel/taste
Double-bond isomerases	Changing the double bond position	Topnote flavour (e.g. (iso)eugenol)
Cis-trans isomerases	*Cis-trans* isomerization	Flavours

7.3.4. Transferases

Transferases are enzymes that transfer a chemical group, such as a methyl or glycosyl group, from one compound (the donor) to another compound (the acceptor). In many cases, the donor is a cofactor that carries the group to be transferred. A relatively large number of transferases is known (about 720), of which about 90 are available.

Many transferases are involved in the biosynthesis of compounds of interest to the food, fine chemicals, and pharmaceutical industries. To date, however, there are no commercial processes involving the deliberate use of transferases. The often expensive cofactor requirement hampers their practical applications, although whole cell conversions might offer a solution to this problem.

A typical research example includes the synthesis of a K^+-channel blocker intermediate (Fig. 7.10). Here, an aminotransferase is used to replace a keto group by an amino function to yield the desired intermediate [20]. Other potential applications include the biosynthesis of flavouring esters. In nature such compounds are often formed by a specific acyl CoA transferase. For practical purposes, however, the alternative, less complicated route, using lipases and esterases in their synthetic mode, seems more appropriate.

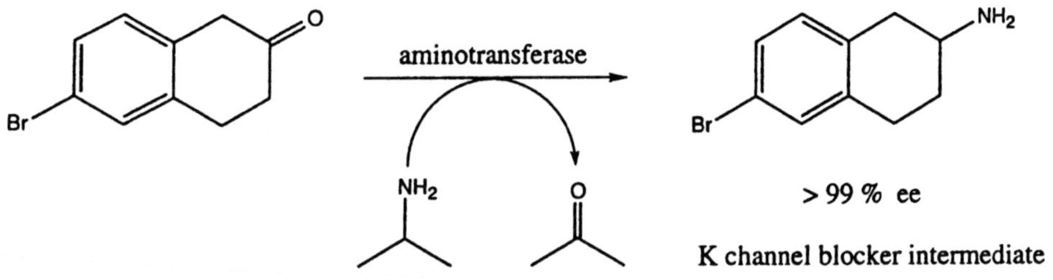

Fig. 7.10. Example of the use of a transferase in the synthesis of a pharmaceutical product (from Ref. [20]).

In principle, transglycosylases could also be applied for the production of chemicals, especially as a tool to selectively protect sensitive groups during a multi-step chemo/biosynthetic procedure. Such an approach could be applied, for example, in the synthesis of (flavouring) phenolic compounds to prevent polymerization during one of the steps.

7.3.5 *Ligases*

Ligases are enzymes that catalyze the joining of two molecules with concomitant hydrolysis of the phosphoric acid anhydride bond in ATP or a similar triphosphate. In nature these enzymes are involved in the biosynthesis of many metabolites. About 80 different ligases are known, of which five are available. Like transferases, ligases seem relatively unsuitable for the *in vitro* production of chemicals, due to the absolute requirement for an expensive cofactor (e.g., ATP). Although sophisticated ATP-regenerating systems have been designed, it is not likely that ligases will be applied in the near future for industrial purposes. At present, ligases can be regarded as a research tool, and in that role they are currently applied successfully in the field of molecular biology.

7.3.6 *Oxidoreductases*

To this class belong all enzymes catalyzing oxidation and reduction reactions, making it a very large class indeed. The substrate that is oxidized is regarded as the hydrogen or electron donor, and the systematic classification is based on 'donor:acceptor oxidoreductase'. The recommended name is 'dehydrogenase', wherever this is possible; 'oxidase' is used only when O_2 is an acceptor. Classification is difficult in some cases, because of the lack of specificity towards the electron or hydrogen acceptor. Over 650 oxidoreductases are known, of which about 90 are available; only a few, however, are currently being applied commercially. Below are depicted the different types of reactions catalyzed by oxidoreductases.

Dehydrogenases

$$SH_2 + D \rightarrow S + DH_2 \tag{7.2}$$

Oxidases

$$SH_2 + O_2 \rightarrow S + H_2O_2 \tag{7.3}$$

Oxygenases

$$\text{Monooxygenases: } S + DH_2 + O_2 \rightarrow SO + D + H_2O \tag{7.4}$$

$$\text{Dioxygenases: } S + O_2 \rightarrow SO_2H \tag{7.5}$$

Peroxidases

$$SH_2 + H_2O_2 \rightarrow S + 2\,H_2O \tag{7.6}$$

$$S + H_2O_2 \rightarrow SO + H_2O \tag{7.7}$$

where S(O) symbolizes the (oxidized) substrate and D the cofactor, such as $NAD^+/NADH$.

Chapter 4.7 gave a detailed description of the steps involved in biological oxidations and introduced some of the important enzymes. At present, only a few of these enzymes, all performing oxidations, are applied in a commercial setting. For the cofactor-requiring reactions, growing-cell conversions are preferred. A nice example of this approach is the oxidation of dipicoline to carboxydipicoline by growing cells of *Pseudomonas putida*. The conversion obtained in this process is >95%, and production rates of 30–40 ton/year have been described [21]. Attractive biocatalytic routes have also been developed for the regio- and stereoselective oxidation of steroids [22]. Another process in the fine chemical industry, which involves permeabilized cells of the yeasts *Pichea pastoris* or *Hansenula polymorpha*, is the >98% selective oxidation of L-lactate to pyruvate by the enzyme glycolate oxidase.

In the food industry, the use of redox enzymes enjoys a longer history, especially for oxidative enzymes that do not require a cofactor. Some examples are listed in Table 7.6; most are still in the developmental stage.

Phenol oxidases belong to the so-called monooxygenases, since they catalyze the incorporation of one atom of oxygen into the substrate. Phenol oxidases are present in all fruit and vegetable products and can affect the flavour and colour by the well-known browning reaction [23]. The enzymatic browning reaction involves the hydroxylation of plant phenolic compounds into the dihydroxy

TABLE 7.6

Examples of oxidative enzymes for food purposes (from Ref. [24])

Type of oxidative enzyme	Purpose	Application example
Phenol oxidases	Enzymatic browning	Fruits, beverages
Lipoxygenases	Hydroperoxidation:	
	–step in aldehyde/alcohol synthesis	Various flavours
	–step in δ-decalactone synthesis	Butter flavour
	–aromatic aldehydes	Vanillin
Glucose oxidases	Acidification	Cheeses
Sulphydryl oxidases	Thiol off-flavour removal	Milk
Alcohol oxidases	Production of aldehydes	Beverages, fruit flavours
Aldehyde oxidases	Off-flavour scavenging	Soybean extracts
Vanillyl alcohol oxidase	Phenolic-derived flavours	Beverages, fruits
Peroxidases	Specific oxidized flavours	Nootkatone
	Demethylation of amines	Methylanthranilate

derivative, which is subsequently oxidized by the same enzyme to the quinone form. In a non-enzymatic reaction, these highly reactive quinones react with, among others, amino acids to generate coloured pigments. In this way tasty phenolic compounds are probably lost, and at the same time flavouring reactive intermediates can be generated.

Lipoxygenases are dioxygenases and are well-known (off-)flavour generators in nature. Their potential in the *in vitro* production of natural flavours has been long recognized. Typical examples include the conversion of polyunsaturated fatty acids into various short to medium chain aldehydes/alcohols [25] or into S(-)-δ-decalactone [26] (butter flavour). Well-known fatty-acid derived flavouring aldehydes/alcohols include (E2)-pentenal (green apple), (E2)-hexenal (leaf aldehyde), (Z3)-hexenol (leaf alcohol), (E2,E6)-nonadienal (cucumber), 1-octen-3-one (field mushroom), (1,Z5)-octadien-3-one (geranium leaves), (1,E3,E5)-undecatriene (balsamic), and (1,E3,Z5,Z8)-undecatetraene (seaweed). In all cases, the first step involves the hydroperoxidation of a specific polyunsaturated fatty acid [27]. Depending on the degree of unsaturation and the regioselectivity of the lipoxygenase, different hydroperoxy compounds are formed, from which the above-mentioned flavours can be derived by subsequent enzymic reactions. For (Z3)-hexenol, linolenic acids are used as a substrate, while for most of the other aldehydes/alcohols higher unsaturated fatty acids are required. For the production of the chiral lactone, linoleic acid is the fatty acid of choice. The production of (Z3)-hexenol has recently been commercialized using plant homogenates (e.g., alfalfa sprouts, green peppers), which are relatively rich in hydroperoxide lyase, the second enzyme required to split the hydroperoxide fatty acid into smaller fragments. (Z3)-Hexenal generated in this way was converted into (Z3)-hexenol using the reductive enzymatic power of baker's yeasts [28]. A very elegant approach has recently been taken by Givaudan-Roure. To unify all three enzymes involved in the formation of (Z3)-hexenol, they have cloned and overexpressed the hydroperoxide lyase from banana in baker's yeast [29]. In this way, they have developed a 'single step' process that produces (Z3)-hexenol in relatively high yields [30].

Other aldehydes/alcohols can be produced following the same principles. For production of C_{10}-lactone, however, a different strategy has to be followed after the formation of the hydroperoxide. It involves the fermentative β-oxidation of the hydroperoxide intermediate by the yeast *Pichia etchellsii*, and the subsequent cyclization of 5-hydroxydecanoic acid to the corresponding S(-)-δ-decalactone. The routes to (Z3)-hexenol and S(-)-δ-decalactone are depicted in Fig. 7.11.

As has been shown by Quest International [31], lipoxygenases also accept a relatively broad spectrum of phenolic compounds as substrates. Of interest to the flavour industry are the lipoxygenase-catalyzed conversions of isoeugenol and coniferyl benzoate from Siam resin to vanillin (4–15% yield, Fig. 7.12). At present, the commercialization of these biotransformations is hampered by the

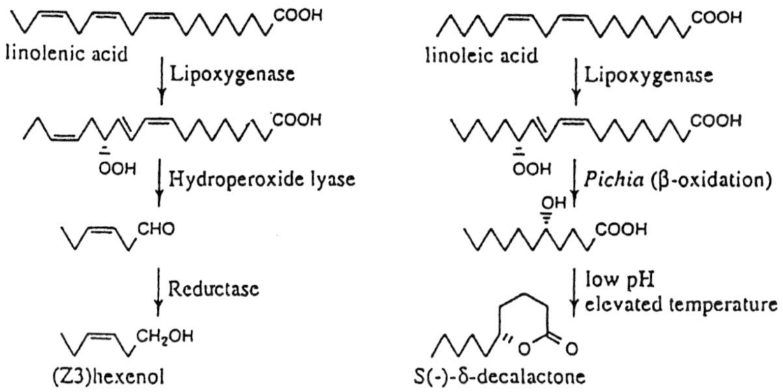

Fig. 7.11. Two examples of the use of lipoxygenase in the production of fatty-acid derived flavours.

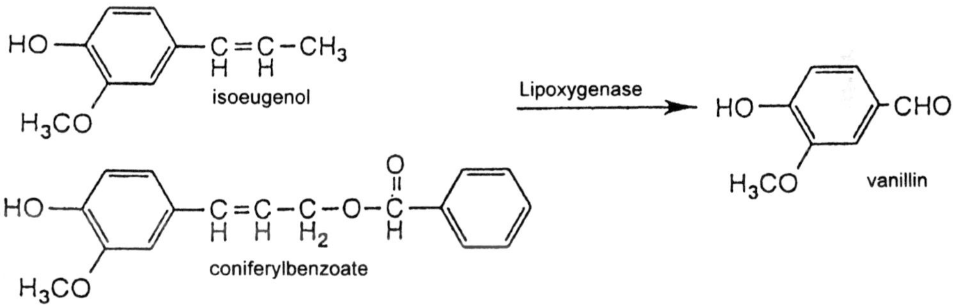

Fig. 7.12. Lipoxygenase-catalyzed production of vanillin.

facts that isoeugenol is not readily available and coniferyl benzoate is difficult to handle in a reactor.

Glucose oxidase is added to food systems mainly to remove traces of glucose (e.g., treatment of egg white prior to dehydration) or oxygen (e.g., prevention of discoloration and/or oxidation), or to produce hydrogen peroxide *in situ* for textural or antimicrobial purposes [32]. The ability to generate hydrogen peroxide is also being applied in certain toothpastes to control the microbiological flora in saliva.

Sulfhydryl oxidases are capable of oxidizing sulfhydryl groups to disulfides in the presence of oxygen. Industrially, the potential of this enzyme lies in its ability to ameliorate the cooked "sulfury" off-note that is formed during the UHT-treatment of milk, and in texturizing food systems by rearrangement of the disulfide linkages between proteins. The former application has been demonstrated on a pilot scale using immobilized enzyme columns [32], while the latter application remains to be proven. Sulfhydryl oxidase from *Aspergillus* is one of those enzymes that have recently become available in sufficient quantities by genetic modifications [33], which will undoubtedly facilitate the quest for (novel) commercial applications.

In general, aldehydes are more potent flavours than their alcohol counterpart. Hence, alcohol oxidases are interesting enzymes for the *in vitro* production of flavouring preparations. Typical examples include the use of methanol oxidase from *Pichia, Hansenula,* and *Candida* [34] for the production of natural acetaldehyde from ethanol. This enzyme is induced during growth on methanol. At the end of the logarithmic growth phase, cells are harvested and incubated with ethanol. In this way, concentrations of about 1.5% natural acetaldehyde can be achieved, which can be concentrated further to the desired application level. The substrate specificity of alcohol oxidases varies substantially from one yeast source to another. Hence, this procedure can also be used to convert other alcohols, such as hexenol and related long-chain alcohols, to their corresponding aldehydes [35].

Aldehyde oxidases may also be applied to improve the aroma profile of certain food products by the conversion of undesirable aldehydes to their corresponding acid in the presence of oxygen. In this way the beany off-flavour from soybean extracts has been successfully 'neutralized' [36].

As an alternative to alcohol and aldehyde oxidases, the corresponding dehydrogenases could in principle be used. A severe drawback, however, is that these dehydrogenases require the expensive cofactor $NAD(P)^+$ instead of (cheap) oxygen as an electron acceptor. Although various sophisticated $NAD(P)^+$-cofactor regenerating systems have been designed and substantial cost reductions have been realized in this way, it is evident that in commercial applications oxidases are preferred over their dehydrogenase counterparts.

A special type of alcohol oxidase is vanillyl alcohol oxidase (VAO) from *Penicillium simplicissimum.* Recently, it has been shown that this stable, flavin-containing enzyme has a very broad substrate specificity and readily converts *para*-substituted phenols into interesting flavour precursors or flavouring compounds (Fig. 7.13). Apart from natural vanillin and coniferyl alcohol, different vinylphenols (e.g., *para*-vinylguaiacol) can be produced from cheap raw materials and oxygen as an electron acceptor. VAO also suggests itself for the generation of flavouring building blocks. To that end, a natural mix of phenolic compounds (e.g., creosote) could be treated with VAO to enrich it with a range of vinylic and aldehydic flavouring substances.

Peroxidases are enzymes that use hydrogen peroxide instead of oxygen as an electron acceptor. These enzymes usually contain heme iron, but examples utilizing a non-heme vanadium centre are also known. Both types occur widely in nature and can be applied to remove excess hydrogen peroxide in food products (e.g., cold sterilization of milk in regions lacking refrigeration), to produce powerful antimicrobial agents such as HOSCN in milk (e.g., lactoperoxidase), to generate bleach (HOCl) *in situ* in washing powders, to texturize

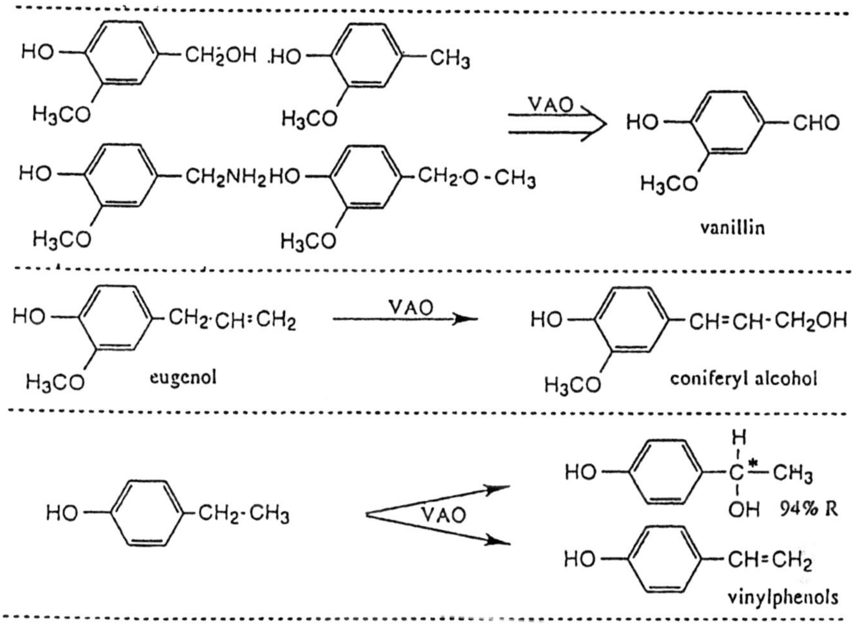

Fig. 7.13. Reactions catalyzed by vanillyl alcohol oxidase (VAO) (from Refs. [37,38]).

food systems by cross-linking polysaccharides and/or proteins, or as a biocatalyst to produce specific flavours [39]. Enzymes currently under investigation for the *in vitro* production of flavours include horseradish peroxidase, soybean peroxidase, and several so-called microperoxidases. These microperoxidases (MPs) consist of a ferriprotoporphyrin IX covalently attached to a small residual oligopeptide chain (see Chapter 4.7). They can be prepared by mild proteolytic digestion of, for example, cytochrome *c*, and exhibit both cytochrome P450-type of oxygen transfer reactions as well as peroxidase type reactions [40].

Reactions catalyzed by (micro)peroxidases include sulfoxidation, N-demethylation, oxidation, and hydroxylation. A typical example of an interesting N-demethylation reaction is the conversion of methyl-N-methylanthranilate from citrus into the concord grape topnote N-methylanthranilate (Fig. 7.14). Horse-

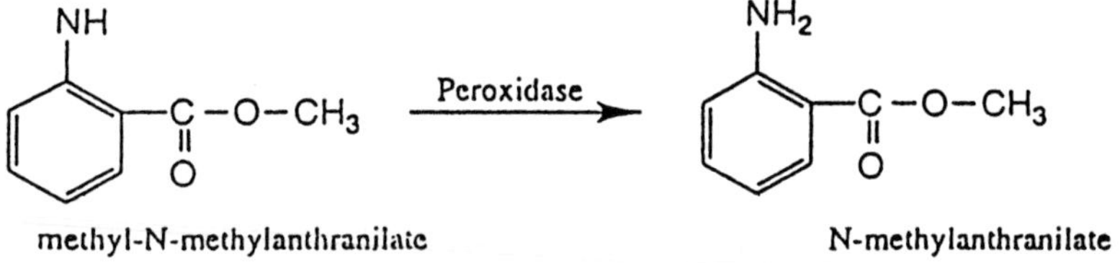

Fig. 7.14 Peroxidase-catalyzed demethylation of methyl-N-methylanthranilate.

radish peroxidase, soybean peroxidase and a microperoxidase have all been shown to catalyze this reaction in the presence of hydrogen peroxide.

Although reductases play an important role in the *in vivo* synthesis of many chemicals (see flavour example in Fig. 7.11), little attention has been paid to this type of biocatalyst. In most cases whole microbial or plant cells are used to perform a bioreductive reaction due to the requirement for (expensive) cofactors. Typical examples include the reduction of certain double bonds in terpenes by plant cells [27,41], the reduction of Massoi lactone to R(+)-δ-decalactone by *Basidiomycetes* and *S. cerevisiae* [28], and the baker's yeast-catalyzed reduction of ketones to (chiral) alcohols [42].

7.4 CONCLUSIONS

Applied biocatalysis is still in its infancy, despite the overwhelming possibilities offered by enzymes, their obvious advantages such as environmental benignity, naturalness, and high selectivity, and the fact that they can operate under mild conditions. There are still many problems to be solved before the field of applied biocatalysis will really come of age and its promise turns into reality. One of the problems is that only a limited number of enzymes is available in sufficient quantities at an acceptable price. The other main problem is the often limited operational stability under industrial conditions. Future prospects are, however, very encouraging. Further exploration of Nature's biocatalytic diversity and tailoring of enzymes by genetic tools will undoubtedly lead to improved, on-the-shelf biocatalysts that exhibit the desired characteristics for practical applications.

REFERENCES

1. P.S.J. Cheetham, in: J.M.S. Cabral, D. Best, L. Boross, J. Tramper (eds.) *Applied BioCatalysis*. Harwood Academic Publishers, Chur, Switzerland, 1994, pp. 47–108.
2. C.B. Wick, *Genetic Eng. News* 15 (2) (1995) 1–22.
3. G.F. Bickerstaff (ed.), *Immobilization of Enzymes and Cells*. Humana Press, 1996.
4. J. Tramper, H. Vermuë, H.H. Beeftink, and U. von Stockar (eds.), *Biocatalysis in Non-conventional Media*. Progress in Biotechnology, Vol. 8. Elsevier, Amsterdam, 1992.
5. L.E.S. Brink and J. Tramper, *Biotechnol. Bioeng.* 27 (1985) 1258–1269.
6. C. Laane, S. Boeren, K. Vos and C. Veeger, *Biotechnol. Bioeng.* 30 (1987) 81–87.
7. R.F. Rekker, 1977. *The Hydrophobic Fragmental Constant*. Elsevier, Amsterdam.
8. J. Tramper, I. Wolters and P. Verlaan, In: C. Laane, J. Tramper and M.D. Lilly (eds.), *Biocatalysis in Organic Media*. Elsevier, Amsterdam, 1987, pp. 311–316.
9. F.W.J.M.M. Hoeks, H. Kulla and H.-P. Meyer, *J. Biotechnol.* 22 (1992) 117–128.
10. O.M. Birch, E. Böhlen, T. Gilligan, F.W.J.M.M. Hoeks, V. Malanik, K. Robins, N.M. Shaw and T. Zimmermann, *Chiral Europe '94*, 1994.
11. V.A. Santos, J., Tramper and R.H. Wijffels, *Biomat., Art. Cells Immob. Biotech.* 21 (1993) 317–322.

12. J.A. Wells and D.A. Estell, *Trends Biol. Sci.* 13 (1988) 291.
13. J. Verwey and E. de Vroom, *Rec. Trav. Chim. Pays Bas* 112 (1993) 66.
14. R.A. Sheldon, *Chirotechnology*. Marcel Dekker, New York, 1993.
15. Y. Isowa, M. Ohmori, T. Ichikawa, K. Mori, Y. Nonaka, K. Kihara, K., Oyama, H. Satoh and S. Nishimura, *Tetrahedron Lett.* 28 (1979) 2611.
16. M.J. Van der Werf, W.J.J. van den Tweel, J. Kamphuis, S., Hartmans and J.A.M. De Bont, *Trends Biotechnol.* 12 (1994) 95.
17. T. Nagasawa, H. Shimizu and H. Yamada, *Appl. Microbiol. Biotechnol.* 40 (1993) 189–195.
18. W.J. Quax, *Trends in Food Sci. Technol.* 4 (1993) 31.
19. J.S. Weisman and P.S. Kim, *Nature* 365 (1993) 185.
20. D.M. Tschaen et al., *J. Org. Chem.* 60 (1995) 4324.
21. A. Kiener, *Chemtech.* 9 (1995) 31.
22. K. Faber, *Biotransformations in Organic Chemistry*. Springer-Verlag, Berlin, 1992.
23. A.M. Mayer and E. Harel, 1991. In: P.F. Fox (ed.), *Food Enzymology*. Elsevier, London, Vol. 2, p. 373.
24. P.F. Fox (ed.), *Food Enzymology*, Vols. 1 & 2. Elsevier Applied Science, London and New York, 1994.
25. J.C. Villettaz and D. Dubourdieu, In: P.F. Fox (ed.), *Food Enzymology*. Elsevier, London, Vol. 2, 1991. p. 427.
26. P.S.J. Cheetham, *Tibtech.* 11 (1993) 478.
27. R.G. Berger, *Aroma Biotechnology*. Springer-Verlag, Berlin, Heidelberg, 1995.
28. Pernod Ricard, FR patent 8912901 and EU patent 481147.
29. A. Muheim, A. Häusler, B. Schilling and K. Lerch, 1999. In: K.A.D. Swift (ed.), *Flavours and Fragrances*, Special Publication No. 214 of the Royal Society of Chemistry, Cambridge, pp. 137–152.
30. P.A. Markus, A.L.J. Peters and R. Roos, 1991. EU patent 542284.
31. P.F. Fox and M.B. Grafferty, In: P.F. Fox (ed.), *Food Enzymology*. Elsevier, London, Vol. 2, 1991. p. 253.
32. J. Maat, W. Musters, H. Stam, P.J. Schaap, P.J. Van der Vonderwoort, J. Visser and J.M. Verbakel, US patent 5,529,926.
33. H. Stam, M. Hoogland and C. Laane, 1998. In: B. Wood (ed.), *Microbiology of Fermented Foods, Vol. 2*. Chapman and Hall, London, pp. 505–543..
34. W.D. Murray and S.J.B. Duff, *Appl. Microbiol. Biotechnol.* 33 (1990) 202.
35. N. Takahashi, R. Sasaki and H. Chiba, *Agric. Biol. Chem.* 43 (1979) 2557.
36. M.W. Fraaije, C., Veeger and W.J.H. van Berkel, *Eur. J. Biochem.* 234 (1995) 271.
37. M.W. Fraaije, F., Drijfhout, G. Meulenbelt and W.J.H. van Berkel, 1996. In: R.M. Buitelaar (ed.), *Proceedings 6th Neth. Biotechnol. Congress* p. PI I1.
38. D.S. Robinson, In: P.F. Fox (ed.), *Food Enzymology*. Elsevier, London, V. 2, 1991. p. 399.
39. A.M. Osman, J. Koerts, M.G. Boersma, S. Boeren, C. Veeger and I.M.C.M. Rietjens, *Eur. J. Biochem.* 240 (1996) 232.
40. F. Drawert and R.G. Berger, R.G. In: 45 Disk Tagung, FK der Ernährungsindustrie Hannover, 1987, p. 93.
41. J. Heidlas et al., *Eur. J. Biochem.* 172 (1988) 633.
42. E. Van den Ban, C. Laane and H. Haaker, *Biotechnol. Bioeng.* (1999) in press.

Applied catalysis

Chapter 8

Catalytic reaction engineering

8.1 INTRODUCTION

Catalytic reaction engineering is a scientific discipline which bridges the gap between the fundamentals of catalysis and its industrial application. Starting from insight into reaction mechanisms provided by catalytic chemists and surface scientists, the rate equations are developed which allow a quantitative description of the effects of the reaction conditions on reaction rates and on selectivities for desired products. The study of intrinsic reaction kinetics, i.e. those determined solely by chemical events, is the basis of catalytic reaction engineering. Another, intimately related, area is the study of the interactions between physical transport and chemical reaction. Such interactions can have pronounced effects on the rates and selectivities obtained in industrial reactors. They have to be accounted for explicitly when scaling up from laboratory to industrial dimensions.

The present chapter is not meant to be exhaustive. Rather, an attempt has been made to introduce the reader to the major concepts and tools used by catalytic reaction engineers. In order to give the reader a feel of the applicability of these concepts and tools, Section 8.2 gives an overview of the most important industrial reactors. Section 8.3 is a review of ideal reactor types. Emphasis is placed on the way mathematical model equations are constructed for each reactor category. Basically, this boils down to the application of the conservation laws of mass, energy and possibly momentum. Section 8.4 presents an analysis of the effect of the finite rate at which reaction species and/or heat are supplied to or removed from the locus of reaction, i.e. the catalytic site. Finally, the material developed in Sections 8.3 and 8.4 is applied to the design of laboratory reactors and the analysis of rate data in Section 8.5.

8.2 INDUSTRIAL REACTORS

Every industrial chemical process, and catalytic processes are no exception, is designed to produce economically and safely a desired product or range of products from a variety of feedstocks. Ideally, one would like to produce safely and at the lowest costs, at a high throughput and the highest yield (100% selectivity and conversion), with minimum energy consumption and no negative environmental impact. In the selection and design of a catalytic conversion process to approach this ideal several factors play a role, which often impose conflicting demands on the process.

This section presents some of the most important reactors used in industry, and their characteristics.

We have based the classification firstly on the type of operation, viz., batch or continuous, and secondly, on whether a solid catalyst is present or not. This classification is not unambiguous. For instance, most continuous-flow reactors discussed in Sections 8.2.2 and 8.2.3 can also be operated in batch mode.

8.2.1 Batch Reactors

Batch reactors (Fig. 8.1) are usually used for the production of high added value products in limited amounts, e.g., fine chemicals and catalysts. They are also used for testing new processes that have not yet been fully developed, and for processes that are difficult to convert to continuous operation, e.g., some polymerization processes. Typical reactor volumes are in the order of 10 m^3 and the reactor content is in the liquid state. Mechanical stirring ensures good mixing as well as sufficient heat exchange with the surroundings and, if necessary, maintains a solid catalyst in suspension.

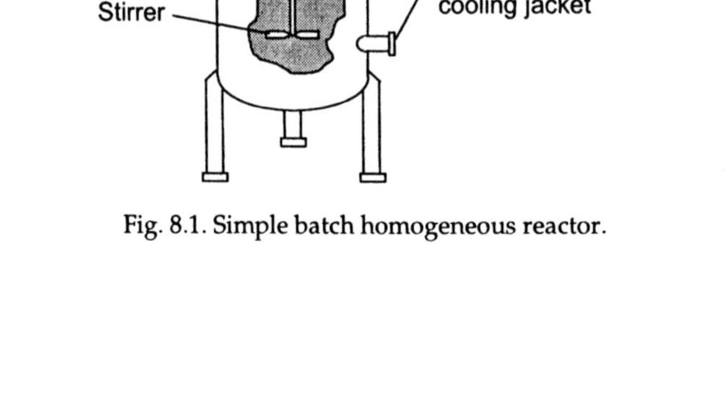

Fig. 8.1. Simple batch homogeneous reactor.

The reactor is charged through the two holes at the top, the reaction is started either by increasing the temperature or by adding a catalyst. At the end of the reaction time the unconverted reactants and the reaction products are discharged through a discharge pipe at the bottom of the reactor.

The batch reactor offers the advantage of high conversions that can be obtained by leaving the reactant in the reactor for long periods of time, but it has the disadvantage of high labour costs per unit production, while large-scale production is difficult. Often various products are manufactured in one plant, a so-called Multi Product or Multi Purpose Plant (MPP), using the same reactors, feed systems, etc. This reduces the investment per product type.

For fast exothermic reactions, temperature control can be a problem. This is often solved by external circulation of part of the reactor content through a heat exchanger, or by adding an internal heat exchange area. Alternatively, semi-batch operation can be applied, i.e., part of a reactant can be fed steadily over time or at certain intervals. This also minimizes the occurrence of unwanted side reactions.

In biocatalysis, temperature control is usually not a problem, because the reactions are generally not accompanied by large heat effects. However, in biocatalysis pH control is an important problem, which can be treated analogously to temperature control.

8.2.2 *Continuous-flow Reactors for Gas–Liquid Reactions (Homogeneous Catalysis)*

In homogeneous catalysis often a reaction takes place between a gaseous reactant and a liquid reactant in the presence of a catalyst that is dissolved in the liquid phase. Examples are carbonylations, hydroformylations, hydrogenations, hydrocyanation, oxidations, and polymerizations. Typically, reactants such as oxygen, hydrogen, and/or carbon monoxide have to be transferred from the gas phase to the liquid phase, where reaction occurs. The choice of reactor mainly depends on the relative flow rates of gas and liquid, and on the rate of the reaction in comparison to the mass and heat transfer characteristics (see Fig. 8.2).

Bubble columns are most widely spread, followed by sparged stirred tanks. Often both are used for a certain application. Stirred tanks and bubble columns have similar characteristics with respect to mass transfer (see Table 8.1). In both reactors the liquid is well mixed. The gas phase in the bubble column shows plug-flow behavior, while in stirred tanks it is well mixed.

Several measures can be taken to control the reactor temperature. For instance, cooling jackets or internal cooling coils can be used. External liquid recirculation through a heat exchanger is another possibility. An elegant solution is to operate at the boiling point of the liquid mixture. In bubble columns and stirred tanks thus near isothermal operation can be achieved by evaporation of one or more of the liquid components present.

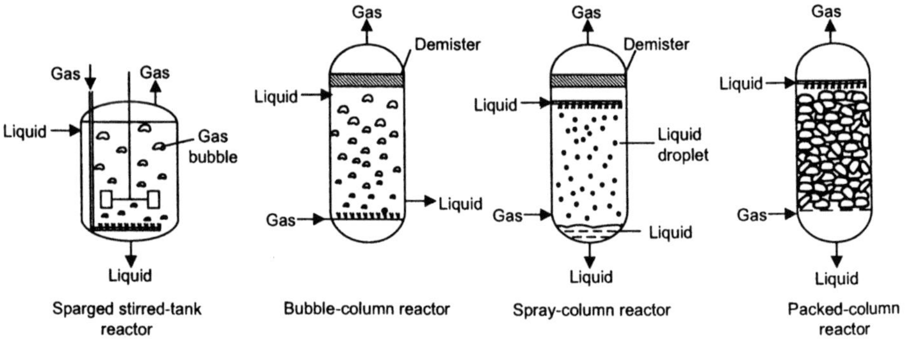

Fig. 8.2. Continuous reactors for gas–liquid reactions.

TABLE 8.1

Mass transfer parameters of gas–liquid reactors. Adapted from [1,2]

Reactor type	Sparged stirred tank	Bubble column	Spray column	Packed column
Liquid holdup, ε_L ($m_L^3\,m_r^{-3}$)	> 0.70	> 0.70	< 0.30	0.05–0.15
Gas hold-up, ε_G ($m_G^3\,m_r^{-3}$)	0.02–0.30	0.02–0.30	> 0.7	0.50–0.80
a_v ($m_i^2\,m_L^{-3}$)	200–2000	100–500	25–200	50–250
k_L ($m_L^3\,m_i^{-2}s^{-1}$)	10^{-4}–$6{\cdot}10^{-4}$	10^{-4}–$6{\cdot}10^{-4}$	$1.5{\cdot}10^{-4}$–$3{\cdot}10^{-4}$	$5{\cdot}10^{-5}$–$3{\cdot}10^{-4}$
β ($m_L^3\,m_f^{-3}$)	10^3–10^4	10^3–10^4	10–40	10–40

The major advantage of the bubble column is the simple design of this type of reactor. Mixing of the liquid is achieved by the upflowing gas bubbles, and, as a consequence, mass transfer rates are limited. Another disadvantage of bubble columns is that they have limited temperature and pressure ranges, while they are also unsuitable for processing viscous fluids.

The major disadvantage of stirred tanks is that much backmixing occurs in both the liquid and the gas phase. A number of stirred tanks can be placed in series in order to narrow the residence time distribution. Mechanical stirring has the advantage that viscous fluids can be handled, but it increases investment and operating costs.

Both the bubble column and the stirred tank can also be used with a catalyst in suspension (see Section 8.2.3).

The spray column and packed column both have low gas-phase pressure drop and are therefore suitable for processes requiring large gas throughputs. They are often used for gas treatment and off-gas scrubbing. For instance, absorption of CO_2 and H_2S from gas streams is often performed in packed columns.

Table 8.1 compares some important parameters of the various gas–liquid reactor types.

The first major decision in the choice of a reactor for gas–liquid reactions taking place in the liquid phase is based on the optimal usage of the total reactor volume, i.e. the choice of the parameter β, which is the ratio of the liquid-phase volume to the volume of the diffusion layer (see Section 8.4.2). When reactions are slow compared to the mass transfer from the gas to the liquid, sparged stirred tanks and bubble columns are preferred, as these reactors have the largest bulk liquid volume. On the other hand, fast reactions for a large part take place in the diffusion layer, so in this case spray columns and packed columns are more suitable.

8.2.3 Continuous-flow Reactors for Solid-catalyzed Reactions

Heterogeneously catalysed reactions are two-, three-, or even more than three-phase operations. Solid catalyst and gaseous and liquid reactants are brought in contact to achieve the desired conversion. Some of the reactor types that are used are briefly presented here for background information with generalized remarks on their advantages and disadvantages.

Gas–solid reactors are the best-known two-phase catalytic reactor types operated continuously. The major reactors for solid-catalyzed gas phase reactions are the fixed-bed, the fluid-bed and the entrained-flow reactor.

The fixed-bed reactor is either operated adiabatically (Fig. 8.3) or isothermally. In the latter case a multitubular reactor system is often chosen where the parallel reactor tubes are surrounded by a heat exchanging fluid. The fixed character of

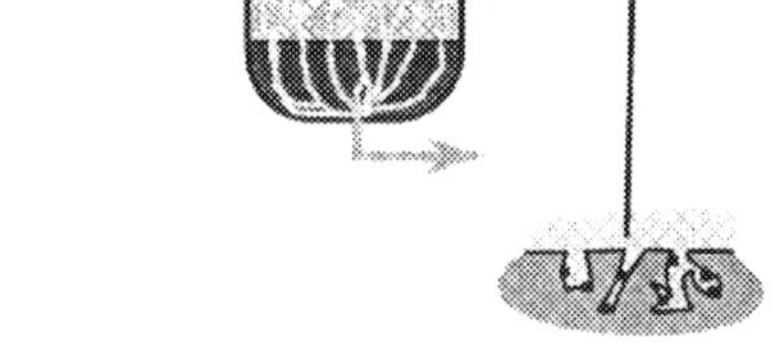

Fig. 8.3. Adiabatic fixed-bed reactor; single fluid phase.

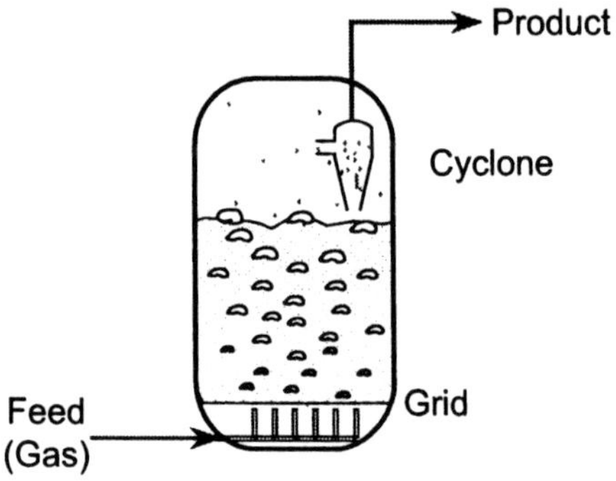

Fig. 8.4. Fluid-bed reactor.

the catalyst bed necessitates a long catalyst life. If deactivation occurs it should be possible to regenerate the catalyst inside the reactor.

For reactions with moderate to strong heat effects, multiple beds or reactors are often placed in series with intermediate heating or cooling.

Advantages of the adiabatic fixed-bed reactor are its simplicity, the high catalyst load per unit volume, little catalyst attrition, and little backmixing. Disadvantages are the large pressure drop, difficult temperature control, and the long diffusion distances.

Some of the many processes using adiabatic fixed-bed reactors with gaseous reactants are the production of methanol and ammonia, and hydrotreating of naphtha.

A reactor system similar to the fixed-bed reactor is the moving-bed reactor, in which the deactivation rate is relatively low, but too high for pure fixed-bed operation. An example of its application is the catalytic reforming processes.

The fluid-bed reactor (Fig. 8.4) is mainly chosen because of its good heat exchange properties and the uniform temperature distribution. Furthermore, due to the fluid-like behavior of the bed, the catalyst can be added and withdrawn during operation for regeneration or replacement in case of fast deactivation. In addition, due to the relatively small particle size, the diffusion distances are short. On the other hand, disadvantages are the occurrence of catalyst attrition and backmixing, and the fact that fluid-bed reactors are difficult to scale up.

The entrained-flow reactor (Fig. 8.5) is used when very short contact times are required, as in the case of highly active catalysts that deactivate fast. In fluid catalytic cracking (FCC) the circulating catalyst also supplies part of the heat for the endothermic reaction. Depending on the catalyst loading one can distinguish dilute and dense phase 'risers'.

Advantages of entrained-flow reactors are the high mass-transfer rates, short contact times, and the possibility of continuous catalyst replacement. Dis-

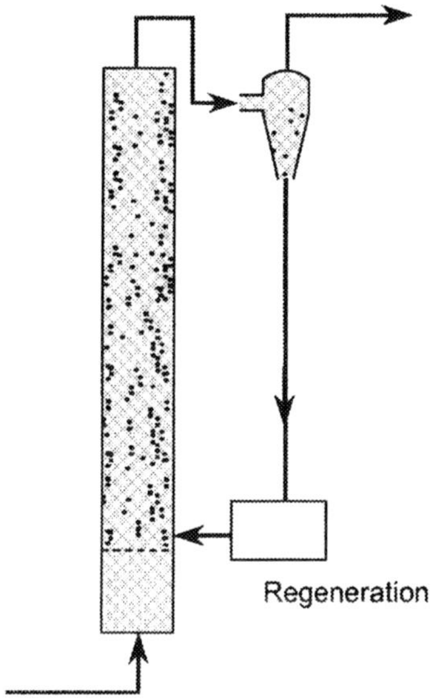

Fig. 8.5. Entrained-flow reactor.

advantages are the occurrence of catalyst attrition, reactor erosion, and the requirement to separate the catalyst from the product.

Reactor types intermediate between fluid bed and entrained flow also exist, like internally circulating bed reactors with an internal riser section and an external annular section of a fluid/moving-bed type.

Relatively new are the monolithic reactor types (shown in Fig. 8.9 for gas/liquid/solid operation), which are mainly used in environmental applications [3]. Examples include the three-way catalyst (TWC) for exhaust gas cleaning of gasoline engines, the selective catalytic reduction catalyst (SCR) for NO_x reduction with ammonia in lean-burn combustion flue gases, the oxidation of VOCs and the decomposition of ozone for air conditioning (airplanes, photocopiers). All have an extremely small pressure drop due to the structured character of the catalyst and can be used at high flow rates. The TWC has small channels (~1 mm) with very thin walls and has high thermostability. The SCR catalyst has relatively large channels (~1 cm) and is tolerant for operation under dusty conditions.

Another example of a reactor with small pressure drop is the parallel-passage reactor for flue gas treatment, which is shown in Fig. 8.6.

Liquid–solid reactors of the fluid bed or slurry type do exist but are much less frequently encountered than their gas–solid analogues. In most cases a gas phase is involved, too, resulting in three-phase operation.

Many applications of *gas–liquid–solid reactors* are found in practice. Selective oxidations and hydrogenations, hydrotreating, hydrations and aminations of a

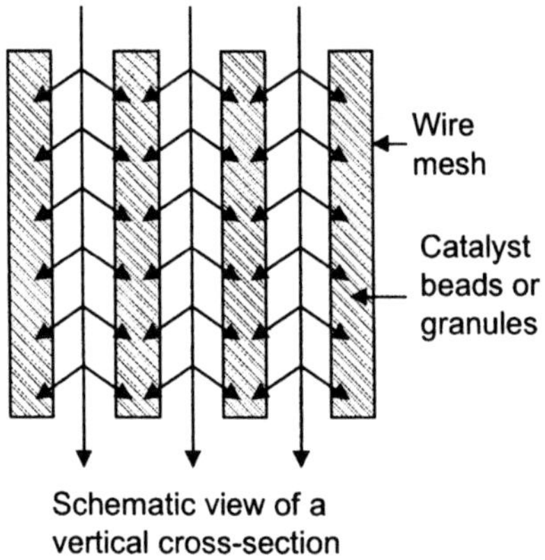

Fig. 8.6. Parallel-passage reactor.

liquid reactant (mixture) catalyzed by a solid catalyst are important industrial examples. Two extremes exist in three-phase operation: the trickle-bed reactor (Fig. 8.7) and the slurry reactor (Fig. 8.8).

In the trickle-bed reactor the gaseous and liquid reactants flow co-currently downward over a packed bed of catalyst. The liquid phase must be distributed such that an even wetting of the catalyst is obtained. Uneven wetting may lead to local hot spots and runaways. The gas phase is the continuous phase. High gas flow rates are used in order to obtain a sufficient heat removal (or supply) rate. Since the liquid covers the catalyst particles, reactant diffusion is much slower than in gas phase operation. The nature of the packed bed requires catalyst bodies of a few millimeters, so the catalyst effectiveness is restricted.

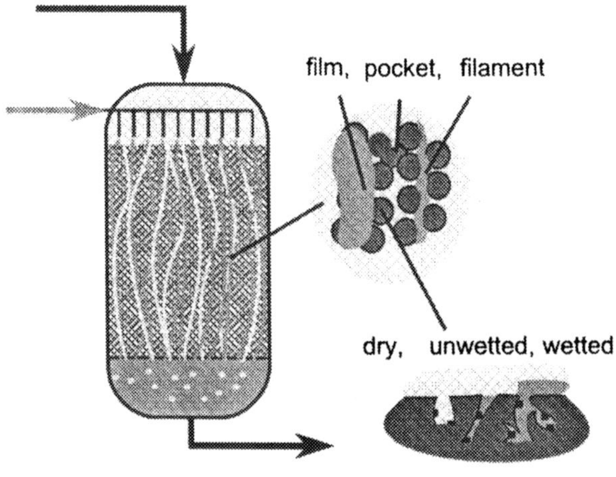

Fig. 8.7. Trickle-bed reactor.

Advantages of trickle-bed reactors are high catalyst load, no need for catalyst separation, no attrition, and very limited backmixing. Disadvantages are lower catalyst effectiveness and selectivity, flow maldistributions, and large pressure drop.

Upflow reactors are sometimes used for small scale testing. The liquid phase is the continuous phase through which the gas bubbles rise and wetting problems are absent. Industrial reactors in which liquid and gas flow upward are rarely found, because it can be difficult to keep the catalyst bed immobile at large flow rates.

Countercurrent operation may offer certain advantages, like the removal of intermediate products to avoid side reactions, to overcome equilibrium limitations (hydrogenation of aromatics) or to avoid product inhibition (hydrodesulfurization). Also, temperature profiles along the reactor length differ which may be used advantageously. The major problem is flooding, i.e., entrainment of the liquid by the gas phase. Larger catalyst particles must be used to allow sufficiently high flow rates, thereby further lowering the catalyst efficiency.

In slurry reactors (Fig. 8.8) small catalyst particles (10–100 μm) are suspended in a liquid phase by mechanical mixing. Stirring also improves the contact between the gas bubbles and the liquid phase and heat exchange with the surroundings. Slurry reactors can be operated in a batch, semi-batch or continuous mode. In the semi-batch mode often the gas-phase is supplied continuously.

The use of small particles results in high catalyst utilization, but the particles are subjected to attrition. Filtration is an energy intensive aftertreatment, therefore an optimum particle size is sought with respect to catalyst effectiveness, settling and mixing speed. In these well-stirred reactors a good heat supply or removal is achieved.

The main advantages of the slurry reactor are the high catalyst utilization, the uniform temperature, and good external mass transfer. Disadvantages are catalyst attrition, the need for catalyst separation, and a high degree of backmixing.

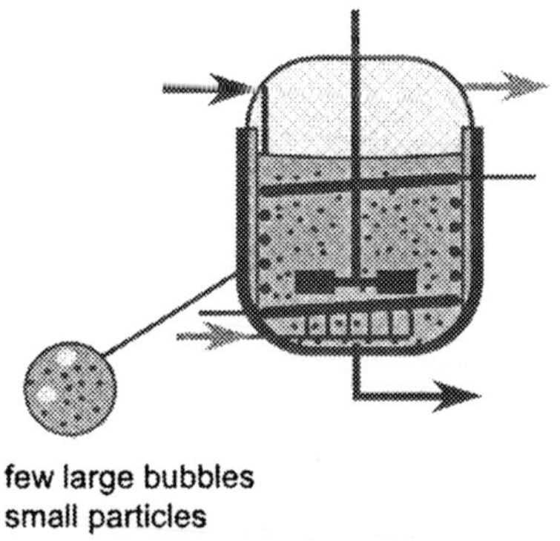

Fig. 8.8. Slurry reactor.

TABLE 8.2

Characteristics of three-phase reactor types (adapted from [1])

Characteristics	Catalyst in suspension		Fixed bed			3-Phase fluid bed
	Bubble column	Stirred tank	Cocurrent down	Cocurrent up	Counter-current	
ε_{cat} ($m_{cat}^3\,m_r^{-3}$)	0.01–0.1	0.01–0.1	0.6–0.7	0.6–0.7	0.5	0.1–0.5
ε_L ($m_L^3\,m_r^{-3}$)	0.8–0.9	0.8–0.9	0.05–0.25	0.2–0.3	0.05–0.1	0.2–0.8
ε_G ($m_G^3\,m_r^{-3}$)	0.1–0.2	0.1–0.2	0.2–0.35	0.05–0.1	0.2–0.4	0.05–0.2
d_p (mm)	≤ 0.1	≤ 0.1	1–5	1–5	>5	0.1–5
a' ($m_s^2\,m_p^{-3}$)	500	500	1000–2000	1000–2000	500	500–1000
k_L ($m_L^3\,m_i^{-2}s^{-1}$)	10^{-4}–$6{\cdot}10^{-4}$	10^{-4}–$6{\cdot}10^{-4}$	$5{\cdot}10^{-5}$–$5{\cdot}10^{-4}$	$5{\cdot}10^{-5}$–$5{\cdot}10^{-4}$	$5{\cdot}10^{-5}$–$5{\cdot}10^{-4}$	10^{-4}–$5{\cdot}10^{-4}$
a_v ($m_i^2\,m_L^{-3}$)	100–400	100–1500	100–1000	100–1000	100–500	100–1000
η (isoth.)	1	1	< 1	< 1	< 1	< 1

In order to reduce attrition, recirculation of the particles can be achieved by the gas bubbles (gas lift) instead of by mechanical agitation (bubble column, see Fig. 8.2). External liquid recirculation offers the possibility of improving the gas–liquid mass transfer by special liquid ejector types. In the case of gas and liquid upflow a three-phase fluidized bed reactor is produced. Here particles of a few millimeters are needed to retain them in the reactor and allow separation from the gas/liquid phase.

Since three-phase operation is industrially very important and since this is the most challenging catalytic operation, Table 8.2 presents a comparison of the main characteristics of three-phase reactor types, while Table 8.3 gives a number of appreciation criteria. These can serve in the choice of a proper reactor type for a specific reaction.

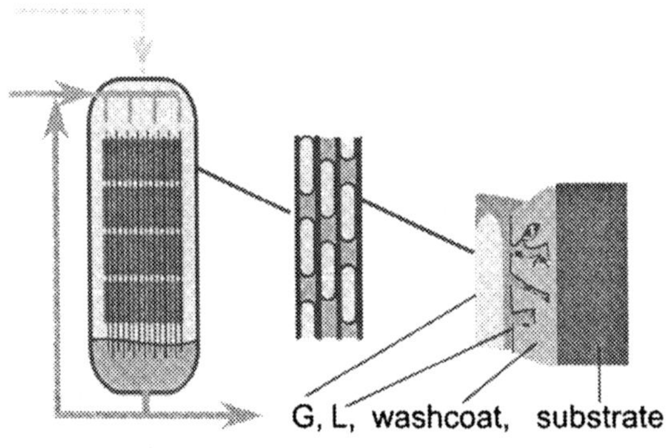

Fig. 8.9. Three-phase monolith reactor.

TABLE 8.3

Comparison of different three-phase reactor types (adapted from [1])

Appreciation criteria	Suspended catalyst	Three-phase fluid bed	Fixed bed
Activity	Highly variable, but often full utilization	Highly variable: inter and intra mass transfer may reduce utilization, especially in fixed bed backmixing –	plug flow +
Selectivity	Generally unaffected by mass/heat transfer	as above backmixing often –	plug flow often +
Stability	Catalyst replacement between batches to overcome rapid poisoning	Possibility of continuous renewal. Must have good attrition resistance	Essential for fixed bed operation. Plug flow may establish a poison adsorption front
Cost	Consumption depends on feed impurities acting as poisons		Necessarily low catalyst consumption
Heat exchange	Fairly easy to achieve	Possibility of heat exchange in reactor	Generally adiabatic operation
Design difficulties	Catalyst separation sometimes difficult: possible problems in pumps due to risks of deposits or erosion		Fairly simple for a downward cocurrent adiabatic bed
Scaling-up	No difficulty: generally limited to batch and relatively small sizes	System poorly known, should be scaled up in steps	Large reactors can be built provided a good liquid distribution is achieved

New reactor types for three-phase operation are still being developed. An example is the application of structured reactors, which may have certain advantages in three-phase operation, and can be operated with co-, cross- as well as with countercurrent flow. A very recent development is the use of monolith reactors (Fig. 8.9) in three-phase operations. Their advantages are the small pressure drop, the good external mass transfer, the short diffusion distance, and the low adiabatic temperature rise. Disadvantages are the higher catalyst costs, importance of liquid distribution, and moderate catalyst load.

Three-phase monoliths are not yet applied industrially (except in one process). Only limited experience exists on laboratory and pilot-plant scale.

8.3 IDEAL REACTORS — MATHEMATICAL DESCRIPTION

For each of the ideal reactor types, *viz.* ideal batch reactor, plug-flow reactor (PFR), and continuous-flow stirred-tank reactor (CSTR), continuity equations or design equations can be derived using mass (or rather molar) balance equations for each species involved.

A molar balance on a species A at any instant in time, t, yields the following general equation:

rate of flow of A into the reactor	+	rate of production of A by chemical reaction	−	rate of flow of A out of the reactor	=	rate of accumulation of A within the reactor	
F_{A0}	+	$R_{v,A}V$	−	F_A	=	dn_A/dt	(8.1)

where: F_{A0} and F_A = molar flow rate of A at reactor inlet and outlet, respectively (mol s^{-1}), $R_{v,A}$ = production rate of A per unit reaction volume (mol m_f^{-3} s^{-1}), V = reaction volume (m_f^3), n_A = the amount of A (mol), t = time (s).

In catalysed reactions it is often more convenient to define the production rate per unit volume of catalyst or even per unit mass of catalyst, instead of the production per unit volume reaction volume. Eqn. (8.2) expresses the relations between the various rate definitions:

$$R_{v,A}V = R_{v,pA}V_p = R_{w,A}W \tag{8.2}$$

with: $R_{v,p,A}$ = production rate per unit volume of catalyst particle (mol m_p^{-3} s^{-1}), V_p = volume of catalyst particle (m_p^3), $R_{w,A}$ = production rate per unit mass of catalyst (mol kg_{cat}^{-1} s^{-1}), W = mass of catalyst in reactor (kg_{cat}).

As in most sections of this chapter we are dealing with heterogeneous catalysis, the production rate will generally be expressed per unit mass of catalyst. The rate of production is negative for reactants and positive for products:

$$R_{w,A} = \nu_A r_w \tag{8.3}$$

where ν_A = stoichiometric coefficient of A (–), and r_w = specific reaction rate of A (mol kg_{cat}^{-1} s^{-1}).

In the next sections, the design equations for the three ideal reactor types will be derived for isothermal conditions. In practice, the heat effects associated with chemical reactions usually result in non-isothermal conditions. The application of the law of conservation of energy leads to the so-called energy balance equation. This derivation is analogous to the derivation of the mass balance equations, and will not be treated here (see for instance, [4,5]). However, it should be noted that under non-isothermal conditions, the energy balance equation should always be solved simultaneously with the corresponding mass balance equation, since the reaction rate depends not only on composition but also on temperature.

8.3.1 Batch Reactor

During batch operation, the concentrations of reactants and formed products change as a function of time. In the case of perfect mixing, i.e., in an ideal batch reactor, both temperature and composition (and if present, also catalyst concentration) are uniform throughout the reactor. Since in a batch reactor there is no inflow or outflow of reactants and products during reaction, a mass balance for species A only contains an accumulation (left) and a production term:

$$\frac{dn_A}{dt} = R_{w,A}\, W \tag{8.4}$$

where: n_A = the amount of A (mol); t = time (s); $R_{w,A}$ = specific production rate of A (mol kg_{cat}^{-1} s^{-1}); and W = mass of catalyst in reactor (kg).

If A is a reactant, its degree of conversion is defined as:

$$X_A = \frac{n_{A0} - n_A}{n_{A0}} \tag{8.5}$$

and, for constant density systems (V = constant):

$$X_A = \frac{C_{A0} - C_A}{C_{A0}} \tag{8.6}$$

which leads to:

$$\frac{dX_A}{dt} = -\frac{1}{C_{A0}}\frac{W}{V} R_{w,A} \tag{8.7}$$

8.3.1.1 Single reaction

For a single reaction and known kinetics the balance for A can be integrated at a given temperature to yield a relation between the batch time and the degree of conversion:

$$\frac{1}{C_{A0}}\frac{W}{V} t = -\int_0^{X_A} \frac{1}{R_{w,A}} dX_A \tag{8.8}$$

In special cases an analytical solution can be derived for Eqn. (8.8). For a single irreversible first-order reaction $A \rightarrow P$:

$$R_{w,A} = \nu_A\, r_w = -r_w = -k_w C_A \tag{8.9}$$

with: k_w = first-order rate constant (m^3 kg_{cat}^{-1} s^{-1}).

TABLE 8.4

CSTR: relations between conversion, X_A, or concentration, C_A, and batch space time, τ_B (for constant density)

r_w (mol kg_{cat}^{-1} s^{-1})	X_A (mol mol^{-1})	C_A (mol m_f^{-3})
k_w	$\frac{k_w \tau_b}{C_{A0}}$	$C_{A0} - k_w \tau_B$
$k_w C_A$	$1 - e^{-k_w \tau_b}$	$C_{A0} e^{-k_w \tau_B}$
$k_w C_A^2$	$\frac{k_w C_{A0} \tau_b}{1 + k_w C_{A0} \tau_b}$	$\frac{C_{A0}}{1 + k_w C_{A0} \tau_b}$

Substitution of Eqns. (8.9) and (8.6) into Eqn. (8.8) and integration then yields:

$$X_A = 1 - e^{-k_w \tau_B} \qquad (8.10)$$

where τ_B is the batch space-time (kg_{cat} m^{-3} s^{-1}):

$$\tau_B = t \frac{W}{V} \qquad (8.11)$$

Table 8.4 lists analytical relations between conversion or concentration and batch space-time for simple reaction kinetics.

8.3.1.2 Multiple reactions

If more than one reaction occurs, it is convenient to define the selectivity for Q as the ratio of the amount of product Q formed to the amount of reactant A converted:

$$S_Q = \frac{n_Q - n_{Q0}}{n_{A0} - n_A} \qquad (8.12)$$

Alternatively, the yield of a product Q relates the amount of Q formed to the amount of A fed:

$$Y_Q = \frac{n_Q - n_{Q0}}{n_{A0}} \qquad (8.13)$$

Note that:

$$Y_Q = S_Q X_A \tag{8.14}$$

If the following reaction occurs:

$$A \xrightarrow{1} Q \xrightarrow{2} P$$

in order to determine the amount of Q produced, a mass balance for Q is required in addition to the balance for A:

$$\frac{dn_A}{dt} = R_{w,A} W = -r_{w1} W \tag{8.15}$$

$$\frac{dn_Q}{dt} = R_{w,Q} W = r_{w1} W - r_{w2} W \tag{8.16}$$

The balance equations form a set of coupled ordinary differential equations, which have to be solved numerically. Analytical integration is possible in special cases only.

Assuming first-order reaction kinetics and constant density leads to:

$$\frac{dC_A}{d\tau_B} = -k_{w1} C_A \tag{8.17}$$

$$\frac{dC_Q}{d\tau_B} = k_{w1} C_A - k_{w2} C_Q \tag{8.18}$$

with τ_B as defined in Eqn. (8.11).

Integration of Eqns. (8.17) and (8.18) with initial conditions:

$$C_A = C_{A0} \quad \text{for} \quad \tau_B = 0 \tag{8.19}$$

$$C_Q = 0 \quad \text{for} \quad \tau_B = 0 \tag{8.20}$$

yields the following analytical expression for C_A and C_Q as a function of time:

$$C_A = C_{A0}\, e^{-k_{w1}\tau_B} \tag{8.21}$$

$$C_Q = \frac{C_{A0}\, k_{w1}}{k_{w2} - k_{w1}} \left(e^{-k_{w1}\tau_B} - e^{-k_{w2}\tau_B}\right) \tag{8.22}$$

Note that the concentration of P follows from the total mass balance.

8.3.2 Plug-Flow Reactor (PFR)

Numerous reactions are performed by feeding the reactants continuously to cylindrical tubes, either empty or packed with catalyst, with a length which is 10 to 1000 times larger than the diameter. The mixture of unconverted reactants and reaction products is continuously withdrawn at the reactor exit. Hence, constant concentration profiles of reactants and products as well as a temperature profile are established between the inlet and the outlet of the tubular reactor (see Fig. 8.10). This requires, in contrast to the batch reactor, the application of the law of conservation of mass over an infinitesimal volume element, dV, or mass element, dW, of the reactor. For a tubular reactor with a fixed catalytic bed, it is more convenient to relate the production rates to the catalyst mass, rather than to the reactor volume.

In contrast to a batch reactor, the existence of a temperature profile does not allow us to consider the mass balances for the reacting components and the energy balance separately. Such a separation can only be performed for isothermal tubular reactors.

In view of the large length-to-diameter ratio of most tubular reactors, the flow through them can in most cases be described adequately as a so-called plug flow. It is assumed that:

- flow occurs in one direction only, i.e. parallel to the axis of the tube;
- radial, i.e. perpendicular on the axis, non-uniformities can be neglected;
- flow occurs by forced convection only.

With these assumptions we can picture the flow as a plug of fluid with uniform properties being pushed through the tube.

In a plug-flow reactor the mass balance for a component A over an infinitesimal reactor element, as defined in Fig. 8.10, yields:

$$(F_A + dF_A) - F_A = R_{w,A}\, dW \tag{8.23}$$

The left-hand side of Eqn. (8.23) corresponds to the net flow of A out of the reactor element which in the steady state has to equal the net production rate of A, i.e. the right-hand side of Eqn. (8.23).

In analogy to the batch reactor, a fractional conversion of a reactant A can be defined as:

$$X_A = \frac{F_{A0} - F_A}{F_{A0}} \tag{8.24}$$

in which F_{A0} is the molar flow rate of A at the reactor inlet (mol s^{-1}).

This leads to the continuity equation for A:

$$\frac{dX_A}{d(W / F_{A0})} = -R_{w,A} \tag{8.25}$$

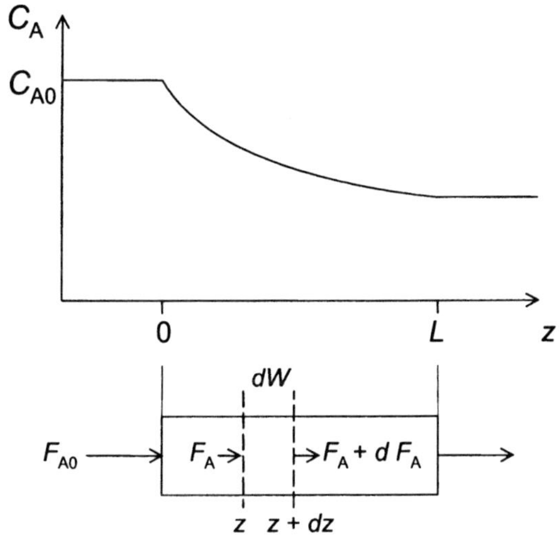

Fig. 8.10. Reactant concentration versus position in a PFR.

Integration results in:

$$\frac{W}{F_{A0}} = -\int_0^{X_A} \frac{1}{R_{w,A}} \mathrm{d}X \tag{8.26}$$

Comparison with Eqn. (8.8) for a batch reactor shows the mathematical analogy between plug-flow and batch reactors. The batch reactor results obtained in the preceding paragraph can be transposed to the plug-flow reactor, provided that a space time τ_0 is defined for the plug-flow reactor as:

$$\tau_0 = \frac{W C_{A0}}{F_{A0}} \qquad (\mathrm{kg_{cat}\ m_f^{-3}\ s}) \tag{8.27}$$

8.3.3 Continuous-Flow Stirred-Tank Reactor (CSTR)

Another type of continuous-flow reactor is found in industry: the Continuous-Flow Stirred-Tank Reactor (CSTR) (see Fig. 8.11).

Mixing can be considered at two length scales. Macromixing relates to the mixing of fluid elements at the scale of the reactor. Ideal macromixing requires that the time scale at which the fluid elements are mixed can be neglected with respect to the time scale at which these elements flow through the reactor. Ideal micromixing requires that the time scale at which reacting molecules are mixed can be neglected when compared to the time scale at which they react. In a perfectly mixed flow reactor (CSTR) the reactor content is uniform at both length scales, both with respect to composition and to temperature. Hence, the mass

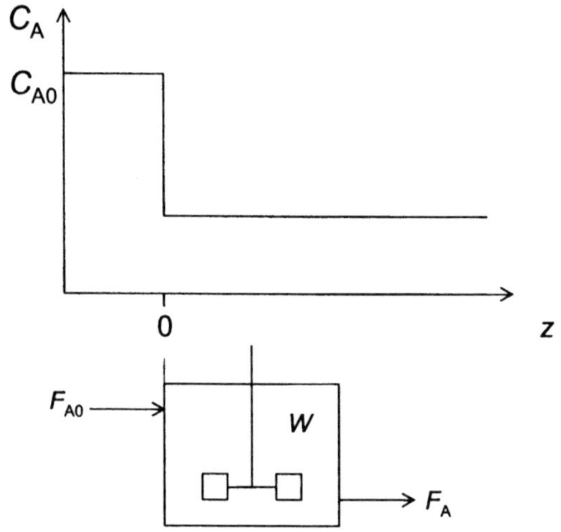

Fig. 8.11. Reactant concentration versus position in a CSTR.

balance for a component can be made over the complete reactor (see Fig. 8.11). At steady state:

$$F_A - F_{A0} = R_{w,A}\, W \tag{8.28}$$

Or, after introduction of the fractional conversion of A, Eqn. (8.24):

$$X_A = -R_{w,A} \frac{W}{F_{A0}} \tag{8.29}$$

Note that the uniform conditions in the reactor equal those at the reactor outlet, which implies that the production rate of A is also determined by the outlet conditions. The mass balance for a component in a CSTR is an algebraic equation, in contrast to the case of reactions in a batch or a plug flow reactor. For a single, irreversible first-order reaction and constant density, Eqn. (8.29) becomes:

$$X_A = k_w\, C_{A0}\, (1 - X_A) \frac{W}{F_{A0}} \tag{8.30}$$

with k_w in $m^3\ kg_{cat}^{-1}\ s^{-1}$ or, after substitution of Eqn. (8.27) and solving for X_A:

$$X_A = \frac{k_w \tau_0}{1 + k_w \tau_0} \tag{8.31}$$

Table 8.5 lists analytical relations between conversion or concentration for some simple reaction kinetics.

TABLE 8.5

CSTR: relations between conversion, X_A, or concentration, C_A, and $\tau_0 = WC_{A0}/F_{A0}$ (constant density) for simple reaction kinetics

r_w (mol kg_{cat}^{-1} s^{-1})	X_A (mol mol^{-1})	C_A (mol m_f^{-3})
k_w	$\dfrac{k_w\tau_0}{C_{A0}}$	$C_{A0} - k_w\tau_0$
k_wC_A	$\dfrac{k_w\tau_0}{1+k_w\tau_0}$	$\dfrac{C_{A0}}{1+k_w\tau_0}$
$k_wC_A^2$	$1+\dfrac{1-\sqrt{1+4k_wC_{A0}\tau_0}}{2k_wC_{A0}\tau_0}$	$\dfrac{-1+\sqrt{1+4k_wC_{A0}\tau_0}}{2k_w\tau_0}$

8.3.4 Comparison of PFR and CSTR

It is worthwhile to compare the conversion obtained in an isothermal plug-flow reactor (PFR) with that obtained in a CSTR for given reaction kinetics. Figure 8.12 shows a fair comparison for irreversible first-order kinetics by plotting the conversion in both reactors as a function of space–time τ_0.

The conversion of A obtained in a PFR is larger than that obtained in a CSTR at equal space-time. This holds for every positive partial reaction order with respect

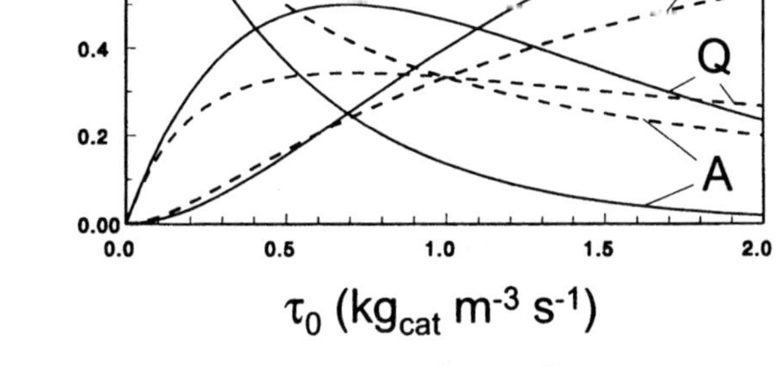

Fig. 8.12. Reactant and product concentrations for A $\xrightarrow{1}$ Q $\xrightarrow{2}$ P versus batch time or space-time τ_0 in a batch reactor or PFR (solid lines) or a CSTR (dashed lines); $k_{w1} = 2$ m_f^3 kg_{cat}^{-1} s^{-1}; $k_{w2} = 1$ m_f^3 kg_{cat}^{-1} s^{-1}; constant density.

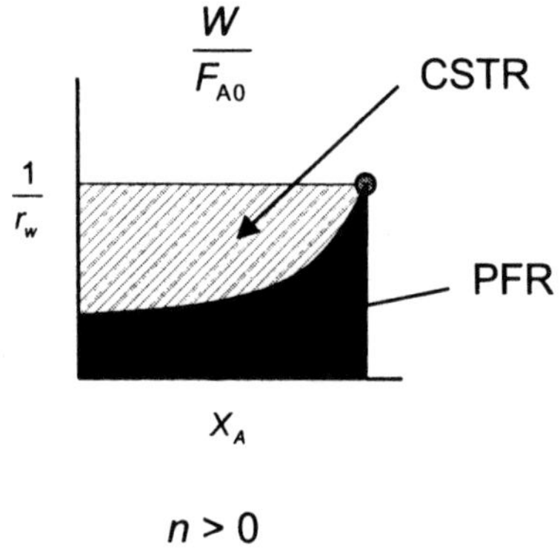

Fig. 8.13. Required W/F_{A0} in a PFR and CSTR for reactions with order $n > 0$.

to A. Similarly, the space-time required for equal conversion is smaller in a PFR than in a CSTR at the same initial concentration C_{A0}. Figure 8.13 illustrates this for an arbitrary reaction with order n greater than zero, using Eqns. (8.26) and (8.29).

For multiple reactions, selectivities and yields enter the picture.

$$S_Q = \frac{F_Q - F_{Q0}}{F_{A0} - F_A} \tag{8.32}$$

Alternatively, the yield of a product Q relates the amount of Q formed to the amount of A fed:

$$Y_Q = \frac{F_Q - F_{Q0}}{F_{A0}} \tag{8.33}$$

Figure 8.12 also compares the evolution of concentrations of the desired intermediate Q and the final product P in the case of two first-order reactions in series in a CSTR with that in a batch or plug-flow reactor. For constant density, the mass balances for the reaction components in a CSTR are:

$$C_A = \frac{C_{A0}}{1 + k_{w1}\,\tau_0} \tag{8.34}$$

$$C_Q = \frac{k_{w1}\,C_{A0}\,\tau_0}{(1 + k_{w1}\,\tau_0)(1 + k_{w2}\,\tau_0)} \tag{8.35}$$

For positive reaction orders with respect to the reactants the intermediate yields obtained in a CSTR are lower than those obtained in a batch or plug-flow reactor.

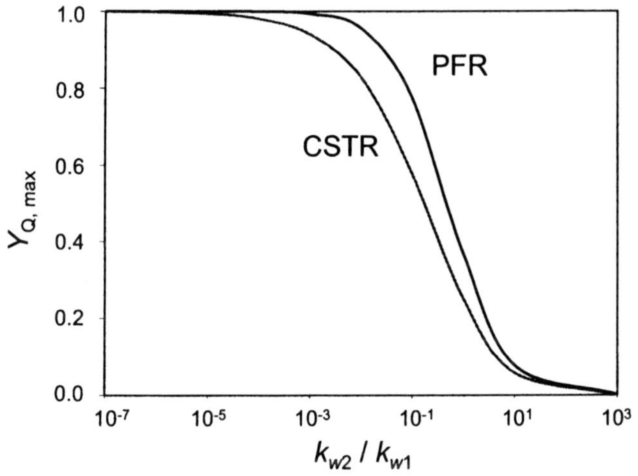

Fig. 8.14. Maximum yield of intermediate in a PFR and a CSTR as a function of k_{w2} / k_{w1}.

The maximum yield for the intermediate Q is a function of k_{w2}/k_{w1} only:

PFR, batch:
$$Y_{Q,max} = \left(\frac{k_{w1}}{k_{w2}}\right)^{\frac{k_{w2}}{k_{w2}-k_{w1}}} \tag{8.36}$$

CSTR:
$$Y_{Q,max} = \frac{1}{\left(\sqrt{\frac{k_{w2}}{k_{w1}}}+1\right)^2} \tag{8.37}$$

Figure 8.14 shows these relationships.

In biocatalysis, selectivity is also of great importance. However, in biocatalysis competition often takes place between parallel reactions: A $\rightarrow$ P and A $\rightarrow$ Q (which is also important in homogeneous enantioselective catalysis), or A $\rightarrow$ P and B $\rightarrow$ Q (kinetic resolution, A and B are enantiomers and so are P and Q). With the latter type of competition, the selectivity in a CSTR is reduced as compared to a PFR, just as in the consecutive reactions of the example. On the other hand, for the parallel reactions of A, the selectivity in a CSTR is the same as in a PFR.

8.4 REACTION COMBINED WITH TRANSPORT

Up to this point it has been assumed that the composition and the temperature are uniform throughout the complete reactor for perfect mixing, or within an infinitesimal volume element for plug flow. These assumptions usually hold for homogeneously catalysed reactions involving a single phase. If the reactants are present in more than one phase, as in gas–liquid reactions, or if a heterogeneous catalyst is involved, it is often necessary to relax the hypothesis of uniformity. Indeed, the presence of several phases requires transport in and/or transfer between these phases to allow reaction to take place.

As the driving force for transport and transfer of mass or heat is provided by concentration or temperature gradients, the latter cannot always be neglected. Consequently, the mass or enthalpy production terms in the reactor equations of the preceding paragraph are not always determined by the chemical (so-called intrinsic) kinetics only, but also by the rates of the physical transport and transfer phenomena. In such a situation the reaction is said to be transport-limited.

8.4.1 Heterogeneous Catalysis

In catalytic reactions mass transfer from the fluid phase to the active phase inside the porous catalyst particle takes place via transport through a fictitious stagnant fluid film surrounding the particle and via diffusion inside the particle. Heat transport to or from the catalyst takes the same route. These phenomena are summarized in Fig. 8.15.

Transport resistances may lead to an under- or overestimation of the local conditions in the catalyst particle, and so to a deviating estimation of the reaction rate. When neglected, the practical consequence is an over-designed, or worse, under-designed reactor.

In general, there are seven different physical and chemical steps starting from a reactant A and ending in a reaction product B via a heterogeneously catalyzed reaction path:

1. Transfer of A from the fluid phase surrounding the catalyst particle, the so-called bulk fluid phase, to the external surface of the particle.

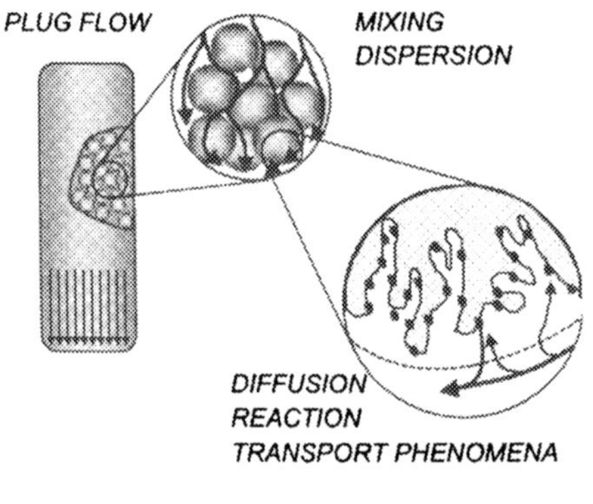

Fig. 8.15. Phenomena in a packed-bed reactor.

2. Transport from the external surface of the particle through the pores towards the active sites on the interior surface.
3. Chemisorption of A on an active site * as A*.
4. Surface reaction of A* into B*.
5. Decomposition of B* (desorption of B from *).
6. Transport of B through the pores towards the external surface of the pellet.
7. Transfer of B from the external surface of the particle to the bulk fluid phase.

Steps 3–5 are strictly chemical and consecutive to each other: Hougen–Watson–Langmuir–Hinshelwood rate equations describing the rate of the purely chemical phenomenon consisting of steps 3–5 have been derived in Chapter 3 on the kinetics of catalysed reactions. In the transport-limited situation the supply of reactant and/or the removal of reaction product will not be sufficiently fast to keep pace with the potential intrinsic rate, and the concentrations of A and B inside the pores will be different from the corresponding concentrations in the bulk of the fluid phase.

The transfer steps 1 and 7 are strictly physical and in series with steps 3–5: the transfer occurs separately from the chemical reaction. The transport steps 2 and 6, however, cannot be separated from the chemical steps 3–5: the transport inside the pores occurs simultaneously with the chemical reaction. In the next sections these two situations are analysed quantitatively.

8.4.1.1 Transport and reaction in series

External mass transfer

The transfer of mass from the bulk of a fluid phase to the external surroundings can be described by the so-called film model (see Fig. 8.16). The film

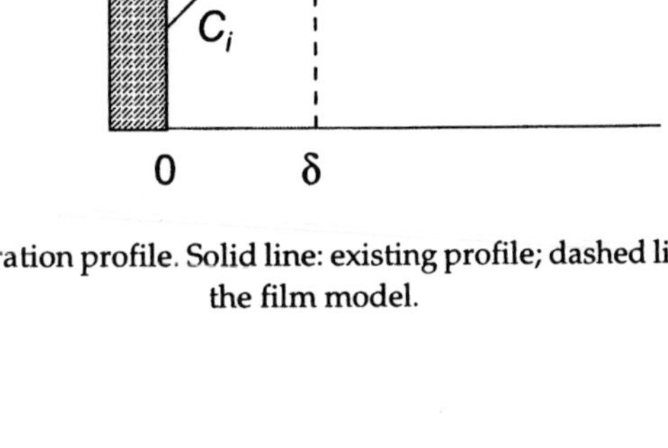

Fig. 8.16. External concentration profile. Solid line: existing profile; dashed line: profile according to the film model.

model assumes the existence of a stagnant layer of thickness δ surrounding the external surface of the catalyst. The complete resistance to mass transfer is located in this layer. Hence, the concentration gradient allowing the transfer is also confined to this layer.

The rate of transfer, expressed as a molar flux N in mol $m_s^{-2}\,s^{-1}$, is proportional to the driving force, which with the film model gives:

$$N = k_f\,(C_b - C_s) \tag{8.38}$$

with: k_f the mass transfer coefficient for A in the film layer surrounding the catalyst particle ($m_f^3\,m_s^{-2}\,s^{-1}$); C_b, C_s the bulk and surface concentration of A, respectively (mol m_f^{-3}). The subscript A is left out from here on for notational convenience. The value of C_s lies between 0 and C_b. This implies that the reaction rate can never exceed $N = k_f C_b$, no matter how active the catalyst is!

At steady state the flux of A towards the external surface of the catalyst particle has to be equal to the rate of disappearance of A, expressed per unit external surface area:

$$N = r_{\mathrm{v,p}}\frac{V_{\mathrm{p}}}{A_{\mathrm{p}}} = r_{\mathrm{v,p}}\frac{1}{a'} = r_{\mathrm{w,p}}\rho_{\mathrm{p}}\frac{1}{a'} \tag{8.39}$$

with $r_{\mathrm{v,p}}$ and $r_{\mathrm{w,p}}$ = reaction rate per unit volume of catalyst particle (mol $m_p^{-3}\,s^{-1}$) and unit mass of catalyst particle (mol $kg_{cat}^{-1}\,s^{-1}$), V_p = volume of catalyst particle (m_p^3), A_p = external surface area of catalyst particle (m_s^2), $a' = A_p/V_p$ = volumetric external surface area ($m_s^2\,m_p^{-3}$), ρ_p = particle density ($kg_{cat}\,m_p^{-3}$). Note: the subscript p has been introduced in the reaction rate to emphasize the fact that we are dealing with one catalyst particle.

For an irreversible first-order reaction (and using the reaction rate per unit volume):

$$r_{\mathrm{v,p}} = k_{\mathrm{v,p}}\,C_s \tag{8.40}$$

Eqn. (8.39) becomes with Eqn. (8.38):

$$k_f\,(C_b - C_s) = \frac{1}{a'}k_{\mathrm{v,p}}\,C_s \tag{8.41}$$

Note that it is the concentration at the external catalyst surface which appears in the first-order rate expression and not the concentration in the bulk of the fluid phase. Solving Eqn. (8.41) for the concentration at the external surface yields:

$$C_s = \frac{k_f\,a'}{k_{\mathrm{v,p}} + k_f\,a'}C_b \tag{8.42}$$

The observed reaction rate is then expressed as:

$$r_{v,p}^{obs} = \left(\frac{1}{k_f\, a'} + \frac{1}{k_{v,p}} \right)^{-1} C_b = \left(\frac{1}{\tau_T + \tau_R} \right) C_b \tag{8.43}$$

which shows that the observed rate is determined by a physical resistance, $\tau_T = 1/k_f a'$, and by a chemical resistance, $\tau_R = 1/k_{v,p}$. τ_T and τ_R can be looked upon as the time scales for transfer and reaction, respectively. The observed rate can also be written as:

$$r_{v,p}^{obs} = \left(\frac{k_{v,p}}{1 + Da} \right) C_b \tag{8.44}$$

with

$$Da = \frac{\tau_T}{\tau_R} = \frac{k_{v,p}}{k_f a'} \tag{8.45}$$

The importance of external mass transfer limitations can now be discussed in terms of *Da*, the Damköhler number. Figure 8.17 shows the concentration profiles that are established at steady state for different values of *Da*. Large values for *Da* correspond to strong mass transfer limitations, in which case the observed reaction rate becomes:

$$r_{v,p}^{obs} = k_f a' C_b \tag{8.46}$$

i.e. the observed reaction kinetics depend only on the mass transfer and, hence, are first order. The observed first-order kinetics are obtained independently of the intrinsic reaction kinetics, i.e. even for higher reaction orders.

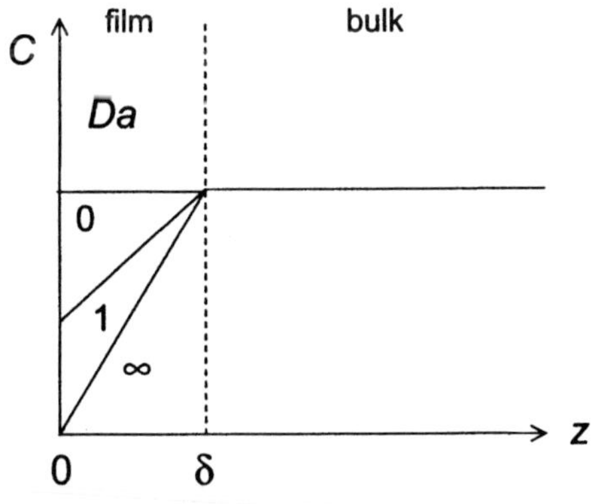

Fig. 8.17. External profiles according to the film model for different values of *Da*.

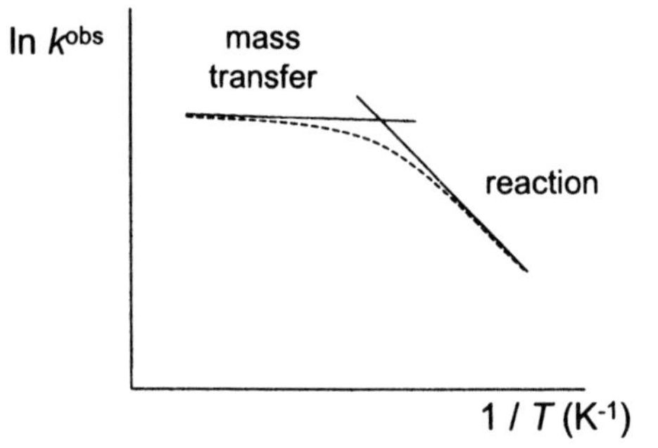

Fig. 8.18. Observed rate coefficient as a function of temperature for transport and reaction in series: external mass transfer (first-order kinetics).

It should also be noted that in this regime the amount of substance converted per unit time is proportional to be the external surface area and not to the catalyst volume, weight or internal surface area.

Small values for *Da* correspond to a situation where mass transfer limitations can be neglected:

$$r_{v,p}^{obs} = k_{v,p}\, C_b \tag{8.47}$$

i.e. the observed kinetics agree with the intrinsic kinetics, because the difference between the concentration in the bulk and at the external surface can be neglected.

Figure 8.18 qualitatively shows the temperature dependence of:

$$k_{v,p}^{obs} = \left(\frac{1}{k_f\, a'} + \frac{1}{k_{v,p}} \right)^{-1} \tag{8.48}$$

Increasing the temperature always leads to the strongly transfer-limited regime, as the mass transfer coefficient is almost independent of temperature.

For positive reaction orders the existence of significant concentration differences between the bulk of the fluid phase and the external surface of the catalyst particle leads to lower reaction rates. This can be expressed by the external effectiveness factor, η_e:

$$\eta_e = \frac{\text{observed reaction rate}}{\text{reaction rate at bulk fluid conditions}} = \frac{r_{v,p}^{obs}}{r_{v,p}(C_b)} = \frac{r_{v,p}(C_s)}{r_{v,p}(C_b)} \tag{8.49}$$

which for intrinsic first-order kinetics can be expressed as:

$$\eta_e = \frac{1}{1 + Da} \tag{8.50}$$

Equation (8.50) also holds for general kinetics, provided that *Da* is generalized to:

$$Da = \frac{\text{reaction rate at bulk fluid conditions}}{\text{maximum mass transfer rate}} = \frac{r_{v,p}(C_b)}{k_f a' C_b} \tag{8.51}$$

Both the calculation of the external effectiveness factor and of the Damköhler number requires knowledge of the intrinsic rate expression, and neither are so-called observables. Therefore, the Carberry number [6] is introduced, which only requires observation of the rate and calculation of the other variables (see also Section 8.5.3.1):

$$Ca = \frac{\text{observed reaction rate}}{\text{maximum mass transfer rate}} = \frac{r_{v,p}^{obs}}{a' k_f C_b} = \eta_e Da = \frac{C_b - C_s}{C_b} \tag{8.52}$$

Hence, small values for *Ca* correspond to the kinetic regime, whereas values of *Ca* close to one correspond to the transfer-limited regime.

External heat transfer

In practice, the effects of heat transfer from the bulk of a fluid phase to the external surface of a solid catalyst particle are more important than the effects of mass transfer. An approach analogous to that of mass transfer, for a single reactant, leads to:

$$-a' h (T_b - T_s) = r_{v,p}^{obs}(-\Delta H_r) \tag{8.53}$$

where ΔH_r = reaction enthalpy (J mol^{-1}, and h = heat transfer coefficient (W m_s^{-2} K^{-1}).

Combination of Eqn. (8.53) with Eqns. (8.38) and (8.39) leads to a relation that allows us to calculate the temperature difference from an observed reaction rate:

$$\frac{T_s - T_b}{T_b} = \frac{k_f C_b (-\Delta H_r)}{h T_b} \frac{C_b - C_s}{C_b} = \beta_e Ca = \frac{\Delta T_{e,max}}{T_b} Ca \tag{8.54}$$

in which *Ca* is the Carberry number, as defined in Eqn. (8.52) and β_e is the external Prater number. It represents the maximum relative temperature difference over the film or, in other words, the ratio of the maximum heat production and heat transfer rates. The external Prater number may reach values up to 3 (corresponding to 100–1000 K) for a gaseous fluid phase. In liquids the Prater number is much smaller.

8.4.1.2 Transport simultaneous with reaction

Internal mass transfer

(a) Effective diffusivity

For most catalysts the fraction of active sites located at the external surface of the catalyst particle can be neglected. The reactant has to be transported through the pores inside the particle to reach the active sites.

The driving force for the transport is provided by a concentration gradient: as the reactant moves further towards the centre of the particle its concentration decreases by reaction. The resistance towards transport mainly originates from collisions of the molecules, either with each other or with the pore walls. The latter dominate when the mean free path of the molecules is larger than the pore diameter. Usually both types of collision are totally random, i.e., the transport mechanism is of the diffusion type. Hence the rate of transport of a species, expressed as a molar flux in mol m_p^{-2} s^{-1}, in the case of equimolar counter-diffusion can be written as:

$$N = -D_e \frac{dC}{dz} \tag{8.55}$$

where z = particle coordinate, defined as the perpendicular distance from the centre of the particle (m_p), and D_e = effective diffusivity of a species A in the particle, m_f^3 m_p^{-1} s^{-1}.

Equation (8.55) is analogous to Fick's law. However, the effective diffusivity has to take account of the fact that the diffusion does not occur in a homogeneous medium, but only through the pores of the particle. This has two consequences. Firstly, pores occupy only a fraction, ε_p, of the particle volume. For an isotopic pore structure it can easily be shown that ε_p also equals the fraction of the area available for the diffusion towards the centre of the particle. Secondly, the driving force for diffusion is based on the concentration difference over a distance between z and (z + dz). In fact the orientation of pores deviates from the direction perpendicular to the external particle surface. This results in a diffusion path through the pores which is longer than dz, and thus, a smaller driving force. This is accounted for by introducing a so-called tortuosity factor, τ_p. These corrections lead to the following expression for the effective diffusivity:

$$D_e = \frac{\varepsilon_p D}{\tau_p} \tag{8.56}$$

Typical values for ε_p are between 0.3 and 0.6, and for τ_p between 2 and 5. In the case of equimolar counterdiffusion D is given by the Bosanquet equation:

$$\frac{1}{D} = \frac{1}{D_m} + \frac{1}{D_k} \tag{8.57}$$

where D_m = molecular diffusivity corresponding to intermolecular collisions ($m_f^2\ s^{-1}$), and D_k = Knudsen diffusivity corresponding to collisions of the molecules with the pore walls ($m_f^2\ s^{-1}$).

Correlations for the determination of D_m can be found in, e.g., Refs. [7,8]. For ideal gases the Knudsen diffusivity is given by:

$$D_k = \frac{2}{3} r_o \sqrt{\frac{8RT}{\pi M}} \tag{8.58}$$

where r_o = average pore radius for unimodal pore radius distribution (m_p), M = molecular mass of species (kg/mol).

(b) Effectiveness factor and Thiele modulus

When the diffusion of a reactant inside the particle is not fast enough to compensate for its disappearance by reaction, a decreasing concentration profile is established in the particle. For positive partial reaction orders with respect to the reactant this leads to lower reaction rates at positions away from the external surface and, hence, to a lower reaction rate when averaged over the complete particle volume. A measure for the degree of internal diffusion limitations is given by the internal effectiveness factor, η_i, defined as:

$$\eta_i = \frac{\text{reaction rate with internal diffusion limitations}}{\text{reaction rate at external surface conditions}} = \frac{\int_{V_p} r_{v,p}(C)\,dV}{V_p r_{v,p}(C_s)} = \frac{r_{v,p}^{obs}}{r_{v,p}(C_s)} \tag{8.59}$$

At steady state, the rate of disappearance of the reactant ($r_{v,p}^{obs}$) must be equal to the flux through the external surface, hence:

$$\eta_i = \frac{N|_{z=L}\, A_p}{r_{v,p}(C_s) V_p} = \frac{D_e \left.\frac{dC}{dz}\right|_{z=L}}{r_{v,p}(C_s)\,/\,a'} \tag{8.60}$$

Obviously, the internal effectiveness factor, η_i, depends on the effective diffusivity, D_e, and kinetic parameters such as the rate coefficient, $k_{v,p}$, but also on the shape of the catalyst particle.

Consider, by way of example, an irreversible first order reaction, A → B, occurring in a particle with slab geometry. As a result of the concentration profile, the reaction rate depends on the position within the catalyst particle. Hence, a mass balance for the reactant has to be made over an infinitesimal slice of the slab. At steady state this leads to:

$$N_{z+dz} - N_z = -k_{v,p} C\,dz \tag{8.61}$$

or

$$\frac{dN}{dz} = -k_{v,p}\,C \tag{8.62}$$

Substitution of Eqn. (8.55) and assuming a constant effective diffusivity yields:

$$D_e \frac{d^2C}{dz^2} = k_{v,p}\,C \tag{8.63}$$

with the boundary conditions:

$$C = C_s \quad \text{at} \quad z = L \text{ (at surface of slab)} \tag{8.64}$$

$$\frac{dC}{dz} = 0 \quad \text{at} \quad z = 0 \text{ (at centre of slab)} \tag{8.65}$$

for a slab with thickness 2 L.

Integration of Eqn. (8.63) yields the concentration profile:

$$\frac{C}{C_s} = \frac{\cosh \phi \left(\frac{z}{L}\right)}{\cosh \phi} \tag{8.66}$$

which depends on a single dimensionless number ϕ, the Thiele modulus:

$$\phi = L\sqrt{\frac{k_{v,p}}{D_e}} \tag{8.67}$$

The Thiele modulus is comparable to the Damköhler number as its square is the ratio of a diffusion (transport) resistance:

$$\tau_D = \frac{L^2}{D_e} \tag{8.68}$$

and a chemical resistance:

$$\tau_R = \frac{1}{k_{v,p}} \tag{8.69}$$

Small values of ϕ correspond to a situation where internal diffusion limitations, and hence, internal concentration gradients, can be neglected, i.e., the observed

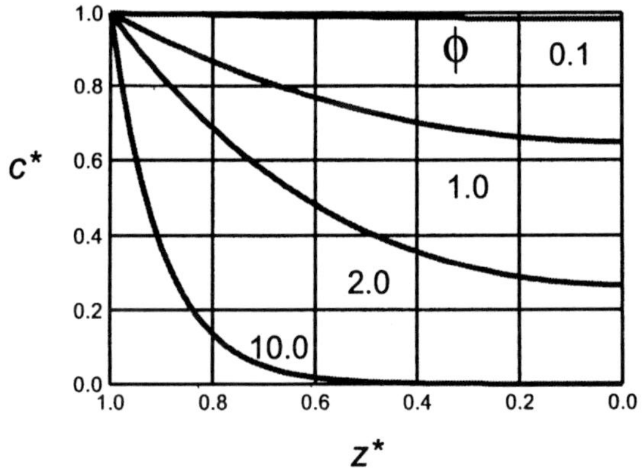

Fig. 8.19. Dimensionless reactant concentration profiles within a catalyst slab for an irreversible first-order reaction, at different values of the Thiele modulus; $C^* = C/C_s$; $z^* = z/L$.

rate agrees with the rate expected from intrinsic reaction kinetics. Small values of ϕ lead to an internal effectiveness factor, η_i, close to 1. Large values of ϕ correspond to strong internal diffusion limitations. In the extreme case, the internal diffusion is so slow that the reactants do not penetrate the particle at all. In this case the reaction is limited to the external surface of the particle. Large values of ϕ lead to an internal effectiveness factor, η_i, close to 0.

Figure 8.19 shows the dimensionless internal concentration profiles calculated using Eqn. (8.66) for different values of the Thiele modulus ($C^* = C/C_s$; $z^* = z/L$).

Knowing the exact internal concentration profile (Eqn. (8.66)) allows us to calculate the molar flux through the external surface of the slab from Eqn. (8.55), and hence the internal effectiveness factor, η_i, from Eqn. (8.60) with $1/a' = L$:

$$\eta_i = \frac{\tanh\phi}{\phi} \tag{8.70}$$

Figure 8.20 shows relation (8.70) on a double logarithmic scale. Two asymptotes can be distinguished:

$$\lim_{\phi \to 0} \eta_i = 1 \tag{8.71}$$

$$\lim_{\phi \to \infty} \eta_i = \frac{1}{\phi} \tag{8.72}$$

From Eqns. (8.72) and (8.67) it follows that for strong internal diffusion limitations, the effectiveness factor is inversely proportional to the particle diameter.

For a particle shape such as a sphere or a cylinder, expressions analogous to (8.70) can be derived using the same procedure as for a slab geometry. For the general case, the linear dimension L of the slab is replaced by:

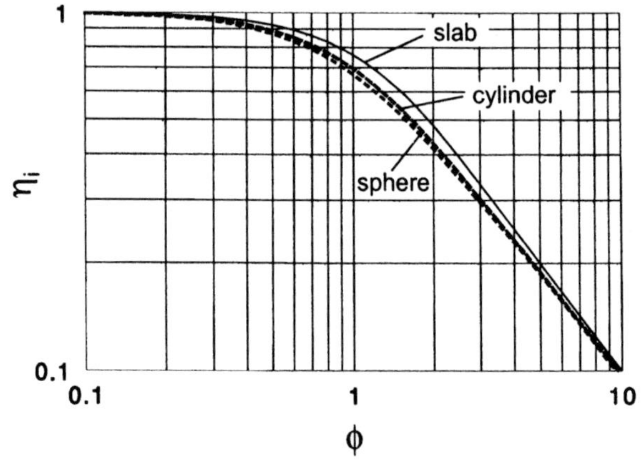

Fig. 8.20. Effectiveness factor for slab, cylinder, and sphere geometry as function of the generalized Thiele modulus.

$$L = \frac{1}{a'} = \frac{V_p}{A_p} \tag{8.73}$$

which leads to a Thiele modulus generalized for particle geometry:

$$\phi = \frac{1}{a'}\sqrt{\frac{k_{v,p}}{D_e}} \tag{8.74}$$

As shown in Fig. 8.20, the asymptotic behavior, Eqns. (8.71) and (8.72), is not affected by the particle shape due to the choice of the characteristic length L.

The above analytical relations have been derived for irreversible first-order kinetics. Arbitrary kinetics do not generally allow an analytical solution. However, a generalization can be made for arbitrary kinetics. Without going into details (see for instance Refs. [9,10]), the following result for a generalized Thiele modulus, which is independent of the reaction kinetics and defined such that the limits expressed in Eqns. (8.71) and (8.72) hold, is obtained:

$$\phi = \frac{1}{a'}\frac{r_{v,p}(C_s)}{\sqrt{2\int_{C_{eq}}^{C_s} D_e r_{v,p}(C)dC}} \tag{8.75}$$

where C_{eq} is the equilibrium concentration, which prevails at the centre plane in case of strong diffusion limitations. For an irreversible reaction C_{eq} equals 0.

Obviously, the use of Fig. 8.20 with the generalized Thiele modulus as defined in Eqn. (8.75) requires knowledge of the rate equation in order to be able to calculate the integral in the denominator. For partial reaction orders varying from one-half to three the deviations from the exact numerical solution are limited to 15%. These deviations are largest at Thiele moduli around one.

For irreversible n^{th} order reactions and assuming a constant D_e, the Thiele modulus at strongly diffusion limited conditions, Eqn. (8.75) becomes:

$$\phi = \frac{1}{a'}\sqrt{\frac{k_{v,p}}{D_e}\left(\frac{n+1}{2}\right)C_s^{n-1}} \tag{8.76}$$

Under these conditions, the observed reaction rate can be expressed as:

$$r_{v,p}^{obs} = \eta_i\, r_{v,p}(C_s) = \frac{r_{v,p}(C_s)}{\phi} = \frac{1}{a'}\left(\frac{2}{n+1}\right)^{1/2}\left(\frac{D_e\, k_{v,p}}{L^2}\right)^{1/2} C_s^{\frac{n+1}{2}} \tag{8.77}$$

This result implies that not the intrinsic kinetics are observed, but the so-called disguised or false kinetics. Consequently, the observed rate is inversely proportional to the particle size, the observed reaction order is $(n + 1)/2$, and the observed activation energy is about half the true activation energy.

During a kinetic study, one measures the observed reaction rate and not the intrinsic reaction rate. Therefore, the following combination has been introduced that yields a procurable quantity, and is referred to as the Wheeler–Weisz modulus [11,12]:

$$\Phi = \eta_i \phi^2 = \frac{\text{observed rate}}{\text{'diffusion rate'}} \tag{8.78}$$

When $\Phi >> 1$, strong diffusion limitations occur. Indeed, large values of the Wheeler–Weisz modulus can only correspond to large values of the generalized Thiele modulus, and hence, to strong internal diffusion limitations.

For an irreversible n^{th} order reaction, and assuming a constant D_e, substitution of Eqns. (8.59) and (8.76) into Eqn. (8.78) leads to:

$$\Phi = \frac{r_{v,p}^{obs}}{(a')^2 D_e C_s}\left(\frac{n+1}{2}\right) \tag{8.79}$$

Note that C_s is not an observable. However, its value may be calculated from Eqn. (8.52). Alternatively, using C_b instead is often justified.

Internal heat transfer

When heat effects are important, one has to account for heat transport, mass transport, and reaction, simultaneously. Heat transport through a catalyst particle can be described by the Fourier law:

$$q = -\lambda_e \frac{dT}{dz} \tag{8.80}$$

in which λ_e is the effective conductivity coefficient in the particle (W m_p^{-1} K^{-1}).

The steady-state enthalpy balance for a slab with thickness $2L$ is then given by:

$$-\lambda_e \frac{d^2T}{dz^2} = (-\Delta H_r)\, r_{v,p}(C,T) \tag{8.81}$$

As the rate expression $r_{w,p}$ obviously depends on the concentrations of the reaction species, a continuity equation has to be considered simultaneously with the enthalpy equation:

$$D_e \frac{d^2C}{dz^2} = r_{v,p}\,(C,T) \tag{8.82}$$

Combination of Eqns. (8.81) and (8.82) with the boundary conditions:

$$C = C_s,\ T = T_s \qquad \text{at} \quad z = L \tag{8.83}$$

$$\frac{dC}{dz} = 0,\ \frac{dT}{dz} = 0 \quad \text{at} \quad z = 0 \tag{8.84}$$

for a first-order reaction yields a simple algebraic relation between the temperature and concentration profile over a particle:

$$\frac{T - T_s}{T_s} = \beta_i \frac{C_s - C}{C_s} \tag{8.85}$$

$$\beta_i = \frac{(-\Delta H_r) D_e C_s}{\lambda_e T_s} = \frac{\Delta T_{i,max}}{T_s} \tag{8.86}$$

β_i is the internal Prater number, which represents the maximum relative temperature difference across the particle, or the ratio of the heat production rate and the heat conduction rate in the particle.

Equations (8.85) and (8.86) hold for any particle geometry. Substitution of typical values for the quantities appearing in the left-hand side of Eqn. (8.86), for exothermal reactions resulting in a typical value for β_i ranging from 0 to 0.3, leads to the conclusion that, for most practical purposes, the internal temperature gradient can be neglected.

8.4.2 Homogeneous Catalysis

Several important homogeneously catalysed reactions involve more than one phase, usually a gas phase and a liquid phase. The gaseous reactant has to be transferred from the gas phase to the liquid phase, where reaction then occurs. The liquid phase contains the homogenous catalyst together with a liquid reactant and often a solvent. The transport from one phase to the other requires a driving force in the form of concentration or temperature gradients. The latter can usually be neglected. Figure 8.21 shows the concentration profiles that are established in a direction perpendicular to the gas–liquid interface.

Several steps can be distinguished:

1. Transport of A from the bulk of the gas phase to the gas–liquid interface.
2. Transfer of A at the interface from the gas to the liquid phase.
3. Transport of A towards the bulk of the liquid.
4. Reaction of A in the liquid.

Steps 1 and 2 are strictly physical and consecutive to each other. In most cases the resistances corresponding to steps 1 and 2 can be neglected when compared to those corresponding to steps 3 and 4. This means that the liquid phase concentration of A at the interface, C_{Ai}^{1}, can be calculated from:

$$C_{Ai}^{1} = \frac{p_A}{H_A} \tag{8.87}$$

where H_A = the Henry coefficient of A (Pa $\mathrm{m_L^3}$ $\mathrm{mol^{-1}}$), and p_A = partial pressure of A in the bulk of the gas phase (Pa).

Steps 3 and 4 cannot, in general, be separated from each other: the transport away from the interface occurs simultaneously with the chemical reaction, just as with transport and reaction inside catalyst particles. The following section offers a quantitative analysis analogous to the analysis in Section 8.4.1.

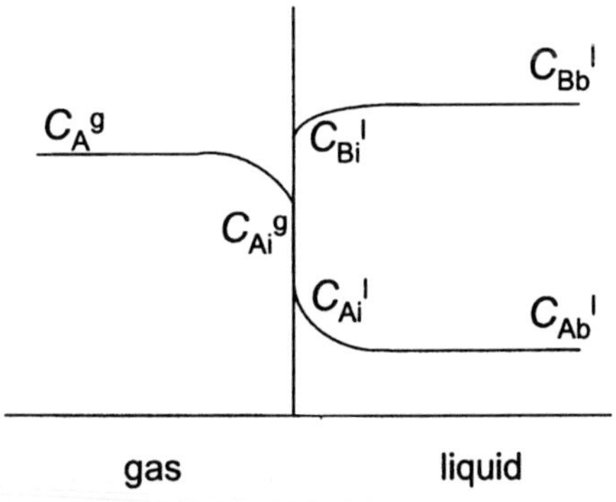

Fig. 8.21. Concentration profile of a reactant in the vicinity of a gas–liquid interface with a homogeneously catalyzed reaction in the liquid phase.

8.4.2.1 Transport simultaneous with reaction

Utilization factor

For positive partial reaction orders with respect to the reactant which has to be transferred, the existence of a concentration profile leads to a reaction rate per unit reaction volume, i.e. liquid volume, which is lower than the volumetric rate that would be obtained in the absence of concentration gradients. The extent of this slowing down of the reaction can be expressed by the effectiveness factor, also called liquid utilization factor in this context, defined for a single reaction as:

$$\eta_L = \frac{\text{reaction rate with transport limitations}}{\text{reaction rate at conditions prevailing at liquid side of interface}} \tag{8.88}$$

It follows from the above definition that for an irreversible first-order reaction, the utilization factor equals:

$$\eta_L = \frac{N_A|_{z=0}\, a_v}{k_v\, C_{Ai}} \tag{8.89}$$

where the superscript, l, has been omitted for notational convenience and with k_v = volumetric first-order rate coefficient (s^{-1}), $a_v = A_L / V_L$ = volumetric interface surface area ($m_i^2\, m_L^{-3}$).

The calculation of the molar flux of A at the interface follows from the film theory.

Film theory

In the film theory the complete resistance to mass transfer from the interface to the bulk of the liquid is confined to a fictitious film along the interface. The bulk of the fluid is considered to be perfectly mixed in the direction perpendicular to the interface, i.e. the concentration of A is uniform beyond the film. Figure 8.22 shows the concentration profile of A according to the film theory.

Furthermore, the only transport mechanism considered in the film is that of molecular diffusion perpendicular to the interface. The diffusion of A in the fictitious liquid film obeys Fick's law:

$$N_A = -D_A \frac{dC_A}{dz} \tag{8.90}$$

where D_A = molecular diffusivity of A ($m_L^2\, m_i^{-1}\, s^{-1}$).

In the absence of reaction — i.e., for purely physical transfer of A from gas to liquid — the flux is independent of z and for a constant diffusivity Eqn. (8.90) can be integrated to:

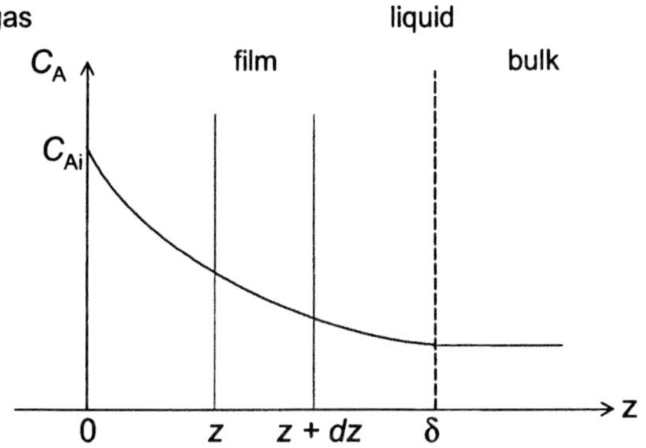

Fig. 8.22. Concentration profile of a reactant A in the liquid phase according to the film theory (δ = film thickness).

$$N_A = \frac{D_A}{\delta}(C_{Ai} - C_{Ab}) \tag{8.91}$$

The film thickness, δ, is unknown, but the mass transfer coefficient, k_L:

$$k_L = \frac{D_A}{\delta} \tag{8.92}$$

can be calculated from correlations available in the literature.

In the presence of reaction, the molar flux of A at a position z within the film follows from the derivative of the concentration profile at that position. The latter follows from a balance for A over an infinitesimal volume element, between z and $z + dz$, of the fictitious film:

$$D_A \frac{d^2 C_A}{dz^2} = -R_{v,A} \tag{8.93}$$

or, for an irreversible first-order reaction:

$$\frac{d^2 C_A}{dz^2} = \frac{k_v}{D_A} C_A \tag{8.94}$$

with boundary conditions:

$$C_A = C_{Ai} \quad \text{at} \quad z = 0 \tag{8.95}$$

$$C_A = C_{Ab} \quad \text{at} \quad z = \delta \tag{8.96}$$

The concentrations which define the boundary conditions, i.e. C_{Ai} and C_{Ab}, follow from mass balances for A in the bulk of the gas and the liquid phase. Only if these phases are perfectly mixed do C_{Ai} or C_{Ab} depend on the position in the reactor.

Integration of Eqn. (8.94) with the boundary conditions (8.95) and(8.96) yields the concentration profile over the film:

$$C_A = \frac{C_{Ai} \sinh\left[Ha\left(1 - \frac{z}{\delta}\right)\right] + C_{Ab} \sinh Ha \frac{z}{\delta}}{\sinh Ha} \tag{8.97}$$

which depends on a single dimensionless number Ha, the Hatta number, completely analogous to the Thiele modulus of Section 8.4.1.2:

$$Ha = \delta \sqrt{\frac{k_v}{D_A}} \tag{8.98}$$

It follows from Eqn. (8.92) that for irreversible first-order kinetics the Hatta number can be calculated from:

$$Ha = \frac{\sqrt{k_v D_A}}{k_L} \tag{8.99}$$

thus avoiding the use of the unknown film thickness.

Figure 8.23 shows the concentration profiles of reactant A calculated using Eqn. (8.97) for various values of the Hatta number.

Application of Fick's law (8.90) allows the calculation of the corresponding molar fluxes, e.g. at the gas–liquid interface:

$$N_A|_{z=0} = \frac{Ha}{\tanh Ha}\left(1 - \frac{C_{Ab}}{C_{Ai} \cosh Ha}\right) k_L C_{Ai} \tag{8.100}$$

Hence, for an irreversible first-order reaction, the utilization factor of the liquid phase follows from Eqn. (8.9):

$$\eta_L = \frac{1}{Sh\, Ha \tanh Ha}\left(1 - \frac{C_{Ab}}{C_{Ai} \cosh Ha}\right) \tag{8.101}$$

with:

$$Sh = \frac{k_L}{a_v D_A} \tag{8.102}$$

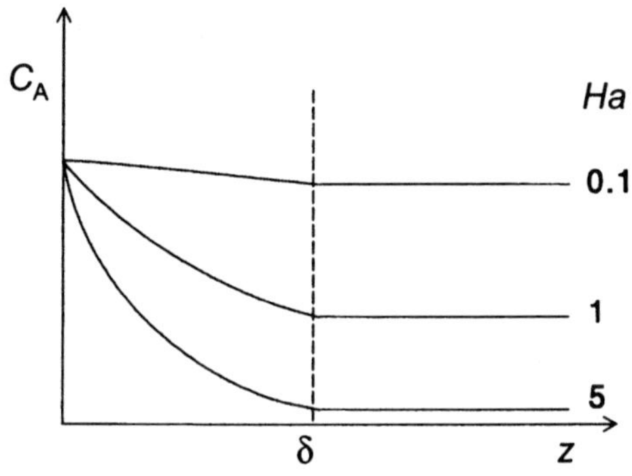

Fig. 8.23. Concentration profiles in the liquid phase with a homogeneous catalyst for irreversible first-order kinetics at different values of the Hatta number. Reactant A has to be transferred from the gas phase.

the dimensionless Sherwood number, the general physical meaning of which is the ratio of the time scale for molecular diffusion to that for the global mass transfer. In this particular situation the Sherwood number also equals the ratio of the total liquid volume, i.e. of the liquid present in the bulk and in the film, to the volume of the liquid in the film. Indeed, substitution of (8.92) in (8.102) leads to:

$$Sh = \frac{1}{a_v \delta} \frac{V_L}{A_L \delta} = \beta \tag{8.103}$$

Equation (8.101) for the utilization factor corresponds with (8.70) for the case of heterogeneous catalysis with diffusional limitations. The analogy between the two expressions is complete when $Sh = 1$, i.e. when the reaction occurs simultaneously with diffusion throughout the complete liquid volume. The presence of a Sherwood number, in addition to the Hatta number, is needed to describe situations where a significant part of the reaction occurs in the bulk of the liquid, i.e. in series with the transport through the film. Such a situation is often encountered. Typical values for the Sherwood number are:

$$10 < Sh < 5000$$

as can be concluded from the values for k_L, a_v and D_A given in Table 8.6.

Typical values for the Hatta number for industrial reactions involving homogeneous catalysis are:

$$10^{-2} < Ha < 1$$

Analogously to the Wheeler–Weisz modulus, Eqn. (8.78), a dimensionless number featuring only observable quantities can be applied to assess the influence of mass transfer limitations of a homogeneously catalysed reaction:

$$\Phi_L = \eta_L \, Ha^2 \tag{8.104}$$

TABLE 8.6

Typical values for the parameters involved in mass transfer from gas to liquid [13]

$k_L (m_L^3\, m_i^{-2}\, s^{-1})$	$5\times 10^{-5} - 5\times 10^{-4}$
$a_v\, (m_i^2\, m_L^{-3})$	$10^2 - 10^3$
$D_A\, (m_L^3\, m_i^{-1}\, s^{-1})$	$10^{-9} - 10^{-8}$

Substitution of Eqns. (8.89) and (8.99) into (8.104), for first-order kinetics, leads to:

$$\Phi_L = \frac{N_A|_{z=0}\, a_v\, D_A}{k_L^2\, C_{Ai}} \tag{8.105}$$

where indeed $N_A|_{z=0}$ is an observable, while D_A, a_v and k_L can be calculated.

Figure 8.24 shows the utilization factor versus the observable Φ_L for some values of *Sh*. In most practical situations a utilization factor close to 1 can be achieved. Then the reaction essentially occurs in the bulk of the liquid phase, and practically no reaction takes place in the film. Hence, homogeneously catalysed reactions are generally performed in reactors with a continuous liquid phase and a dispersed gas phase. The gas transfer rate can often be kept sufficiently high by simply bubbling the gas through the liquid (see Section 8.2, bubble column). For highly exothermic reactions, mechanical stirring may be used to improve heat removal from the reactor towards the internal heat exchangers or the cooling jacket.

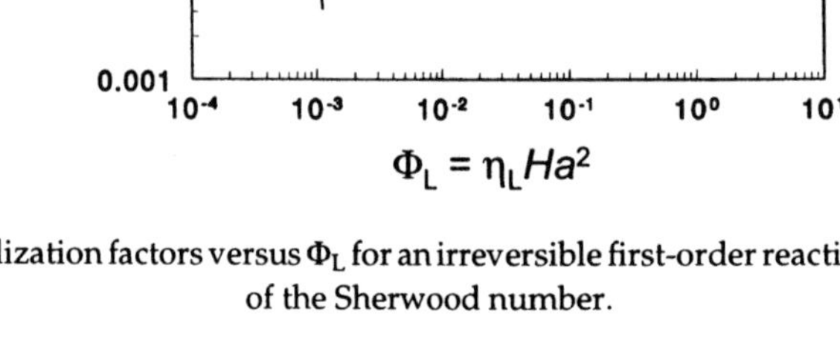

Fig. 8.24. Liquid utilization factors versus Φ_L for an irreversible first-order reaction at different values of the Sherwood number.

Multiple reactions

When multiple reactions occur, one not only has to account for decreasing reaction rates as a result of transport limitations, but also for the effects on selectivity and yields.

For the following irreversible reaction with first-order kinetics:

$$A \xrightarrow{1} Q \xrightarrow{2} P$$

in Section 8.3.3 it has been shown that in an ideal (i.e. homogeneous, and without mass transfer limitations) continuous-flow stirred-tank reactor the maximum yield of Q is given by:

$$Y_{Q,\max} = \frac{1}{\left(\sqrt{\dfrac{k_2}{k_1}} + 1\right)^2} \tag{8.37}$$

This section deals with a situation typical for homogeneously catalysed reactions. A is absorbed from the fluid phase 1 (gas) and reacts to Q in the fluid phase 2 (liquid). Q may then either desorb into the gas phase or react further to P. Application of the film model leads to the concentration profiles shown in Fig. 8.25.

Analytical expressions for the concentration profiles shown can be obtained by integrating the continuity equations for A and Q with the boundary conditions corresponding to Fig. 8.25 excluding terms for transport by convection or any other mechanism. The yield of Q then follows from:

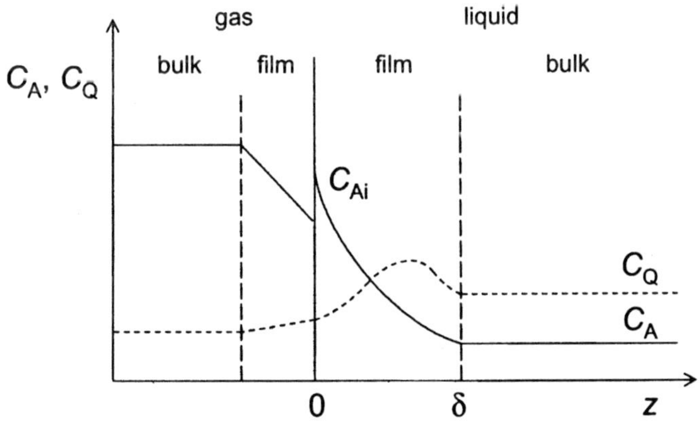

Fig. 8.25. Concentration profiles corresponding to $A \rightarrow Q \rightarrow P$ according to the film model as applied by Bridgwater [14].

$$Y_Q = \frac{D_Q \left.\dfrac{dC_Q}{dz}\right|_{z=0}}{-D_A \left.\dfrac{dC_A}{dz}\right|_{z=0}} \tag{8.106}$$

and can be shown to be determined not only by the dimensionless groups characterizing the absorption-reaction of A, Eqns. (8.99) and (8.103), but also by:

$$\lambda_1 = \frac{k_L D_Q}{k_G H_Q D_A} \tag{8.107}$$

which, for $D_Q \approx D_A$, equals the ratio of the liquid-side mass transfer coefficient to the gas-side mass transfer coefficient of Q, and:

$$C_{Qi} = \frac{p_Q|_{z=0}}{H_Q C_A} \tag{8.108}$$

i.e. the concentration of Q at the liquid side of the interface in equilibrium with the gas-phase concentration relative to the concentration of A, and finally by:

$$\lambda_2 = \frac{k_{v2} / D_Q}{k_{v1} / D_A} \tag{8.109}$$

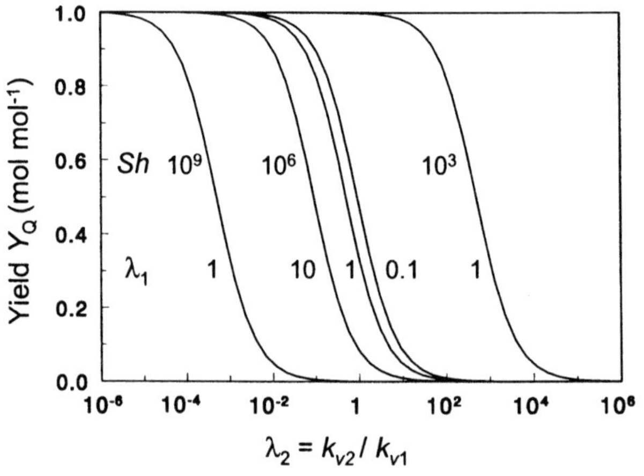

Fig. 8.26. Yield of Q versus $\lambda_2 = k_{v2}/k_{v1}$ ($A \xrightarrow{1} Q \xrightarrow{2} P$, first-order kinetics: $Ha_A = 10^{-3}$, $C_{Qi} = 0$, $D_Q/D_A = 1$) for different values of Sh and λ_1. A is transferred from the gas to the liquid phase [14].

which, for $D_Q \approx D_A$, determines the intrinsic chemical yield that can be obtained, viz. Eqn. (8.37). Bridgwater calculated the yield of Q for different combinations of Sh, λ_1 and λ_2. Figure 8.26 shows results obtained for a situation not unlikely to be encountered in homogeneously catalysed reactions involving two fluid phases.

8.5 EXPERIMENTAL DETERMINATION OF REACTION KINETICS

8.5.1 Scope

The ultimate goal of catalysis, and catalyst development in particular, is the maximization of the rate of a desired reaction. Despite considerable progress in the last few years, the formulation of new catalysts is still very much a matter of trial and error, requiring substantial experimental effort. An initial screening of candidate catalysts should result in a reliable first selection. The most promising formulations should be investigated in more detail. Finally, scaling up of the catalytic reaction towards the industrial application should be envisioned. Such a scale-up starts from the knowledge of detailed reaction kinetics, i.e. rate equation(s) describing the effects of process conditions, such as temperature and concentrations, on the reaction rate(s). Detailed reaction kinetics also allow optimization of existing processes. From a more fundamental point of view, experimental rate data can provide information which, combined with spectroscopic techniques, improves insight into the catalytic reaction mechanisms and hence provides guidelines for further improvement.

The principles underlying kinetic experiments do not depend upon the specific objectives aimed at. No matter what the purpose, accurate and easily interpretable data are required. Hence, this section relates to catalyst screening and development, as well as to detailed reaction kinetics either for design or for fundamental purposes.

8.5.2 Reactors

The selection and design of a reactor for bench-scale kinetic experiments should be considered case by case. It is important to stress, however, that one should not try to build a bench-scale replica of what is believed to be or is the industrial reactor. Industrial reactors are designed to operate a process in a profitable way, which is not the case for experimental reactors. In industrial reactors heat, mass and momentum transport has to occur in an economically justifiable way, leading in general to temperature, concentration and/or pressure gradients inside the reactor. Also, the hydrodynamics can be rather complicated. Fluidized beds, bubble columns and trickle-flow reactors require model equations that involve several physical parameters, besides the intrinsic kinetic parameters. Empirical correlations allowing the calculation of these physical parameters are available. The validity region of these correlations is often limited to the hydrodynamic conditions prevailing during industrial operation, however. Hence, the only straightforward and accurate access to rate data is provided by laboratory reactors where care has been taken that the effects of transport phenomena on the measured rates can be neglected (see also Ref. [9]). Section 8.5.3 discusses the means to ascertain the absence of transport limitations on the scale of the catalyst particle. Also, the hydrodynamics should be ideal in laboratory reactors: either plug flow or perfect mixing should be established. Laboratory reactors should be operated isothermally and isobarically. If this cannot be achieved, at least the temperature profiles in the reactor should be measured. Finally, operating with continuous flow as opposed to batch operation, allows the variation of the conversion in an independent way, with proper account being taken of catalyst deactivation.

Solid-catalyzed reactions are generally examined in continuous-flow reactors, while homogeneously catalyzed reactions (including reactions catalyzed by non-immobilized biocatalysts) are usually studied in batch reactors. An important difference between chemical catalysis and biocatalysis is that it is relatively easy to study the biocatalyst either as a homogeneous or a heterogeneous catalyst, as desired.

The following sections discuss the two commonly preferred laboratory reactors for solid-catalyzed reactions, viz., the plug-flow reactor (PFR) and the continuous-flow stirred-tank reactor (CSTR).

8.5.2.1 Plug-flow reactor (PFR)

Plug-flow reactors are often used to investigate heterogeneously catalyzed reactions. Typically 0.1–10 g of catalyst with a particle diameter smaller than 1 mm is loaded in a tube of 1 cm diameter and a few dm long. A central thermocouple well allows the measurement of the temperature inside the catalyst bed.

The assumptions underlying plug flow and the corresponding continuity equations have been given in Section 8.3.2.

For a catalytic reaction the continuity equation for a reactant provides access to the corresponding net production rate through the measurement of the reactant conversion versus the space–time W / F_{A0} followed by differentiation:

$$R_{w,A} = -\frac{d X_A}{d(W / F_{A0})} \tag{8.110}$$

The conversion can be determined by a measurement of the flow rates, Eqn. (8.24). This usually involves a chromatographic analysis of the reactor outlet stream. A complete kinetic analysis requires the determination of X_A versus

W/F_{A0} curves over the range of temperatures, total pressures and inlet compositions of interest.

The production rates obtained after differentiation are then regressed by a postulated rate equation on parameters such as preexponential factors and activation energies. This method is termed the differential method since the fit of the rate expression is tested to the data directly, without any integration. This method is useful in relatively complicated situations. However, the required differentiation of X_A versus W/F_{A0} data can be a source of errors. To avoid this, the same set of data can be analyzed by the so-called integral method, i.e., by selecting a rate equation and integrating it (Eqn. (8.26)). The integral method can only test one particular rate expression at a time, while the differential method can be used to develop a rate equation to fit the data [16].

In some cases it is possible to perform experiments at conversions low enough to neglect the effect of conversion on the rate. In other words: the rate is constant throughout the plug-flow reactor. Equation (8.110) can then be integrated to:

$$R_{w,A} = -\frac{X_A}{W/F_{A0}} \tag{8.111}$$

which shows that measurement of the conversion can give direct access to the rate. A plug-flow reactor operated in such a way that Eqn. (8.111) holds, is called a differential plug-flow reactor.

The plug-flow reactor has the following advantages for laboratory purposes:

- simple set-up, basically a tube with packed bed of catalyst particles;
- variation of W/F_{A0} by changing the amount of catalyst W or the initial molar flow rate F_{A0} allows a broad range of conversions to be covered.

However, it also has disadvantages:

- low flow rates can lead to concentration and temperature gradients over the stagnant layer surrounding the catalyst particles;
- conversions are measured rather than rates;
- care must be taken to ensure plug flow;
- isothermicity is difficult to achieve for reactions with strong heat effects.

Concerning the plug-flow requirement, the following rules of thumb can be applied for $Re_p > 10$:

$$\frac{d_t}{d_p} > 10 \tag{8.112}$$

$$\frac{L_b}{d_p} > 50 \tag{8.113}$$

where d_p = particle diameter (m_p), d_t = inner tube diameter (m_r), and L_b = length of the bed (m_r).

Fulfilment of Eqn. (8.112) means that the effect of the reactor wall on the flow pattern can be neglected. Deviation from plug flow can also occur by superposition on the plug flow of a diffusion-like transport mechanism in the axial direction. This can be neglected provided Eqn. (8.113) is satisfied. A more detailed analysis is necessary for $Re_p < 10$.

Deviations from isothermicity can be suppressed by diluting the catalyst bed with inert particles up to volumetric ratios of inert to catalytic material of 10 to 1. Larger dilution rates can cause deviations from plug flow because all the catalyst particles are no longer contacted by the reactant to the same extent (see also Ref. [9]).

Radial temperature gradients are more critical than axial gradients. In order to limit the effects of heat transport limitations on the observed rates to 5% it is sufficient that for a single reaction with the following kinetics:

$$r_w = A_0 e^{-E_a/RT} f(C) \tag{8.114}$$

the following criterion holds [15]:

$$\left| \frac{E_a}{RT_w} \frac{r_w^{obs} \rho_{cat} (-\Delta H_r)(1-\varepsilon_b)(1-b)\frac{d_t^2}{4}}{\lambda_{eb} T_w} \right| \left(\frac{1}{8} + \frac{1}{Bi_{hw}} \frac{d_p}{d_t} \right) < 0.05 \tag{8.115}$$

where $1-\varepsilon_b$ = bed fraction of the reactor volume ($m_b^3\ m_r^{-3}$); $1-b$ = catalyst particle fraction of the bed ($m_p^3\ m_b^{-3}$); T_w = internal wall temperature (K); and:

$$Bi_{h_w} = \frac{h_w d_p}{\lambda_{eb}} \tag{8.116}$$

is the thermal Biot number, the ratio of the radial heat transport resistance through the bed to the heat transfer resistance at the wall, where h_w = heat transfer resistance at the wall (W m^{-2} K^{-1}); and λ_{eb} = effective radial thermal conductivity of the bed (W m^{-1} K^{-1}).

A complete derivation of Eqn. (8.115) would take us too far from the central theme of this chapter. Basically, a perturbation approach is followed, i.e. the rate equation is linearized with respect to the temperature in the vicinity of the inner wall temperature T_w by a Taylor expansion, neglecting the higher order terms:

$$r_w(T) = r_w(T_w)\left(1 + \frac{T - T_w}{T_w} \frac{E_a}{RT_w}\right) \tag{8.117}$$

Correlations for h_w and λ_{eb} can be found in the literature. Values of 10^3 W m^{-2} K^{-1} for h_w and of 1 W m^{-1} K^{-1} for λ_{eb} are typical. Thermal Biot numbers are generally around 1. It follows from Eqn. (8.115) that much dilution with inert particles and small tube diameters favour radial isothermicity. Also, an inert with a good thermal conductivity should be used whenever possible.

8.5.2.2 Continuous-flow stirred-tank reactor (CSTR)

Perfectly mixed flow reactors are often used for kinetic studies when a liquid phase is involved. Perfect mixing, as defined in Section 8.3, is obtained by mechanical stirring. Homogeneous catalysts are continuously fed in parallel with the reactants. Catalyst recovery is not a concern during kinetic studies. Heterogeneously catalyzed reactions can be investigated by suspending the catalyst particles in the liquid phase in so-called slurry reactors (see Fig. 8.8). The particles, with a diameter that is usually smaller than 100 μm, can be maintained in the reactor by membrane filters. The heterogeneous catalyst can also be packed in a fixed bed over which the fluid stream is internally recirculated, as shown schematically in Fig. 8.27.

External recirculation, also represented schematically in Fig. 8.27, is an alternative to achieve perfect mixing, which is convenient when only gas streams are involved. The minimum external recirculation ratio depends on the reaction kinetics, but typically ratios higher than 10 are required to achieve perfect mixing [9]. Experiments performed with a CSTR give direct access to the net production rates of the species involved by solving the corresponding mass balances:

$$R_{w,A} = -\frac{X_A}{W / F_{A0}} \qquad (8.118)$$

Hence, the differential method of kinetic analysis can be applied.

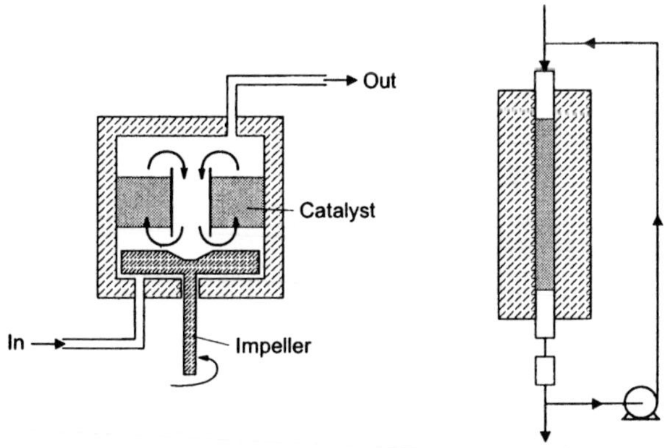

Fig. 8.27. CSTR with fixed bed and internal recirculation (Berty reactor) or external recirculation.

The CSTR has the following advantages in laboratory operation:
- isothermicity is easily achieved;
- concentration and temperature gradients over the stagnant film surrounding the catalyst particles can be neglected;
- rates are directly measured.

Disadvantages are:
- rates are measured at outlet conditions, viz. Fig. 8.11, which cannot be adjusted in a straightforward manner;
- stagnant zones in the reactor or the external recirculation lines can lead to non-ideal behaviour.

8.5.3 Transport Disguises at Particle Scale

In order to obtain accurate and easily interpretable rate data, it is not sufficient to ensure either plug flow or perfect mixing, i.e. ideality at reactor scale. For heterogeneously catalyzed reactions in particular, care has to be taken that heat and/or mass transport to or from the catalytically active site does not interfere with intrinsic rate measurement. The transport phenomena involved have been analyzed in Section 8.4. This section provides criteria to assess whether or not transport disguises occur and discusses how to avoid them.

8.5.3.1 External transfer

Assessment of external transfer effects on measured rates requires the knowledge of the corresponding transfer coefficients.

(a) Determination of mass and heat transfer coefficients

Various correlations exist for mass and heat transfer coefficients in terms of dimensionless numbers. Table 8.7 surveys the most appropriate ones for laboratory fixed-bed reactors [8,17–19]. The Sherwood number, *Sh*, and the Nusselt number, *Nu*, express the ratio of total mass transfer to diffusive mass transfer, and the ratio of total heat transfer to conductive heat transfer, respectively. Values of k_f and h for gases in laboratory systems range from 0.1 to 10 $m_f^3\ m_s^{-2}\ s^{-1}$ and 100–1000 W $m_s^{-2}\ K^{-1}$, respectively.

For laboratory slurry reactors the following correlation can be used to calculate the mass transfer coefficient [20]:

$$Sh = 2 + 0.4\, Re^{1/4}\, Sc^{1/3} \tag{8.119}$$

with *Sh* and *Sc* as defined in Table 8.7, and *Re*:

$$Re = \frac{N_p d_I^5 N_I^3 d_p^4 \rho_f^3}{\mu_f^3 V_L} \tag{8.120}$$

where: d_I = diameter of stirring blade (m); N_I = stirring frequency (s^{-1}); V_L = liquid volume (m_L^3); N_P = power number (typically 5 in laboratory slurry reactors); and d_p = catalyst particle diameter, $6 \cdot 10^{-5}$ m < d_p < 10^{-3} m_p.

A typical value for the mass transfer coefficient, k_f, is $10^{-3}\, m_f^3\, m_s^{-2}\, s^{-1}$, which is usually sufficient to allow neglect of the effects of mass transfer from the bulk of the liquid to the external particle surface on the measured rates.

The heat transfer coefficient can be calculated from the mass transfer coefficient by means of the Chilton–Colburn analogy, also called the *j* factor analogy. The *j* factors are defined as:

$$j_D = Sh\, Re^{-1}\, Sc^{-1/3} \tag{8.121}$$

$$j_H = Nu\, Re^{-1}\, Pr^{-1/3} \tag{8.122}$$

with *Nu* and *Pr* as defined in Table 8.7.

TABLE 8.7

Correlations for the calculation of mass and heat transfer coefficients in packed beds [8,17–19]

Mass transfer	Heat transfer	Range of validity
$Sh = \frac{k_f \cdot d_p}{D}$	$Nu = \frac{h \cdot d_p}{\lambda_f}$	$Re_p = \frac{\rho_f \cdot u \cdot d_p}{\mu_f}$
$Sc = \frac{\mu_f}{\rho_f \cdot D}$	$Pr = \frac{\mu_f \cdot c_{pf}}{\lambda_f}$	
Gases		
$Sh = \frac{0.357}{\varepsilon_b} \cdot Re_p^{0.641} \cdot Sc^{1/3}$	$Nu = \frac{0.428}{\varepsilon_b} \cdot Re_p^{0.641} \cdot Pr^{1/3}$	$3 < Re_p < 2000$
$Sh \approx 0.07 \cdot Re_p$	$Nu \approx 0.07 \cdot Re_p$	
Liquids		
$Sh = \frac{0.250}{\varepsilon_b} \cdot Re_p^{0.69} \cdot Sc^{1/3}$	$Nu = \frac{0.300}{\varepsilon_b} \cdot Re_p^{0.69} \cdot Pr^{1/3}$	$55 < Re_p < 1500$
$Sh = \frac{1.09}{\varepsilon_b} \cdot Re_p^{1/3} \cdot Sc^{1/3}$	$Nu = \frac{1.31}{\varepsilon_b} \cdot Re_p^{1/3} \cdot Pr^{1/3}$	$0.0016 < Re_p < 55$

In case of perfect Chilton–Colburn analogy, both factors are equal, $j_D = j_H$, and h follows from k_f:

$$h = \rho_f \, c_{pf} \left(\frac{Sc}{Pr}\right)^{2/3} k_f \tag{8.123}$$

Typical values for the heat transfer coefficient are around 10^4 W m_s^{-2} K^{-1}. Heat transfer effects rarely affect rates measured in slurry reactors.

Naturally, when one or more reactants are fed as a gas, the possible effects of the transfer from the gas phase to the liquid phase must also be considered. Correlations are available for the calculation of the corresponding transfer coefficients. Gas–liquid transfer effects can be easily diagnosed experimentally, however. The absence of gas–liquid transfer effects can be assumed if the observed specific rate and selectivities are independent of the stirring frequency and/or of the catalyst concentration in the range of conditions investigated.

(b) Criteria for the absence of transport limitations (fixed-bed reactors)

Criteria are usually derived so that deviations from the ideal situation are not larger than 5%. In order for external mass transfer limitations to be negligible, for an isothermal, n^{th} order irreversible reaction in a spherical particle, a criterion for the Carberry number can be derived, which assures that the observed rate does not deviate more than 5% from the ideal rate:

$$Ca = \frac{d_p r_{w,p}^{obs} \rho_p}{6 k_f C_b} < \frac{0.05}{|n|} \tag{8.124}$$

Substitution of the calculated value for the mass transfer coefficient in Eqn. (8.124) allows us to assess whether or not mass transfer effects can be neglected.

Figure 8.28 schematically depicts a way to determine experimentally whether external mass transfer can be neglected. Transfer limitations occur when the conversion at fixed space–time (W/F_{A0}) depends on the flow rate. The test is not very sensitive, however. This is caused by the fact that the mass transfer coefficient depends only slightly on the flow rate at the low Reynolds numbers prevailing in laboratory fixed-bed reactors.

The assessment of the heat transfer effects requires linearization of the rate expression, Eqn. (8.114) around the bulk temperature T_b:

$$r_{w,p}^{obs}(T_s) = r_{w,p}(T_b)\left(1 + \frac{T_s - T_b}{T_b}\frac{E_a}{RT_b}\right) \tag{8.125}$$

Combination of Eqn. (8.125) with the energy balance for a spherical particle:

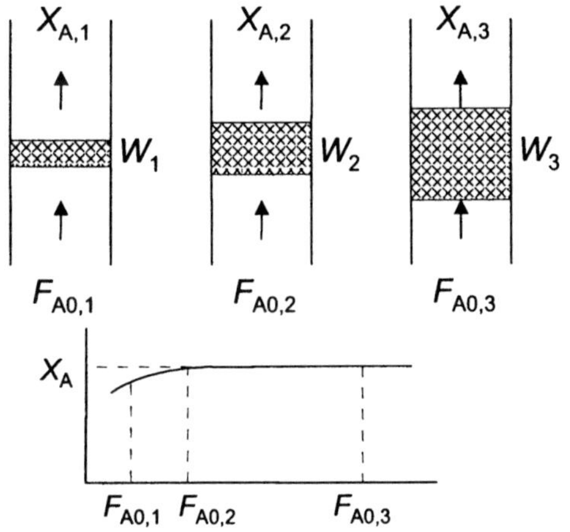

Fig. 8.28. Experimental tests to verify the absence of external mass transfer effects.

$$-\frac{6}{d_{\mathrm{p}}} h(T_b - T_s) = r_{\mathrm{w,p}}^{\mathrm{obs}}(-\Delta H_{\mathrm{r}}) \tag{8.126}$$

yields:

$$r_{\mathrm{w,p}}^{\mathrm{obs}}(T_s) = r_{\mathrm{w,p}}(T_b)\left(1 + \frac{d_{\mathrm{p}} r_{\mathrm{w,p}}^{\mathrm{obs}} \rho_{\mathrm{p}}}{6 k_{\mathrm{f}} C_b} \frac{(-\Delta H_{\mathrm{r}}) k_{\mathrm{f}} C_b}{h T_b} \frac{E_{\mathrm{a}}}{R T_b}\right) = r_{\mathrm{w,p}}(T_b)\,(1 + Ca\,\beta_{\mathrm{e}}\,\gamma_b) \tag{8.127}$$

If the deviation between the observed rate and the rate at the bulk temperature (ideal rate) has to be less than 5%, the following criterion is obtained:

$$|Ca\,\beta_{\mathrm{e}}\,\gamma_b| = \left|\frac{d_{\mathrm{p}} r_{\mathrm{w,p}}^{\mathrm{obs}} \rho_{\mathrm{p}}}{6 k_{\mathrm{f}} C_b} \frac{(-\Delta H_{\mathrm{r}}) k_{\mathrm{f}} C_b}{h T_b} \frac{E_{\mathrm{a}}}{R T_b}\right| < 0.05 \tag{8.128}$$

Note that criterion (8.128) requires knowledge of the apparent activation energy of the reaction. Criterion (8.128) holds whether internal diffusion limitations exist or not. When this criterion is compared to criterion (8.115) concerning radial heat transport limitations in the catalyst bed, it can be concluded that the latter criterion is more critical unless:

$$\frac{\gamma_b}{\lambda_{eb}} (1 - \varepsilon_b)(1 - b)\, d_{\mathrm{t}} < 5.3 \tag{8.129}$$

which is rather unusual.

8.5.3.2 *Internal transfer*

Internal diffusion limitations can in general be avoided easily in laboratory slurry reactors as the minimum particle diameter is determined by the maximum pore size of the membrane filter used to maintain the catalyst in the reactor. Membrane filters allowing the use of particles with a diameter as small as 1 μm are commercially available.

In fixed-bed reactors, however, the catalyst particle diameter has to be at least 0.1 mm in order to avoid an excessive pressure drop over the bed. With such a diameter internal diffusion effects can be important. For a 5% tolerance level, the criterion for the absence of internal diffusion limitations is:

$$\eta_i = 1 \pm 0.05 \tag{8.130}$$

However, since no intrinsic kinetics are measured, one cannot determine whether criterion (8.130) is satisfied. Introduction of the Wheeler–Weisz modulus, Eqn. (8.78), and series expansion, leads to the Weisz–Prater criterion to assess the importance of internal diffusion limitations. For an irreversible n^{th} order reaction and for a spherical particle, this criterion is:

$$\Phi = \frac{\rho_{\mathrm{p}} r_{\mathrm{w,p}}^{\mathrm{obs}} d_{\mathrm{p}}^2}{36 D_{\mathrm{e}} C_s} \left(\frac{n+1}{2} \right) < 0.15 \tag{8.131}$$

In this case the experimental determination of the rate and the knowledge of the reaction order is sufficient to assess the importance of internal diffusion limitations. Note that only $n > -1$ is meaningful.

An experimental test to verify the absence of significant concentration gradients inside the catalyst particle is based on the inverse proportional relation between the effectiveness factor and the particle diameter for strong internal diffusion limitations. Hence, a measured rate that is independent of the particle size indicates that internal diffusion limitations can be neglected. Care should be taken to avoid artefacts. External heat transfer effects also depend on particle size and for exothermic reactions might balance the internal diffusion effects. Furthermore, if the catalyst particle consists of a support with a non-uniformly distributed active phase, crushing and sieving to obtain smaller particles is hazardous.

Concerning internal temperature gradients, a criterion which guarantees that the deviation of the observed volumetric rate from the isothermal rate is less than 5% is given in an analogous way by [15, 21]:

$$|\Phi \beta_i \gamma_s| = \left| \frac{\rho_{\mathrm{p}} r_{\mathrm{w,p}}^{\mathrm{obs}} d_{\mathrm{p}}^2}{36 D_{\mathrm{e}} C_s} \left(\frac{n+1}{2} \right) \frac{(-\Delta H_{\mathrm{r}}) D_{\mathrm{e}} C_s}{\lambda_{\mathrm{p}} T_s} \frac{E_{\mathrm{a}}}{R T_s} \right| < 0.1 \tag{8.132}$$

For all practical purposes internal temperature gradients can be neglected in laboratory reactors.

NOTATION

a'	exterior surface area of particles per unit volume	$m_s^2\ m_p^{-3}$
a_v	gas–liquid interfacial surface area	$m_i^2\ m_L^{-3}$
A_p	external surface area of a catalyst particle	m_s^2
A_0	pre-exponential factor	depending on reaction
b	inert bed fraction	$m_{in}^3\ m_b^{-3}$ ($m_b^3 = m_{in}^3 + m_p^3$)
Bi_h	Biot number for heat transport ($= \frac{hd_p}{\lambda_e}$)	
c_p	specific heat capacity	$J\ kg^{-1}\ K^{-1}$
C_A	molar concentration of A	$mol\ m_f^{-3}$
C_A^*	dimensionless molar concentration of A	
Ca	Carberry number ($= \eta_e\ Da$)	
d_p	catalyst particle diameter	m_p
d_t	inner tube diameter	m_r
d_I	diameter of stirring blade	m
D_A	equimolar diffusion coefficient of A	$m_f^3\ m^{-1}\ s^{-1}$
D_{Ak}	Knudsen diffusion coefficient of A	$m_f^3\ s^{-1}$
D_{Am}	molecular diffusion coefficient of A	$m_f^3\ s^{-1}$
D_{eA}	effective diffusion coefficient of A	$m_f^3\ m_p^{-1}\ s^{-1}$
Da	Dahmköhler number $\left(= \frac{\tau_T}{\tau_R} = \frac{k_{w,p}\,\rho_p}{k_f a'}\right)$	
E_a	activation energy	$J\ mol^{-1}$
F_A	molar flow rate of A	$mol\ s^{-1}$
h	heat transfer coefficient	$W\ m^{-2}\ K^{-1}$
H_A	Henry coefficient of A	$Pa\ m_L^3\ mol^{-1}$
Ha	Hatta number $\left(= \delta\sqrt{\frac{k_v}{D_A}}\right)$	
ΔH_r	reaction enthalpy	$J\ mol^{-1}$
j_D	j-factor for mass transfer Chilton–Colburn ($= Sh\ Re^{-1}\ Sc^{-1/3}$)	
j_H	j-factor for heat transfer Chilton–Colburn ($= Nu\ Re^{-1}\ Sc^{-1/3}$)	
k_f	mass transfer coefficient (fluid film-to-particle)	$m_f^3\ m_s^{-2}\ s^{-1}$
k_G	mass transfer coefficient in gas	$mol\ m_i^{-2}\ s^{-1}\ Pa^{-1}$
k_L	mass transfer coefficient in liquid	$m_L^3\ m_i^{-2}\ s^{-1}$
k_v	volumetric rate coefficient	depends on reaction
k_w	rate coefficient per unit mass of catalyst	depends on reaction
L	length	m

L_b	length of bed	m_r
M_A	molar mass of A	kg mol^{-1}
n	reaction order	
n_A	amount of A	mol
N_A	molar flux of A	mol m^{-2} s^{-1}
N_p	power number	
N_I	stirring frequency	s^{-1}
Nu	Nusselt number $\left(= \frac{h d_p}{\lambda_f}\right)$	
p_A	partial pressure of A	Pa
p_t	total pressure	Pa
Pr	Prandtl number $\left(= \frac{c_{pf} \mu_f}{\lambda_f}\right)$	
q	heat flux	W m^{-2}
r_o	average pore radius	m_p
r_v	volumetric reaction rate	mol m_f^{-3} s^{-1}
$r_{v,p}$	reaction rate per unit particle volume	mol m_p^{-3} s^{-1}
r_w	specific reaction rate (catalyst bed)	mol kg_{cat}^{-1} s^{-1}
$r_{w,p}$	specific reaction rate (one particle)	mol kg_{cat}^{-1} s^{-1}
R	gas constant	J mol^{-1} K^{-1}
R_v	volumetric production rate	mol m_f^{-3} s^{-1}
$R_{v,p}$	production rate per unit volume of catalyst particle	mol kg_{cat}^{-1} s^{-1}
R_w	specific production rate	mol kg_{cat}^{-1} s^{-1}
Re	Reynolds number in stirred tank $\left(= \frac{N_p d_I^5 N_I^3 d_p^4 \rho_f^3}{\mu_f^3 V_L}\right)$	
Re_p	Reynolds number based on particle diameter $\left(= \frac{\rho_f u d_p}{\mu_f}\right)$	
S_A	selectivity	mol mol^{-1}
Sc	Schmidt number $\left(= \frac{\mu_f}{\rho_f D}\right)$	
Sh	Sherwood number $\left(= \frac{k_f d_p}{D}\right)$	
t	time	s
T	temperature	K
T_w	internal wall temperature	K
V	reaction volume	m^3

V_L	liquid volume	m_L^3
W	catalyst mass	kg_{cat}
X_A	conversion of A	
Y_A	yield on A	$mol\ mol^{-1}$
z	distance coordinate in one direction	m
z^*	dimensionless distance coordinate	

Greek symbols:

β	liquid phase volume/volume of diffusion layer	$m_L^3\ m_f^{-3}$
β_e	external Prater number	
β_i	internal Prater number	
γ	dimensionless activation energy (= E_a/RT)	
δ	film thickness	m_f
ε_b	void fraction of the bed	$m_f^3\ m_r^{-3}$
ε_{cat}	fraction of reactor occupied by catalyst particles	$m_p^3\ m_r^{-3}$
ε_G	fraction of reactor occupied by gas	$m_G^3\ m_r^{-3}$
ε_L	fraction of reactor occupied by liquid	$m_L^3\ m_r^{-3}$
ε_p	void fraction of catalyst particle	$m_f^3\ m_p^{-3}$
η_i	internal catalyst effectiveness factor	
η_e	external catalyst effectiveness factor	
η_L	liquid utilization factor	
λ_c	effective conductivity coefficient	$W\ m^{-1}\ K^{-1}$
λ_1	ratio of the liquid side mass transfer coefficient to gas side mass transfer $\left(= \dfrac{k_L D_Q}{k_G H_Q D_A}\right)$	
λ_2	rate constant ratio $\left(= \lambda_2 = \dfrac{k_{v2}/D_Q}{k_{v1}/D_A}\right)$	
μ	dynamic viscosity	$kg\ m^{-1}\ s^{-1}$
ν	stoichiometric coefficient	
ρ	density	$kg\ m^{-3}$
τ_0	space–time in continuous-flow reactor	$kg_{cat}\ m_r^{-3}\ s$
τ_B	batch space–time	$kg_{cat}\ m_r^{-3}\ s$
τ_D	characteristic time for diffusion	s
τ_p	tortuosity factor	
τ_R	characteristic time for reaction	s
τ_T	characteristic time for mass transfer	s

ϕ Thiele modulus $\left(= L \sqrt{\frac{k_{w,p}\rho_p}{D_e}} \right)$

Φ Wheeler–Weisz modulus ($= \eta_i \phi^2$)

Subscripts:

A,*B*,..	with respect to *A*,*B*,...
b	bulk; also bed
e	external; also effective
eq	equilibrium
f	fluid; also film
G	gas
i	interface; also internal
L	liquid
p	particle
r	reactor; also radial
s	at exterior surface
S	solid
t	tube; also total
u	surrounding
v	volume
w	mass; also wall
0	initial or inlet condition

Superscripts:

o	initial; also standard
obs	observed

REFERENCES

1 P. Trambouze, H. van Landeghem, and J.-P Wauquier, *Chemical Reactors Design/Engineering/Operation*, Editions Technip, Paris, 1988, pp. 439–475.
2 R. Krishna and S.T. Sie, *Chem. Eng. Sci.*, 49 (24A) , (1994) 4029–4065.
3 A. Cybulski and J.A. Moulijn (Eds.), *Structured catalysts and reactors*, Marcel Dekker, New York, 1998, pp. 1–14.
4 H. Scott Fogler, *Elements of Chemical Reaction Engineering*, Prentice-Hall, London, 1986, pp. 349–451.
5 G. F. Froment and K.B. Bischoff, *Chemical Reactor Analysis and Design*, Second Edition, Wiley, New York, 1990, pp. 293–304.

6 J.J. Carberry, *Chemical and Catalytic Reaction Engineering*, Mc. Graw-Hill,New York, 1976, pp. 205–208.
7 L.K. Doraiswamy and M.M. Sharma, *Heterogeneous Reactions: Analysis, Examples and Reactor Design, Vol. 1: Gas–Solid and Solid–Solid Reactions*, Wiley, New York, 1984 pp. 52–54.
8 C.N. Satterfield, *Heterogeneous Catalysis in Industrial Practice*, Second Edition, McGraw-Hill, New York, 1992, pp. 471–540.
9 F. Kapteijn and J.A. Moulijn, in: *Handbook of Heterogeneous Catalysis*, Volume 3, G. Ertl, H. Knözinger and J. Weitkamp (Eds.), VCH, Weinheim, 1997, pp. 1359–1376.
10 H.H. Lee, *Heterogeneous Reactor Design*, Butterworth, Boston, 1985, pp. 95–150.
11 J.J. Carberry, in *Catalysis: Science and Technology*, Vol. 8, J.R. Anderson and M. Boudart (Eds.), Springer-Verlag, Berlin, 1987, 131–171.
12 G. F. Froment and K.B. Bischoff, *Chemical Reactor Analysis and Design*, Second Edition, Wiley, New York, 1990, pp. 166–171.
13 P. Trambouze, H. van Landeghem, and J.-P Wauquier, *Chemical Reactors Design/Engineering/Operation*, Editions Technip, Paris, 1988, p. 195.
14 J. Bridgwater, *Chem. Eng. Sci.*, 22, 1967, 185–191.
15 D.E. Mears, *J. Catal*, 20, 1971, 127–131.
16 O. Levenspiel, *Chemical Reaction Engineering*, Second Edition, Wiley, New York, 1972, p. 42.
17 P. Trambouze, H. van Landeghem, and J.-P Wauquier, *Chemical Reactors Design/Engineering/Operation*, Editions Technip, Paris, 1988, pp. 385–438.
18 A. Cybulski, M.J. van Dalen, J.W. Verkerk and P.J. van den Berg, *Chem. Eng. Sci.*, 30, (1975) 1015–1018.
19 L.K. Doraiswamy and M.M. Sharma, *Heterogeneous Reactions: Analysis, Examples and Reactor Design, Vol. 1: Gas–Solid and Solid–Solid Reactions*, Wiley, New York, 1984 pp. 220–252.
20 Y. Sano, N. Yamaguchi and T. Adachi, *J. Chem. Eng. Jpn.*, 7, 1974, 255–261.
21 J.A. Moulijn, A. Tarfaoui and F. Kapteijn, *Appl. Catal*, 12, 1991, 1–12.

Chapter 9

Preparation of catalyst supports, zeolites and mesoporous materials

9.1 INTRODUCTION

Catalysts usually consist of two or more components in which the various constituents are assembled in the desired structure and shape. In this chapter we will consider the synthesis of a set of building blocks for the catalyst. The building blocks considered are support materials (silica, alumina and carbon), zeolites and mesoporous materials. Often the catalyst assembly process involves the addition of an active (second) component onto these building blocks. A case in point is the emplacement of a metal onto the support materials in order to obtain so-called supported catalysts. Note that the synthesis of supported catalysts is discussed separately in Chapter 10.

Support materials are vehicles for the active phase. They exercise several functions, among which are the maximization of the surface area of the active phase by providing a large area over which it may be spread and allowing the catalyst to be cast into the form of coarse bodies suitable for use in technical reactors. The active phase usually constitutes between 0.1 and 20% of the total catalyst mass and is normally in the form of very small crystallites, e.g. 1 nm Pt crystals or 50 nm Ag particles. The support is often supposed to be catalytically inactive by itself, however in partnership with the active phase it can participate in the total reaction in an important way (e.g. so-called bi-functional catalysis).

Typical substances that find wide use as high-area supports include silica gel and γ-alumina, which can be obtained with surface areas in the range 100–800

m^2/g. Materials used as low surface area supports (~ 1 m^2/g) include α-alumina and mullite (alumina–silica). These oxidic support materials are in fact ceramics with good thermal stability, which is relevant to obtain thermostable catalysts (cf. Fig. 10.1). Carbon is an attractive carrier because it is chemically inert in (aqueous) liquids, which allows one to use it in liquid-phase catalysis for fine chemicals synthesis. The support materials mentioned are dealt with in Sections 9.2–9.4.

Microporous and, more recently, mesoporous solids comprise a class of materials with great relevance to catalysis (cf. Chapters 2 and 4). Because of the well-defined porous systems active sites can now be built in with molecular precision. The most important catalysts derived from these materials are the acid zeolites. The acid site is defined by the crystalline structure and exhibits great chemical and steric selectivities for catalytic conversions, such as fluid catalytic cracking and alkane isomerization (cf. Chapter 2). In Section 9.5 we discuss the synthesis of zeolites and, briefly, of mesoporous solids.

It is not easy to make general statements about the preparation of industrial catalysts, but in many cases one can distinguish between the chemical operations to assemble the chemical components, and the fabrication steps in which the catalyst is made into the desired shape. The shaping procedures, unfortunately, are very empirical in nature. A descriptive summary of shaping methods used most frequently is given in Section 9.6.

9.2 PREPARATION OF SILICA SUPPORTS

9.2.1 Preparation of Silica Gel

Soluble silicates are commonly used as starting material. The key variables in these solutions are:

1. the alkali metal;
2. the ratio of SiO_2 to alkali metal oxide expressed as a molar ratio;
3. the concentration of solids.

Sodium silicates are produced as glasses having SiO_2:Na_2O molar ratios between 1.6 and 3.9. These are sold in lump or pulverized form, partly hydrated powders and as concentrated solutions. For the production of precipitated silicas a solution of sodium silicate with a ratio of about 3.25 is usually used, rather than lower ratios, since less acid is required for the neutralization of alkali per unit of silica, and this ratio is available at low cost [1]. In aqueous solution this sodium silicate not only contains monomeric silicate species (SiO_4^{4-}) but also dimeric ($Si_2O_7^{6-}$), trimeric ($Si_3O_{10}^{8-}$) tetrameric ($Si_4O_{12}^{8-}$) and higher oligomeric species, as for example, the double-four ring octameric species ($Si_8O_{20}^{8-}$), see Scheme 9.1.

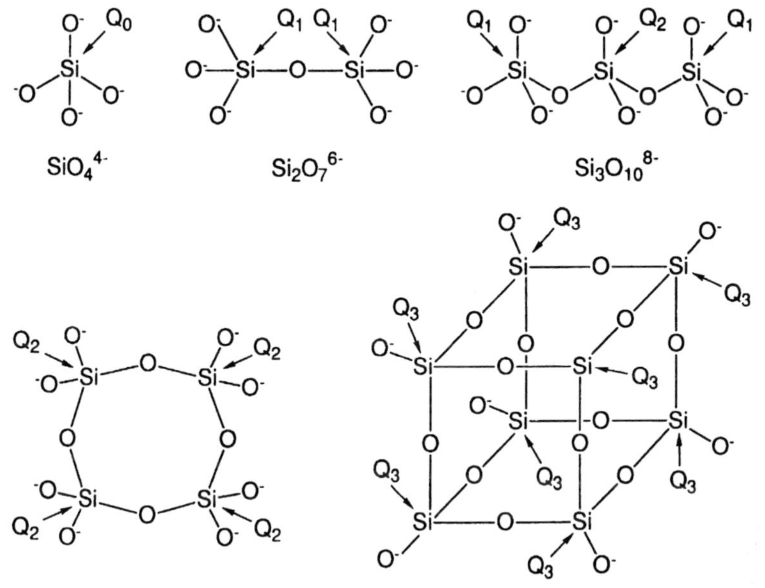

Scheme 9.1.

A powerful technique for investigating these different species is ^{29}Si-NMR (see Fig. 9.1). Different lines can be observed in the ^{29}Si-NMR spectrum of a silicate solution corresponding to the differently positioned ^{29}Si nuclei in the (poly)silicate ions. The highest values for the chemical shift ($\delta = -71.5$ ppm with respect to tetra methyl silane) are found for the monomeric silicate units while the resonances of fully condensed silicon atoms (i.e. Si–$(O–Si)_4$: tetrasiloxy silane) are to be found at the lowest values for the chemical shift ($\delta = -110$ ppm).

In order to distinguish between the different signals a nomenclature based on the number, n, of siloxane bridges (Si–$(O–Si)_n$: $n = 0, 1, 2, 3$, or 4) is introduced (Q_0, Q_1, Q_2, Q_3 and Q_4, respectively).

The formation of silica gel from these silicate solutions proceeds according to the following steps:

- Firstly, when lowering the pH of the solution by addition of an acid (usually H_2SO_4), silanol groups are formed according to the following equilibrium

$$\equiv Si–O^- + H_3O^+ \rightleftharpoons \equiv Si–OH + H_2O$$

- The silanol groups so formed have a strong tendency to polymerize in such a way that polymers are formed with a maximum of siloxane (Si–O–Si) bonds and a minimum of uncondensed Si–OH groups. The main polymerization reaction is:

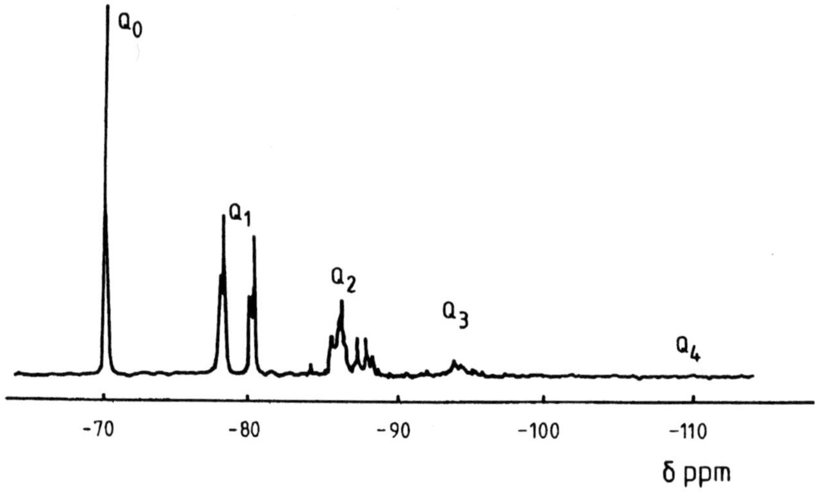

Fig. 9.1. Typical ^{29}Si-NMR spectrum of a potassium silicate solution $SiO_2/K_2O = 1.0$ [2].

$$\equiv Si{-}O^- + HO{-}Si\equiv \rightleftharpoons \equiv Si{-}O{-}Si\equiv + OH^-$$

- At the earliest stage of polymerization this leads to the formation of ring structures, the cyclic tetramer, for example, followed by addition of monomers to these and linking together of the cyclic oligomers to larger three-dimensional molecules. These condense internally to the most compact state with ≡Si–OH groups remaining on the outside.
- The resulting spherical units are, in effect, the nuclei that develop into larger particles. Because small particles are more soluble than larger ones, and since not all particles have the same size, the particles grow in average size and diminish in numbers as the smaller ones dissolve and the silica is deposited upon the larger ones (Ostwald ripening).
- Above pH 6 or 7 and up to pH 10.5, where silica begins to dissolve as silicate, the silica particles are negatively charged and repel each other. Therefore they do not collide, so that particle growth continues without aggregation. However, if salt is present at a concentration greater than 0.2–0.3 molar, as when sodium silicate is neutralized with sulphuric acid, the charge repulsion is reduced and aggregation and gelling occur.
- At low pH the silica particles bear very little ionic charge and thus can collide and aggregate into chains and then gel networks.

The successive steps in polymerization from monomer to large particles and gels have been represented schematically by Iler as in Fig. 9.2.

The silica gels obtained in this way are mixed with a solution of the salts formed by the neutralization of the alkali by the acid (usually a sodium sulphate solution). These salts can be removed by filtering and washing the gel. The silica gel

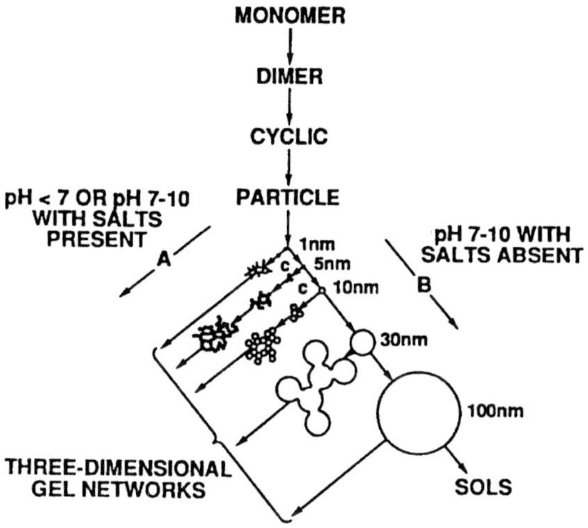

Fig. 9.2. Schematic representation of the formation of silica gel.

obtained in this way, in which the pores are filled with water, is called the *aquagel*. From this a *xerogel* is obtained when the water is removed by evaporation.

The structure of the xerogel will be compressed and the porosity reduced at least to some degree by the surface tension forces as the water is removed. An *aerogel* is a special type of xerogel from which the liquid has been removed in such a way as to prevent any collapse or change in the structure as liquid is removed. Specifically, this is done by heating the liquid-filled gel, usually an *alcogel*, in an autoclave to above the critical point of the liquid, so that there is no liquid–vapour interface, and releasing the vapour.

All silicas dried from water at less than 150°C have a fully hydroxylated surface in which the surface structure terminates in silanol groups. This is readily wettable by water and water-soluble organic molecules.

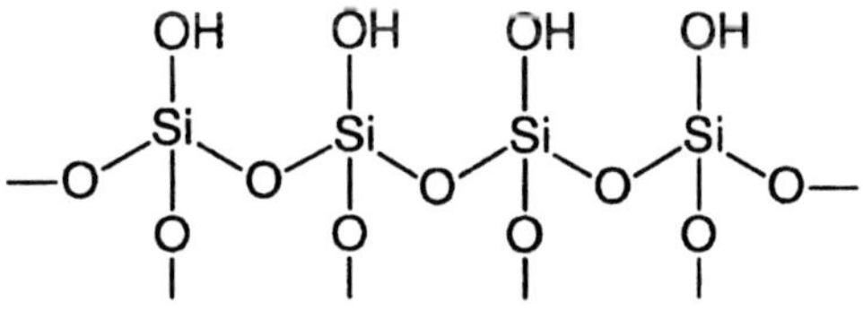

According to Böhm the concentration of ≡Si–OH groups on the silica surface is rather constant, at a value of 6.6 OH groups per nm^2 [3].

Commercial silica gels for use as catalyst bases are, for example, produced by AKZO Nobel and Davison, having specific surface areas between 100 and 800 m^2/g and pore volumes between 0.3 and 2.0 ml/g. These silica gels contain as main impurities S (sulphate), Na, Al, Fe and Ti.

9.2.2 Silica Precipitation from Vapour: Pyrogenic Silica

Very pure silica can be obtained by the oxidation or hydroxylation of purified $SiCl_4$ at high temperature.

$$SiCl_4 + O_2 \rightarrow SiO_2 + 2Cl_2$$

$SiCl_4$ can be burned in oxygen to yield SiO_2 and Cl_2, but simpler equipment can be used when $SiCl_4$ is mixed with CH_4 and burned to produce fine SiO_2 and HCl.

$$SiCl_4 + CH_4 + 2O_2 \rightarrow SiO_2 + CO_2 + 4HCl$$

The silicas made by this process are commonly known by their trademarks, 'Aerosil' produced by Degussa in Europe, and 'Cabosil' produced by Cabot in America.

The structure of these pyrogenic silicas has been discussed by Barby [4], particularly with reference to their specific surface area. It was concluded that the initially condensed particles are only about 1 nm in diameter and that these are so closely packed (high coordination number) to secondary particles of 10 to 30 nm that only a small amount of nitrogen can penetrate the micropores between them. Thus the secondary particles are the ones that are commonly identified in electron micrographs and which determine the specific surface area. They are the 'primary' particles in the voluminous aggregate structure and have a low coordination number of about 3 (see Fig. 9.3). Because of the low level of impurities this type of silica is often used as catalyst support in fundamental studies.

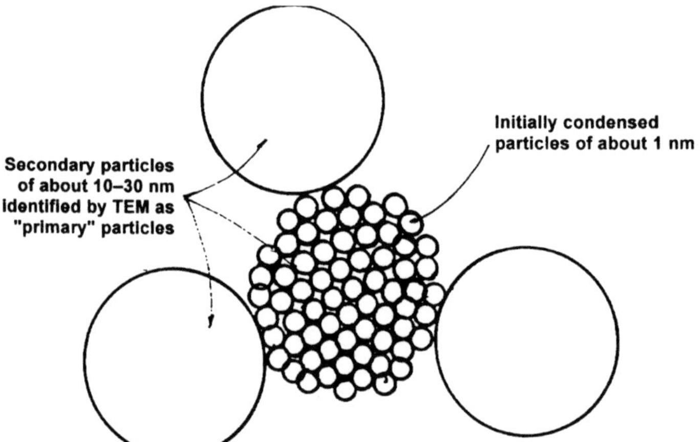

Fig. 9.3. Schematic representation of the structure of pyrogenic silica.

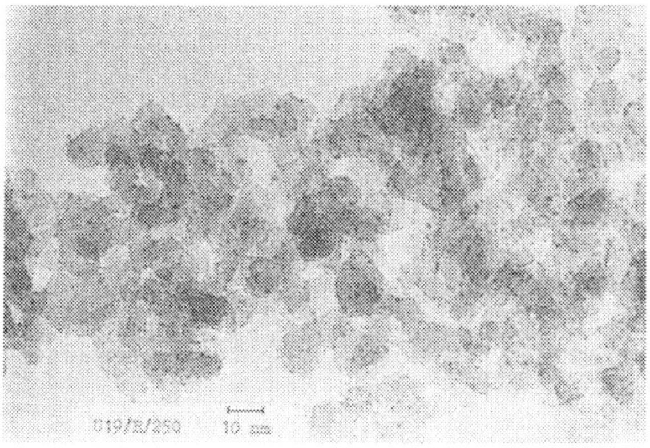

Fig. 9.4. TEM micrograph of Rh/SiO_2 containing 4%w Rh, previously reduced at 250°C [5].

The structure of a silica support is illustrated by showing a transmission electron microscopy (TEM) picture of a Rh/SiO_2 catalyst [5]. In Figure 9.4 we observe particles with medium contrast of about 10 nm which are the primary support particles with a surface area of ~300 m^2/g. The darkest, smallest dots are the Rh particles of ~1 nm size.

9.3 PREPARATION OF ALUMINA SUPPORTS

9.3.1 Preparation of γ-Al_2O_3 and η-Al_2O_3

Because of the amphoteric character of aluminium, it is soluble in both acidic and basic solvents. In acidic solutions below pH = 2 aluminium will be present as solvated Al^{3+} ions, while in basic solutions above pH = 12 it is present in the form of aluminate ions (AlO_4^{5-}).

Starting from an aqueous acidic Al^{3+} solution (for example an aluminium sulphate solution) precipitation occurs if the pH of the solution is increased above about pH = 3 by addition of a base. The first precipitate is a gel-like substance in which minute crystals of boehmite (AlO(OH)) are present. If this is filtered without aging and then calcined at temperatures up to 600°C an X-ray amorphous material is obtained. The material remains amorphous until after firing to temperatures greater than 1100°C. α-Al_2O_3 is formed at higher temperatures.

If the initial microcrystalline boehmite gel slurry is aged at 40°C it is converted into bayerite, a crystalline form of $Al(OH)_3$. If this product is filtered, dried and then calcined, a compound designated as η-Al_2O_3 is formed. Calcination at yet higher temperatures produces another compound θ-Al_2O_3, which is converted into α-Al_2O_3 at temperatures exceeding 1100°C.

The bayerite may also be converted by further aging of the slurry at higher temperatures (80°C) and pH about 8 into a much more crystalline form of boehmite, referred to as crystalline boehmite. When this precipitate is heated, after filtration and washing, it becomes another set of compounds, γ-Al_2O_3 and δ-Al_2O_3, which are very similar but not identical to η-Al_2O_3 and θ-Al_2O_3. Upon heating to temperatures higher than 1100°C, the compound γ-Al_2O_3 becomes α-Al_2O_3. If the bayerite is aged at high pH, gibbsite is formed. This is converted to χ-Al_2O_3 and κ-Al_2O_3 as its dehydrated forms, materials that are less important for the use as catalyst supports.

It is also possible to start the precipitation from a basic AlO_4^{5-} solution by addition of an acid. At pH below about 11 precipitation of a gel-like bayerite occurs, which can be treated as indicated above.

Finally it is also possible to obtain a crystalline boehmite precipitate directly, by simultaneous addition of a basic and an acidic solution to a reactor to obtain precipitation in the pH range between 6 and 8.

The different routes for the formation of the various aluminas are shown schematically in Fig. 9.5.

The important compounds for use as catalyst supports are γ-Al_2O_3 and η-Al_2O_3, particularly the former. They represent supports with high surface area (15–300 m^2/g) and high thermal stability from which the surface acidity can be controlled.

9.3.2 *Structure of γ-Al_2O_3 and η-Al_2O_3*

Both γ-Al_2O_3 and η-Al_2O_3 are members of a class of binary oxides in which the oxygen packing is of the cubic close packing type, with the cations situated partly in tetrahedral and partly in octahedral positions. These compounds are called spinels after the mineral spinel which has the composition $MgAl_2O_4$. In spinel Mg^{2+} occupies tetrahedral and Al^{3+} occupies octahedral positions. The unit cell of a cubic close packed structure is shown in Fig. 9.6.

The O^{2-} ions form a face centred cube (fcc) containing 4 O^{2-} ions. As indicated in the figure, each cube contains four octahedral positions, from which two are occupied by Al^{3+} and eight tetrahedral positions, from which one is occupied by Mg^{2+}. Schematically this can be indicated by the formula

$$\underset{\text{tetr.}}{Mg}\ \underset{\text{oct.}}{[Al_2]}\ O_4$$

in which the cations in octahedral positions are placed between brackets.

The X-ray diagrams of γ-Al_2O_3 and η-Al_2O_3 are similar to that of spinel. Because of this similarity and the presence of H atoms in the structures the formulas of the two aluminas are given formally as:

$$\gamma Al_2O_3\text{: } H_{1/2}\ Al_{1/2}\ [Al_2]\ O_4 \text{ and } \eta Al_2O_3\text{: } Al\ [H_{1/2}\ Al_{3/2}]\ O_4$$

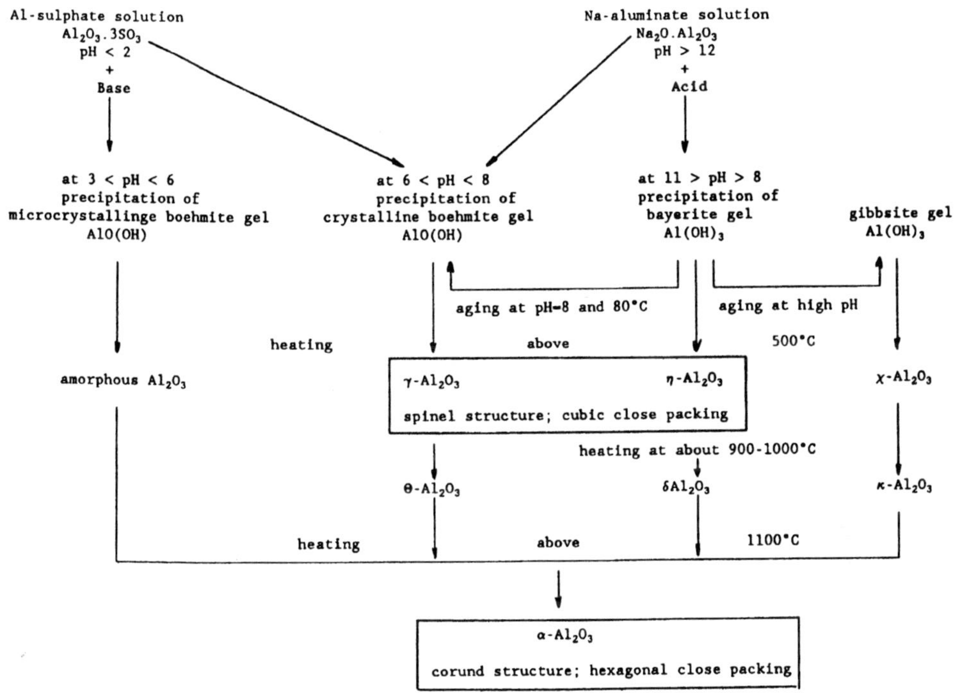

Fig. 9.5. Schematic representation of the formation of the different forms of alumina.

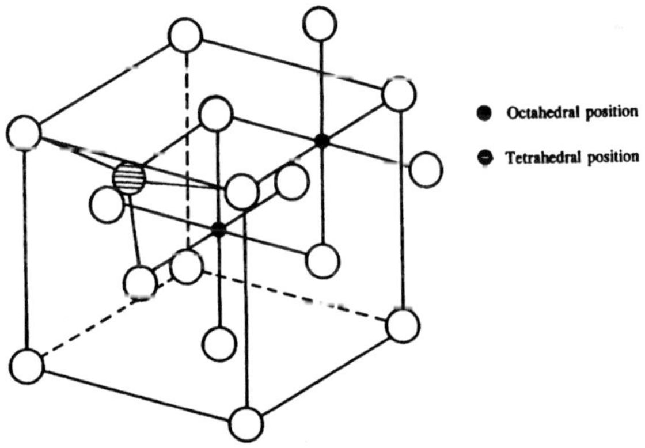

Fig. 9.6. Spinel structure.

in which some of the Al^{3+} ions occur in tetrahedral positions. In all probability the protons are not located in tetrahedral or octahedral positions, as suggested here, but at the surface in OH groups. One in every eight O^{2-} ions is therefore at the surface as OH^-, which means that the crystals are small and their surfaces consist largely of OH^- groups. These conclusions are in good agreement with the observation that the surface areas of γ-Al_2O_3 and η-Al_2O_3 are high (~ 250 m^2/g) and that these structures contain a relatively large amount of 'bonded water'.

The important structural characteristic of alumina in catalysis is the surface, and since alumina occurs in the form of lamellae it is most probable that only one type of surface plane is predominant. According to Lippens [6], this is the (111) plane for η-Al_2O_3 and the (110) plane for γ-Al_2O_3. These differences are important to the applicability of the alumina in catalytic processes; γ-Al_2O_3 appears better suited to hydrodesulphurization catalysts, whereas η-Al_2O_3 appears better suited to reforming catalysts. The η-Al_2O_3 is the more acidic of the two because of the higher density of Al^{3+} in tetrahedral sites.

9.4 CARBON SUPPORTS

Carbon supports are relevant because of their high surface area and good chemical stability in liquid media. Another advantage relates to the facile reclaim of noble metals from used catalysts, which can be achieved by combusting the carbon. Fine chemicals synthesis in the liquid phase often benefits from carbon-supported catalysts.

The most important forms of carbon are summarized in Table 9.1. We will elaborate on the two materials for catalyst support, viz. activated carbon which is commonly used and carbon fibrils, a relatively new material with potential as catalyst support.

Activated carbon — sometimes referred to as charcoal — is obtained by the pyrolysis of natural or synthetic organic polymer material. Mostly natural material is used, which can range from coconut shells to wood. Following the pyrolysis,

TABLE 9.1

Characteristics of carbon support materials with indicative figures on surface areas readily obtainable

Type of carbon	Origin	Surface area (m^2/g)	Pore texture
Activated carbon	Pyrolyis of natural materials	500-1200	micro-, meso-, macro-pores
Carbon black	Partial oxidation of hydrocarbons	10-400	meso-pores
Graphite	High-temperature treatment of carbon	0.1-300	meso- and macro-pores
Carbon fibrils *	Decomposition of CH_4 on small metal particles	100-300	meso-pores

*Often referred to as carbon nanotubes.

activation consists of controlled oxidation at, e.g., 900°C. The activation step increases the accessibility of the carbon surface. In the activated state active carbon still contains a large range of elements, such as hydrogen, oxygen, nitrogen and — important for catalysis — sulphur. Furthermore, inorganic residues are present. The natural origin of the raw material next to the presence of 'foreign elements' in the active carbon often leads to cumbersome reproducibility of the material properties. Furthermore, the very broad pore size distribution in combination with both hydrophobic and hydrophilic areas in the material, often makes metals emplacement to produce supported catalysts an art rather than a science. The favourable costs and the high surface area today nevertheless make active carbon the support of choice.

Quite recently, researchers have realized that carbon fibres grown from either methane or carbon monoxide on small metal particles may be used as an attractive support material. The growth of carbon fibres from C_1 gases on metals has been known for a long period of time. This growth is an important mechanism of catalyst deactivation in steam reforming of methane and Fischer Tropsch chemistry (cf. Chapter 5). In the late eighties and early nineties a number of research groups [7,8] realized that the controlled growth of carbon fibres provides a means of obtaining graphitic material with a unique texture. The TEM photograph in Fig. 9.7 shows a skein of carbon fibres grown from methane and an Ni/Al_2O_3 catalyst at 600°C. A high-resolution TEM micrograph of a fibre grown on nickel shows that the orientation of the graphite planes is highly organized (Fig. 9.8). This type of fibre is referred to as the fishbone or herringbone type. Using iron particles one can obtain fibres with the graphite plane parallel to the fibre axis. Carbon fibres treated with nitric acid, to enhance the number of anchoring carboxylic groups, give rise to excellent dispersions of Pd metal in a Pd/C catalyst [9].

9.5 SYNTHESIS OF ZEOLITES AND MESOPOROUS MATERIALS

9.5.1 Introduction

As zeolite structures are built from SiO_4^{4-} and AlO_4^{5-} tetrahedra these primary building blocks must also be present in the synthesis mixture and, because aluminate ions are only stable at high pH values, zeolite synthesis always takes place from basic solutions. An important factor determining the synthesis conditions is the Si/Al ratio of the zeolite to be synthesized. The higher the Si/Al ratio and so the lower the Al content the more difficult is the synthesis and the more severe the conditions that are needed. Therefore, to illustrate the different aspects of zeolite synthesis the preparation of four types of zeolites with increasing Si/Al will be described in this chapter. Successively, these will be:

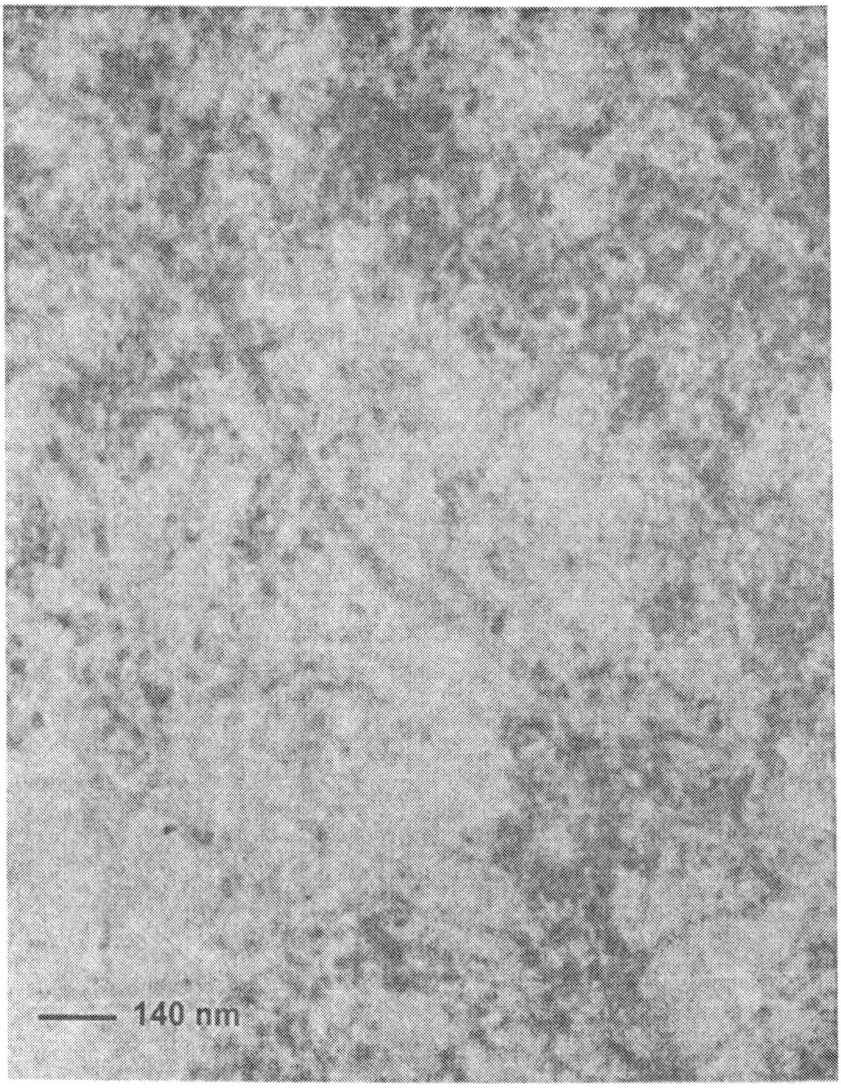

Fig. 9.7. Carbon fibres grown from methane over a Ni/Al_2O_3 catalyst at 600°C. Please note the metallic particles at the end of the fibres.

zeolite A	or	LTA	with	Si/Al = 1	
zeolite Y	or	FAU	with	Si/Al ≈ 2.5	
Mordenite	or	MOR	with	Si/Al ≈ 5	and
zeolite ZSM-5	or	MFI	with	Si/Al > 12	

However, before these specific preparation methods are described, we first discuss the subsequent events occurring in the course of a zeolite synthesis.

Generally the first step is the preparation of the reaction mixture at low temperature (< 60°C). The different ingredients are mixed in this step, which in

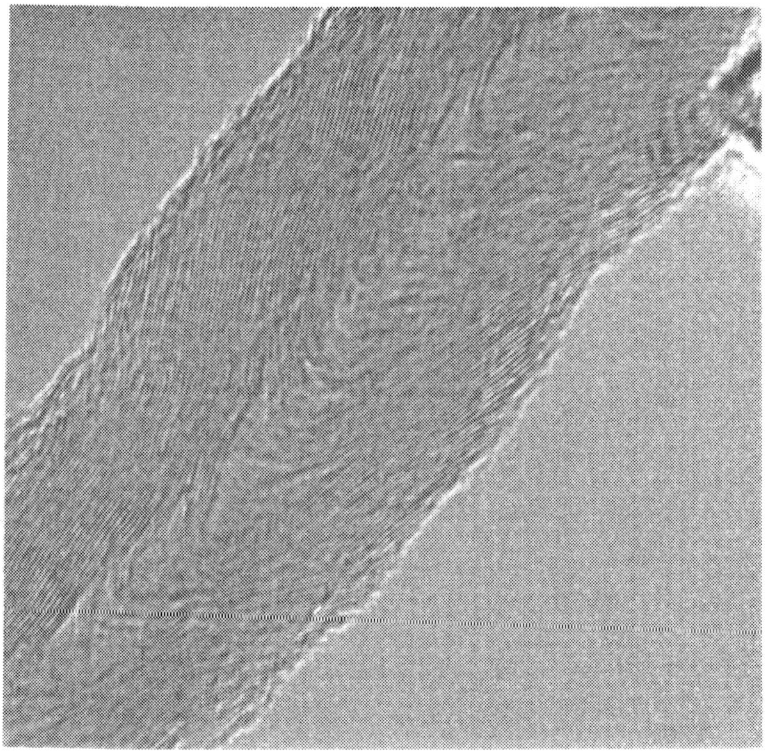

Fig. 9.8. High-resolution TEM micrograph of fishbone type carbon fibre. Note the graphite lattice planes with distances in between of 0.34 nm [Courtesy J.W. Geus].

most cases results in the formation of the so-called synthesis gel. In this gel silicate and aluminate monomers and oligomers in solution are in equilibrium with condensed silicate and aluminate units in the gel phase. In some cases a digestion period is necessary to reach this equilibrium.

In the second step the temperature of the synthesis mixture is raised to the crystallization temperature (generally between 100 and 200°C). During this process continued dissociation of silicate and aluminate oligomers occurs, resulting in an increasing concentration of monomeric silicate and aluminate species in the solution.

Once the high crystallization temperature is reached, zeolite crystallization will start after an induction period. During this induction period the dissolution of the gel phase continues, leading to the formation of clusters of polysilicates or aluminosilicates that become stable above a certain critical dimension (e.g. about 10 Å for zeolite A and about 20 Å for zeolite ZSM-5), and crystallization commences. The course of nucleation and crystallization of zeolites can generally be described by a characteristic S-shaped crystallization curve, in which the amount of crystalline material is plotted against crystallization time. Depending on the

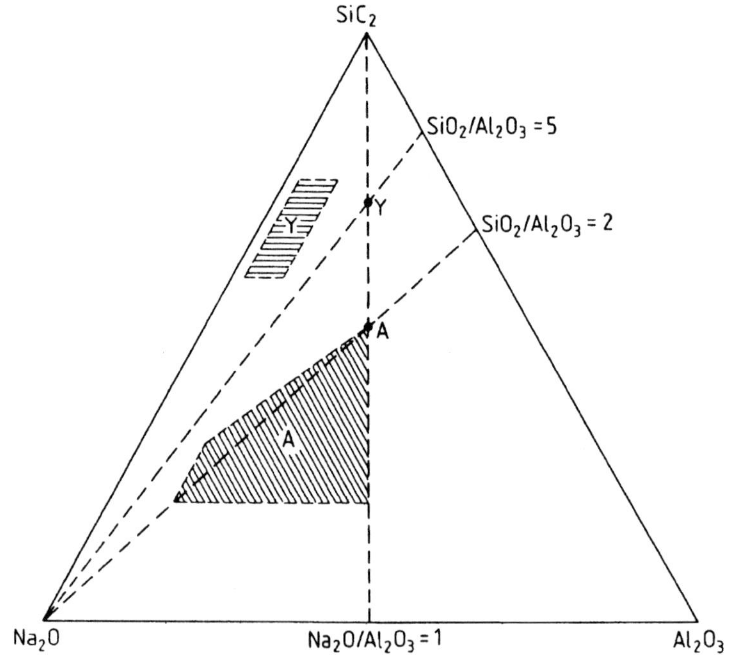

Fig. 9.9. Reaction composition diagram for the synthesis of zeolite A and zeolite Y.

type of zeolite, crystallization will be completed after several hours to several days. The crystals must then be separated from the mother liquor by decantation, filtration or centrifugation, followed by washing with water. The final step in zeolite preparation is the drying and calcination procedure, after which the zeolite void volume is free for the different applications.

9.5.2 *Synthesis of Zeolite A*

Zeolite A or LTA (Linde Type A) with a chemical composition of $Na_2O.Al_2O_3.2SiO_2$ can be synthesized from sodium aluminate–sodium silicate gels. Typical gels are prepared from aqueous solutions of $Na_2O.Al_2O_3$, $Na_2O.3.25SiO_2$ (water-glass) and NaOH. The starting reactant mixture composition can be illustrated by diagrams of the type given in Fig. 9.9. The anhydrous composition of the gel is plotted on a mole percent basis in triangular coordinates. The concentration (H_2O content) of the gel, which cannot be shown in this type of diagram, is also important, as is the nature of the starting materials.

A typical gel composition, taken from Ref.[10], is:

$$\text{mol ratio} — Na_2O : Al_2O_3 : SiO_2 : H_2O = 2.50 : 1.00 : 1.95 : 170$$
$$\text{weight ratio} — 2.5{\times}62 : 1.0{\times}102 : 1.95{\times}60 : 170{\times}18 = 155 : 102 : 117 : 3060$$

corresponding to a solids concentration of about 11%.

Crystallization is performed at 80–100°C and takes about 4–24 h, after which the zeolite can be obtained by filtering and washing with water. Typical crystallite sizes obtained by this method are between 1 and 10 μm.

9.5.3 *Synthesis of Zeolite Y*

For the synthesis of zeolite Y it is necessary to add extra SiO_2 in the form of a reactive silica gel or silica sol to the synthesis mixture [11] to obtain the right composition. A typical gel composition in this case is:

$$\text{mol ratio— } Na_2O : Al_2O : SiO_2 : H_2O = 4.00 : 1.00 : 10.0 : 160$$
$$\text{weight ratio — } 4{\times}62 : 1{\times}102 : 10{\times}60 : 160{\times}18 = 248 : 102 : 600 : 2880$$

corresponding to a solids concentration of about 25%.

The gel prepared from colloidal silica is heterogeneous on a molecular scale, containing relatively large units of silica particles with molecular weights of 1,000,000 to 500,000 and ranging in size from 10 to 20 nm. Therefore, room temperature equilibration or aging of the gel is necessary before the crystallization at 100–120°C can be started. In 2–4 days a highly crystallized product can be obtained which, after filtering and washing, can be transformed into the acidic form by ion exchange.

As zeolite Y is not stable in low pH solutions, however, it is not possible to introduce H^+ by direct exchange with Na^+ using mineral acid solutions. Generally two methods can be considered to solve this problem:

(i) Ion exchange with NH_4^+ followed by decomposition of NH_4^+ into H^+ and NH_3 by heating above about 350°C; or
(ii) Ion exchange by divalent or trivalent metal ions followed by a heat treatment to obtain hydrolysis. For this purpose rare earth ions (RE^{3+}) are usually used.

9.5.4 *Synthesis of Mordenite*

Synthesis of low silica mordenite is possible from alkaline mixtures, but in order to obtain high silica mordenite with a chemical composition of $Na_2O.Al_2O_3.10SiO_2$ it is necessary to add an organic modifier. Because of the low Al content, the zeolite framework loses its polar properties and can no longer be stabilized by the adsorption of hydrated alkali ions. Instead, positively charged organic molecules are needed which, besides charge compensation, also have nonpolar interactions with the zeolite framework. Amines are often used for this purpose, and for mordenite synthesis diethylamine (DEA) is used in particular.

A typical gel composition is:

$$\text{mol ratio — } Na_2O : Al_2O_3 : SiO_2 : DEA : H_2O = 1.00 : 1.00 : 6.60 : 20 : 300$$
$$\text{weight ratio — } 1{\times}62 : 1{\times}102 : 6.6{\times}60 : 20{\times}73 : 300{\times}18 = 62 : 102 : 396 : 1400 : 5400$$

corresponding to a solids concentration of about 7.5%. Crystallization is performed at 175°C under autogenous pressure and takes 2–4 days. After filtering and washing, the mordenite in its sodium form is obtained, which can be transformed in the acidic form by ion exchange of Na^+ with H^+, which is usually done via ammonium exchange and subsequent heating.

9.5.5 Synthesis of ZSM-5

The synthesis of ZSM-5 and other (very) high silica zeolites becomes possible when, instead of alkali hydroxides, tetraalkylammonium hydroxides are used to obtain the high pH necessary for zeolite synthesis. In the case of ZSM-5 tetrapropylammonium hydroxide must be used [12].

A typical gel composition is

$$\text{mol ratio — } Na_2O : Al_2O_3 : SiO_2 : TPAOH : H_2O = 1.25 : 1.00 : 30 : 18 : 800$$
$$\text{weight ratio—} 1.25{\times}62:1{\times}102:30{\times}60:18{\times}203 : 800{\times}18 = 77.5:102:1800:3654:14400$$

corresponding to a solid concentration of about 10%. Crystallization is performed at 150–175°C under autogenous pressure and takes 2–5 days. After filtering and washing the zeolite is obtained with the tetrapropylammonium ion still present in the zeolite framework (1 TPA per pore intersection). It is for this reason that the tetrapropylammonium ion is also indicated as template for this particular zeolite.

To remove the template it is necessary to decompose it, which can be done by heating it above about 550°C. What is obtained is ZSM-5 in which some Na^+ is still present. Because of the low alumina content the acid stability is so good that this can be exchanged in mineral acid solution to obtain H-ZSM-5. The synthesis mentioned above can also be performed in the absence of Na_2O and Al_2O_3. In that case a purely silica form of ZSM-5 is obtained which is also called silicalite 1.

9.5.6 Synthesis of mesoporous Al-MCM-41 [13,14]

The mesoporous materials discussed here comprise silicates and aluminosilicates that are formed from synthesis conditions comparable to zeolites. The main difference with the latter is the use of supra-molecular assemblies of surfactant molecules, e.g. alkyltrimethylammonium ions, as the structure directing agents. Due to the large dimensions of these spherical or cylindrical micelles, the framework of the silicates is not so well crystallized as in the case of zeolites. This causes lower framework stability as well as weaker Brönsted acidity. Upon calcination, large pores of uniform diameter (say 4–10 nm, depending on the surfactants used) become accessible.

A typical laboratory synthesis of Al-MCM-41 runs as follows [15]. A mixture is made of 200 g solution containing 26 wt% hexadecyltrimethylammoniumhydroxide, 2 g Alumina (Catapal), 100 g tetramethylammonium silicate solution (10% SiO_2 and TMA/SiO_2 =1) and 25 g fumed silica. The mixture is transferred into a teflon-lined stainless steel autoclave. The temperature is held at 150°C for 48 h. The material is recovered, washed with water and calcined at 550°C for 1 h in flowing N_2.

9.6 SHAPING OF CATALYST BODIES

9.6.1 Introduction

Forming or shaping of supports and catalysts is an important step in the preparation procedure of a commercial catalyst. The catalyst must have a prescribed shape and size for the given chemical reactor. This shape and size is determined by the application of the catalyst and the type of the reactor in which it will be used. Table 9.2 gives a rough survey of the different shapes and the reactors in which they are used.

For a fixed bed reactor it is very important that the pressure drop over the bed is as low as possible. This condition is usually fulfilled by using pellets, extrudates or spheres with a diameter greater than 3 mm. A fixed bed can have a height of ten meters or more. For this reason the catalyst particles in a fixed bed must have a high mechanical strength, otherwise the particles in the lower part of the bed will break under the weight of the upper half of the catalyst bed. In the riser reactor there is a continuous transport of catalyst spheres. Here it is necessary that these catalyst particles are as smooth and strong as possible because the constant movement can degrade the catalyst very rapidly. In the

TABLE 9.2

Different types of catalyst shapes

Shape	Size	Type of reactor
Extrudate	d = 1–5 mm l = 3–30 mm	Fixed bed reactor
Pellet	d = 3–10 mm h = 3–10 mm	Fixed bed reactor
Granule Bead	d = 1–20 mm d = 1–5 mm	Fixed bed reactor
Sphere	d = 1–5 mm	Fixed bed reactor
Sphere	d = 20–100 μm	Fluid bed reactor Slurry reactor Riser reactor

fluidized bed reactor the catalyst bed is fluidized and the catalyst particles are in constant vigorous movement during which the particles constantly impinge on each other. These particles must thus have a very high attrition resistance, otherwise fines will be formed and blown out of the reactor. Different techniques are used for shaping the catalysts into the shapes mentioned in Table 9.2. In the following paragraphs we will deal with these techniques in more detail [16–18].

9.6.2 Spray Drying

Spray drying is a technique which is used to produce spherical materials. It can be used as an intermediate process for drying and producing free-flowing powders, or as a real shaping technique producing attrition-resistant spherical particles with a diameter between 10 and 100 μm.

These particles are used in fluidized beds. Different types of spray dryers are used. (The difference is the way in which the solution is sprayed or atomized.) In a spray dryer a hydrogel or sol solution is sprayed through nozzles into a heated zone. Depending on the temperature, either drying or both drying and calcination occur as the drops fall. Two different types of spray dryers are shown in Fig. 9.10. The difference between the two types is the way in which the small droplets are formed or atomized. For example, the nozzle through which the liquid is blown may have very small holes or may be a rapidly rotating wheel. There are also spray dryers with a nozzle blowing upwards, like a fountain. The type of liquid atomizer is very important for the maximum attainable particle size and the broadness of the particle size distribution.

The spray of spherical droplets consisting of sol or gel particles will grow together, forming agglomerates as the liquid in the droplets evaporates during the process. Spray drying for the production of fluid bed catalyst will start with gel forming materials. These are materials which shrink and coalesce on drying or which are film-forming materials. After drying and calcination these materials form good attrition-resistant spheres. If non-gel-forming or non-coalescing-type materials are spray dried, gels or film formers may be used as matrices for the more crystalline materials. For example, bismuth molybdate can be included in silica gel and then spray dried. This silica gel is added in the form of colloidal silica. Another example is zeolites in a matrix of alumina gel.

The process itself is more art than science. There is hardly any information in the open scientific literature. Most publications have to do with spray drying of ceramic powders (Ref. [19] and references therein). There are also some standard books about spray drying [20]. Important process parameters are the viscosity of the liquid, the solids content of the suspension, the film-forming characteristics, the type of atomizer, the temperature, the rotation speed of the wheel, gas velocity, etc.

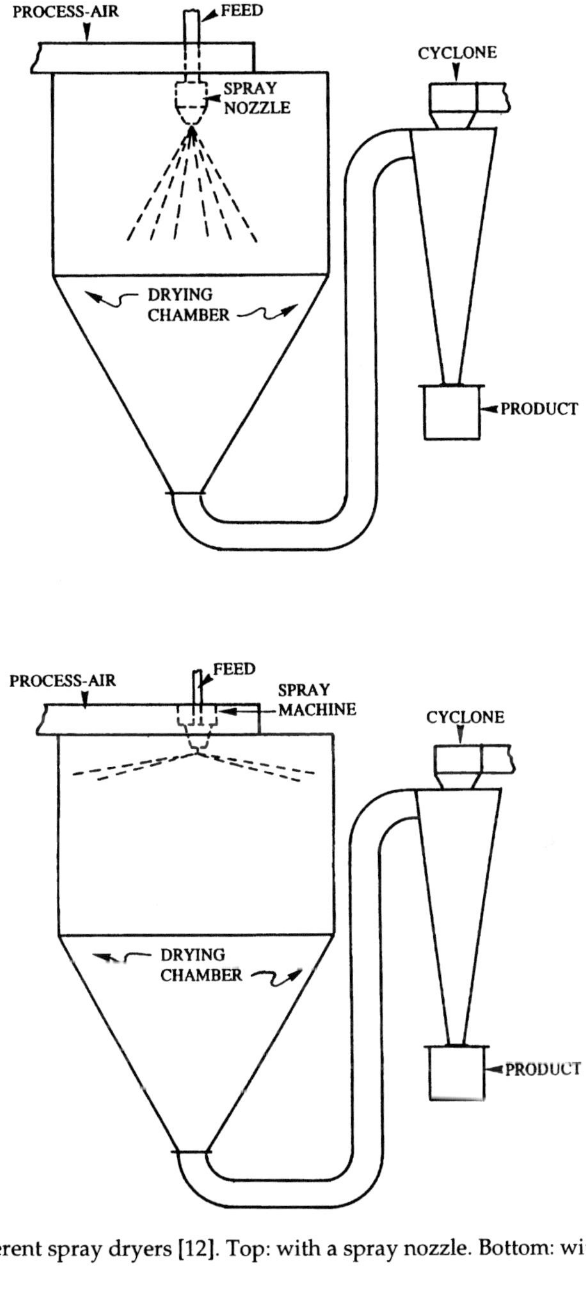

Fig. 9.10. Two different spray dryers [12]. Top: with a spray nozzle. Bottom: with a rotating wheel.

9.6.3 *Granulation*

Granules are particles with a more or less spherical shape. The diameter of these particles can be between 2–30 mm. Granulation is widely used in other fields of

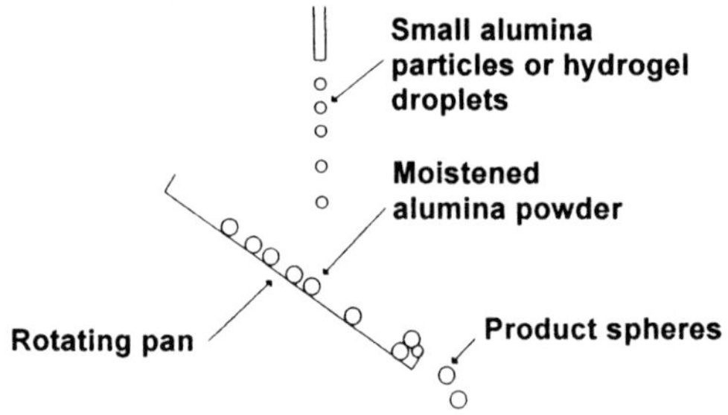

Fig. 9.11. Production of spheres using the rotating dish method [17].

technology, if spherical particles are needed. The principle of the shaping method is best described by the snowball effect. A round dish (see Fig. 9.11) is used, which is rotating about an inclined axis, the angle of inclination of which is variable. Small particles are fed into the dish. At the same time a cohesive slurry is sprayed onto the small particles. The particle surface becomes wet and granules begin to grow. There are many mechanisms for growth and which mechanism predominates will depend on many factors, such as the size, strength, plasticity and surface wetting of the granules, the degree of agitation and the amount of fines in the powder bed. The basic mechanisms are:

- nuclei creation
- nuclei growth
- coalescence of nuclei
- crushing of smaller granules
- layering of fragments onto larger granules

As a result of all these mechanisms the small particles rotating in the dish grow layer by layer and become larger, more or less spherical particles. As the process is moisture dependent, the size and strength can be controlled by controlling the rate of liquid input and by the fineness of the spray. By adjusting the inclination, granules of the desired size automatically leave the granulator dish.

There is a lot of art in this technique, too, but there are also several publications in the open literature about the principles of this shaping process. The basic mechanisms which play a role are described in the classic review by Rumpf [21]. Newitt and Conway-Jones [22] identified four types of liquid bonding mechanisms which depend on the relative amount of liquid present in the wetted powder. Rumpf calculated the strength of the bond between the particles in the wetted powder.

After drying, the green spheres are calcined to obtain a higher mechanical strength. It is most important in the process to control the parameters which are responsible for making spherical particles. Since the spheres grow layer by layer, it is important to prevent the formation of onion-like material, the layers of which can be peeled away very easily. To prevent this, binders are used to improve the adherence of the layers after calcination. In other fields of application, e.g. the production of iron ore pellets, cement is used to give a very strong bonding. In the case of catalysts the use of binders is restricted. If there is poor adherence the spheres are very weak and it will also be impossible to impregnate these spheres with a water-soluble salt.

9.6.4 *Extrusion*

Extrusion is the most commonly applied shaping technique for catalysts or catalyst supports. Many types of extrusion equipment are used, some of which are very sophisticated, but the process is in principle very simple. A wet paste of the catalyst is fed to a screw transport system and the screw forces the paste through small holes in a die plate at the end of the extruder (see Fig. 9.12). This die can have holes in the form of a circle, rings, ovals, stars, three-lobed joined rings and many other shapes, so extrudates of different shapes and sizes can be made. As the ribbon of the extruded paste emerges from the hole in the die plate it begins to dry and harden sufficiently to maintain its shape. The ribbon is either cut into prescribed lengths by a knife rotating outside the end plate or allowed to break up in small pieces of irregular length when the material is dried and calcined in a drier or calciner.

The extruder can be fed with a paste or a dry powder. In the case of a dry powder the extruder will have a kneading section where a liquid is added to the powder and the powder is kneaded to a paste. In the other case the paste is made in a separate kneader before feeding it into the extruder. The rheological properties of the paste are very important, so the formulation of the paste is the most important part of this shaping process. Peptising agents such as nitric acid are used to deagglomerate the particles in the powder. This step also helps to govern the mesoporosity of the end product [23–25]. The peptization influences the flow behaviour of the paste. The porosity of the starting powder and the end

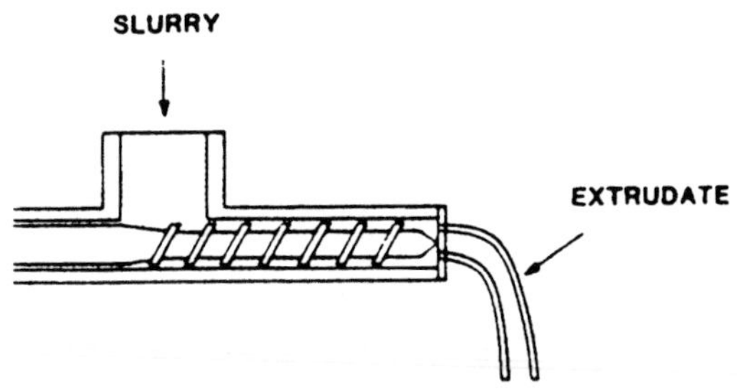

Fig. 9.12. Production of extrudates [17].

product determine the quantity of fluid which must be added to the powder. The high pressure in the extruder will diminish the pore volume of the extrudate during the extrusion process, so the pore volume of the extrudate will be less than the pore volume of the starting powder. The volume of fluid/gram of powder which is added to the powder must be equal to or slightly more than the pore volume of the end product. If it is much more there will be no ribbon coming out of the extruder but a fluid. If it is much less the paste will stick in the extruder and it will not be easy to get the dry paste out.

The shape of the holes in the die, and especially the shape of the entrance of these holes is also very important in this shaping process. The material used for the die plate like teflon, nylon and stainless steel, can also play an important role in the extrusion process.

Sometimes additives are used, e.g.
- compounds for improving the rheological behaviour of the paste;
- deflocculants to improve the peptization and to increase the content of solids;
- compounds to bind the water coming out of the pores during the extrusion;
- combustible material to increase the macroporosity.

This type of shaping is also an art and is the proprietary technology of the catalyst manufacturer. Thus little is known in the open literature. However, there are more publications about the extrusion of ceramics and clays [26].

If the extrusion is performing well the particles formed are very regular, hard and uniform. The extruder can produce great quantities of variously shaped products and, as a consequence, the extrusion process is relatively cheap in comparison with the pelletization method of making catalyst shapes. The mechanical strength is less than that of the pellets so the extrudates are less resistant to abrasion, but they do have better characteristics from the standpoint of porosity and freedom from lubricant. Furthermore, many different shapes and sizes are possible.

9.6.5 Oil-Drop Method/Sol–Gel Method

Spheronising processes for particles in the millimetre range may be classified into methods which rely on the build-up of smaller particles into spheres by means of a rolling, or 'snowball' technique (see granulation), and those which form spheres by individually shaping particles. The production costs are low for the first technique, but the spheres are irregular and there is a broad particle size distribution. The spheres manufactured by individual shaping are more expensive, but their shape and particle size are very regular and they have a higher mechanical strength. In general, catalysts or supports of this type are produced by the so-called 'oil-drop' or 'sol–gel' method; i.e. by suspending drops of an aqueous liquid in a water-immiscible liquid such as silanol or another oil so as to form spherical drops. The gelation of the drops is brought about by a change in pH.

The spheres are then aged for a sufficient time so that the drops can be withdrawn, washed, dried and calcined.

The principle of the sol–gel process is as follows. A sol is prepared by peptising a well suited alumina powder with an acid. The powder particles or agglomerates are stabilized by a positively charged layer on the surface of these particles. An increase in pH allows the hydroxyl ions to neutralize the positive ions on the surface of the particles. These particles will grow together to larger particles and gelation will occur. The sol–gel process can be carried out in several ways:

- For small spheres the sol in the spheres is gelated by an external ammonia solution through which the sol droplets fall after passing the oil layer. The gelating process in the droplets continues during the aging step to confer sufficient rigidity on the spheres [27].
- For larger spheres the sol itself contains compounds like urea or hexamethylenetetramine which decompose in the heated oil (ca 90°C) giving hydroxyl ions which will increase the pH [27,28] and cause gelation.

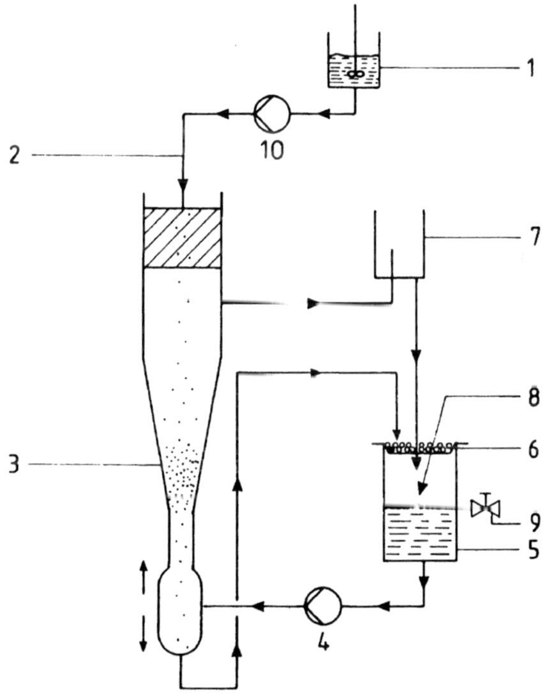

Fig. 9.13. Diagram of a set up for continuous sol–gel preparation of alumina spheres. (1) sol preparaton; (2) dropping of sol into layer of petrol ether; (3) gelation reactor filled with a NH_4OH solution; (4) circulation pump; (5) supply vessel for the gelation fluid; (6) particle separation; (7) overflow, maintaining a certain level in (3); (8) addition of fresh NH_4OH solution; (9) purge; (10) sol pump.

- Another method has been developed by Labofina [29] in which organic monomers are used which polymerize in the heated oil, giving sufficient rigidity to the spheres so that they may be handled with ease during the ensuing steps.

Sol–gel processes have been used for the preparation of silica [30] and alumina supports [31]. A description is given of a sol–gel process in Fig. 9.13 [28].

REFERENCES

1 R.K. Iler, *The Chemistry of Silica*, Wiley, 1979.
2 P.W.J.G. Wijnen, A spectroscopic study of silica gel formation from aqueous silicate solutions, Ph.D. Thesis, TU Eindhoven, 1990.
3 H.P. Böhm, *Adv. Catal.*, 16 (1966) 220.
4 D. Barby, Silicas, in: G.D. Parfitt and K.S.W. King (Editors), *Characterization of Powder Surfaces*, Academic Press, New York, 1976.
5 K.P. de Jong, J.H.E. Glezer, H.P.C.E. Kuipers, A. Knoester and C.A. Emeis, *J. Catal.* 124 (1990) 520.
6 B.C. Lippens, Ph.D. Thesis, TU Delft, 1961.
7 M.S. Hoogenraad, R.A.G.M. van Leeuwarden, G.J.B. van Breda Vriesman, A. Broersma, A.J. van Dillen en J.W. Geus in: *Preparation of Catalysts VI*, G. Poncelet et al. (Editors), Elsevier (1995), p. 263.
8 N.M. Rodriguez, A. Chambers and R.T.K. Baker, *Langmuir* 11 (1995) 3862.
9 B.L. Mojet, M.S. Hoogenraad, A.J. van Dillen, J.W. Geus and D.C. Koningsberger, *J. Chem. Soc., Faraday Trans.*, 93 (1997) 4371.
10 Henkel GmbH, Dutch Patent Application 74 06306 (1974).
11 D.W. Breck, U.S. Patent 3,130,007 (1964).
12 R.J. Argauer and G.R. Landolt, U.S. Patent 3,702,886 (1972).
13 J.S. Beck et al., *J. Am. Chem. Soc.* 114 (1992) 10834.
14 S.B. McCullen et al., in: Th.J. Pinnevaia and M.F. Thorpe (Editors), *Access in Nanoporous Materials*, Plenum Press, New York (1995), p. 1–11.
15 S. Biz and M.L.Occelli, *Catal. Rev.- Sci. Eng.* 40 (1998), 329.
16 A.B. Stiles, *Catalyst Manufacture*, Chemical Industries 14, Marcel Dekker, New York, 1983.
17 J.T. Richardson, *Principles of Catalyst Development*, Plenum Press, New York, 1989.
18 'Compaction of Particulate Solids', Course Manual, University of Bradford, 9–11 April 1980.
19 S.J. Lukasiewicz, *J. Am. Ceram. Soc.*, 72 (4) (1989) 617–624.
20 K. Masters, *Spray Drying* (4th edn.), Wiley, New York, 1985.
21 H. Rumpf, in: W.A. Knepper (Editor), *Agglomoration*, Interscience, 1962.
22 D.M. Newitt and J.H. Conway-Jones, *Trans. Inst. Chem. Eng.*, 36 (1958) 422.
23 W. Stoepler and K.K. Unger, in: G. Poncelet, P. Grange and P.A. Jacobs (Editors), *Preparation of Catalysts III*, Studies in Surface Science and Catalysis Vol. 16, Elsevier, Amsterdam, 1983, pp. 643–651.
24 K. Jiratova, L. Janácek and P. Schneider, *ibid.* pp. 653–663.
25 A. Damme and K.K. Unger, in: B. Delmon, P. Grange, P.A. Jacobs and G. Poncelet (Editors), *Preparation of Catalysts IV*, Studies in Surface Science and Catalysis Vol. 31, Elsevier, Amsterdam 1987, pp. 343–351.

26 J.J. Benbow, S.H. Jazayeri and J. Bridgewater, in: G.J. Messing, E.R. Fuller and H. Hauser (Editors), *Cer. Trans.*, Vol. IIB, Am. Cer. Soc., pp. 624–634.
27 M.E.A. Hermans, *Sc. of Cer.*, 5 (1970) 523–538.
28 H. Barton *et al.*, in: B. Delmon, P.A. Grange and G. Poncelet (Editors), *Preparation of Catalysts IV*, Studies in Surface Science and Catalysis Vol. 31, Elsevier, Amsterdam, 1987, pp. 103–111.
29 R.M. Cahen, J.M. Andre and H.R. Debus, in: B. Delmon, P. Grange, P. Jacobs and G. Poncelet (Editors), *Preparation of Catalysts II*, Studies in Surface Science and Catalysis Vol. 3, Elsevier, Amsterdam, 1979, pp. 585–594.
30 T.G. Spek, Eur. Patent 0067459 (1982).
31 M.A. Day, Eur. Patent 0115927 (1984).

Chapter 10

Preparation of supported catalysts

10.1 INTRODUCTION

The preparation of supported catalysts is, as the saying goes, still much more an art than a science. Usually, after the catalyst for a particular purpose has been identified, its manufacture is optimized through varying experimentally easily accessible parameters (pH, *T*, loading of the active phase, additives) in a previously established basic recipe. It is not that the fundamental chemistry involved is not known, but the best catalyst-preparation recipes are generally so complicated, chemically speaking, that no precise description of the reactions taking place during the manufacturing process can be given. Nevertheless, the scientific basis of catalyst preparation has received quite a lot of attention in recent years, and in a few cases a rather detailed understanding of the various steps in the preparation procedure has been achieved. Since the aim of the present work is to emphasize molecular aspects wherever possible, it is these cases that will receive most attention, more than perhaps would be warranted on the basis of their (industrial) importance.

It is worth mentioning again that supported catalysts are often applied, because they combine a relatively high dispersion (amount of active surface) with a high degree of thermostability of the catalytic component, cf. Fig. 10.1. The support, which is itself usually not catalytically active (but it may be), is a thermostable, highly porous material onto which the active component is applied. Frequently used support materials are γ-Al_2O_3, SiO_2, active C, and, to a lesser extent, TiO_2. Supported catalysts can be prepared basically in two ways: (i) selective removal of a component from a non-porous phase containing (a precursor of) the active component(s) and the support, e.g. a coprecipitate; and (ii)

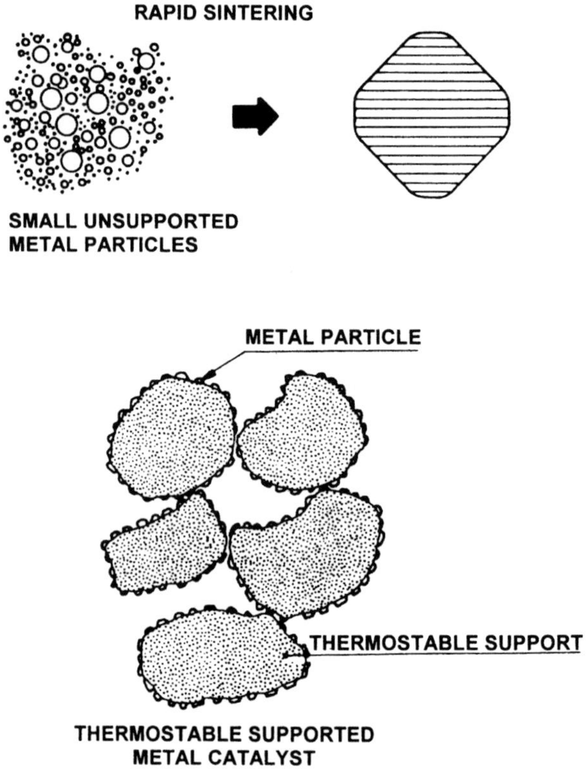

Fig. 10.1. A thermostable carrier prevents sintering of the active material.

separate application of (a precursor of) the catalytically active material(s) onto a pre-existing support, e.g. by impregnation or precipitation. Which one will be chosen in a particular case depends, among other things, on the loading one wants to achieve: when applying cheap metals and oxides one typically strives for maximum active surface area per unit volume, in which case the selective removal strategy may well be advantageous; however, the very expensive noble metals are employed only at low loadings, where the aim is to prepare very small particles having almost all their atoms at the surface (maximum utilization), and impregnation is then the method of choice. In this chapter, we will first discuss route (i), and then go on to discuss route (ii). In the latter case, the interaction of the catalyst precursor and the support surface turns out to be of crucial importance. To understand this interaction one needs some understanding of the surface chemistry of support materials, and this is provided in Section 10.3.1. We note in passing, that some attempts have recently been made to prepare supported catalysts via direct sol-gel synthesis, but since this area is still in its infancy, it will not be discussed here.

10.2 SELECTIVE REMOVAL

Many industrially used, high-loading supported catalysts are produced by selective removal of mainly water and oxygen, and, sometimes, carbon dioxide, from a non-porous precursor material. An old, but still very important example is the ammonia synthesis catalyst. The starting material of this catalyst is a melt of magnetite (Fe_3O_4), alumina, and potassium oxide produced in an electric furnace in a non-oxidising atmosphere at about 1500°C. Utilization of magnetite, or a corresponding mixture of iron (II) and iron (III) oxides, is required to dissolve the alumina. The alumina and potassium oxide content is low, viz. of the order of 2 to 5 wt.%. After a (difficult) crushing to particles (granules) of the required size, the catalyst is reduced and passivated. The selective removal of oxygen during reduction leads to highly porous, mechanically strong particles. The high iron content causes the iron surface area per unit volume to be high; but the relatively narrow pores in the resulting catalyst lead to diffusion limitations. Collapse of the structure during reduction and at reaction temperatures (400–500°C) is prevented by the alumina acting as a spacer. Here, the alumina, due to its minor presence, is not referred to as a support, but as a structural promoter. Some authors assume that K_2O is reduced to metallic K, which facilitates adsorption of N_2. Other authors feel that K_2O is not reduced during ammonia synthesis. The strongly basic K_2O is required to desorb ammonia at elevated pressures, which accounts for the fact that K_2O is only required at pressures above about 70 bar. Although the production of this catalyst is rather expensive, its excellent properties have ensured it a dominant position ever since its invention in about 1910.

Other catalysts used in large-scale processes are produced starting from coprecipitates. Instances are the Ni/alumina (methane) steam reforming catalyst and Cu(oxide)–Zn oxide/alumina catalysts applied in the low-temperature CO shift reaction and in methanol synthesis. The ions of the active component(s) and of the support are precipitated together as hydroxides or hydroxy salts (e.g. hydroxy carbonates), using NaOH or $NaHCO_3$ as the precipitating agent. After filtering and drying, the solid is processed into pellets, calcined and reduced (alternatively the solids are pelletized after calcination). It is essential that the ions of the active material and the support are intimately mixed already in the precipitation step to achieve a homogeneous distribution of the active particles in the final catalyst, i.e. they have to precipitate simultaneously. Usually this requires variations of pH to be prevented, since this would result in the preferential precipitation of one component. Interestingly, with nickel and aluminium ions coprecipitation can result in a hydrotalcite, a mixed basic carbonate, while copper and zinc ions can also react to form a hydrotalcite. Indeed, strictly speaking, real coprecipitation only occurs when a `double salt', i.e. a material with fixed stoichiometry and characterized by a single solubility product, is produced; most often one has simultaneous precipitation and co-aggregation. Removal of CO_2 and water during calcination and of oxygen during reduction leads to the porous catalyst. In the Ni/Al_2O_3 case the calcination temperature should not be too high to prevent spinel formation, which would reduce the reducibility of the nickel component drastically.

It is fair to say, however, that this manufacturing route is not easy to control. The extent of mixing of the constituents depends on the details of the precipitation procedure (order and mode of addition of the different solutions, temperature, aging period of precipitate in the mother liquor, filtering and washing procedures). The size of the active particles, their distribution, as well as the porous structure and the mechanical strength of the finished (reduced) catalyst pellets vary considerably with the extent of mixing of the components established in the precipitation step. Scaling-up of the coprecipitation reaction, asking for the efficient mixing of large volumes of solutions, is hard. Also, the structure of the solid is profoundly changed during calcination and reduction. The properties of the resulting catalyst depend on variables such as water vapour pressure and the local thermal history. With large volumes of catalyst to be thermally treated, it is difficult to control these variables and to prevent considerable variations over the catalyst bed. Production of supported catalysts by coprecipitation is analogous to building and pushing over a brick wall, the pushing over being a metaphor for the drastic change in the structure of the solid during the removal of water, carbon dioxide and oxygen. While it is already difficult to build the wall reproducibly by coprecipitation, it is well nigh impossible to push it over in an entirely controlled way. Nevertheless, it is often not possible to establish a reasonably uniform, dense distribution of active particles by other methods, while the intimate interactions fostered during coprecipitation may result in higher activity and/or stability of the active metal crystallites, and several industrially important catalysts are still produced from coprecipitates.

10.3 APPLICATION ON A SEPARATELY PRODUCED SUPPORT

In this section we will discuss extensively the processes of precipitation and impregnation. The phenomena of adsorption and ion-exchange will be subsumed under the latter heading. Each preparation route has its advantages and disadvantages, and often a preference for one technique over the other is a matter of compromise. The active component can be applied on powder supports or on preshaped porous carrier bodies, e.g. extrudates. In the former case, pore-diffusion limitations that would give uneven active-phase distributions are virtually absent, but in the latter case one has better control of pore-size distribution and handling is easier. Active components deposited in large amounts onto powders change their mechanical and surface properties, so that subsequent pelleting or extrusion may be more difficult. In the laboratory, deposition onto powders is often practised, but commercial manufacturers find preshaping on the whole more practical.

10.3.1 Support Surface Chemistry

10.3.1.1 Oxides

The surface of an oxide offers an approaching active-phase precursor various sites with which to interact, viz. different types of hydroxyl groups, roughly classified as acidic, neutral and basic, and coordinatively unsaturated (c.u.s.) metal sites (Lewis-acid centres). The OH groups can be distinguished with IR spectroscopy: the lower its OH stretching frequency, for a given oxide, the more acidic its character. Some representative IR spectra (O–H stretch region) of the more important oxidic support materials γ-Al_2O_3, SiO_2 and TiO_2, are shown in Fig. 10.2. Roughly speaking, one can distinguish in the γ-Al_2O_3 OH spectrum three band systems, which can be assigned to specific surface OH configurations: The high-frequency band is due to structures of type I, while the middle and low-frequency bands are due to structures of types II and III, respectively.

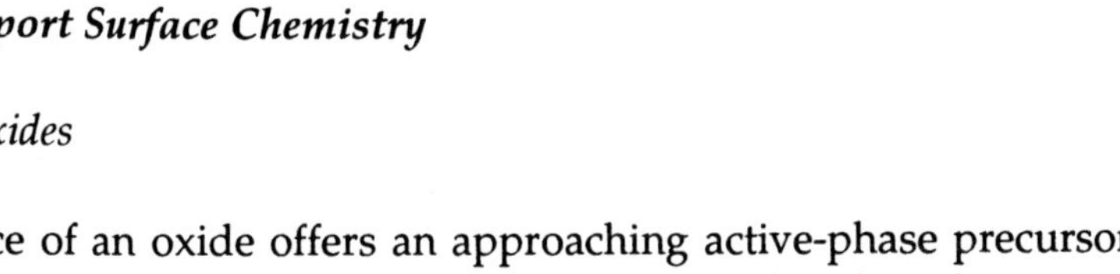

Similarly, for TiO_2, whose spectrum consists of two band systems, the high-frequency one corresponds to basic, i.e. single-bonded, hydroxyl groups, while the other corresponds to more acidic bridged hydroxyls. The anionic character of the type I hydroxyls on alumina and titania is exemplified by the fact that they preferentially exchange with fluoride ions:

$$M\text{–}OH + F^- \rightarrow M\text{–}F + OH\text{–} \qquad (10.1)$$

In their turn, the acidic hydroxyl groups on TiO_2 are preferentially methylated using such agents as methyl iodide and p-toluenesulphonic acid. The exchange of Cd^{2+} ions also involves primarily these groups, as one would expect:

$$Ti\text{–}OH + Cd^{2+} \rightleftharpoons Ti\text{–}OCd^+ + H^+ \qquad (10.2)$$

The OH stretch region of the IR spectrum of SiO_2 typically consists of a strong sharp band at about 3740 cm^{-1} and a broader feature at lower wavenumbers (cf. Fig. 10.2b). The sharp band can be ascribed to isolated (A), and presumably also

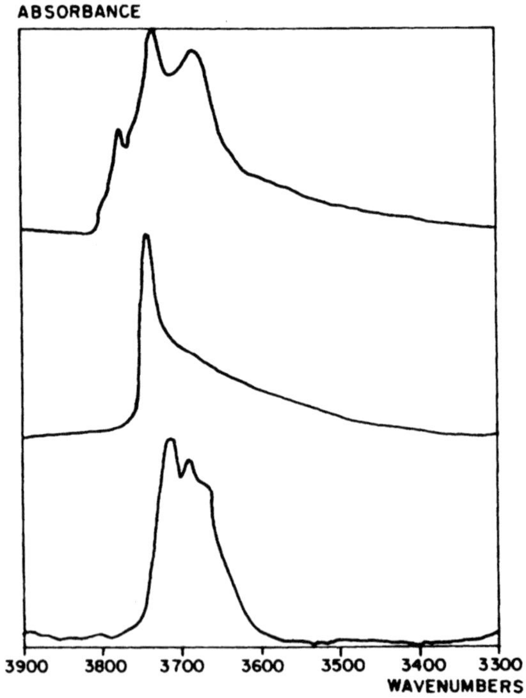

Fig. 10.2. IR spectra (OH stretch region) of, from top to bottom, γ-Al_2O_3, SiO_2 and TiO_2.

to geminal (B) hydroxyls, while the broader feature is due to vicinal (C) hydroxyl groups, which show internal hydrogen bonding:

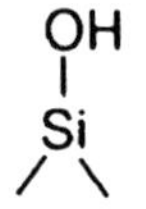
(A)

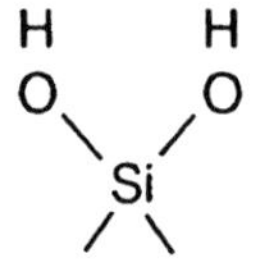
(B)

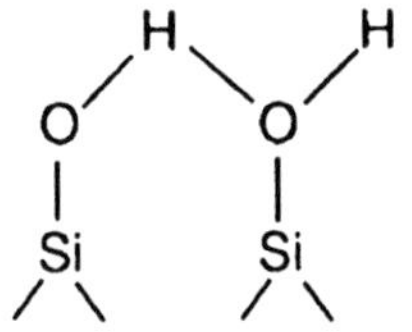
(C)

Chemically speaking, however, these Si–OH groups are difficult to distinguish, and for most purposes one can regard them as forming a homogeneous population. We note in passing that 'internal' hydroxyls (i.e. existing in the bulk, rather than on the surface of the silica) are sometimes present and contribute to the broad feature in the IR spectrum.

When an oxide is brought into contact with an aqueous solution, the surface hydroxyl groups enter into the following equilibria:

$$M\text{–}OH + H^+ \rightleftharpoons [M\text{–}OH_2]^+, \text{ or } M\text{–}OH + H_2O \rightleftharpoons [M\text{–}OH_2]^+ + OH^- \quad (10.3)$$

$$M\text{–}OH + OH^- \rightleftharpoons M\text{–}O^{(-)} + H_2O \quad (10.4)$$

where the basic OH groups will be primarily involved in reaction (10.3) and acidic ones in reaction (10.4). The extent of surface charging, and its sign, obviously depends on the pH of the solution with which the surface is in contact. The pH at which the net surface charge is zero is called the pzc (= point of zero charge). Of the oxides discussed here, γ-Al_2O_3 is the most basic and has a pzc of about 8, TiO_2 is intermediate with a pzc of 5–6, while SiO_2 is the most acidic with a pzc near 2. Above the pzc, the surface of a particular oxide is negatively charged, and will attract cations, while below the pzc it is positively charged, and anionic species will be able to adsorb, under formation of an outer-sphere (reaction (10.5)), or an inner-sphere (reaction (10.6)) complex:

$$[M\text{–}OH_2]^+ + X^- \rightleftharpoons [M\text{–}OH_2]\,X \tag{10.5}$$

$$[M\text{–}OH_2]^+ + X^- \rightleftharpoons M\text{–}X + H_2O \tag{10.6}$$

A good example of outer-sphere complex formation is the adsorption of $Co(CN)_6^{[3-]}$ ions on γ-Al_2O_3. As can be seen from Fig. 10.3, lowering the pH leads to a concomitant increase in the amount of anion adsorbed, while neutralising the solution again induces the $Co(CN)_6^{[3-]}$ ions to desorb again. The process is perfectly reversible, which is not usually the case for inner-sphere complex formation.

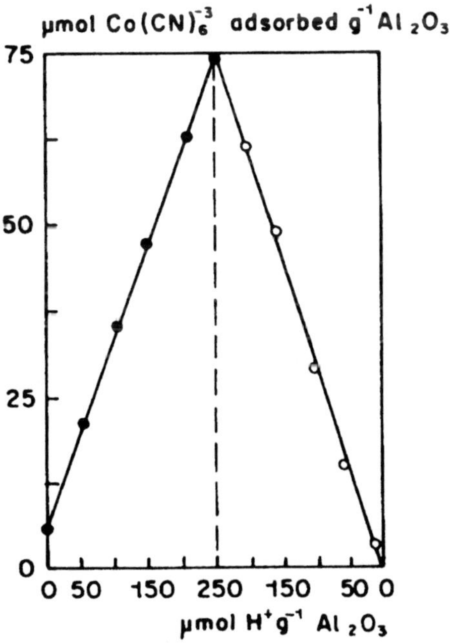

Fig. 10.3. Adsorption of cobalt cyanide ions on γ-Al_2O_3 as a function of surface protonation. ●, addition of HNO_3; o, addition of KOH.

In this chapter, the adsorption of catalyst precursors on oxidic surfaces will be discussed in terms of a chemical-interaction model, i.e. allowing the formation of inner-sphere complexes, but it is important to realize that other models are being advocated in the literature as well: on the one hand, one has the physical adsorption models, allowing only the formation of outer-sphere complexes, and on the other, it is proposed that during 'adsorption' one really has a reaction between the precursor and the support to form a new phase or phases. We will meet the latter situation in our discussion of deposition–precipitation (Section 10.3.3), but we will disregard it when discussing impregnation chemistry (Section 10.3.2).

10.3.1.2 Active carbon

The surface of active carbons contains a rather wide variety of oxygen-containing groups, which have not yet been completely analyzed. Such acidic surface groups as carboxylic acid and phenolic hydroxyl groups are often present. However, the main feature of (active) carbons is that they have weak anionic-exchange properties. The structure of the site responsible for this feature has not been elucidated until now; a recent suggestion is presented in Fig. 10.4. Another important property of (active) carbons is that they can fairly strongly adsorb aromatic molecules, a property that in catalyst preparation is sometimes made use of when cations complexed with (an) aromatic ligand(s) are employed as catalyst precursors. Carbon-supported transition-metal porphyrins, catalysts for the electrochemical reduction of dioxygen, for example, can be simply prepared by letting the chelate adsorb onto the carbon from a suitable solvent.

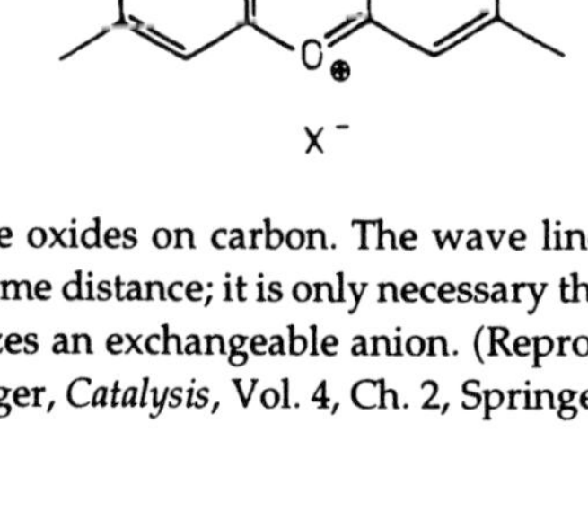

Fig. 10.4. Model of basic surface oxides on carbon. The wave line indicates that the two oxygen functions can be separated by some distance; it is only necessary that a suitable resonance structure can be constructed. X^- symbolizes an exchangeable anion. (Reproduced by permission from H.P. Boehm and H. Knözinger, *Catalysis*, Vol. 4, Ch. 2, Springer-Verlag, Berlin, 1983.)

10.3.2 Impregnation

In impregnation, a solution of a metal salt of sufficient concentration to give the desired loading is added to the support, after which the system is aged, usually for a short time only, dried, and calcined. With a preshaped support an 'incipient wetness' (also called 'dry' or 'pore volume') impregnation is generally used, in which an amount of solution is added which is just sufficient to fill up the pore volume of the support particles, or slightly less. With powdered supports a volume of solution substantially larger than the pore volume can, and often is, applied ('wet' impregnation). Unless the catalyst precursor is strongly adsorbed on the support, it is important to stir the slurry continuously during drying in order to get a homogeneous distribution of the precursor over the support.

Adsorption effects in wet impregnation can usually be described in terms of the surface charging model, Eqns. (10.5) and (10.6). As an example, Fig. 10.5 shows the adsorption of the tetraamine Pt cation on silica and alumina as a function of pH; this is a classical route to highly dispersed Pt-on-silica. As expected, in view of its lower pzc, the extent of adsorption is higher for silica, and the adsorption increases with increasing pH. A pH of about 9 is the practical limit, since above that value the silica and alumina supports start to dissolve appreciably. In the case of zeolites, one introduces the catalyst precursor by ion exchange if at all possible. The problem here is, to keep the metal particles inside the crystallites during calcination and reduction. Various detailed procedures have been developed to deal with this problem. In the remainder of this section we will focus on dry impregnation of preshaped carriers — we will first have a

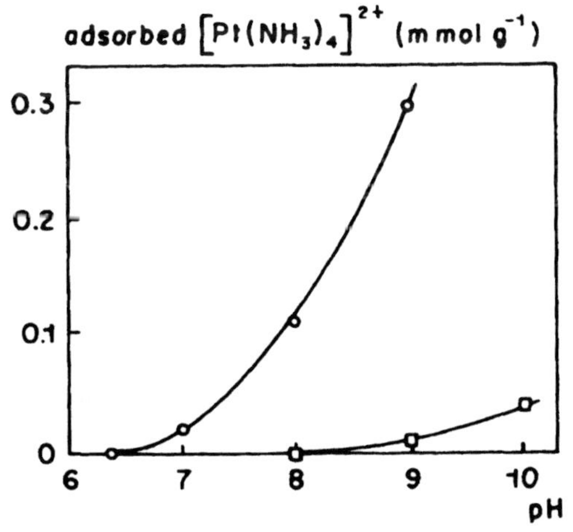

Fig. 10.5. Adsorption from solution containing $[Pt(NH_3)_4]^{2+}$ (as chloride) onto silica gel (Davison 70, 370 $m^2 g^{-1}$) and γ-alumina (Alcoa F-20, 204 $m^2 g^{-1}$) as a function of pH. Room temperature. ○, silica gel; □, γ-alumina. After H.A. Benesi, R.M. Curtis and H.P. Studer, *J. Catal.*, 10 (1968) 328.

look at three examples in which adsorption plays an important role, and then go on to discuss what can happen if the precursor–support interaction is weak or absent, and what tricks there are to counteract any undesirable effects.

10.3.2.1 Supported noble-metal catalysts

As there are no qualitative differences between the various noble metals as far as the chemistry involved in placing them on the various supports is concerned, we will limit our discussion to just one of them, viz. platinum.

Alumina-supported Pt catalysts are conventionally made via impregnation of γ-Al_2O_3 extrudates with an aqueous solution of chloroplatinic acid, H_2PtCl_6. The acid adsorbs strongly on the alumina surface and, in view of its being relatively irreversible, can be described as involving an inner sphere complex according to:

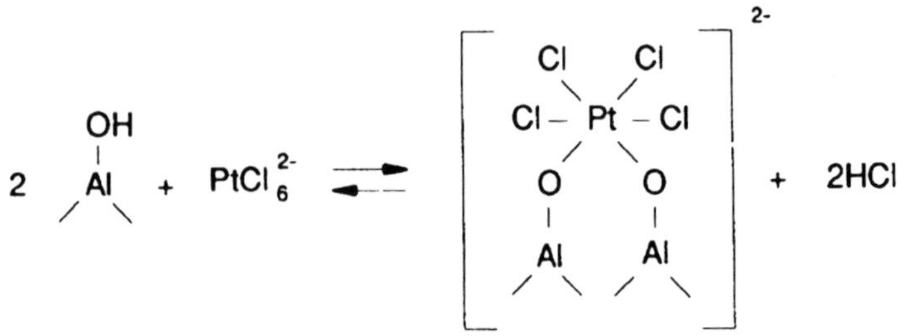

(10.7)

The Al–OH groups involved in this reaction are likely to be the basic ones, since the adsorption reaction is probably to be thought of as analogous to the hydrolysis of the $PtCl_6^{(2-)}$ species that occurs in aqueous solution at higher pH values. The fact that the platinum adsorption capacity of a typical γ-alumina is of the order of 1.5 $\mu mol/m^2$, while the amount of basic OH groups is typically 3 to 3.5 $\mu mol/m^2$, as determined by titration with MoO_2(acetylacetonate)$_2$, fits in nicely with this idea.

Usually, one aims at a Pt loading considerably lower than that corresponding to saturation, viz. 1 wt.% or less, which corresponds to 0.2 $\mu mol/m^2$ or less, the surface area of alumina being typically around 250 m^2/g. Impregnation with an aqueous solution containing the required amount of H_2PtCl_6, however, does not lead to a homogeneous Pt distribution: because of the strong interaction between $PtCl_6^{(2-)}$ ions and the alumina surface, and since the adsorption reaction is very fast, Pt will be deposited on all the adsorption sites these ions encounter on their way in from the outer surface of the extrudate, until the solution is completely depleted. This results in supported catalyst bodies with the platinum concentrated in the outer rim, as visualized in Fig. 10.6 (egg-shell distribution). Sometimes one wants such a distribution, to be sure, e.g. when the reaction to be catalysed is severely diffusion limited, but more often a homo-

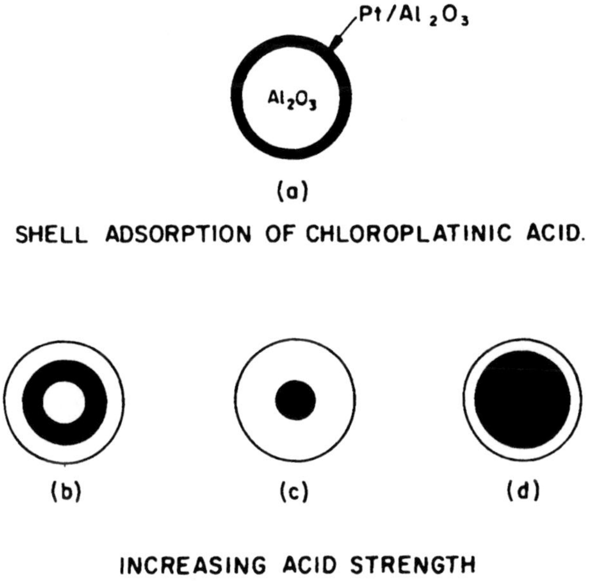

Fig. 10.6. Pellet concentration profiles with adsorbing acids. (b)–(d): see text.

geneous distribution is desired. This can be achieved by adding HCl to the impregnation solution, which adsorbs on the same sites as the chloroplatinic species according to

$$\underset{/\ \backslash}{\overset{\text{OH}}{\overset{|}{\text{Al}}}} + \text{HCl} \rightleftharpoons \underset{/\ \backslash}{\overset{\text{Cl}}{\overset{|}{\text{Al}}}} + H_2O \tag{10.8}$$

This competitive adsorption drives the platinum deeper into the extrudates, and when sufficient HCl is added, the $PtCl_6^{(2-)}$ adsorption reaction is moderated to such an extent that a reasonably homogeneous distribution is obtained. One can also add acids that adsorb even more strongly than chloroplatinic acid, e.g. oxalic or citric acid, with which Pt profiles such as shown in Fig. 10.6 can be prepared. Profile (b) can be useful when a strongly adsorbing poison is present in the stream to be treated, e.g. Pb in automobile exhaust gas, while profiles (c) and (d) could be advantageous e.g. in diffusion-limited consecutive reactions.

After impregnation a Pt/Al_2O_3 catalyst is dried and calcined to produce supported Pt-oxide species (some Cl is retained on the alumina), which can then be reduced, usually at relatively mild temperatures (about 300–350°C) in hydrogen to produce small Pt particles (about 10 Å).

In certain applications, e.g. liquid-phase hydrogenations and H_2/O_2 fuel cells, carbons (active carbon or carbon black) are preferred as support material, because they are chemically rather inert (not attacked by strongly acidic or caustic solutions) and, important for fuel cells, electrically conducting. Carbon-supported Pt catalysts are normally prepared in the same way as their alumina-supported counterparts, i.e. via impregnation with an aqueous solution of chloroplatinic acid. Because of the anion-exchange capacity of carbons, there is an interaction, albeit rather weak, between the $PtCl_6^{(2-)}$ ions and the support which is sufficient to largely prevent migration of the Pt species during drying (see below), but not strong enough to induce large inhomogeneities in Pt distribution, obviating the addition of HCl to the impregnation solution. Alternatively, the carbon can be mildly oxidized to produce surface carboxyl groups and the Pt brought on through ion-exchange. As oxidizing agent, one usually employs concentrated nitric acid, or sodium hypochlorite, and it can be advantageous to activate unreactive carbons like high surface area graphite first by treatment in flowing oxygen at 500°C.

10.3.2.2 Disproportionation catalysts

In the disproportionation (metathesis) of olefins, i.e. the reaction $2R_1\text{–}CH{=}CH\text{–}R_2 \rightarrow R_1\text{–}CH{=}CH\text{–}R_1 + R_2\text{–}CH{=}CH\text{–}R_2$, a MoO_3/Al_2O_3 catalyst is often applied. Its preparation involves impregnation of γ-Al_2O_3 extrudates with an aqueous solution of ammonium heptamolybdate (AHM) or an ammoniacal solution of ammonium dimolybdate (ADM).

The interaction between AHM and an γ-Al_2O_3 surface has been extensively studied. Two adsorption reactions take place. In the first instance, HM ions react with the basic OH groups according to

$$Mo_7O_{24}^{(6-)} + 8\ Al\text{–}OH \rightarrow 4\ (Al)_2MoO_4 + 3\ MoO_4^{(2-)} + 4\ H_2O \qquad (10.9)$$

That only the basic OH groups are involved here follows from FTIR measurements: only the high-frequency OH band (cf. Fig. 10.2a) disappears from the IR spectrum of the alumina support. The tetrahedral coordination of Mo in the adsorbed phase has been verified by EXAFS and Raman spectroscopies (Mo is octahedrally coordinated in AHM). Reaction (10.9) is followed by a (reversible) physisorption of AHM molecules on the c.u.s. Al sites in the alumina surface. When the impregnation is effected with the ammoniacal molybdate solution, which contains only $MoO_4^{(2-)}$ ions at a pH of 9–10, no interaction initially occurs between the solute and the alumina surface: the pH being appreciably higher than the pzc, the basic OH groups are unreactive and the molybdate ions are not attracted to the negatively charged surface. Upon drying, however, the ammonia escapes in preference to water, so that the pH inside the pores drops, which in turn leads to the formation of HM ions according to

$$7\ MoO_4^{(2-)} + 4\ H_2O \rightleftharpoons \quad + 8\ OH^- \qquad (10.10)$$

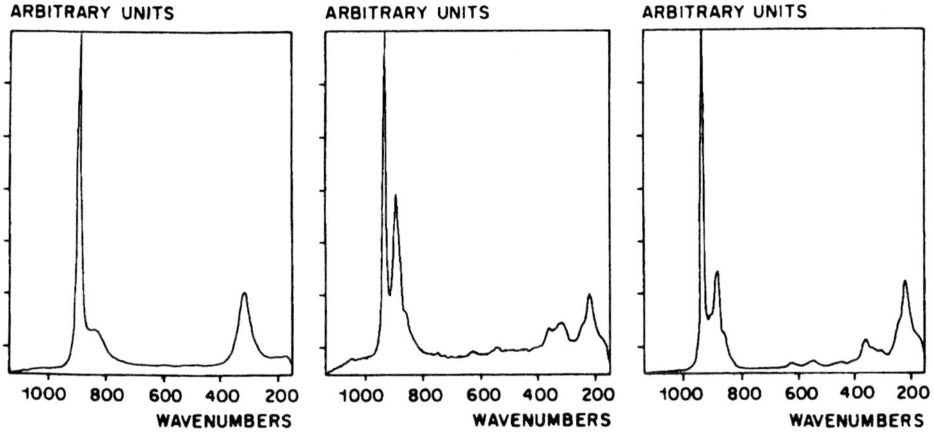

Fig. 10.7A. Ammoniacal Mo solution, after evaporation of (a) 1/3, (b) 2/3, and (c) all of the solution. Raman *in-situ* cell. Spectrum (a) is identical to $(NH_4)_2MoO_4$ aq. Spectrum (c) is identical to that of solid AHM.

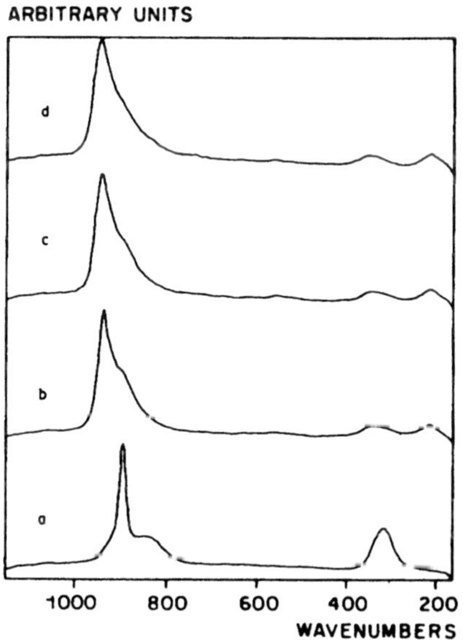

Fig. 10.7B. *In-situ* Raman spectrum of γ-Al_2O_3 impregnated with ammoniacal Mo solution ($\rightarrow$ 10.5 %w Mo). (a) Wet catalyst, and after (b) 20, (c) 40, and (d) 60 min drying at 120°C. Spectrum (d) is identical to that of dried AHM/γ-Al_2O_3 of the same Mo loading.

this equilibrium starting to shift to the right hand side at pH around 7. The occurrence of this process in the catalyst pores has been verified with Raman spectroscopy (Fig. 10.7). The HM ions, thus formed, then engage in the same adsorption reactions as described above, with this difference that at the drying temperature (70–110°C) the physisorption reaction takes precedence over reaction (10.9).

It has been found that maximum performance is reached at just that loading which corresponds to the adsorption capacity of the alumina carrier employed.

In the case of functionalized olefins, i.e. where R_1 and/or R_2 are not simple alkyl chains, it is better to use a Re_2O_7/Al_2O_3 catalyst. Such catalysts can be prepared by pore-volume impregnation with an aqueous solution of ammonium perrhenate. It is observed that the turnover frequency, i.e. the activity per Re atom, increases considerably with Re_2O_7 loading up to monolayer coverage of the alumina surface (about 18 wt.% Re_2O_7 for an alumina of 200 m^2/g). A theory explaining this behaviour has been proposed, postulating that at low loadings the ReO_4^- ions react with the basic OH groups, and at higher loadings with neutral and acidic ones, only in the last case leading to active, because electron-deficient, Re centres capable of complexing a double bond.

In the dried catalyst Re is mainly present as small $(NH_4)ReO_4$ crystallites, which upon calcination spread over the alumina surface to form isolated ReO_4 units, as has been established by Raman spectroscopy. Studying the evolution of the alumina OH bands with IR, Fig. 10.8, shows that up to 3 wt.% Re_2O_7 the OH groups are not too much affected, the ReO_4^- ions presumably reacting with the c.u.s. Al sites, leading to totally inactive sites. With increasing Re_2O_7 loading, it is

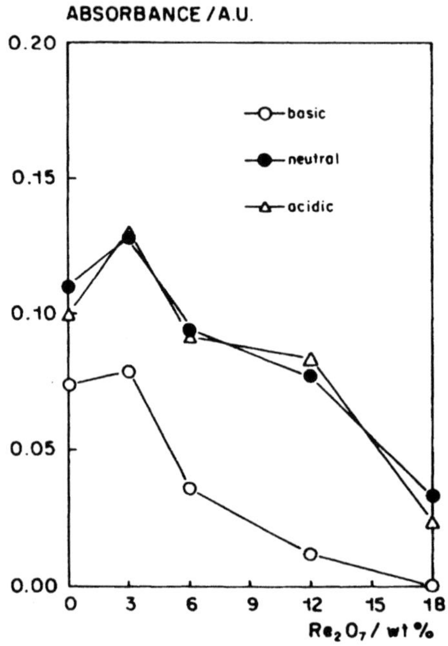

Fig. 10.8. The absorbances of the three different types of OH groups as a function of the Re_2O_7 loading of $Re_2O_7/\gamma\text{-}Al_2O_3$ catalysts.

indeed the high-frequency band that is preferentially reacted away, thus verifying the above postulate.

10.3.2.3 Supported Co(Ni)Mo hydrotreating catalysts

Alumina-supported Co- and Ni-promoted molybdenum sulphide hydrotreating catalysts are the main workhorses in many refineries and have, therefore, attracted a lot of attention from catalytic chemists. They are usually prepared via co-impregnation, i.e. pore-volume impregnation with both Mo and the promoter atom present in solution. After drying and calcining, the catalyst manufacture is complete, but it has to be sulphided before use. Traditionally, this is done *in situ* with the feed to be treated, spiked with some extra sulphur (as, e.g., CS_2), but nowadays one can buy presulphided catalysts, which can be converted into the active form much faster, thus limiting the loss of production days of the hydrotreatment unit. The active phase consists of small MoS_2 particles, 10–15 Å in diameter, with promoter atoms decorating their edges. It is generated from the molybdenum species that adsorb on the alumina carrier during impregnation and drying, as Fig. 10.9 makes plausible — the details of the adsorption reactions will be discussed below. The problem with non-adsorbed Mo is that during calcination it tends to react with the promoter atoms present to form $Co(Ni)MoO_4$, which unfortunately is not a good precursor to the active phase. Interacting with the subjacent alumina stabilizes the $Mo–O_x$ species against this reaction, leading to a two-dimensional surface phase from which highly dispersed MoS_2 can be generated. The promoter atoms should also be present in this surface phase, but in the absence of long-range ordering it has been impossible to locate them precisely.

There are two main co-impregnation solutions, one being ammoniacal and the other containing phosphoric acid. The chemistry involved when applying the former is quite similar to that described above for the MoO_3/Al_2O_3 disproportionation catalyst. The point is that the presence of Co or Ni ions does not appear to influence the molybdate equilibria and adsorption reactions to any great extent. On the other hand, the concentration profiles of Co/Ni and Mo over the alumina extrudates are always identical, indicating that adsorption or precipitation of the molybdate species entrains adsorption or precipitation of the Co/Ni species. Indeed, when a precipitate develops in an ammoniacal impregnation solution, it does contain both promoter and Mo ions. The precise chemistry involved is as yet obscure, however.

In general, concentrated Mo solutions are difficult to handle, especially when they also contain Co or Ni ions, as they tend to be rather unstable. The situation can be much improved by adding phosphoric acid, because it reacts with molybdate to form a variety of, usually extremely soluble, molybdophosphates, such as $H_6P_2Mo_5O_{23}$ and $H_3PMo_{12}O_{40}$. The other side of the coin is that the presence of

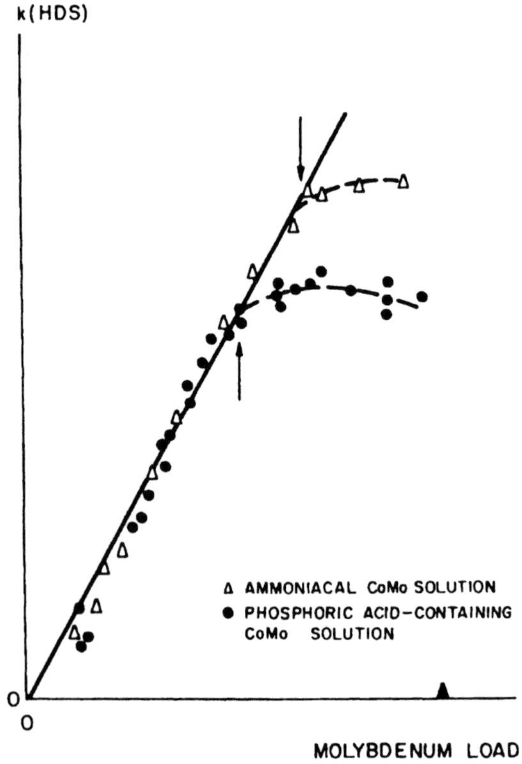

Fig. 10.9. Relationship between HDS activity and CoMo loading. Fixed γ-Al_2O_3 support, fixed Co/Mo atomic ratio. ▲ indicates monolayer coverage of the support by Mo; → indicates Mo-adsorption capacity of the γ-Al_2O_3 applied when using the impregnation solutions indicated.

phosphate ions considerably complicates the Mo adsorption chemistry. As far as it is known today, what happens when an alumina is impregnated with a phosphoric acid-containing co-impregnation solution is that part of the alumina surface is converted into an $AlPO_4$ surface phase. The formation of this phase does not start with the exchange of a basic hydroxyl group for a $H_2PO_4^-$ ion, as might have been expected, but rather with the exchange of a bridging OH. Aluminium-phosphate formation has the following consequences: (i) some loss in support surface area, and (ii) a decrease in the Mo-adsorption capacity of the support. Polyoxomolybdates do adsorb on $AlPO_4$ via its P–OH groups, e.g. according to

$$2P_s - OH + Mo_7O_{24}^{6-} \rightarrow (P_s - O)_2Mo_5O_{15}^{2-} + 2\,MoO_4^{2-} + H_2O \qquad (10.11)$$

but the level of adsorption is lower than on the original alumina. Moreover, the patches of alumina left in the impregnated particles are somewhat less reactive than in the absence of neighbouring AlPO patches, perhaps for steric reasons.

The Mo adsorption capacity for both the ammoniacal and the phosphoric acid routes of a particular alumina can be estimated on the basis of the adsorption reactions found to obtain in the two cases, and the numbers derived for the alumina used as a support for the catalysts figuring in Fig. 10.9 are indicated in that figure by arrows.

Since a somewhat higher active phase loading can be achieved in the ammoniacal as against the phosphoric acid route, one may wonder why P-containing catalysts are still in use. Well, here we have to draw a distinction between hydrodesulphurization (HDS) and hydrodenitrogenation (HDN). In the former process phosphorus has no catalytic role to play, so that its presence is of no advantage, and indeed, as the handling of non-P impregnation solutions has improved, we have witnessed a gradual replacement of $CoMoP/Al_2O_3$ HDS catalysts by their $CoMo/Al_2O_3$ counterparts. In HDN, however, P acts as a promoter, which more than offsets the slight loss in the amount of active phase, and for this application $NiMoP/Al_2O_3$ remains the catalyst of choice.

As to calcination temperatures, it is advisable to keep them relatively low, say about 723 K. At significantly higher T, the promoter ions tend to go subsurface (incipient spinel formation) and so are lost from the surface, and eventually from the active phase. Also, in the case of a P-containing catalyst, further surface-area loss may occur due to migration of the phosphate anions into the bulk of the alumina.

Although most attention has been paid, naturally, to alumina-supported Co(Ni)Mo catalysts, other supports have been looked at as well. The adsorption of polyoxo molybdates on titania can be described in the same terms as that on alumina: a reaction with basic surface OH groups, followed by physisorption on c.u.s. Ti sites. Since freshly calcined titania contains a relatively small amount of surface hydroxyls, practically a full monolayer coverage can be achieved via the latter adsorption reaction. Classical impregnation routes have not been very successful in the case of silica, however, as adsorption of molybdates occurs only to a limited extent owing to a lack of basic OH groups and Lewis-acid sites on the silica surface. Reasonable Mo loadings at reasonable dispersions can be prepared on active carbon carriers, making use again of their weak anion-exchange capacity; such catalysts should not be calcined, however.

10.3.2.4 Consequences of the absence of precursor–support interaction

In the case where there is no interaction between the catalyst precursor and the support surface to fix the former to the latter, the dynamics of drying become of prime importance. Briefly, upon drying impregnated porous particles (extrudates, pellets, spheres), evaporation of the solvent (almost always water), starts at the external surface, resulting in menisci at the pore mouths and setting up a suction pressure. At normal drying rates, and employing simple aqueous solutions, the liquid lost by vaporization in smaller pores is supplied from the connecting larger pores due to the capillary pressure difference. Thus, the evaporation takes place mainly at the pore mouths of the smaller pores while solution flows from the interior of the particle towards them. The impregnation liquid is by this means concentrated near the mouths of the small pores at the outer surface, until (super)saturation occurs and the catalyst precursor crystallizes or precipitates, which results in the pore-mouths becoming even smaller and so remaining preferentially wetted, etc. The end result is a segregation of the catalyst precursor at the particle periphery (egg-shell distribution). Here it is assumed, that the concentration differences, caused by the localized evaporation, are not counterbalanced by diffusion, in which case (super)saturation and crystallization would occur in all still solution-filled pores simultaneously. This is a safe assumption, diffusion being a rather slow process, unless the drying rate is extremely low. Nevertheless, the picture given here is somewhat simplified, and part of the catalyst precursor does usually remain in the interior of the support particle, especially when the pore-size distribution is very sharp (entailing small capillary pressure differences, i.e. relatively little liquid movement). It has also been observed that during drying voids in the liquid arise in the pore system located at inner parts of the support bodies. During this stage of the drying process, the so-called funicular state, the liquid phase is covering the (internal) surface of the support by a continuous layer. If this layer breaks up (giving rise to the so-called pendular state), some fraction of the active component remains in the interior of the support. It is worth bearing in mind that not only is the distribution of the catalyst precursor over the support body inhomogeneous, its particle-size distribution is also often very broad.

As is implicit in the above, the transport of the catalyst precursor to the outer rim of the support body can be suppressed by (i) using very high drying rates, e.g. through applying micro-wave heating, and (ii) increasing the viscosity of the impregnation solution, e.g. by adding a polymeric substance (preferably one that does not increase the viscosity so much at room temperature, so that the impregnation rate is not seriously affected, but only at temperatures in the drying regime). In either case, the rate of removal of water is made higher than the rate of liquid capillary flow, so that the evaporation front moves smoothly inward into the particle, and the solute precipitates concomitantly. It is to be borne in mind, however, that the use of polymeric additives may affect the morphology of the precipitated active phase.

It has also been observed that employing, at near-saturation, a catalyst precursor that does not crystallize at all but rather forms a coating on the support surface, can lead to a homogeneous distribution of small particles — a case in point being the application of an iron EDTA complex at alkaline pH to produce, after calcination, small (2–5 nm) Fe_2O_3 particles homogeneously distributed through SiO_2 extrudates.

10.3.3 Deposition–Precipitation

As coprecipitation methods are not easy to control and reproduce, and impregnation techniques cannot always be made to yield the desired active-phase distribution, loading and/or dispersion, it is worthwhile considering alternative methods. One of these is deposition–precipitation, which will be discussed in this section.

10.3.3.1 Basics of the method

Traditionally, the objective of precipitation is to obtain a well-dispersed metal hydroxide (or carbonate) phase on a support through its precipitation from an aqueous metal solution onto a support powder by adding a base. The support powder is suspended in the metal salt solution. Upon addition of, say, NaOH, the pH increases strongly and metal hydroxide species are generated. When their concentration exceeds the (super)solubility limit, metal hydroxide particles can nucleate and grow. The trick is to let the nucleation happen on the support surface rather than in the solution, which would result in large crystallites and an inhomogeneous metal distribution. To prevent rapid nucleation and growth in the bulk of the solution, the precipitant has to be added slowly and under vigorous mixing conditions to prevent the formation of regions of high local pH. The best strategy would appear to be to inject the alkali solution slowly through a hypodermic syringe below the surface of the liquid, while the suspension is vigorously agitated. Highly efficient mixing is difficult to achieve on a large scale, however, so this condition poses an up-scaling problem.

To establish a very uniform distribution of small active particles over a support, and to get around the mixing problem, the procedure of deposition–precipitation was developed. For the method to work well it is essential, as for precipitation in general, that the support facilitates the nucleation of an active precursor (cf. below). In the classical embodiment of the method, a metal salt solution containing urea is well mixed with the support powder at room temperature. Then, the temperature is raised to 70–90°C, where the urea slowly hydrolyses according to

$$CO(NH_2)_2 + 3\,H_2O \rightarrow 2\,NH_4^+ + CO_2 + 2\,OH^- \qquad (10.12)$$

The hydroxyl ions are produced homogeneously throughout the solution, thus obviating the need for strong mixing. Since, in the presence of an active precursor–support interaction, nucleation of the catalyst precursor (metal hydroxide species in the present example) at the surface of the support proceeds at lower concentrations than those needed for precipitation in the bulk of the solution, precipitation exclusively on the support surface can be achieved. This

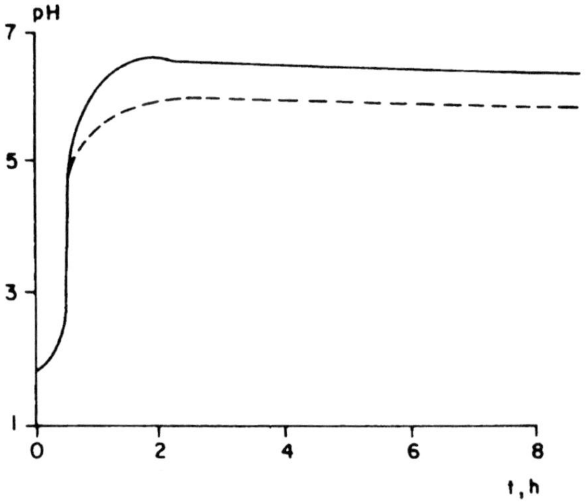

Fig. 10.10. Records of pH as a function of time in the absence (——) and presence (- - - -) of silica during the precipitation of Mn(II). After K.P. de Jong, *Stud. Surf. Sci. Catal.*, 63 (1991) 19.

procedure has been used with success to produce a number of different catalysts. An example to illustrate the basics of the method is shown in Fig. 10.10: precipitation of manganese(II) from a homogeneous solution by raising the hydroxyl ion concentration is represented by the evolution of the pH as a function of time during the hydrolysis of urea. Without suspended silica, the pH of the manganese nitrate solution passes through a maximum. Before nucleation in the bulk of the solution can proceed, the concentration has to be increased above the concentration corresponding to the solubility of larger crystallites (supersaturation limit). After nucleation, the pH level drops to that corresponding to the solubility product of manganese(II) hydroxide. With suspended silica, on the other hand, the nucleation barrier is absent, and the precipitation on the silica surface proceeds smoothly.

Apart from raising the pH level, precipitation can be induced in a variety of other ways. Anionic species, for example, can be deposited on the surface of suspended carriers by decreasing the pH level. This procedure has been used for vanadium(V) and Mo(VI). Oxidation at a pH level where the ions of the lower valency are soluble and the oxidized species insoluble, can also be utilized to precipitate from a homogeneous solution. Iron(II/III) and Mn(III/IV) are cases in point. Oxidation can be effected by dissolved agents like nitrate, or electrochemically. Reduction to insoluble ionic compounds has been applied with Cr, Cu, and Mo. Reduction to the metal has also been practised, and appears to work especially well with noble metals. Decomplexing is another possibility: e.g. destruction of complexing EDTA by hydrogen peroxide has been employed to produce supported catalysts.

After the deposition–precipitation step has been completed, the solid is filtered, washed and dried. Before activation and use, it has to be shaped (e.g. pelleted), and, in most cases, calcined.

10.3.3.2 Deposition–precipitation on silica powders

Deposition-precipitation is often practised with silica as the support. Especially suitable is aerosil silica, which consists of very small non-porous spheres, so that the precipitation process is not affected in any way by diffusion processes. It is well known that most hydrolysed metal species have a high affinity for the silica surface, thus fulfilling the condition for obtaining surface precipitation only. In the colloid-chemical literature, the initial adsorption of the (partially) hydrolysed metal ions with a silica surface is often described in terms of a surface-complexation model, involving negatively charged surface sites, which exist on silica at pH above 2 (= pzc of silica), and positively charged metal species:

$$-\overset{|}{\underset{|}{\mathrm{Si}}}-\mathrm{O}^- + \mathrm{M(OH)}^{(n-1)+} \rightarrow -\overset{|}{\underset{|}{\mathrm{Si}}}-\mathrm{O}-\mathrm{M(OH)}^{(n-2)+} \qquad (10.13)$$

On the other hand, there is evidence that the pH may affect the reactivity of the hydroxyl groups, in that OH^- ions have been found to catalyse the condensation of two hydroxyl groups on different silica particles to water and an oxygen bridge. In analogy, an alternative description of the initial reaction between a hydrolysed metal species and the silica surface would be

$$-\overset{|}{\underset{|}{\mathrm{Si}}}-\mathrm{OH} + \mathrm{HO}-\overset{|}{\underset{|}{\mathrm{M}}}- \rightarrow -\overset{|}{\underset{|}{\mathrm{Si}}}-\mathrm{O}-\overset{|}{\underset{|}{\mathrm{M}}}- + \mathrm{H_2O} \qquad (10.14)$$

A careful study of Cu(II) precipitation on SiO_2 from different salts has shown that the latter description is in fact quite satisfactory.

It has been established that the pH needs to be above about 4 for reaction (10.14) to proceed. Thus, in the case of Fe(III), where nucleation already starts at pH around 2, the silica surface is not able to compete, and $Fe(OH)_3$ particles are generated in solution. These particles do not subsequently react with the silica surface upon further increasing the pH, resulting eventually in a separate, rather than supported, iron (hydr)oxide phase. However, iron(III) can be deposited on SiO_2 by injecting a weakly acidic Fe(III) solution into a silica suspension, whose pH is kept above 5.

To produce Fe(oxide)/SiO_2 particles one could use a solution of Fe(II), in which case air has to be excluded to prevent its oxidation to Fe(III). Iron (II) starts to react markedly above pH = 4.8 (urea, 90°C). In this case, the reaction is not

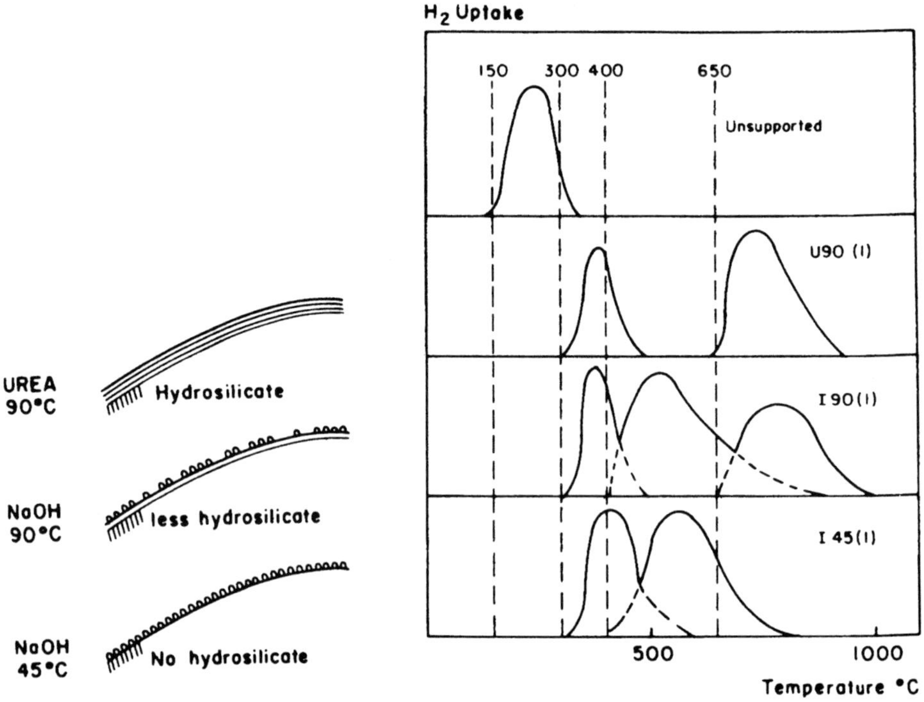

Fig. 10.11. Left: structures resulting from deposition–precipitation of Fe(II) by different procedures. Right: temperature-programmed reduction of Fe oxide deposited on silica by hydrolysis of urea (U 90(1)), injection at 90°C (I 90(1)), and injection at 45°C. For comparison bulk Fe oxide is included.

limited to Eqn. (10.14), but a bulk hydrosilicate is formed. Upon performing an injection experiment at 45°C, it is observed that the reaction with the support is less extensive at that temperature. As the slightly higher pH at the injection point brings about the formation of a less reactive iron species, attack of the support is less marked than in the urea case even at 90°C. The structures obtained in the three different experiments are indicated in Fig. 10.11. The different extents of hydrosilicate formation are reflected in the temperature-programmed reduction (TPR) experiments, as can be seen in Fig. 10.11. Previous air-drying partially oxidizes the Fe(II). As interaction with silica stabilizes Fe(II), the supported Fe samples show a separate reduction step to Fe(II), which is not displayed by bulk Fe oxide. The iron hydrosilicate obtained in urea precipitation at 90°C is fairly stable and is reduced only above 650°C. The Fe(II) precipitated at 45°C is more loosely bonded and its reduction starts already at 450°C, and injection at 90°C leads to an intermediate situation. The different extents of reaction of the support can also be visualized through transmission electron microscopy: the carrier loaded at 45°C shows the silica particles covered by very small Fe oxide particles, while the carrier has reacted to thin hydrosilicate platelets at 90°C (urea).

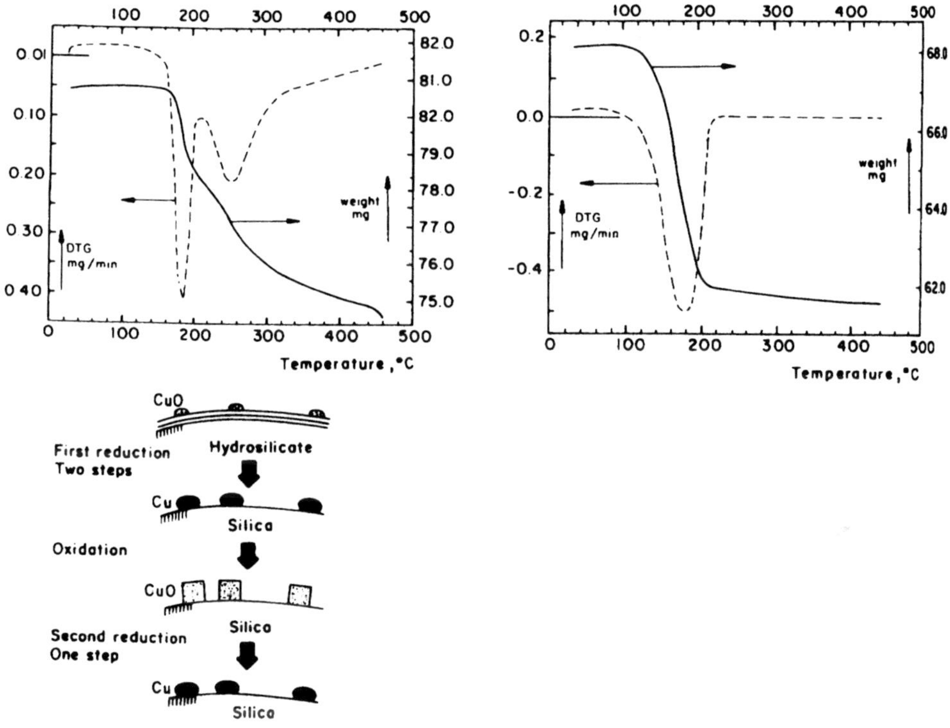

Fig. 10.12. Thermogravimetric reduction of Cu deposited on silica from basic Cu nitrate. Both the weight and the rate of the decrease in weight are represented. Top left: first reduction. Top right: second reduction. Bottom: structure of catalyst before and after reduction, reoxidation and second reduction.

Hydrosilicate formation is also in evidence in the Cu(II)–SiO_2 system. Via precipitation from a homogeneous solution one can obtain highly dispersed copper oxide on silica (cf. above, Fig. 10.10, where it should be noted that the Cu case is more complicated than the Mn one in that intermediate precipitation of basic salts can occur). Reaction to copper hydrosilicate is evident from temperature-programmed reduction. As shown in Fig. 10.12 the freshly dried catalyst exhibits reduction in two peaks, one due to Cu(II) (hydr)oxide and the other, at higher temperature, to Cu(II) hydrosilicate. Reoxidation of the metallic copper particles leads to Cu(II) oxide, and subsequent reduction proceeds therefore in one step. The water resulting from the reduction of the oxide does not produce significant amounts of copper hydrosilicate, in contrast to what usually happens in the case of nickel.

The activity of the thus prepared Cu(oxide)/SiO_2 catalyst in the oxidation of carbon monoxide is represented in Fig. 10.13. It can be seen that the activity of the fresh catalyst before reduction is rather low. In spite of the very good dispersion (high surface area), the conversion at e.g. 180°C is much lower than that of

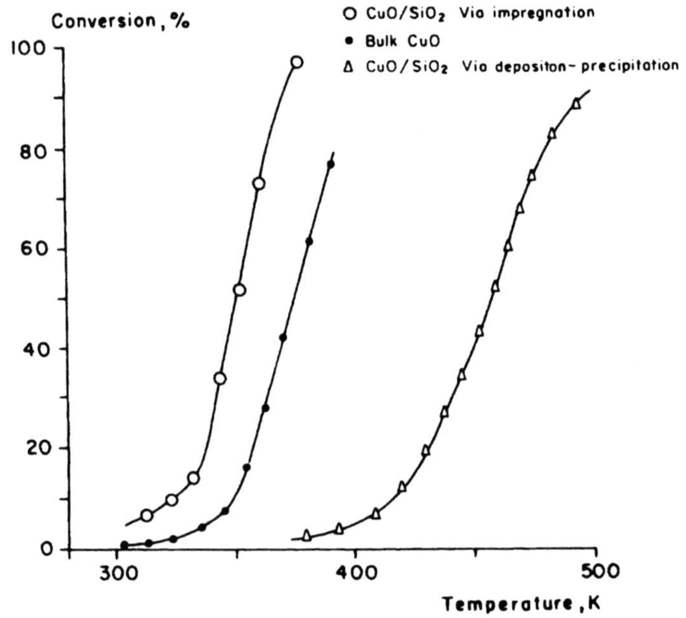

Fig. 10.13. CO-oxidation activity of freshly calcined Cu(II)-oxide catalysts. CuO loading in the SiO_2-supported catalysts is about 28 wt.%.

unsupported copper oxide and the less well dispersed CuO/SiO_2 prepared via impregnation. However, reduction and reoxidation leads to a much higher activity, cf. Fig. 10.14. Apparently, then, the activity of copper ions in a copper hydrosilicate is much lower than that of those in an oxide. As the dispersion remains very high in the transformation of the one phase into the other, a high catalytic activity can therefore be achieved.

Nickel reacts even more strongly than copper with silica; indeed, silica completely reacts with nickel(II) ions precipitating at 90°C to nickel hydrosilicate, provided sufficient nickel is supplied. As a result, the structure or the support completely disappears from the electron micrographs, only the hydrosilicate platelets being in evidence. Deposition–precipitation of Ni onto γ-Al_2O_3 leads to the formation of a Feitknecht compound.

The technique of deposition-precipitation has been continuously improved. One of the latest improvements has been the use of electrochemical procedures both to dissolve and to precipitate the active precursor. The adjustment of the valency of the element to be applied on the support to establish a sufficiently strong interaction with it has continued to broaden the range of application of the deposition–precipitation procedure considerably. Recent instances are the application of vanadium(III) and molybdenum(IV) on silica.

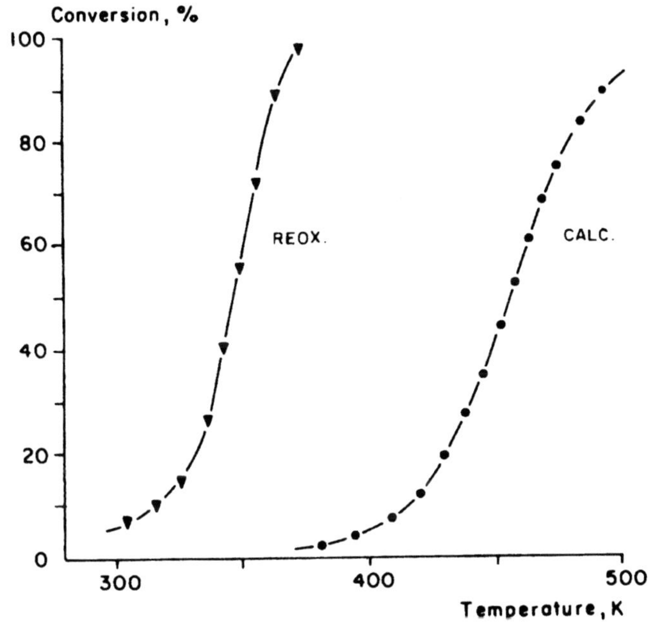

Fig. 10.14. Effect of reduction and reoxidation on the CO-oxidation activity of a CuO/SiO_2 catalyst prepared by deposition-precipitation.

10.3.3.3 Deposition-precipitation onto preshaped carrier bodies

As remarked above, it is often desirable to apply a catalyst precursor directly onto a preshaped support. An obvious way to proceed would be to pore-volume impregnate the carrier bodies with the required amounts of metal salt and urea (or, alternatively, nitrite) in an aqueous solution and heat the impregnated particles to 70–90°C. This can actually be made to work, as has been shown recently in the case of $Ni/\gamma\text{-}Al_2O_3$, but the preparation of highly loaded catalysts would appear to be problematic with this route.

The deposition on about 1 mm porous SiO_2 granules via urea hydrolysis has been studied for $Mn^{(2+)}$. As expected, the rate determining step is the urea hydrolysis step itself. In these conditions, the metal profile over the granules is determined by the relative rates of diffusion of hydrolysed Mn species in the pores and of reaction of these species with the SiO_2 surface. This is relatively well described with the Thiele concept, as is evident from Fig. 10.15. The rate of the surface reaction is lower for $Rh^{(3+)}$ and $Ni^{(2+)}$ than for $Mn^{(2+)}$, which should lead to flatter metal profiles. However, a certain concentration gradient inside the carrier body cannot be avoided when applying this preparation method to larger support particles. Whether these concentration gradients are acceptable or not will depend on the catalyst and its specific application.

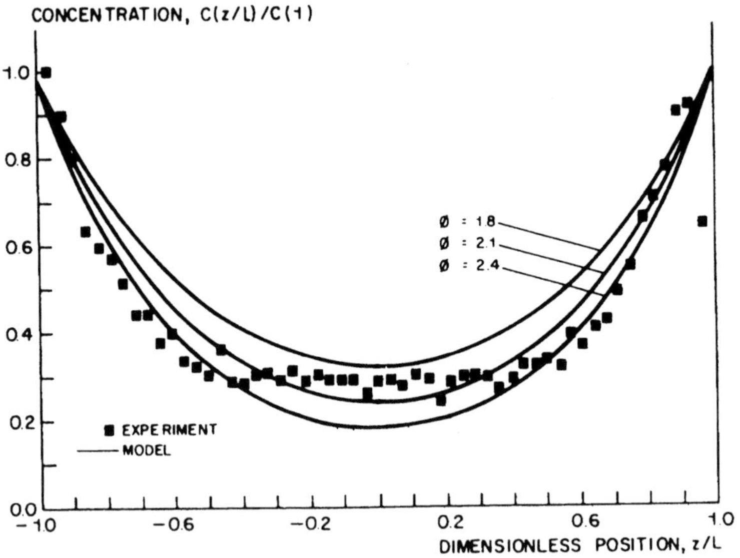

Fig. 10.15. Distribution of manganese over a representative silica granule from experiment U34. The drawn lines have been calculated for different values for the Thiele modulus (Φ). After K.P. de Jong, *Stud. Surf. Sci.Catal.*, 63 (1991) 19.

FURTHER READING

James T. Richardson, *Principles of Catalyst Development*, Plenum Press, New York, 1989, Ch. 6.

C. Marcilly (and J.P. Franck), La preparation des catalyseurs, *Rev. Inst. Fr. Petrole*, 39 (1984) 189–208, 337–364.

J.A. Schwarz, C. Contescu, and A. Contescu, Methods for the preparation of catalytic materials, *Chem. Rev.*, 95 (1995) 477–510.

H.P.Boehm and H.Knozinger, Nature and estimation of functional groups on solid surfaces, in: *Catalysis*, Vol.4, Springer-Verlag, Berlin etc., 1983, pp. 39–207.

J.A.R. van Veen, A method for the quantitative determination of the basic OH groups on an alumina surface, *J. Colloid. Interface Sci.*, 121 (1988) 214–219.

J.A.R. van Veen, An enquiry into the surface chemistry of TiO_2(anatase), *Z. Phys. Chem. NF*, 162 (1989) 215–229.

J.A. Schwarz, C. Contescu, and J. Jagiello, Binding Properties of Hydroxyl Groups on Substrates in Aqueous Environments: Their Relationship to Catalyst Preparation, *Catalysis*, Vol. 11, Royal Soc. Chem., 1994, Ch. 4.

A.V. Neimark, L.I. Kheifetz and V.B.Fenelonov, Theory of preparation of supported catalysts, *Ind. Eng. Chem. Prod. Res. Dev.*, 20 (1981) 439–450.

M. Kotter, Herstellung von Trankkatalysatoren als Verfahrenstechnische Aufgabe, *Chem.-Ing.-Tech.*, 55 (1983) 179–185.

S.-Y. Lee and R. Aris, The distribution of active ingredients in supported catalysts prepared by impregnation, *Catal. Rev.-Sci. Eng.*, 27 (1985) 207–340.

M. Komiyama, Design and preparation of impregnated catalysts, *ibid.*, 341–372.

O.A. Sceiza, A.A. Castro, D.R. Ardiles and J.M. Parera, Modeling of the impregnation step to prepare supported Pt/Al_2O_3 catalysts, *Ind. Eng. Chem. Fundam.*, 25 (1986) 84–88.

J.R. Regalbuto *et al.*, A Scientific Description of Pt Adsorption onto Alumina, *Stud. Surf. Sci. Catal.*, 118 (1998) 147–156.

M. Sibeijn, R. Spronk, J.A.R. van Veen, and J.C. Mol, IR studies of Re_2O_7 metathesis catalysts supported on (phosphated) alumina, *Catal. Lett.*, 8 (1991) 201–208.

J.A.R. van Veen *et al.*, Chemistry of (phospho-)molybdate adsorption on alumina surfaces, *J. Phys. Chem.*, 94 (1990) 5275–5285.

X. Carrier, J.F. Lambert, and M. Che, Ligand-Promoted Dissolution in the Preparation of MoO_x/γ-Al_2O_3 Catalysts: Evidence for the Formation and Deposition of an Anderson-type Alumino Heteropolymolybdate, *J. Am. Chem. Soc.*, 119 (1997) 10137–10146.

J.W. Geus, Production and Thermal Pretreatment of Supported Catalysts, *Preparation of Catalysts III*, Elsevier, Amsterdam, 1983, pp. 1–32.

L.M. Knijff, P.H. Bolt, R. van Yperen, A.J. van Dillen, J.W. Geus, Production of nickel-on-alumina catalysts from preshaped bodies; K.P. de Jong, Deposition–precipitation onto pre-shaped carrier bodies — possibilities and limitations; P.J. van den Brink, A. Scholten, A. van Wageningen, M.D.A. Lamers, A.J. van Dillen, J.W. Geus, The use of chelating agents for the preparation of iron oxide catalysts for the selective oxidation of hydrogen sulfide, all in *Preparation of Catalysts V*, Elsevier, Amsterdam, 1991.

P.K. de Bokx, W.B.A. Wassenberg, and J.W. Geus, Interaction of Nickel Ions with γ-Al_2O_3 Support during Deposition from Aqueous Solutions, *J. Catal.*, 104 (1987) 86–98.

J.B. d'Espinose de la Caillerie, M. Kermarec, and O. Clause, Impregnation of γ-Al_2O_3 with Ni(II) or Co(II) ions at Neutral pH: Hydrotalcite-Type Coprecipitate Formation and Characterization, *J. Am. Chem. Soc.*, 117 (1995) 11471–11481.

Catalyst characterization

Chapter 11

Catalyst characterization with spectroscopic techniques

11.1 INTRODUCTION

Catalyst characterization plays an important role in catalysis. No fewer than 78% of the roughly 250 papers presented at the Ninth International Congress on Catalysis (Calgary, Canada, 1988) contained at least some results on the catalyst(s) obtained by characterization techniques. The remaining 22% of the papers dealt with reactions over essentially uncharacterized catalysts [1]. Figure 11.1 shows which techniques are often used in catalysis.

This chapter briefly describes the concepts of the techniques that are most frequently used for catalyst characterization. The intention is to give the newcomer an idea of what these methods tell about a catalyst. References to more extensive treatments are given at the end of this chapter [2–7]. Physisorption is dealt with in a separate chapter (Chapter 13) and two examples of temperature programmed techniques are treated in some detail in Chapter 12.

11.1.1 *Aim of Catalyst Characterization*

The catalytic properties of a surface are determined by its composition and structure *on the atomic scale*. Thus, it is not sufficient to know that a surface consists of a metal and a promoter, say iron and potassium, but it is essential to know the exact structure of the iron surface, including defects, steps, etc., as well as the exact location of the promotor atoms. Thus, from a fundamental point of view, the ultimate goal of catalyst characterization should be to look at the surface atom by atom, and under reaction conditions. At present, this is only occasionally possible in highly simplified model systems such as the well defined

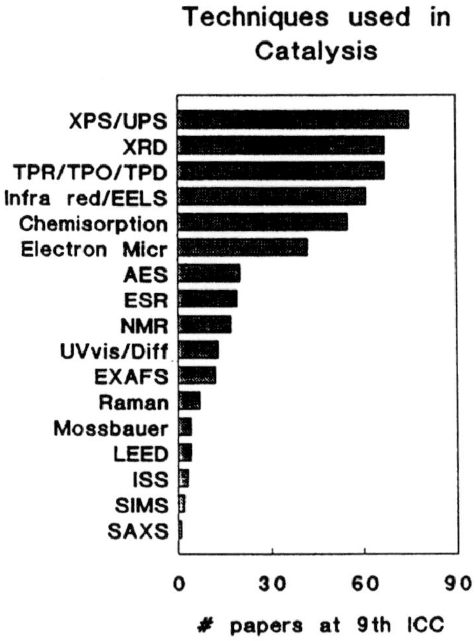

Fig. 11.1. Bar diagram showing how often certain techniques have been used in the papers at the Ninth International Congress on Catalysis [1].

surfaces of single crystals, or the needle shaped tips used in field emission studies [8].

The industrial point of view of catalyst characterization is different. Here the main interest is to optimize or produce an active, selective, stable and mechanically robust catalyst. In order to accomplish this, tools are needed which identify those structural properties that discriminate efficient from less efficient catalysts. In principle, all spectroscopic information that helps to achieve this is welcome. Empirical relations between the factors that govern catalyst composition, particle size and shape, and pore dimensions on one side, and catalytic performance on the other can be extremely useful in developing catalysts.

Simplifying, one could say that catalyst characterization in industrial research deals with the materials science of catalysts on a more or less mesoscopic scale, whereas the ultimate goal of fundamental catalytic research is to characterize the surface of a catalyst on the *atomic* scale.

11.2 TECHNIQUES

There are many ways to obtain information on the physico chemical properties of materials. Figure 11.2 presents a scheme from which almost all techniques can be derived. Spectroscopies are based on some type of excitation, represented by the

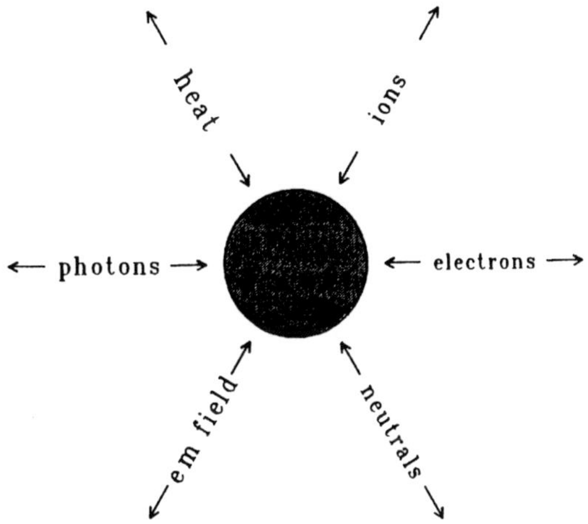

Fig. 11.2. Possibilities for catalyst investigations, each inward arrow stands for an excitation, each outward arrow indicates how the information is to be detected.

ingoing arrow in Fig. 11.2, to which the catalyst responds, symbolized by the outgoing arrow. For example, one can irradiate a catalyst with X-rays and study how the X-rays are diffracted (X-ray diffraction, XRD), or one may study the energy distribution of electrons that are emitted from the catalyst due to the photoelectric effect (X-ray photoelectron spectroscopy, XPS). One can also heat up a spent catalyst and look at what temperatures reaction intermediates and products desorb from the surface (temperature programmed desorption, TPD). We will now discuss some of the most important techniques for investigating catalysts.

11.2.1 X-Ray Diffraction (XRD)

According to Fig. 11.1, XRD is one of the most frequently applied techniques in catalyst characterization. X-rays have wavelengths in the Å range, are sufficiently energetic to penetrate solids and are well suited to probe their internal structure. XRD is used to identify bulk phases and to estimate particle sizes [9].

An X-ray source consists of a target which is bombarded with high energy electrons. The emitted X-rays arise from two processes. Electrons slowed down by the target emit a continuous background spectrum of Bremsstrahlung. Superimposed on this are characteristic, narrow lines. The Cu Kα line, for example, arises because a primary electron creates a core hole in the K shell, which is filled by an electron from the L shell (Kβ: the K-hole is filled from the M-shell, etc.) under emission of an X-ray quantum. The process is called X-ray fluorescence. It is not only the basis for X-ray sources but is also encountered in electron microscopy, EXAFS and XPS.

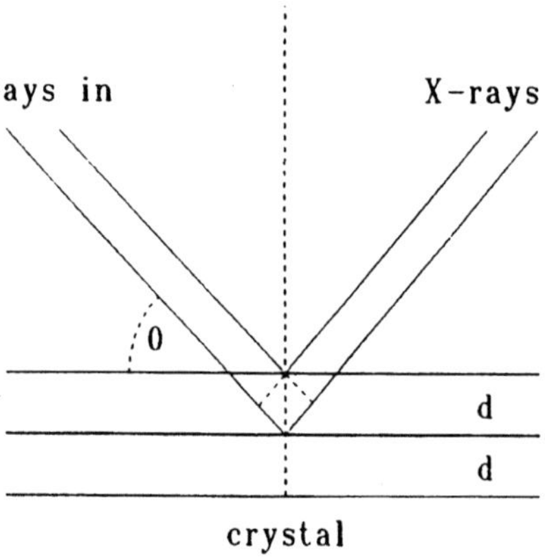

(a)

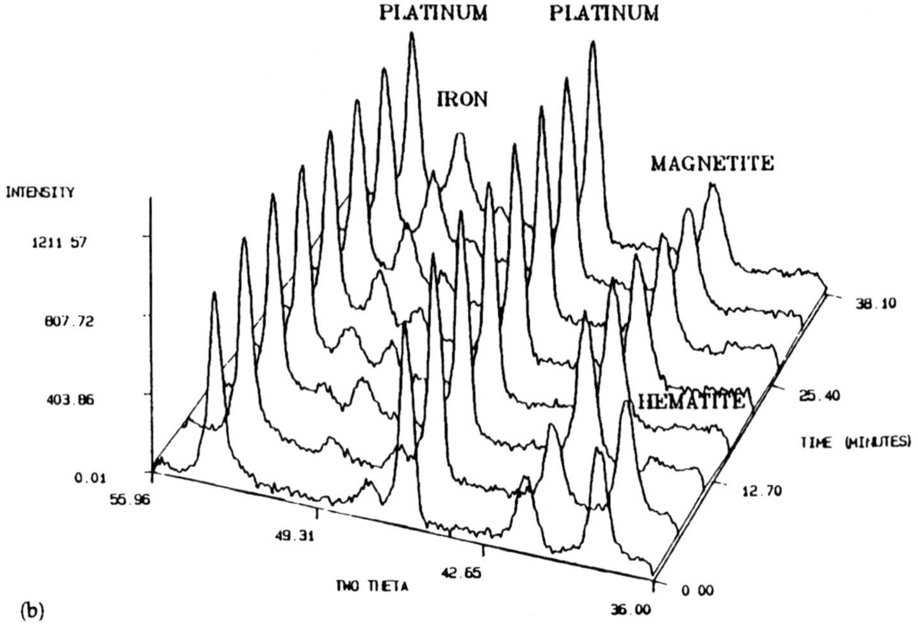

(b)

Fig. 11.3. Top: X-rays reflected by a series of atoms in a lattice plane interfere constructively in directions given by Bragg's Law. Bottom: XRD shows the reduction of alumina-supported iron oxide at 675 K (from Jung and Thomson [10]).

X-ray diffraction is the elastic scattering of X-ray photons by atoms in a periodic lattice. The scattered monochromatic X-rays that are in phase give constructive interference. Figure 11.3a illustrates how diffraction of X-rays by crystal planes allows one to derive lattice spacings, d, by measuring the angles, 2θ, under which constructively interfering X-rays with wave length, λ, leave the crystal, by using the Bragg relation:

$$n\,\lambda = 2d \sin\theta; \qquad n = 1,2,... \tag{11.1}$$

The XRD pattern of a powdered sample is measured with a stationary X-ray source (usually Cu Kα) and a movable detector, which scans the intensity of the diffracted radiation as a function of the angle 2θ between the incoming and the diffracted beams. When working with powdered samples, an image of diffraction lines occurs because a small fraction of the powder particles will be oriented such that by chance a certain crystal plane (*hkl*) is at the right angle with the incident beam for constructive interference.

Diffraction patterns can be used to identify the various phases in a catalyst. An example is given in Fig. 11.3b, where XRD is used to follow the reduction of alumina-supported iron oxide at 675 K as a function of time. The initially present α-Fe_2O_3 (haematite) is partially reduced to metallic iron, with Fe_3O_4 (magnetite) as the intermediate. The diffraction lines from platinum are due to the sample holder [10].

The width of diffraction peaks carries information on the dimensions of the reflecting planes. Diffraction lines from perfect crystals are very narrow. For crystallite sizes below 100 nm, however, line broadening occurs due to incomplete destructive interference in scattering directions where the X-rays are out of phase. The Scherrer formula relates crystal size to line width:

$$\langle L \rangle = \frac{K\lambda}{\beta \cos\theta} \tag{11.2}$$

in which β is the peak width and K a constant (often taken unity). The linear dimension $\langle L \rangle$ is a volume-averaged thickness of the crystallites, measured in a direction normal to the reflecting planes.

X-ray line broadening provides a quick but not always reliable estimate of the particle size. Better procedures are based on line-profile analysis with Fourier transform methods. The average size is obtained from the first derivative of the cosine coefficients and the distribution of particle sizes from the second derivative. The reader is referred to Cohen and Schwartz [9] for more details.

XRD may be very common in catalysis, but it does have disadvantages. Because XRD is based on interference between reflecting X-rays from lattice planes, the technique requires samples which possess sufficient long range order. Amorphous phases and small particles give either broad and weak diffraction lines or no diffraction at all, with the consequence that if catalysts contain particles with a size distribution, XRD may only detect the larger ones. In unfortunate cases, the diffraction lines from the metal may overlap with those from the support. Finally, the surface region is where catalytic activity resides, but this part of the catalyst is virtually invisible for XRD.

11.2.2 Electron Microscopy

Electron microscopy is a rather straightforward technique to determine the size and shape of supported particles. It can also reveal information on the composition of the particles, for example by detecting the X-rays which are produced by the interaction of the electrons with matter, or by analyzing the way how the electrons are diffracted [11].

Electrons have characteristic wavelengths in the range 0.1–1 nm (visible light: 400–700 nm), and come close to seeing atomic detail. Figure 11.4 summarizes what happens with the high energy (100–400 keV) electrons when the primary beam hits the sample.

- Depending on sample thickness a fraction of the electrons passes through the sample without suffering energy loss. As discussed later, these electrons are used in transmission electron microscopy (TEM).
- Electrons are diffracted by particles with the favourable orientation towards the beam, enabling one to obtain crystallographic information.
- Electrons can collide with atoms in the sample and be scattered back; back scattering is more effective when the mass of the atom increases.
- X-rays and Auger electrons are formed in the relaxation of core-ionized atoms, as discussed in the sections on AES and XPS.
- Electrons excite characteristic vibrations in the sample which can be studied by analyzing the energy loss suffered by the primary electrons.

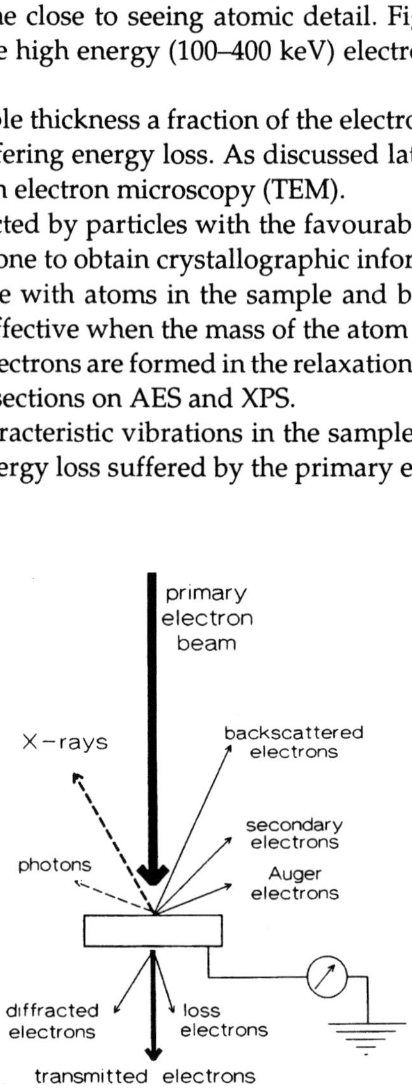

Fig. 11.4. The interaction of the primary electron beam with a sample in electron microscopy produces a wealth of detectable signals.

- Many electrons loose energy in a cascade of consecutive inelastic collisions. These are called secondary electrons; most of these had their last loss process in the surface region just before leaving the sample.
- The emission of a range of photons from UV to infrared is called cathodo-luminescence and is caused by the recombination of electron-hole pairs. Thus, the interaction of the primary beam with the sample provides a wealth of information on morphology, crystallography and chemical composition.

Transmission electron microscopy (TEM) uses transmitted and diffracted electrons. A TEM instrument is very similar to an optical microscope, if one replaces optical by electromagnetic lenses. In TEM, a high intensity primary electron beam passes through a condenser to produce parallel rays which impinge on the sample. As the attenuation of the beam depends on the density and the thickness, the transmitted electrons form a two-dimensional projection of the sample mass, which is subsequently magnified by the electron optics to produce a so-called bright field image. The dark field image is obtained from the diffracted electron beams, which are slightly off angle from the transmitted beam. For supported catalysts, dark field pictures show enhanced contrast for supported metals. Typical operating conditions of a conventional TEM instrument are 100 keV electrons, 10^{-6} mbar vacuum, 0.5 nm resolution and a magnification of 300,000. High resolution microscopes show atomic resolution and magnifications of 1,000,000.

Scanning electron microscopy (SEM) is carried out by rastering a narrow electron beam over the surface and detecting the yield of either secondary or back-scattered electrons as a function of the position of the primary beam. Contrast is caused by the orientation: parts of the surface facing the detector appear brighter than parts of the surface with their surface normal pointing away from the detector. The secondary electrons have low energies (10–50 eV) and originate from the surface region of the sample. Back-scattered electrons come from deeper and carry information on the composition of the sample, because heavy elements are more efficient scatterers and appear brighter in the image. Dedicated SEM instruments have resolutions of about 5 nm. Simple versions of SEM with micron resolution are often available on Auger electron spectrometers, for the purpose of sample positioning. The main difference between SEM and TEM is that SEM sees contrast due to the topology of a surface, whereas TEM projects all information in a two-dimensional image, which, however, is of sub nanometer resolution. A STEM instrument combines the two modes of electron microscopy.

As illustrated by Fig. 11.4, an electron microscope offers additional possibilities for analyzing the sample. Diffraction patterns (spots from a single-crystal particle and rings from a collection of randomly oriented particles) enable one to identify crystallographic phases as in XRD. Emitted X-rays are characteristic for an element and allow for a determination of the chemical composition of a selected part of the sample (typical dimension 10 nm). This technique is called electron microprobe analysis (EMA, EPMA) or, referring to the usual mode of detection, energy dispersive analysis of X-rays (EDAX or EDX). Also the Auger electrons carry information on sample composition, as do the loss electrons. The latter are potentially informative on the low Z elements, which have a low efficiency for X-ray fluorescence.

For studying supported catalysts, TEM is the commonly applied form of electron microscopy. Detection of supported particles is possible provided that there is sufficient contrast between particles and support. This may impede applications of TEM on supported oxides. Determination of particle sizes or of distributions therein rests on the assumption that the size of the imaged particle is truly proportional to the size of the actual particle and that the detection probability is the same for all particles, independent of their dimensions. Semi *in situ* studies of catalysts are possible by coupling the instrument to an external reactor. After evacuation of the reactor, the catalyst can be transferred directly into the analysis position, without seeing air [12]. For reviews of TEM on catalysts we refer to Refs. [11,12].

11.2.3 Temperature Programmed Techniques

Temperature programmed reduction (TPR), oxidation (TPO), desorption (TPD), sulphidation (TPS) and reaction spectroscopy (TPRS) form a class of techniques in which a chemical reaction is monitored while the temperature increases linearly in time. These techniques are applicable to real catalysts and single crystals and have the advantage that they are experimentally simple and inexpensive in comparison to many other spectroscopies. Interpretation on a qualitative base is rather straightforward [13].

The basic set up for TPR, TPO and TPD consists of a reactor and a thermal conductivity detector to measure the hydrogen content in TPR/TPD or the oxygen content in TPO, of the gas mixture before and after reaction. More sophisticated TP equipment contains a mass spectrometer for the detection of reaction products.

TPR and TPO patterns of silica-supported rhodium, iron, and iron–rhodium catalysts are shown in Fig. 11.5 [14]. These catalysts were prepared by pore volume impregnation from aqueous solutions of iron nitrate and rhodium chloride. Note the difference in reduction temperature between the noble metal rhodium and the non-noble metal iron. The bimetallic combination reduces largely in the same temperature range as the rhodium catalyst does, indicating that rhodium catalyzes the reduction of the iron. This forms evidence that rhodium and iron are well mixed in the fresh catalyst. The TPR patterns of the freshly prepared catalysts consist of two peaks, one coincides with that of the TPR pattern of the fully oxidized catalyst (right panel of Fig. 11.5) and can thus be

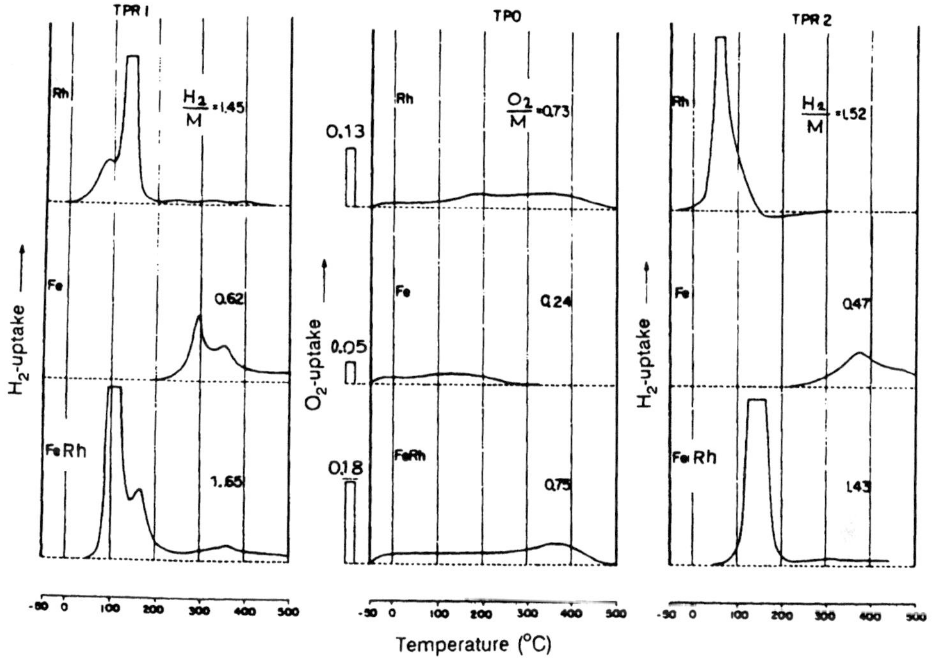

Fig. 11.5. TPR/TPO of silica-supported Rh, Fe, and Fe–Rh catalysts. The left TPR curves have been measured from freshly impregnated catalysts, the right ones after complete oxidation in the TPO experiment (from van't Blik and Niemantsverdriet [14]).

assigned to the reduction of the metal oxide. The remaining peak of the first TPR is assigned to the reduction of the metal salt precursors, iron nitrate and rhodium chloride.

The area under a TPR or TPO curve represents the total hydrogen consumption, and is commonly expressed in terms of moles of H_2 consumed per mol of metal atoms, H_2/M. The ratios of almost 1.5 for rhodium in Fig. 11.5 indicate that rhodium was present in the form of Rh^{3+}. For iron, the H_2/M ratios are significantly lower, indicating that this metal is only partially reduced.

TPR provides a quick characterization of metallic catalysts. It gives information on the phases present after impregnation and on the eventual degree of reduction. For bimetallic catalysts, TPR patterns often indicate if the two components are mixed or not. A more detailed description of TPR and TPS is given in Chapter 12. Temperature programmed desorption (TPD) is popular for determining concentrations of species adsorbed on the surfaces of single crystals and of real catalysts. In principle one can also determine the energy of the adsorption bond, however, only after applying rather involved computations [15]. We refer to Ref. [16] for a critical evaluation of TDS analysis.

11.2.4 Surface Spectroscopy

A technique becomes surface sensitive if the radiation or particles to be detected can travel no more than a few atomic distances through the solid. Low energy electrons and ions are particularly suitable for investigating the surface region. The consequence of working with electrons and ions is that experiments have to be done in vacuum (at least 10^{-6} mbar) otherwise the particles leaving the surface would not be able to reach a detector. The reason that surface spectroscopies are preferably applied under ultra high vacuum (pressures in the 10^{-10} mbar range) is that, in order to obtain meaningful results on the intrinsic properties of a clean surface, or of the surface with a certain adsorbed gas, contamination by residual gases in the measurement chamber should be prevented. For the characterization of catalysts the requirements on vacuum are less stringent and pressures on the order of 10^{-8} mbar are quite acceptable.

11.2.4.1 The electron spectroscopies: XPS, UPS, AES [17,18]

X-ray photoelectron spectroscopy (XPS) is based on the photoelectric effect: an atom absorbs a photon of energy hν; next, a core or valence electron with binding energy E_b is ejected with kinetic energy (Fig. 11.6):

$$E_k = h\nu - E_b \tag{11.3}$$

Routinely used X-ray sources are Mg Kα ($h\nu$ = 1253.6 eV) and Al Kα ($h\nu$ = 1486.3 eV). In XPS one measures the intensity of photoelectrons $N(E)$ as a function of their kinetic energy E_k. The XPS spectrum is a plot of $N(E)$ versus E_k, or, more often, versus binding energy E_b. Figure 11.7a shows the XPS spectrum of a freshly impregnated alumina-supported rhodium catalyst.

Photoelectron peaks are labelled according to the quantum numbers of the level from which the electron originates. An electron coming from an orbital with main quantum number n, orbital momentum l (0,1,2,3,.. indicated as s, p, d, f, ..) and spin momentum s (+1/2 or −1/2) is indicated as nl_{l+s}. Examples are Rh $3d_{5/2}$, Rh $3d_{3/2}$, O $1s$, Fe $2p_{3/2}$, Pt $4f_{7/2}$. For every orbital momentum $l > 0$ there are two values of the total momentum: $j = l + 1/2$ and $j = l - 1/2$, each state filled with $2j$ + 1 electrons. Hence, most XPS peaks come in doublets and the intensity ratio of the components is $(l + 1)/l$. In case the doublet splitting is too small to be observed (as in practice with Si $2p$, Al $2p$, Cl $2p$), the subscript $l + s$ is omitted.

Because a set of binding energies is characteristic for an element, XPS can be used to analyze the composition of samples. Almost all photoelectrons used in laboratory XPS have kinetic energies in the range of 0.2 to 1.5 keV, and probe the outer layers of the catalyst. The mean free path of electrons in elemental solids depends on the kinetic energy. Optimum surface sensitivity is achieved with

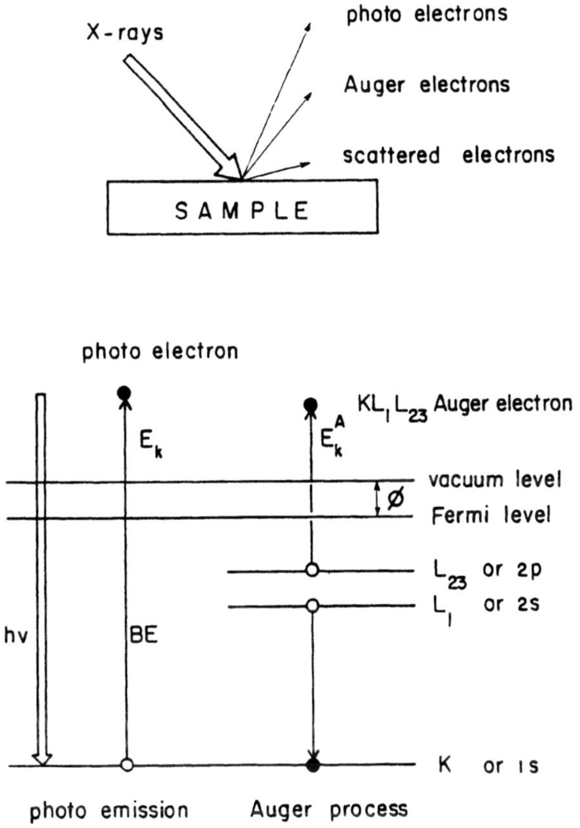

Fig. 11.6. Photoemission and Auger decay: an atom absorbs an incident X-ray photon with energy h and emits a photoelectron with kinetic energy $E_k = h\nu - E_b$. The excited ion decays either by the indicated Auger process or by X-ray fluorescence.

electrons at kinetic energies of 50–250 eV, where about 50% of the electrons come from the outermost layer.

Binding energies are not only element specific but contain chemical information as well: the energy levels of core electrons depend on the chemical state of the atom. Chemical shifts are typically in the range 0–3 eV. Figure 11.7b shows that the binding energy of the Rh $3d_{5/2}$ peak of $RhCl_3$ in the XPS spectrum of a freshly impregnated catalyst is some 2.5 eV higher than that of rhodium metal in the spectrum of the reduced catalyst [19]. The reason is that the electrons of the Rh^{3+} ion feel a higher attractive force from the nucleus than those of the neutral atom. In general, the binding energy increases with increasing oxidation state and, for a fixed oxidation state, with the electronegativity of the ligands, as the series $FeBr_3$, $FeCl_3$, FeF_3 in Table 11.1 illustrates.

Note that XPS measures *binding energies*. These are not necessarily equal to the energies of the orbitals from which the photoelectron is emitted. The difference is

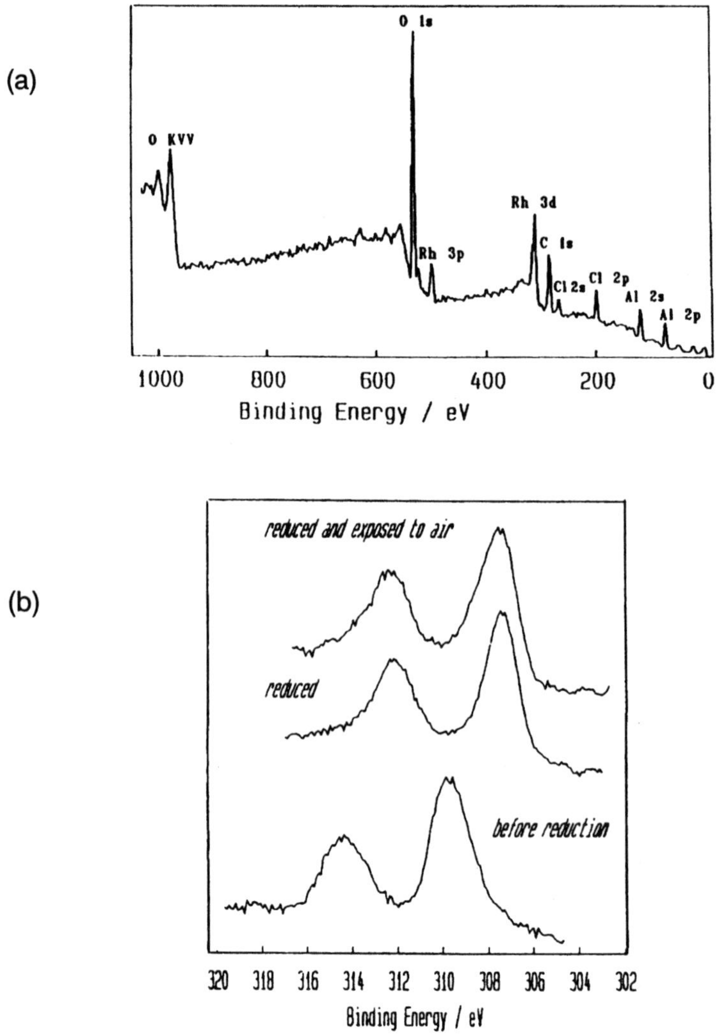

Fig. 10.7. XPS spectra of Rh/Al_2O_3 prepared by impregnation from a $RhCl_3$ solution, (a) survey spectrum of low resolution, (b) Rh 3d spectra at high resolution (from Borg et al. [19]).

caused by the reorganization of the remaining electrons when an electron is removed from an inner shell. Thus the binding energy of the photoelectron contains both information on the state of the atom before photoionization (the initial state) and on the core-ionized atom left behind after the emission of an electron (the final state). Fortunately, it is often correct to interpret binding energy shifts as those in Fig. 11.7b in terms of initial state effects. The charge potential model elegantly explains the physics behind such binding energy shifts, by means of the formula:

TABLE 11.1

Binding energies of Fe $2p_{3/2}$ electrons in several components [20]

compound	E_b (eV)
iron metal	706.7
FeO	710.0
Fe_2O_3	710.7
$Fe(CO)_5$	709.4
$FeBr_3$	710.0
$FeCl_3$	711.1
FeF_3	714.0

$$E_b^i = kq_i + \sum_j \frac{q_j}{r_{ij}} + E_b^{\mathrm{ref}} \tag{11.4}$$

in which E_b^i is the binding energy of an electron from an atom i, q_i the charge on the atom, k a constant, q_j the charge on a neighbouring atom j at distance r_{ij} from atom i, and E_b^{ref} a suitable reference. The first term indicates that the binding energy goes up with increasing positive charge on the atom from which the photoelectron originates. In ionic solids, the second term counteracts the first, because the charge on a neighbouring atom will have the opposite sign. Because of its similarity to the lattice potential in ionic solids, the second term is often referred to as the Madelung sum.

Although XPS is predominantly used for studying surface compositions and oxidation states, the technique can also yield information on the dispersion of supported catalysts, as illustrated by the example on ZrO_2/SiO_2 catalysts, see Fig. 11.8a. [21]. The spectrum of a catalyst made by a new preparation process starting from zirconium ethoxide exhibits a significantly higher Zr intensity than those of catalysts prepared by the usual impregnation from a zirconium nitrate solution. Because the escape depth of the electrons in XPS is limited to a few nanometers, a higher Zr/Si intensity ratio (at the same zirconium content) indicates that the ZrO_2 phase is better dispersed.

One can use the Zr/Si intensity ratio to investigate the thermal stability of the catalysts. As Fig. 11.8b shows, the impregnated catalysts show a decrease in Zr/Si ratio at relatively low calcination temperatures, indicating that the zirconium phase loses dispersion. The catalyst prepared via the new ethoxide route is much better resistant to sintering. Dispersion values can be calculated from the Zr/Si ratio, by using a model, for example the one published by Kuipers et al. [22]. When applied to the spectra of Fig. 11.8, Kuipers' model indicates that the

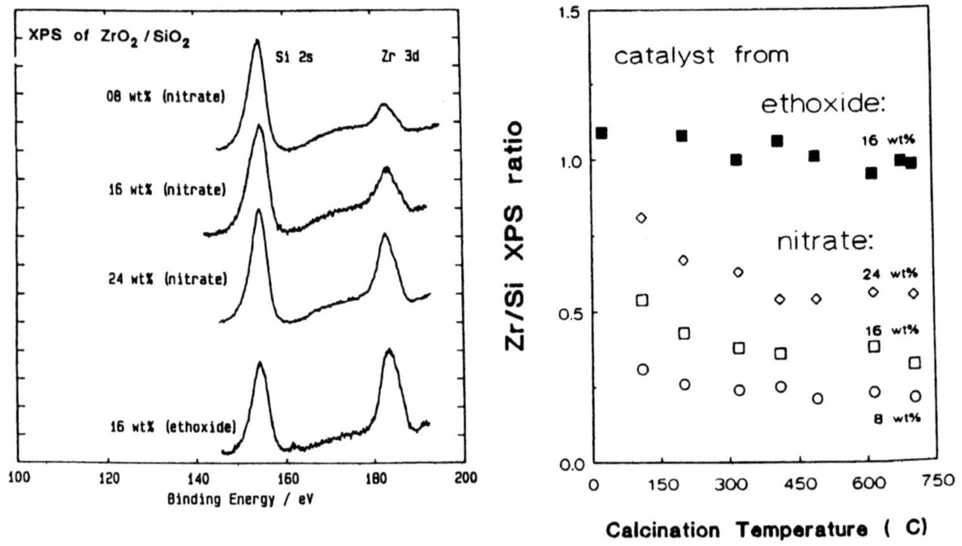

Fig. 11.8. Left: XPS spectra of calcined ZrO_2/SiO_2 catalysts; right: Zr/Si intensity ratios indicate that the catalyst prepared from zirconium ethoxide is much better dispersed than the ones from nitrate (from Meijers et al. [21]).

ZrO_2 dispersion of the three impregnated and calcined catalysts is 5 to 15%, whereas the dispersion of the catalyst from ethoxide is around 75%.

An experimental problem in XPS (and in other electron or ion spectroscopies as well) is that electrically insulating samples may charge up during measurement, because photoelectrons leave the sample. The potential which the sample acquires is determined by the photoelectric current of electrons leaving the sample, the current through the sample holder (usually grounded) towards the sample and the flow of Auger and secondary electrons from the source window onto the sample. Due to the positive charge on the sample, all XPS peaks in the spectrum shift by the same amount to higher binding energies. More serious than the shift itself is that different parts of the catalysts may acquire slightly different amounts of charge. This phenomenon, called differential charging, gives rise to broadening of the peaks, which leads to less resolution, see Fig. 11.9. In fact, the broad Zr 3*d* signal in the spectra of Fig. 11.8 is actually a doublet (similar to the Rh 3*d* signal in Fig. 11.7b) but is not measured as such due to differential charging.

Correcting for charge shifts is done by using the binding energy of a known compound. In SiO_2-supported catalysts one uses the binding energy of the Si 2*p* electrons which should be 103.4 eV. If nothing else is available, one can use the always present carbon contamination with a C 1*s* binding energy of 284.4 eV. The use of a floodgun, which sprays low energy electrons onto the sample, and

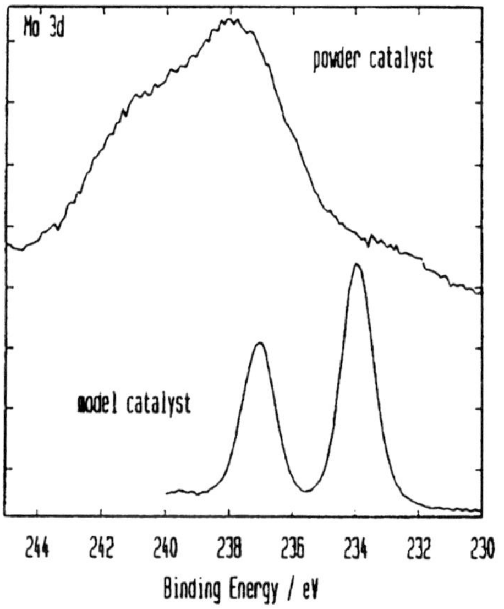

Fig. 11.9. Effect of charging on the Mo 3d signal in the XPS spectrum of a MoO_3/SiO_2 catalyst. Charging is absent in a conducting model catalyst [23].

special sample mounting techniques in which powders are pressed in a conducting indium foil, may alleviate the charging problems to some extent.

Sensitive materials, such as metal salts or organometallic compounds used as catalyst precursors, may decompose during XPS analysis, particularly when a standard X-ray source is used. Heat and electrons from the source may cause damage to samples. In such cases a monochromatic XPS offers a solution [24].

Ultraviolet photoelectron spectroscopy (UPS) differs from XPS in that UV light (He I, 21.2 eV; He II 40.8 eV) is used. At these low exciting energies, photoemission is limited to valence electrons. Hence, UPS spectra contain important chemical information. At low binding energies UPS probes the density of states (DOS) of the valence band (but images it in a distorted way in a convolution with the unoccupied states). At slightly higher binding energies (5–15 eV), occupied molecular orbitals of adsorbed gases may become detectable. UPS also provides a quick measure of the macroscopic work function, ϕ, the energy separation between the Fermi and the vacuum level: $\phi = h\nu - W$, where W is the width of the spectrum. UPS is a typical single crystal technique applied in surface science, but is rarely used on catalysts. The main reason is that all elements contribute peaks to the valence band region. As a result, the UPS spectra of supported catalysts, which contain at least three elements and often more, are rather complicated.

After the emission of a photoelectron in XPS, the atom is left behind as an ion with a hole in one of its core levels. This is an unstable state. Deexcitation of the excited ion occurs via X-ray fluorescence or via an Auger transition (Fig. 11.6).

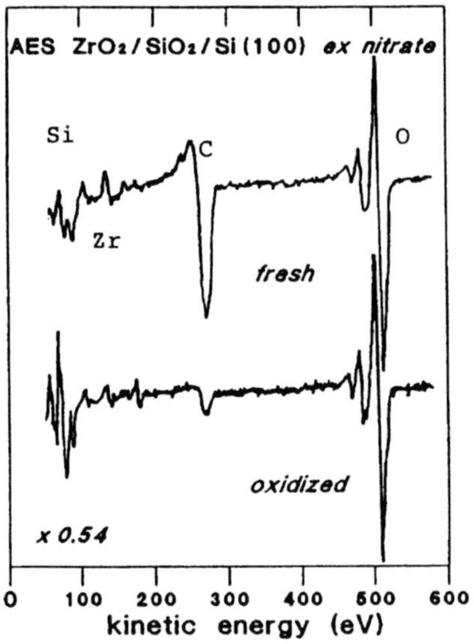

Fig. 11.10. Auger spectra of a conducting $ZrO_2/SiO_2/Si(100)$ model catalyst after impregnation and calcination (from Eshelman et al. [25]).

Therefore, XPS spectra contain peaks due to Auger electrons, which have the characteristic property that they occur at fixed kinetic energies, characteristic of the element from which they are emitted.

In Auger electron spectroscopy (AES), core holes are created by exciting the sample with a beam of electrons (1–5 keV). The Auger electrons appear as small peaks on a high background of secondary electrons scattered by the sample from the incident beam. To enhance the visibility of the Auger peaks, spectra are usually presented in the derivative dN/dE mode. Figure 11.10 shows the Auger spectra measured from a flat, conducting model for a SiO_2-supported ZrO_2 catalyst, in which the signals of Zr, Si, O and C are clearly visible [25]. Note the decrease in intensity of the carbon impurity upon oxidation, as well as the appearance of a new impurity, chlorine, visible in the spectrum at about 180 eV.

The strong point of AES is that it provides a quick measurement of the surface region of conducting samples. For many catalytically relevant elements (C, Cl, S, Pt, Ir, Rh) the corresponding Auger electrons have energies in the range of 100–300 eV where the mean free path of the electrons is close to its minimum. For those elements, AES is considerably more surface sensitive than XPS. A disadvantage of AES is that the intense electron beam easily causes damage to sensitive materials (polymers, insulators, adsorbate layers). Charging of isolating samples also presents serious problems. This is the main reason that AES, although of great importance in surface science and materials science, is not often used on supported catalysts.

In conclusion, XPS is among the most frequently used techniques in catalysis. The advantages of XPS are that it readily provides the composition of the surface region and that it can also distinguish between chemical states of one element. XPS is becoming an important tool for studying the dispersion of active phases over the support. The related techniques UPS and AES are very useful in surface science but, with a few exceptions for AES, less suitable for the characterization of catalysts.

11.2.4.2 The ion spectroscopies: SIMS, LEIS, RBS

When a beam of low energy (≤ 5 keV) ions impinges on a surface, the ions can penetrate the sample and loose their energy in a series of inelastic collisions with the target atoms. This may lead to the emission of secondary ions and neutrals from the surface (SIMS, SNMS). Another possibility is that the ions scatter elastically from the surface atoms (low energy ion scattering) or, when high energy ions are used, from atoms deeper in the sample (Rutherford back scattering). In addition to this, ions can be neutralized by taking up an electron from the sample.

Secondary ion mass spectrometry (SIMS) is by far the most sensitive surface technique, but also the most difficult to quantify [26,27]. When a surface is exposed to a beam of ions (Ar^+, 0.5–5 keV), energy is deposited in the surface region of the sample by a collisional cascade. Some of the energy will return to the surface and stimulate the ejection (desorption) of atoms, ions, and multi atomic clusters. In SIMS, positive or negative secondary ions are detected directly with a quadrupole mass spectrometer. In the more recently developed SNMS, the secondary neutrals are ionized, for example by electron impact ionization, before they enter the mass spectrometer.

Figure 11.11 shows the positive SIMS spectrum of a silica-supported zirconium oxide catalyst precursor, freshly prepared by a condensation reaction between zirconium ethoxide and the hydroxyl groups of the support [21]. Note the simultaneous occurrence of single ions (H^+, Si^+, Zr^+) and molecular ions (SiO^+, $SiOH^+$, ZrO^+, $ZrOH^+$, ZrO_2+). Also the isotope pattern of zirconium is clearly visible. Isotopes are important in the identification of peaks, because all peak intensity ratios must agree with the natural abundances. In addition to the peaks expected from zirconia on silica mounted on an indium foil, the spectrum of Fig. 11.11a also contains peaks from Na^+, K^+, and Ca^+. This is typical for SIMS: sensitivities vary over several orders of magnitude and elements such as the alkalis are already detected when present in trace amounts.

SIMS is strictly speaking a destructive technique. In the dynamic mode, used for making concentration depth profiles, several tens of monolayers are removed per minute. In static SIMS, however, the rate of removal corresponds to one monolayer per hour or more, implying that the surface structure does not change during the measurement (between seconds and minutes). In this case one can be

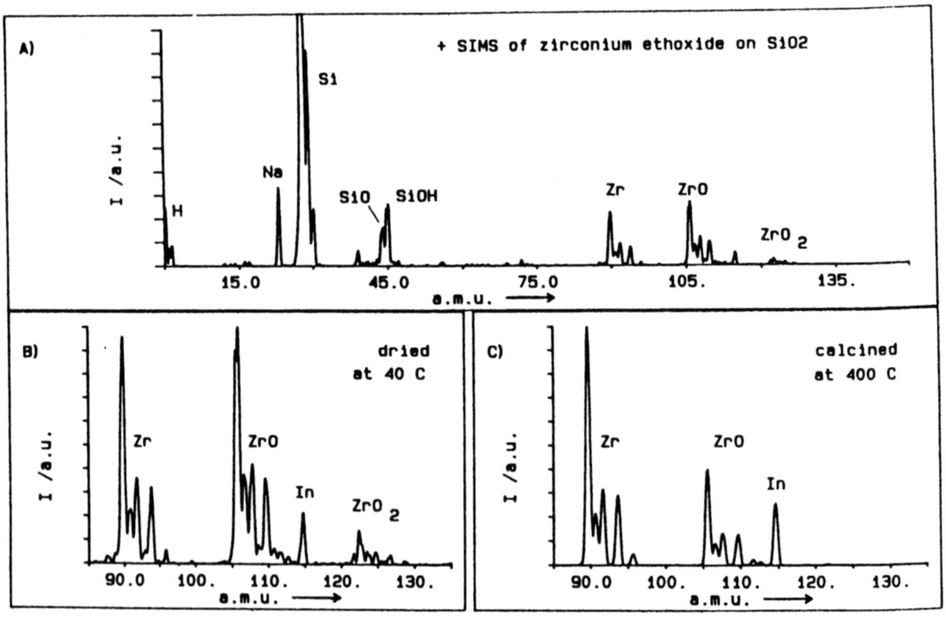

Fig. 11.11. Positive SIMS spectra of a ZrO_2/SiO_2 catalyst, (A,B) after preparation from $Zr(OC_2H_5)_4$ and (C) after calcination in air at 775 K (from Meijers et al. [21]).

sure that the molecular ion fragments are truly indicative of the chemical structure on the surface.

The advantages of SIMS are its high sensitivity (detection limit of ppm for certain elements), its ability to detect hydrogen and the emission of molecular fragments which often bear tractable relationships with the parent structure on the surface. Disadvantages are that secondary ion formation is a poorly understood phenomenon and that quantization is often difficult. A major drawback is the matrix effect: secondary ion yields of one element can vary tremendously with its chemical environment. This matrix effect and the elemental sensitivity variation of five orders of magnitude across the periodic table make quantitative interpretation of SIMS spectra of technical catalysts extremely difficult.

In spite of quantization problems, SIMS can give useful information if one uses appropriate reference materials. The spectra in Fig. 11.11b and c indicate that the relative intensities of the ZrO_2^+, ZrO^+ and Zr^+, contain information on the chemical environment of the zirconium. Spectrum 11b of the zirconium ethoxide (O:Zr = 4:1) shows higher intensities of the ZrO_2^+ and ZrO^+ signals than the calcined ZrO_2 (O:Zr = 2:1) does. The way to interpret this information is to compare the spectra of the catalysts with reference spectra of ZrO_2 and zirconium ethoxide [21].

For single crystals, matrix effects are largely ruled out and excellent quantization has been achieved by calibrating SIMS yields by means of other techniques such as electron energy loss spectroscopy or thermal desorption spectroscopy [26]. Here SIMS offers the challenging perspective to monitor reactants, intermediates and products of catalytic reactions in real time while the reaction is in progress.

SNMS does not suffer from the matrix effects that are associated with ionization. Here the sensitivity for a certain element is mainly determined by its sputter yield. Sputtering is relatively well understood, and consequently, excellent quantization of SNMS has been demonstrated. SNMS is a relatively new technique; it has not yet been fully exploited on catalysts.

In low energy ion scattering (LEIS, also called ion scattering spectroscopy, ISS) a beam of noble gas ions with energy of a few keV scatters elastically from a solid surface. The energy of the outgoing ion is determined by energy and momentum conservation and reveals the mass of the target atom from which it scattered. When the ion energy detector is placed at an angle of 90$ with respect to the incident beam, one has:

$$\frac{E}{E_o} = \frac{M_{at} - M_{ion}}{M_{at} + M_{ion}} \tag{11.5}$$

in which E_o is the energy of the incident ion, E its energy after scattering, M_{ion} the mass of the incident ion and M_{at} the mass of the atom with which the ion collides. Hence, the energy spectrum of the back-scattered ions is equivalent to a mass spectrum of the surface atoms [7]. Figure 11.12 shows LEIS spectra of an alumina-supported lanthanum oxide catalyst. In agreement with the expression for the energy, Figure 11.12 illustrates that the primary ions loose less energy to

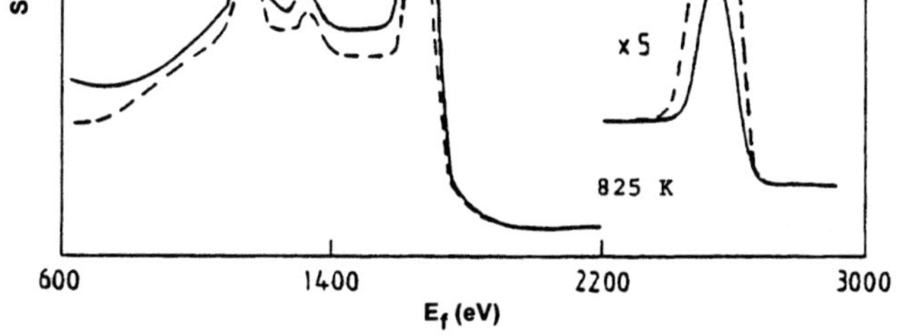

Fig. 11.12. LEIS spectra of La_2O_3/Al_2O_3 catalysts after calcination at 825 and 1325 K, taken with a 3 keV He^+ beam indicate that lanthanum oxide spreads over alumina at high temperatures (from van Leerdam et al. [28]).

heavy atoms than to light atoms. The La dispersion is reflected by the intensity ratio La/Al of the corresponding peaks in the LEIS spectrum: the La/Al ratio is low when lanthanum oxide occurs in particles, and high when it forms a monolayer. This is similar as in the ZrO_2/SiO_2 example we discussed in the XPS section, but with the difference that the information now comes from the outer one or two layers only. The spectra in Fig. 11.12 indicate that La_2O_3 spreads over the support at high temperatures [28].

Two factors determine the intensity of the scattered beam: the scattering cross section for the incident ion–target atom combination and the neutralization probability of the ion in its interaction with the solid. It is the latter quantity that makes LEIS surface sensitive: 1 keV He ions have a neutralization probability of about 99% in passing through one layer of substrate atoms. Hence, the majority of ions that reach the detector must have scattered off the outermost layer. At present, there is no simple theory which describes the scattering cross section and the neutralization probability adequately. Satisfactory calibration procedures by use of reference samples exist, however.

The fact that LEIS provides quantitative information on the outer layer composition of multicomponent materials makes this technique a powerful tool for the characterization of catalysts. The somewhat limited mass resolution may, however, impede certain applications. For example, horizontal neighbours in the heavier metals range of the periodic table (e.g. Pt and Ir, Pd and Ag) are not distinguished. An interesting feature of LEIS is that atoms in the second layer can be distinguished from those in the first by a broadening of the peaks to lower energies, because an ion scattered in the second layer has a considerable chance to loose an additional amount of energy on its way back through the first layer. (Note the high step in the background on the low energy side of the Al peak of the support, in Fig. 11.12.) In addition, concentration depth profiles of the constituents can be investigated by using the sputtering action of the ion beam.

In Rutherford back scattering, RBS, one uses a primary beam of high energy H^+ or He^+ ions (1–5 MeV), which scatter from the nuclei of the atoms in the target. A fraction of the incident ions is scattered back and is subsequently analyzed for energy. As in LEIS, the energy spectrum represents a mass spectrum, but this time it is characteristic for the interior of the sample [7]. The technique has successfully been applied to determine the concentration of, for example, rhodium in model systems where the rhodium is present in sub monolayer quantities on thin, conducting oxide films [19].

11.2.5 *Infrared Spectroscopy*

The most common application of infrared spectroscopy in catalysis is to identify adsorbed species and to study the way in which these species are chemisorbed on the surface of the catalyst. Sometimes the infrared spectra of adsorbed probe molecules such as CO and NO give valuable information on the adsorption sites that are present on a catalyst. We will first summarize the theory behind infrared absorption [29].

Molecules possess discrete levels of rotational and vibrational energy. Transitions between vibrational levels occur by absorption of photons with frequencies ν in the infrared range (wavelength 1–1000 micrometer, wave numbers 10000–10 cm^{-1}, energy differences 1240–1.24 meV). The C–O stretch vibration, for example, is at 2143 cm^{-1}. For small deviations of the atoms in a vibrating diatomic molecule from their equilibrium positions, the potential energy $V(r)$ can be approximated by that of the harmonic oscillator:

$$V(r) = \frac{1}{2} k (r - r_{eq})^2 \tag{11.6}$$

in which k is the force constant of the vibrating bond, r the distance between the two atoms, and r_{eq} the distance at equilibrium. The corresponding vibrational energy levels are

$$E_n = \left(n + \frac{1}{2}\right) h\nu; \quad \nu = \frac{1}{2\pi}\sqrt{\frac{k}{\mu}}; \quad \frac{1}{\mu} = \frac{1}{m_1} + \frac{1}{m_2} \tag{11.7}$$

in which $n = 0, 1, 2, \ldots$, h is Planck's constant, ν the frequency, 0 m_1 and m_2 are the masses of the vibrating atoms and μ is the reduced mass. Allowed transitions in the harmonic approximation are those for which the vibrational quantum number changes by one unit. A general selection rule for the absorption of a photon is that the dipole moment of the molecule must change during the vibration.

The harmonic approximation is only valid for small deviations of the atoms from their equilibrium positions. The most obvious shortcoming of the harmonic potential is that the bond between two atoms cannot break. With physically more realistic potentials, such as the Lennard-Jones or the Morse potential, the energy levels are no longer equally spaced and vibrational transitions with $\Delta n > 1$ are no longer forbidden. Such transitions are called overtones. The overtone of gaseous CO at 4260 cm^{-1} (slightly less than $2 \times 2143 = 4286$ cm^{-1}) is an example.

The simple harmonic oscillator picture of a vibrating molecule has important implications. First, knowing the frequency, one can immediately calculate the force constant of the bond. Note from Eqn. (11.6) that k, as coefficient of r^2, corresponds to the curvature of the interatomic potential and not primarily to its depth, the bond energy. However, as the depth and the curvature of a potential usually change hand in hand, it is often allowed to take the infrared frequency as an indicator for the strength of the bond. Second, isotopic substitution can be useful in the assignment of frequencies to bonds in adsorbed species, because frequency shifts due to isotopic substitution (of for example D for H in adsorbed ethylene, or OD for OH in methanol) can be predicted directly.

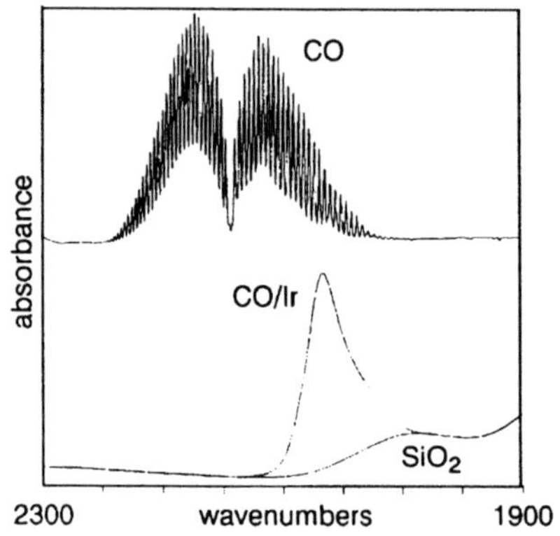

Fig. 11.13. The infrared spectrum of gas phase CO shows rotational fine structure, which is absent in the spectrum of CO adsorbed on an Ir/SiO_2 catalyst [30].

Molecules in the gas phase have rotational freedom, and the vibrational transitions are accompanied by rotational transitions. For a rigid rotor which vibrates as a harmonic oscillator the expression for the available energy levels is:

$$E = \left(n + \frac{1}{2}\right)h\nu + \frac{h^2}{8\pi^2 I}\, j(j+1) \tag{11.8}$$

where j is the rotational quantum number and I the moment of inertia of the molecule. Here a third selection rule applies: $\Delta j = \pm 1$. Note that because of this third selection rule the purely vibrational transition is not observed in gas phase CO. Instead, two branches of rotational side bands appear, see Fig. 11.13 for CO.

Upon adsorption, the molecule loses its rotational freedom and now only the vibrational transition is observed, however, at a different frequency, see Fig. 11.13. In the case of CO, three factors contribute to this shift:

- mechanical coupling of the CO molecule to the heavy substrate increases the CO frequency by some 20 to 50 cm^{-1},
- the interaction between the C–O dipole and its image in the conducting, polarizable metal weakens the CO frequency by 25–75 cm^{-1},
- the formation of a chemisorption bond between CO and the substrate alters the distribution of electrons over the molecular orbitals and weakens the CO bond.

Thus, strictly speaking, it is not correct to interpret the frequency difference between adsorbed and gas phase CO in terms of chemisorption bond strength only.

Infrared frequencies are characteristic for certain bonds in molecules and they can often be used to identify chemisorbed species on surfaces. Investigation of adsorbed species in relation to their behaviour in catalytic reactions is the main field of application for infrared spectroscopy, but falls outside the scope of this chapter. However, infrared spectra of adsorbates can sometimes also be used to recognize sites on the surface of a catalyst, as the following two examples show.

Sulphided Mo and Co–Mo catalysts, used in hydrotreating reactions, contain Mo as MoS_2. This compound has a layer structure consisting of sandwiches, each of a Mo layer between two S layers. The chemical activity of MoS_2 is associated with the edges of the sandwich where the Mo is exposed to the gas phase; the basal plane of the S^{2-} anions is largely unreactive. The infrared spectrum of NO on a sulphided Mo/Al_2O_3 catalyst (Fig. 11.14a) shows two peaks at frequencies which agree with those observed in organometallic clusters of Mo and NO groups [31]. NO on sulphided Co/Al_2O_3 also gives two infrared peaks (Fig.

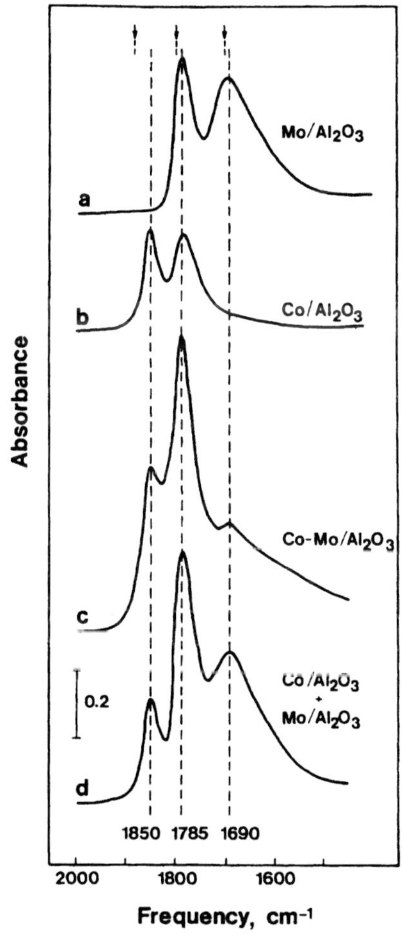

Fig. 11.14. Infrared spectra of NO probe molecules on sulphided Mo, Co, and Co–Mo hydrodesulphurization catalysts allow for titration of Co and Mo sites in the Co–Mo catalyst (from Topsøe and Topsøe [31]).

11.14b), but at different frequencies as observed on MoS_2. These results suggest that NO can be used as a probe to titrate the number of Co and Mo sites in the Co–Mo/Al_2O_3 catalyst. Figure 11.14c confirms that this idea works. Moreover, comparison of the intensities of the NO/Mo infrared signals on Mo/Al_2O_3 and Co–Mo/Al_2O_3 reveals that the presence of Co decreases the number of Mo atoms that are accessible to NO. This means that Co most probably decorates the edges of the MoS_2 sandwiches, because the edges constitute the adsorption sites for NO [31].

Infrared spectroscopy is also an important tool to investigate hydroxyl groups on the surfaces of oxidic catalyst supports and zeolites [32]. These OH groups may possess positive, zero or negative charge and are called acidic, neutral or basic, respectively. In solution these groups exchange with metal ion complexes (OH^- with a negative ion complex, H^+ with a cation) or they provide sites where ions of opposite charge adsorb by electrostatic interaction.

As an example, let us take the (100) surface of TiO_2 in the crystallographic form of anatase. The surface is overall neutral, but possesses charged sites: in the absence of water, it contains Ti^{4+} cations with a net excess charge of +2/3, and two types of oxygen atoms, one with zero excess charge and the other with a net charge of –2/3 [33]. As a result, heterolytic dissociation of water occurs, in which the protons adsorb on the oxygen atoms with negative excess charge, leading to a bridging OH group of acidic character (excess charge +1/3), and the OH^- group attaches to the Ti ion, giving a terminal OH group of basic character (excess charge –1/3). The latter can exchange with negative ions such as $\{RhCl_x(OH)_y\}^-$ or F^-.

The different hydroxyl groups (see Fig. 11.15) have different vibrational properties and can be distinguished by their O–H stretch frequencies in infrared spectroscopy.

As Fig. 11.15A shows, the OH stretch region of a spectrum of TiO_2 contains at least three peaks. The one at about 3730 cm^{-1} disappears when the surface is exchanged with F^- ions and corresponds to basic OH groups. The other peaks around 3670 and 3630 cm^{-1} correspond to neutral and acidic hydroxyl groups, respectively [34]. Similar correlations exist for the O–H stretch frequencies of OH groups on alumina supports [32].

In catalyst preparation, one can determine the relative contributions of various hydroxyl groups before and after application of the active phase onto the support. In this way Sibeijn et al. [35] established that rhenium oxide attached to acidic sites of the alumina support exhibit higher activity for the metathesis of olefins than rhenium oxide on neutral or basic sites. As, however, rhenium species preferentially exchange with basic hydroxyls, one needs to go to loadings above a certain value (6 wt% for an alumina of 200 m^2/gram) before the catalyst exhibits appreciable activity [35].

Another application of O–H infrared spectra is to determine to what extent the support is covered by the active phase by monitoring the total signal intensity of

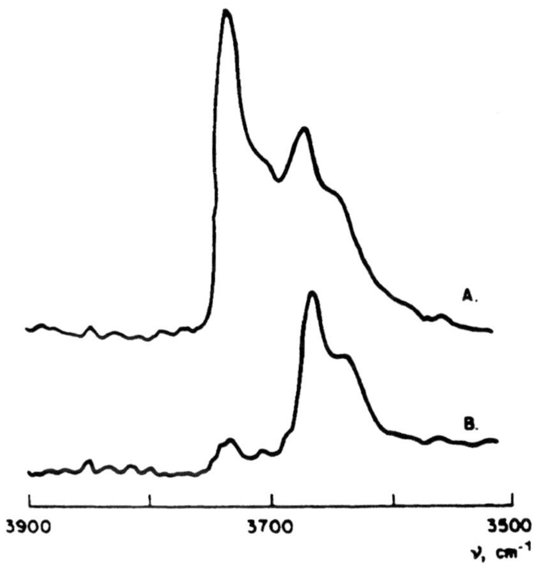

Fig. 11.15. Infrared spectra of the OH stretch region of TiO_2, (A) before and (B) after exchange with F^- confirm that the peak around 3720 cm^{-1} is due to basic $OH^{\delta-}$ groups (from van Veen et al. [34]).

the hydroxyl groups as a function of catalyst loading. One should note, however, that conclusions can only be drawn on a qualitative base, because (i) the extinction coefficients of the various OH groups are unknown, (ii) the infrared signals are easily perturbed by adsorbed water, and (iii) the signal intensities can depend on temperature and concentration. For quantitative results on hydroxyl groups, infrared studies should be combined with wet chemical titrations, as described by van Veen and coworkers [34].

In summary, infrared spectroscopy is very popular for studying the adsorption of gases on supported catalysts or for studying the decomposition of infrared active catalyst precursors during catalyst preparation. Infrared spectroscopy is an *in situ* technique which is applicable in transmission or diffuse reflection mode, on real catalysts.

11.2.6 Extended X-Ray Absorption Fine Structure (EXAFS)

EXAFS is an X-ray absorption technique which gives detailed local structure information, such as the type, number and distance of neighbouring atoms [36]. As in XPS, the basic process is the photoelectric effect: a photon is absorbed by an atom or ion and an electron is emitted from an inner shell.

An EXAFS spectrum is a plot of the X-ray intensity transmitted by the sample, as a function of the energy $E = h\nu$ of the monochromatic X-rays. Each time E reaches the threshold for photoemission of a core electron with binding energy

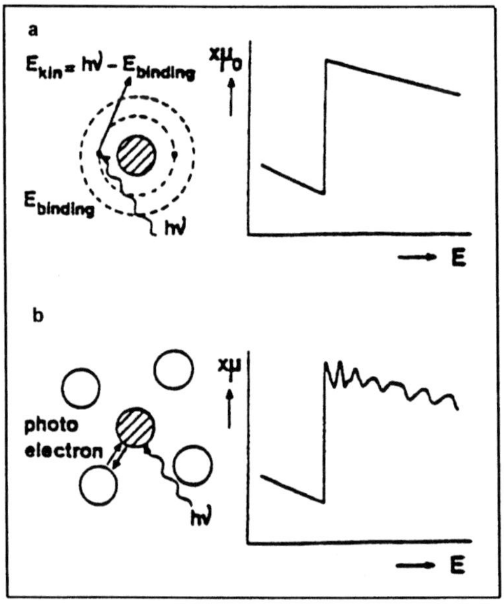

Fig. 11.16. Absorption of X-rays as a function of photon energy *hv* by a free atom and by an atom in a lattice. The spherical electron wave from the central atom is scattered back by the neighbouring atoms, which leads to interference, which is constructive or destructive depending on the wavelength of the electron wave (or the kinetic energy of the electron) and the distance between the atoms. As a result, the X-ray absorption probability is modulated and the spectrum shows fine structure which represents the EXAFS spectrum [37].

E_b, the absorption cross section increases dramatically (Fig. 11.16). In a solid or a liquid, the photoelectron with kinetic energy $E_k = h\nu - E_b$ can be scattered back from neighbouring atoms to the atom where it originated. There it affects the probability that a photon can be absorbed. As electrons are waves, the backscattered wave interferes with the original outgoing wave. Depending on the kinetic energy of the electron, which, in its turn, depends on the energy of the incoming X-rays, the interference will be constructive or destructive. The interference pattern shows up as the fine structure in the spectrum up to several hundred eV above the absorption edge.

In a monoatomic solid, each shell gives rise to a characteristic interference. This is expressed as follows in the EXAFS function $\chi(k)$:

$$\chi(k) = \sum_j A_j\,(k) \sin\,(2kr_j + \phi_j\,(k))\,; \quad k = \frac{2\pi}{h}\sqrt{2mE} \tag{11.9}$$

$$A_j\,(k) = \frac{N_j}{kr_j^2}\,S_o^2(k)\,F_j\,(k)\,e^{-2k^2\sigma_j^2}\,e^{-2r_j\,/\,\lambda(k)} \tag{11.10}$$

where k is the wave number, m the mass of an electron, E is kinetic energy, j labels the coordination shells around the absorbing atom, r_j is the average distance between the central atom and the atoms in the jth coordination shell and $\phi(k)$ is the phase shift. The amplitude function A contains the information on the coordination numbers N, on lattice dynamics and on structural disorder. The back-scattering amplitude F is element specific, S is a correction for relaxation processes in the absorbing atom and for multielectron excitations. These processes cause the outgoing electron to have a lower energy than $h\nu - E_b$. Exponential terms describe the effect of lattice vibrations and of structural disorder by a mean squared displacement σ^2, as well as the attenuation of the electron wave when it travels through the solid, similarly as in XPS, UPS and AES.

The EXAFS function becomes understandable if we look at the Fourier Transform of $\chi(k)$, which resembles a radial distribution function (the probability of finding an atom at a distance r from the absorbing atom):

$$\theta_n(r) = \frac{1}{\sqrt{2\pi}} \int_{k_{min}}^{k_{max}} k^n \chi(k) e^{2ikr} dk \qquad (11.11)$$

The function $\theta_n(r)$ represents the probability of finding an atom at a distance r. The transform is weighted with either k or k^3, to emphasize the role of light or heavy atoms, respectively. In principle, the Fourier transform becomes more accurate when the k interval is larger, but in practice the signal to noise ratio of the spectrum sets the limit for k.

A straightforward Fourier transform of the EXAFS signal does not yield the true radial distribution function. First, the phase shift causes each coordination shell to peak at the incorrect distance. Second, due to the element specific back-scattering amplitude, the intensity may not be correct. Third, coordination numbers of distant shells will be too low mainly because of the term $1/r^2$ in the amplitude (10.10) and also because of the small inelastic mean free path of the photoelectron. The appropriate corrections can be made, however, when phase shift and amplitude functions are derived from reference samples or from theoretical calculations. Figure 11.17 illustrates the effect of phase and amplitude correction on the EXAFS of a Rh foil [38]. Note that unless the sample is that of a single element, N is a fractional coordination number, i.e. the product of the real coordination number and the concentration of the element involved. Also, the EXAFS information is an average over the entire sample. As a consequence, meaningful data on supported catalysts are only obtained when the particles have a monodisperse size distribution.

Figure 11.18 shows an application of EXAFS on sulphided catalysts as used in hydrotreating processes [39]. The results in the corresponding table have been obtained by computer fitting. The great disadvantage of EXAFS is that data analysis is complicated and time consuming; considerable expertise is required

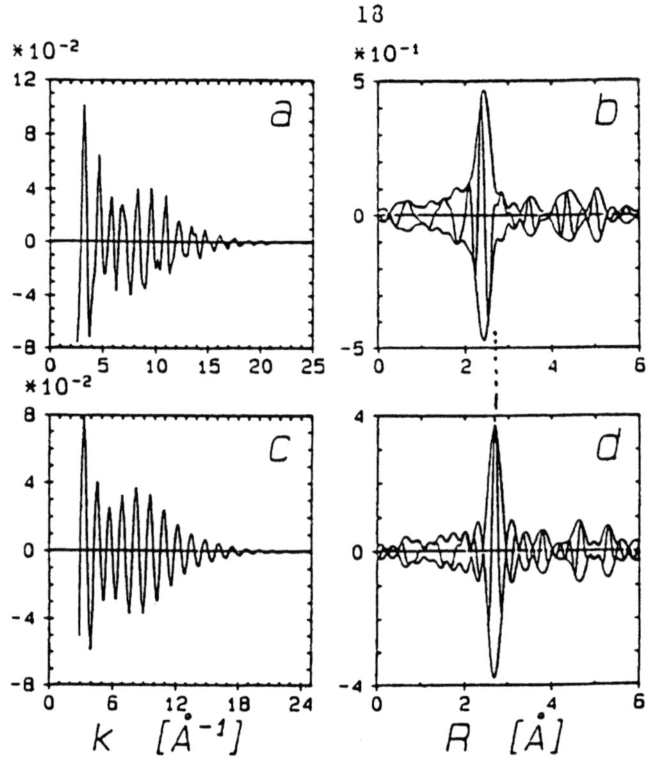

Fig. 11.17. EXAFS and Fourier transforms of rhodium metal: (a) measured EXAFS spectrum, (b) the uncorrected Fourier transform, (c) the inverse Fourier transform of the main contribution in (b), and (d) the phase and amplitude corrected Fourier transform [38].

to avoid ambiguities. However, if the analysis is carried out properly, one obtains structure information on the atomic scale.

11.2.7 Mössbauer Spectroscopy

Although relatively little used in catalysis, Mössbauer spectroscopy has given important information on the state of iron and cobalt in Fischer–Tropsch and hydrodesulphurization catalysts. Mössbauer spectroscopy provides the oxidation state, the magnetic field and the lattice symmetry of a number of elements such as iron, tin, iridium, and cobalt, and can be applied *in situ*. We will first describe the theory behind the Mössbauer effect and explain how a nuclear technique gives information on the state of atoms.

The Mössbauer effect (named after its discoverer, Rudolf L. Mössbauer, who demonstrated the effect in 1957) can in short be described as the recoil free emission and resonant absorption of gamma rays by nuclei [40]. The source contains nuclei in the first excited state. Figure 11.19 shows how excited iron

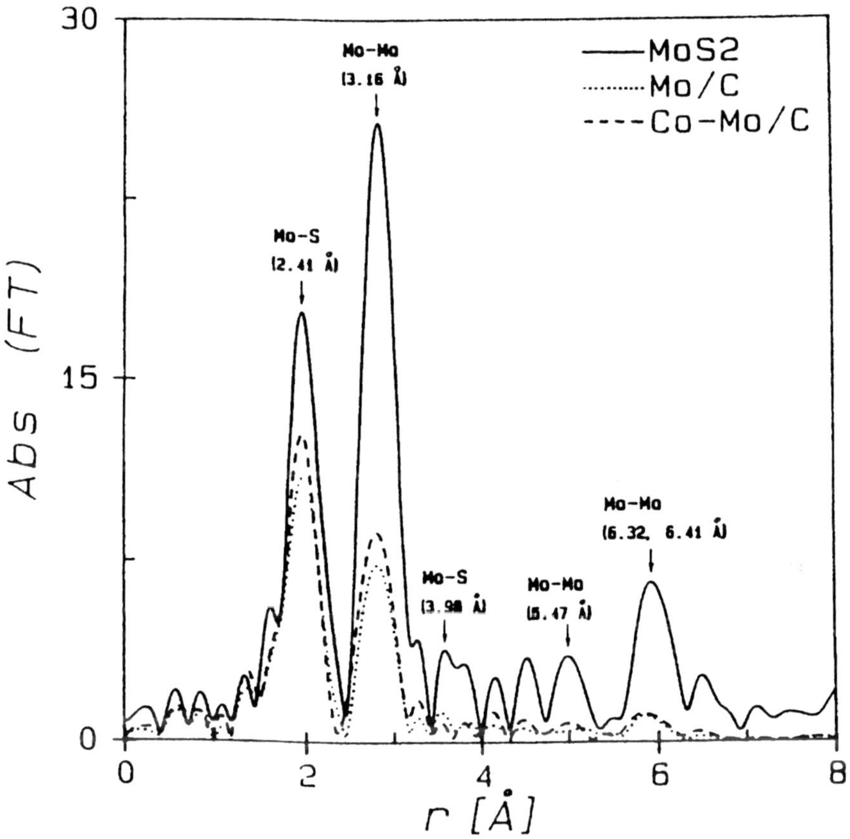

	Sholl	N	R(A)
MoS_2	Mo–S	6	2.41
	Mo–Mo	6	3.16
Mo/C	Mo–S	5.2±0.5	2.41±0.01
	Mo–Mo	2.7±0.3	3.15±0.01
Co–Mo/C	Mo–S	5.5±0.5	2.40±0.01
	Mo–Mo	3.2±0.3	3.13±0.01

Fig. 11.18. Magnitude of the Fourier Transform of the Mo K edge EXAFS spectrum of carbon-supported, sulphided Co–Mo and Mo catalysts, along with EXAFS parameters (from Bouwens et al. [39]).

nuclei are obtained in a nuclear decay process starting from cobalt. The 14.4 keV transition is the one that is normally used for Mössbauer spectroscopy in iron (or, more precise, in the ^{57}Fe isotope, which has a natural abundance of about 2%). The absorber contains iron with nuclei in the ground state. These nuclei could in

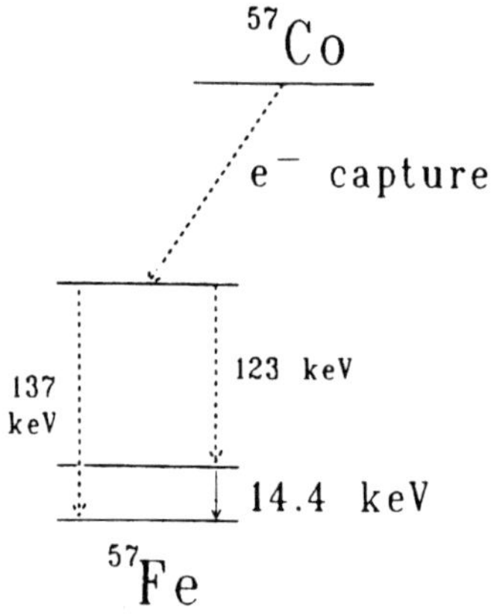

Fig. 11.19. Mössbauer spectroscopy in ^{57}Fe: the 14.4 keV source radiation is generated in a nuclear decay process starting from ^{57}Co.

principle absorb the radiation from the source, if their nuclear levels were separated by exactly the same energy difference as in the source. This is, in general not the case, because nuclear levels can be shifted or splitted by the so-called hyperfine interactions between the nucleus and its environment, as we will discuss in a moment. In order to cover all possible transitions between the shifted and splitted nuclear levels, one must be able to vary the energy of the source radiation. This is done by using the Doppler effect. If the source has a velocity v towards the absorber, the energy of the gamma radiation is given by

$$E(v) = E_0\left(1 + \frac{n}{c}\right) \tag{11.12}$$

in which E_0 is the energy of the stationary source and c the velocity of light. A velocity range of –10 to +10 mm/s is sufficient to cover all the shifts that can occur in iron.

Gamma rays involved in nuclear transitions are quanta of extremely narrow range in energy. It is this spectral precision that enables the detection of the so-called hyperfine interactions, the minute changes in the nuclear energy levels which result from a change in chemical state or in magnetic or electric surroundings of the atom to which the nucleus belongs.

Three hyperfine interactions play a role; they are illustrated in Fig. 11.20.

(1) The isomer shift (IS), is the consequence of the Coulomb interaction between the positively charged nucleus and the negatively charged s-electrons. The isomer shift is therefore a measure of the s-electron density at the nucleus, and yields useful information on the oxidation state of the iron in the absorber. Figures 11.20a and b show how the levels in the absorber have shifted with

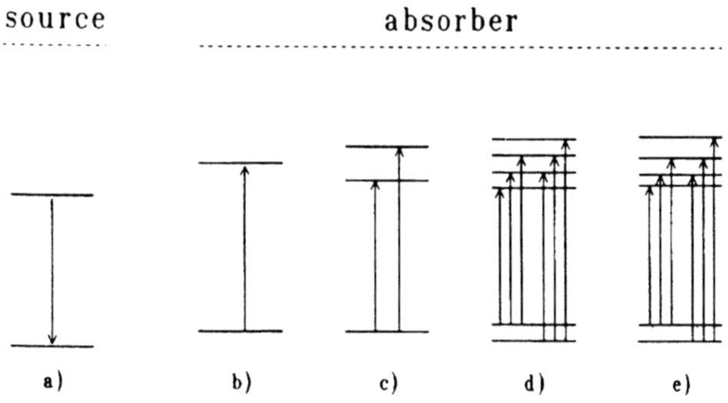

Fig. 11.20. Hyperfine interactions shift and split the nuclear levels of iron (see text) and thus make the state of the nucleus dependent on the state of the atom. Each transition corresponds to a peak in the Mössbauer spectrum, see e.g. Fig. 10.21.

respect to those of the source, because the two have a different oxidation state. An example of a single line spectrum is fcc iron, as in stainless steel or in many alloys with noble metals.

(2) The electric quadrupole splitting is caused by the interaction of the electric quadrupole moment with an electric field gradient. The latter can be caused by asymmetrically distributed electrons in incompletely filled shells of the atom itself and by charges on neighbouring ions. The splitting occurs only in the excited level (Fig. 11.20c) and is proportional to the magnitude of the electric field gradient at the nucleus. The spectrum contains two peaks, the so-called quadrupole doublet. This type of spectrum is often observed with highly dispersed iron (III) oxides, see the top spectrum in Fig. 11.21.

(3) Magnetic hyperfine splitting or the so-called Zeeman effect arises from the interaction between the nuclear magnetic dipole moment and the magnetic field H at the nucleus. This interaction splits both nuclear levels and removes all degeneracy. From the eight possible transitions only six are allowed (Fig. 11.20d) and the spectrum contains six equidistant peaks, called sextet or sextuplet. The separation between the peaks in the spectrum is proportional to the magnetic field at the nucleus. The best known example is that of metallic or bcc iron with a magnetic hyperfine field of 33 T (330 kOe), visible in the third spectrum of Fig. 10.21 as the main contribution.

(4) If the nucleus feels both a magnetic field and an electric field gradient, and the electric quadrupole interaction is small, then the excited levels shift as indicated in Fig. 11.20e. The result is that only the inner four lines of the sextet are equidistant. This type of spectrum is measured from bulk Fe_2O_3, with a magnetic hyperfine field of 51.5 T and a small quadrupole shift on the absorption peaks of 0.10 mm/s.

The Mössbauer effect can only be detected in the solid state. The reason is that the absorption and emission events must occur without energy losses due to recoil effects. The fraction of the absorption and emission events without exchange of recoil energy is called the recoilless fraction, *f*. It depends on temperature and on the energy of the lattice vibrations: *f* is high for a rigid lattice, but low for surface atoms [41].

A Mössbauer spectrometer consists of a radioactive ^{57}Co source on a transducer that continuously scans the desired velocity range, an absorber consisting of the catalyst and a detector to measure the intensity of the gamma radiation transmitted by the absorber as a function of the source velocity. This is the common mode of operation, called Mössbauer absorption spectroscopy, sometimes abbreviated as MAS. It is also possible to fix the ^{57}Co containing source and move a single-line ^{57}Fe absorber, in order to investigate Co-containing catalysts. This technique, called Mössbauer emission spectroscopy (MES), has successfully been applied to study Co–Mo hydrodesulphurization catalysts [42].

Mössbauer spectroscopy provides phase identification, determination of oxidation states, and incidentally structure information and particle size. A little used application is to follow in real time the kinetics of phase transitions (carburization, reduction) in catalysts by monitoring the intensities of a few selected peaks in a single velocity experiment. Examples of applications on catalysts have recently been reviewed [43].

Figure 11.21 illustrates how Mössbauer spectroscopy reveals the identity of the major iron phases in a supported iron catalyst after different treatments. The top spectrum belongs to a fresh Fe/TiO_2 catalyst, i.e. after impregnation and drying. The quadrupole doublet has an isomer shift which shows that iron is present in the ferric state, probably as small particles or a dispersed layer of an oxide or oxyhydroxide. After reduction in H_2 at 675 K the catalyst consists mainly of metallic iron, as evidenced by the sextet, along with some unreduced iron, which gives rise to two doublet contributions of Fe^{2+} and Fe^{3+} in the centre. The overall degree of iron reduction, as reflected by the relative area under the bcc-ion sextet, is high. Fischer–Tropsch synthesis at 575 K in CO and H_2 converts the metallic iron to the Hägg carbide, χ-Fe_5C_2. Almost all unreduced iron is now present as Fe^{2+}. Exposure of the carburized catalyst to the air at room temperature leaves most of the carbide phase unaltered but oxidizes the ferrous to ferric iron [44].

The example is typical for many applications of Mössbauer spectroscopy in catalysis: a catalyst undergoes a certain treatment; next its Mössbauer spectrum is measured *in situ* at room temperature. However, if the catalyst contains highly dispersed particles, the measurement of spectra at cryogenic temperatures becomes advantageous. First, because the recoil-free fraction of surface atoms increases substantially at temperatures below 300 K [41]. Second, because spectra of small particles which behave superparamagnetically at room temperature become magnetically split and therefore more informative at the temperatures of liquid nitrogen, 78 K, or liquid helium, 4.2 K [40,43,45].

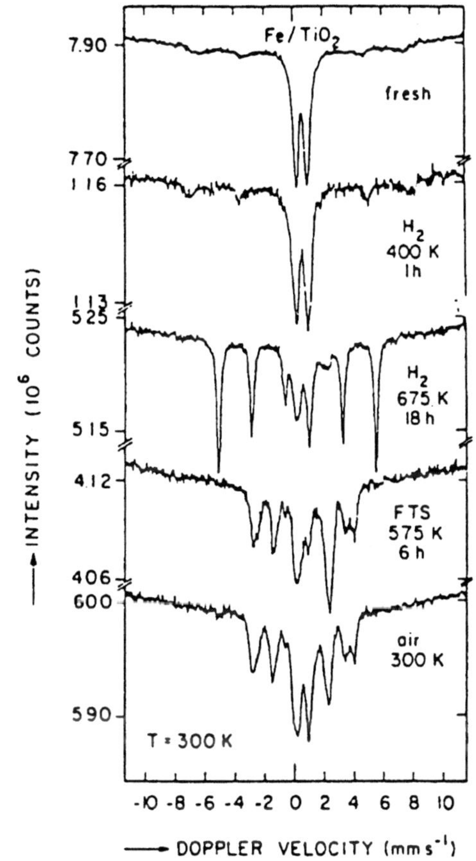

Fig. 11.21. Mössbauer spectra of an Fe/TiO_2 catalyst after different treatments (from van der Kraan et al. [44]).

11.3 CONCLUDING REMARKS

In this chapter we have limited ourselves to the most common techniques in catalyst characterization. Of course, there are many other spectroscopies which can and have been used to study catalysts, such as nuclear magnetic resonance (NMR), electron spin resonance (ESR), Raman spectroscopy, calorimetry, thermogravimetric or thermomagnetic analysis, or small-angle X-ray scattering (SAXS). Also many powerful surface science techniques, such as low energy electron diffraction (LEED), electron energy loss spectroscopy (EELS), thermal desorption spectroscopy (TDS), or scanning tunnelling microscopy (STM), have not been discussed. These methods are invaluable tools for the investigation of well defined surfaces and species adsorbed thereon, and have greatly contributed to new insights in catalysis.

11.3.1 Research Strategies

In principle, two approaches can be adopted for fundamental investigations of the relations between catalytic properties on one hand, and catalyst composition and structure on the other. The first is to model the catalytic surface, for example with that of a single crystal. By using the appropriate combination of surface spectroscopies, the desired characterization on the atomic scale is certainly possible in favourable cases. The disadvantage, however, is that although one may be able to study the catalytic properties of such samples under realistic conditions (pressures of 1 atm or higher), most of the characterization is necessarily carried out in ultra high vacuum, and not under reaction conditions.

The second approach is to study real catalysts with in situ techniques such as Infrared and Mössbauer spectroscopy, EXAFS and XRD, under reaction conditions, or, as is more often done, under a controlled environment after quenching of the reaction. The *in situ* techniques, however, are not sufficiently surface specific to yield the desired atom by atom characterization of the surface. At best they determine the composition of the particles. The situation is schematically represented in Fig. 11.22.

The dilemma is thus that investigations of real catalysts under relevant conditions by *in situ* techniques give little information on the surface of the catalyst, and that the techniques which are surface sensitive can often only be applied on model surfaces under vacuum. Bridging the gap between UHV and high pressures and between the surfaces of single crystals and of real catalysts is an important issue in catalysis.

Another point which concerns the relevance of spectroscopic research in catalysis is the following. Both catalysis and spectroscopy are disciplines which demand considerable expertise. For instance, the state of a catalyst depends often critically on the method of preparation, pretreatment or even on its environment. It is therefore essential to investigate a catalyst under carefully chosen, relevant conditions and after the proper treatment. On the other hand, no spectroscopy is really simple. The correct interpretation of spectra of catalysts requires ex-

	REAL CATALYST	SINGLE CRYSTAL
REACTION CONDITIONS	XRD, TP techniques, IR, Raman spectroscopy, EXAFS, Mössbauer spectroscopy, ESR, NMR	IR, TP techniques
VACUUM	XPS, SIMS, SNMS, LEIS, RBS, TEM	All surface science techniques

Fig. 11.22. Classification of spectroscopic research in catalysis.

perience based on practice and a sound theoretical background in spectroscopy, physical chemistry and solid state physics. Hence, intensive cooperation between experts in spectroscopy and experts in catalysis is probably the best way to warrant meaningful and correctly interpreted results.

Finally, it is good to realize that, although many techniques undoubtedly reveal usable information on catalysts, the information desirable from a fundamental point of view can often not be obtained. In this situation it is a good strategy to combine all techniques which tell us at least something about the catalyst. The catalysis literature contains several examples where this approach has been remarkably successful.

REFERENCES

1 M.J. Phillips and M. Ternan (Editors), *Proc. 9th Int. Congr. on Catalysis*, Calgary 1988, The Chemical Institute of Canada, Ottawa, 1988.
2 W.N. Delgass, G.L. Haller, R. Kellerman and J.H. Lunsford, *Spectroscopy in Heterogeneous Catalysis*, Academic Press, New York, 1979.
3 J.M. Thomas and R.M. Lambert (Editors), *Characterisation of Catalysts*, Wiley, Chichester, 1980.
4 J.L.G. Fierro (Editor), *Spectroscopic Characterization of Heterogeneous Catalysts*, Elsevier, Amsterdam, 1990.
5 R.B. Anderson and P.T. Dawson (Editors), *Experimental Methods in Catalytic Research*, Academic Press, New York, 1976.
6 J.W. Niemantsverdriet, *Spectroscopy in Catalysis, an Introduction*, VCH, Weinheim, 1993.
7 L.C. Feldman and J.W. Mayer, *Fundamentals of Surface and Thin Film Analysis*, North Holland, Amsterdam, 1986.
8 G.A. Somorjai, *Chemistry in Two Dimensions: Surfaces*, Cornell University Press, Ithaca, 1980.
9 J.B. Cohen and L.H. Schwartz, *Diffraction from Materials*, Academic Press, New York, 1977.
10 H. Jung and W.J. Thomson, *J. Catal.* 128 (1991) 218.
11 F. Delannay, Chapter 3 in F. Delannay (Editor), *Characterization of Heterogeneous Catalysts*, Dekker, New York, 1984.
12 R.T.K. Baker, *Catal. Rev. - Sci. Eng.* 19 (1979) 161.
13 N.W. Hurst, S.J. Gentry, A. Jones and B.D. McNicol, *Catal. Rev. - Sci. Eng.* 24 (1982) 233.
14 H.F.J. van 't Blik and J.W. Niemantsverdriet, *Appl. Catal.* 10 (1984) 155.
15 J.L. Falconer and J.A. Schwarz, *Catal. Rev. - Sci. Eng.* 25 (1983) 141.
16 A.M. de Jong and J.W. Niemantsverdriet, *Surface Sci.* 233 (1990) 355.
17 G. Ertl and J. Küppers, *Low Energy Electrons and Surface Chemistry*, VCH, Weinheim, 1985.
18 D. Briggs and M.P. Seah (Editors), *Practical Surface Analysis by Auger and X-Ray Photoelectron Spectroscopy*, Wiley, New York, 1983.
19 H.J. Borg, L.C.A. van den Oetelaar, L.J. van IJzendoorn, and J.W. Niemantsverdriet, *J. Vac. Sci. Technol.*, A 10 (1992) 2737; H.J. Borg, L.C.A. van den Oetelaar and J.W. Niemantsverdriet, *Catal. Lett.* 17 (1993) 81.
20 C.D. Wagner, W.M. Riggs, L.E. Davis, J.F. Moulder, and G.E. Muilenberg, *Handbook of X-ray Photoelectron Spectroscopy*, Perkin-Elmer Corporation, Eden Prairie, 1979.

21 A.C.Q.M. Meijers, A.M. de Jong, L.M.P. van Gruijthuijsen and J.W. Niemantsverdriet, *Appl. Catal.* 70 (1991) 53.
22 H.P.C.E. Kuipers, H.C.E. van Leuven and W.M. Visser, *Surf. Interface Anal.* 8 (1986) 235.
23 Spectra measured by H. Korpik and A.M. de Jong, unpublished.
24 J.C. Muijsers, J.W. Niemantsverdriet, I.C.M. Wehman-Ooyevaar, D.M. Grove and G. van Koten, *Inorg. Chem.*, 31 (1992) 2655.
25 L.M. Eshelman, A.M. de Jong and J.W. Niemantsverdriet, *Catal. Lett.*, 10 (1991) 201.
26 H.J. Bord and J.W. Niemantsverdriet, Chapter 11 in J.J. Spivey and S.K. Agarwal (Editors), *Catalysis*, Specialist Periodical Report, Vol. 11, The Royal Society of Chemistry, Cambridge, 1994.
27 J.C. Vickerman, A. Brown and N.M. Reed (Editors), *Secondary Ion Mass Spectrometry, Principles and Applications*, Clarendon, Oxford, 1989.
28 G.C. van Leerdam, H.H. Brongersma, I.M.M. Tijburg and J.W. Geus, *Appl. Surface Sci.*, 55 (1992) 11.
29 A. Fadini and F.-M. Schnepel, *Vibrational Spectroscopy*, Ellis Horwood Ltd, Chichester, 1989.
30 Spectra measured by L.M.P. van Gruijthuijsen and J.H.M.C. van Wolput, unpublished.
31 N.Y. Topsøe and H. Topsøe, *J. Catal.* 84 (1983) 386; *J. Elec. Spectr. Relat. Phenom.* 39 (1986) 11.
32 H. Knözinger and P. Ratnasamy, *Catal. Rev. - Sci. Eng.* 17 (1978) 31.
33 R.A. van Santen, *Theoretical Heterogeneous Catalysis*, World Scientific, 1991.
34 J.A.R. van Veen, F.T.G. Veltmaat and G. Jonkers, *J. Chem. Soc., Chem. Commun.* (1985) 1656.
35 M. Sibeijn, R. Spronk, J.A.R. van Veen and J.C. Mol, *Catal. Lett.* 8 (1991) 201.
36 D.C. Koningsberger and R. Prins (Editors), *X-ray Absorption*, Wiley, New York, 1987.
37 F.B.M. van Zon-Duivenvoorden, Thesis, Eindhoven University of Technology, 1988
38 J.H.A. Martens, Thesis, Eindhoven, 1988
39 S.M.A.M. Bouwens, R. Prins, V.H.J. de Beer, and D.C. Koningsberger, *J. Phys. Chem.* 94 (1990) 3711.
40 J.A. Dumesic and H. Topsøe, *Adv. Catal.* 26 (1977) 121.
41 J.W. Niemantsverdriet, A.M. van der Kraan and W.N. Delgass, *J. Catal.* 89 (1984) 138.
42 H. Topsøe and B.S. Clausen, *Catal. Rev. - Sci. Eng.* 26 (1984) 395.
43 J.W. Niemantsverdriet, *Hyperfine Interactions* 47 (1989) 219.
44 A.M. van der Kraan, R.C.H. Nonnekens, F. Stoop and J.W. Niemantsverdriet, *Appl. Catal.* 27 (1986) 285.
45 P.H. Christensen, S. Mørup and J.W. Niemantsverdriet, *J. Phys. Chem.* 89 (1985) 4898.

Chapter 12

Catalyst characterization and mimicking pretreatment procedures by temperature-programmed techniques

12.1 INTRODUCTION

Although in theory a catalyst is supposed to facilitate reactions without changing itself, it is a well-known fact that during its life profound changes occur. During start-up the fresh catalyst will adapt to the reaction conditions and if no precautions are taken, the performance changes drastically as a function of time. After start-up, a kind of steady state is usually observed, followed by a decay that eventually necessitates regeneration or complete substitution of the aged catalyst with a fresh catalyst load. For an optimal process the catalyst performance should be as constant as possible. It is not surprising that for every individual process a tailored procedure is called for. This can best be illustrated with a few examples.

In reduction processes catalysed by supported metals the catalyst might be prepared by impregnation of an alumina support with nickel nitrate. In the steady state the system will not contain nitrate ions, nor will it be in the oxidic state. In practice, a calcination step is carried out and reduction of the resulting oxidic nickel phase is carried out *ex situ* (by the manufacturer) or *in situ* (during start-up). As reduction occasionally requires temperatures far exceeding the temperature that is required in the catalytic process, it is often practical to carry out the reduction as a manufacturing step, followed by a so-called passivation step (controlled mild oxidation). In the reactor the oxygen that was chemisorbed during passivation can easily be removed at relatively mild reduction conditions. Many of these treatment steps involve temperature-programmed contacting in the right gaseous atmosphere.

An example where changes in pretreatment steps are very drastic is in hydrotreatment. The most applied catalysts are based on a combination of molybdenum and cobalt sulphides. They are manufactured in the oxidic state, and under reaction conditions these oxides are not stable and are transformed into the sulphidic state. In practice, the oxidic catalyst is presulphided, either in the reactor (by adding H_2S or a compound that is easily converted into H_2S), or *ex situ* by the manufacturer. These presulphiding procedures often involve temperature-programmed reactions.

A third example is the water–gas-shift reaction catalysed by iron catalysts. The active phase is an intermediate iron oxide (Fe_3O_4), distinctly different from the manufactured catalyst (Fe_2O_3). So, mild reduction should be carried out. In optimising the pretreatment procedure it should be realized that over-reduction is to be avoided because of the danger of carbon deposition and methane formation (highly exothermic). It has been found that a well-controlled reduction in a H_2/H_2O mixture is possible, whereas the use of H_2 and steam separately should be avoided.

In all these examples temperature-programmed reaction steps have to be optimized. This can be done by the techniques described in this chapter.

Temperature-programmed reduction (TPR) is a useful technique for the characterization of metal oxide catalysts. During a TPR experiment, the catalyst under investigation is placed in a fixed-bed reactor and exposed to a reducing mixture that continuously perfuses the catalyst bed, while the temperature is increased according to a linear temperature programme. The difference between the inlet and outlet concentration of the gas mixture is measured as a function of time using a thermal conductivity detector. The resulting TPR profile contains qualitative information on the oxidation state of the reducible species present and, in that sense, it is a fingerprint. The technique is intrinsically quantitative and the information obtained is of a kinetic nature and, as a consequence, directly correlated with catalytic behaviour. On the other hand, information on the structure of the species present is less straightforward than for instance that obtained by spectroscopic methods. Temperature-programmed techniques are thus complementary to spectroscopic ones and the two methods are applied in combination.

Undoubtedly, more information can be extracted from a more thorough analysis of the TPR data. To stimulate this an example of a detailed modelling will be presented.

Temperature-programmed sulphiding (TPS) is another member of a family of techniques in which catalysts are characterized by subjecting them to a temperature programme under specified conditions. The applicability of TPS is highlighted by an example in this chapter.

12.2 APPLICATION OF TPR

In TPR oxidic species are reduced following the general reduction path:

$$MO + H_2 \rightarrow M + H_2O \tag{12.1}$$

It should be noted that TPR is not limited to the reduction of oxides. Many species react with hydrogen and can be studied by TPR. Examples are the study of 'coke' deposits on catalysts:

$$\text{coke} + H_2 \rightarrow \text{hydrocarbons and water} \tag{12.2}$$

and the reduction of sulphides:

$$MS + H_2 \rightarrow M + H_2S \tag{12.3}$$

The major part of this chapter is devoted to TPR of oxides but other systems are occasionally discussed.

12.3 THERMODYNAMICS

The enthalpy change, ΔG of reaction (12.1) is:

$$\Delta G = \Delta G^o + RT \ln\left(\frac{p_{H_2O}}{p_{H_2}}\right) \tag{12.4}$$

ΔG^o is positive for a number of oxides [1] such as V_2O_5, ZnO, and Fe_2O_3 (see Fig. 12.1). However, the reduction of these oxides is still thermodynamically feasible because the ratio of partial pressures in Eqn. (12.4) becomes sufficiently small at elevated temperature and due to the high total flow rate the water formed is constantly swept out of the system.

12.4 APPARATUS

A flow scheme of a TPR instrument [2] is given in Fig. 12.2. The reactor containing the catalyst is situated in a furnace equipped with a temperature programmable controller. The effluent of the reactor is analyzed by one or more suitable detectors. A TCD is most commonly used for the detection of hydrogen consumption. In order to obtain a quantitative signal, the H_2O produced during reduction is trapped or removed before the gas mixture reaches the TCD detector. In general, more than one detector should be used. It is convenient also to

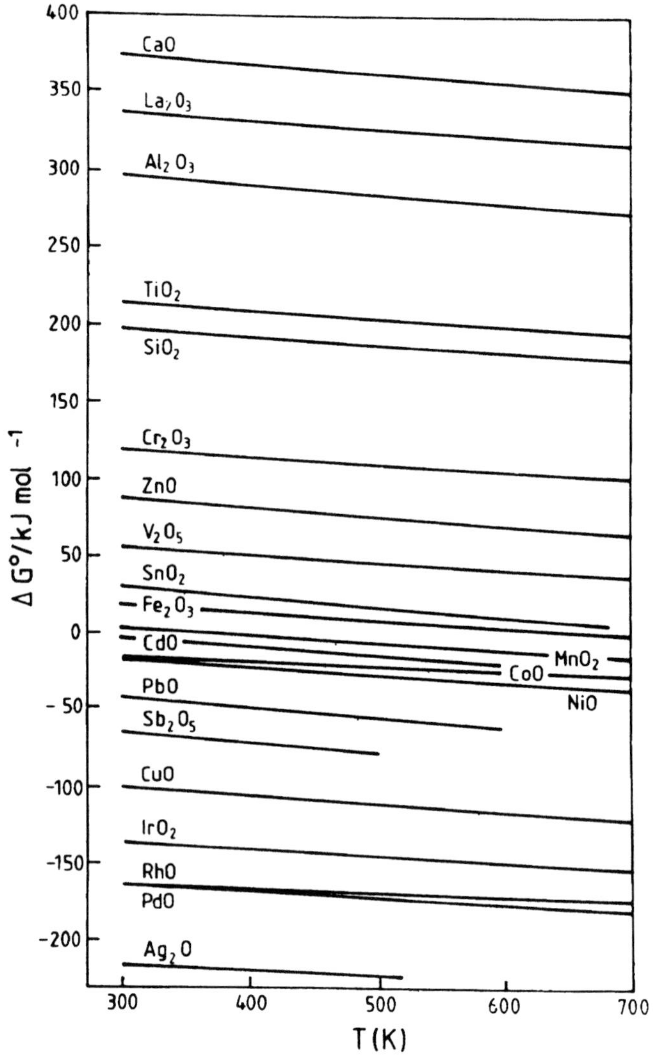

Fig. 12.1. Standard free energy, ΔG^o, as a function of temperature for the process: $MO + H_2 \rightarrow M + H_2O$.

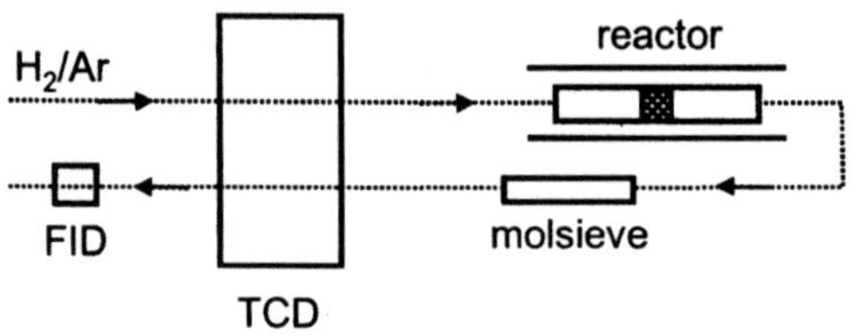

Fig. 12.2. Schematic of temperature-programmed reduction apparatus.

use an FID detector because porous catalysts nearly always contain adsorbed carbonaceous material, which leads to additional hydrogen consumption. The reducing mixture consists of H_2 and an inert gas. The latter should be really inert and therefore argon is well suited while N_2 is not, since it can react with hydrogen over the catalyst under investigation. The detector can easily be calibrated by performing a TPR run with a well defined material, for which often CuO is used that reduces to Cu. An alternative method is to use the injection of a well known amount of the inert gas, e.g. Ar, in the TPR feed mixture and measure the detector's response. In this way the consumption of hydrogen is simulated. This is a very rapid, convenient and reproducible method. For TPS the use of MoO_3 is recommended.

12.5 EXAMPLE 1: TEMPERATURE-PROGRAMMED REDUCTION (TPR) OF CoO/Al_2O_3

The CoO/Al_2O_3 catalyst of this example [3] was prepared by pore volume impregnation of Al_2O_3 with $Co(NO_3)_2$ followed by drying and calcination (heating in air). It appeared that calcination conditions are critical. For instance, the colour of the sample depends strongly on the calcination temperature.

Figure 12.3 gives the TPR patterns of a series of catalysts calcined at different temperatures [3]. Clearly, they differ profoundly and reduction takes place over a wide range of temperatures. Figure 12.4 gives TPR patterns of reference compounds used for identification [3]. CoO and Co_3O_4 have relatively sharp reduction peaks at 610 and 590, respectively. Apparently, $CoAl_2O_4$ is very stable to reduction. The explanation is that in this spinel the CoO bonds are strongly polarized by the neighbouring Al ions. In Co_3O_4 no Al^{3+} ions are present and, as a consequence, reduction is easy. This picture is supported by the behaviour of $CoMoO_4$. Figure 12.5 [4] shows TPR patterns of Co_3O_4, $CoAl_2O_4$ and $CoMoO_4$. In comparison with Al ions Mo ions are soft and it may be expected that the CoO bonds are less polarized. This explains the intermediate reduction rate of $CoMoO_4$.

When we compare these patterns with those of the series of catalysts, it is clear that $CoAl_2O_4$ is formed at high calcination temperature. This was confirmed by XRD. The very intense peak around 500 K of the sample calcined at the lowest temperature is due to the decomposition of nitrate groups. The patterns show that this peak only disappears when calcination is carried out above 650 K.

The sharp peak indicated in Fig. 12.3c is due to Co_3O_4 crystallites. They disappear at higher temperatures because they react with the support. Figure 12.3 also clearly shows peaks due to the hydrogenation of carbonaceous material. This is proven by the fact that they occur both in the TCD and the FID signal. The other peaks have been assigned to three other species, i.e. a Co^{3+} spinel and two surface compounds. It is clear from the peak area that Co^{3+} species must be

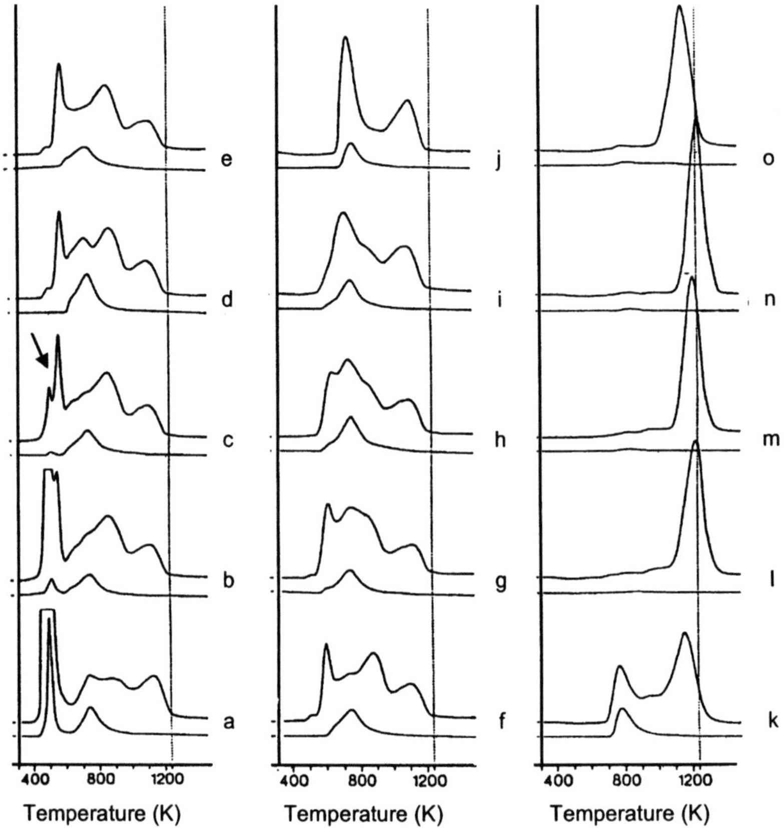

Fig. 12.3. TPR patterns of 9.1% CoO/Al_2O_3 catalysts calcined at various temperatures. Calcination temperature (K):380 (a),(575) (b), 625 (c), 675 (d), 725 (e), 775 (f), 825 (g), 875 (h), 900 (i), 925 (j), 975 (k), 1025 (l), 1075 (m), 1175 (n), 1290 (o), the upper and lower part of each pattern represents the TCD and FID signal, respectively.

present, besides Co^{2+} species. The structure of the compounds is not completely clear but the amounts of reducible species are known.

This example illustrates the potential of TPR. Quantitative information is obtained on the speciation of a complex catalyst system. On the other hand, additional techniques are needed to get a molecular picture. Furthermore, from the results major elements are obtained for a recipe for calcination and reduction treatments. For instance, calcination should take place at conditions such that nitrates are completely removed: the temperature should exceed 650 K. Furthermore, in general, Co ions should not diffuse into the support. As a consequence, a temperature of 900 K should not be exceeded (see Fig. 12.3; above a calcination temperature of 925 K the high temperature peak increases strongly due to the

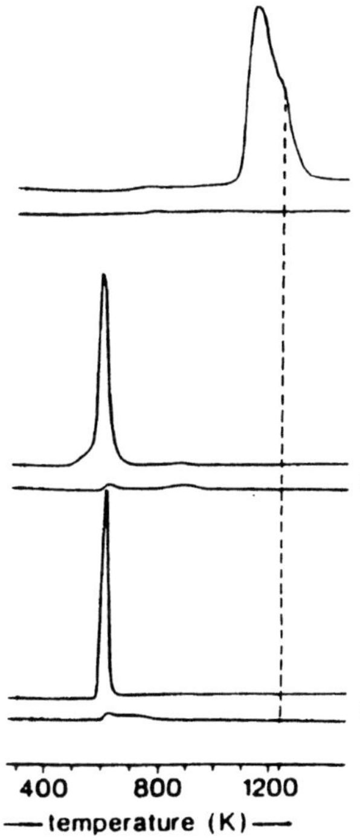

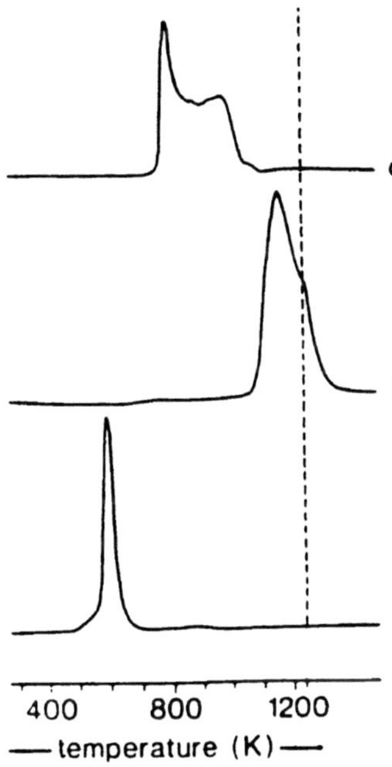

Fig. 12.4 (*left*). TPR patterns of reference materials. (a) CoO; (b) Co_3O_4; (c) $CoAl_2O_4$. The upper and lower part of each pattern represents the TCD and FID signal, respectively.

Fig. 12.5 (*right*). The influence of hard and soft ions on the TPR patterns (10 K/min) of Co-containing crystalline compounds. (a) Co_3O_4; (b) $CoAl_2O_4$; (c) $CoMoO_4$. Only the TCD signals are given.

formation of $CoAl_2O_4$). When Co-metal is desired, the right reduction temperature can be read directly from TPR traces. In the latter case, of course, extensive research is required in order to get information on the influence of the crucial variables, the major ones being heating rate and partial pressure of H_2O. In the last part of this chapter an example illustrates the way of extracting kinetic data from the TPR profiles.

12.6 EXAMPLE 2: TEMPERATURE-PROGRAMMED SULPHIDING (TPS) OF MoO_3/Al_2O_3

Alumina supported molybdenum oxide catalysts are extensively used in hydrotreating reactors. Under reaction conditions the oxides are not stable because they are transformed into molybdenum sulphides. The sulphiding step was investigated by carrying out a TPS study. TPS is strictly analogous to TPR, except for the gas mixture which is here a H_2/H_2S mixture.

A priori the following reactions might be expected:

$$MO + H_2 \rightarrow M + H_2O \quad (12.5)$$

$$MO + H_2S \rightarrow M + H_2O + S \quad (12.6)$$

$$MO + H_2S \rightarrow MS + H_2O \quad (12.7)$$

$$M + H_2S \rightarrow MS + H_2 \quad (12.8)$$

The first reaction is reduction, as in TPR. Elementary sulphur is formed in the

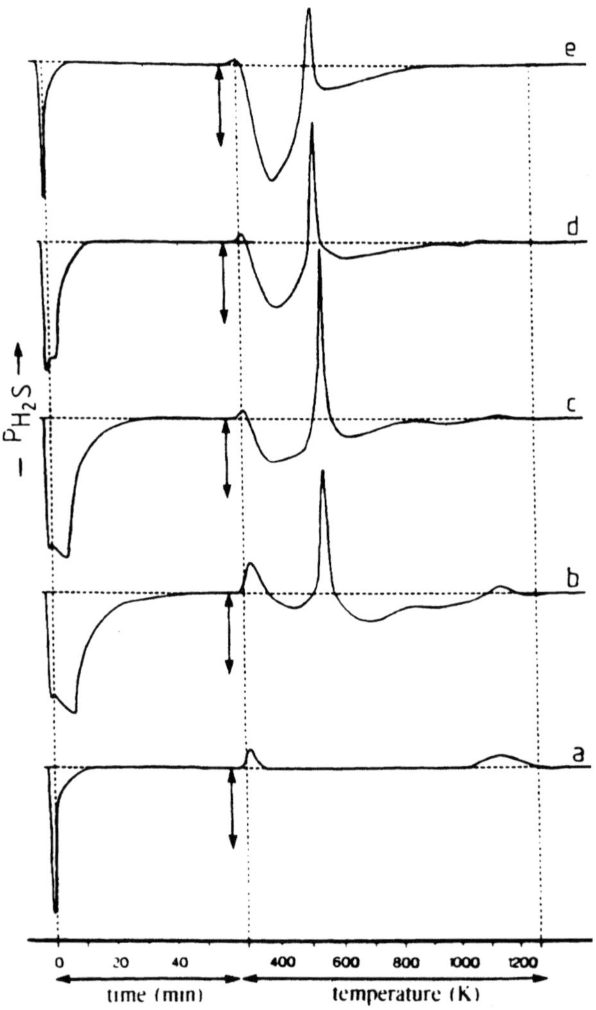

Fig. 12.6. TPS patterns (H_2S) of MoO_3/Al_2O_3 with varying Mo content, pretreated in Ar at room temperature. (a) 400 mg, carrier; (b) 800 mg, 0.5 atom/nm^2; (c) 400 mg, 1.0 atom/nm^2; (d) 200 mg, 2.2 atom/nm^2; (e) 100 mg, 4.5 atom/nm^2. The 50% conversion level of H_2S is indicated by a double-headed arrow.

second reaction. The third reaction will be referred to as 'oxygen–sulphur exchange'. The last reaction is an oxidation.

Figure 12.6 [4] shows the TPS patterns for a series of catalysts with different loadings. The bottom curve represents the carrier. At room temperature extensive uptake of H_2S is observed. In order to measure quantitatively, an isothermal stage of ca. 1 hour is introduced at the start of the experiment.

The unloaded support adsorbs some H_2S at room temperature, which is desorbed at the start of the temperature program. The small peak at the high temperature is attributed to impurities in the carrier.

All catalysts adsorb a large amount of H_2S at room temperature. From the fact that the colour of the catalyst changes during this stage we conclude that at least partial sulphiding takes place. At the beginning of the temperature program some H_2S desorption is observed followed by continued sulphiding. It is striking that in all cases a sharp H_2S production peak exists superimposed on the sulphiding pattern.

In order to obtain information on the stoichiometry of the reaction steps, a mass spectrometer was used, besides a TCD. The results for a catalyst of intermediate loading are given in Fig. 12.7. It is clear that the pattern is explained by an O–S exchange over the whole sulphiding window with hydrogenation of S superimposed on it.

On a molecular scale these results are interpreted in the scheme shown in Fig. 12.8 [4]. For simplicity the molybdenum oxide layer is represented by a tetrahedral structure. From other results and considerations of minimal steric hindrance it is concluded that the pathway via complexes III and VI is the dominant one. These patterns reveal the basic chemistry of sulphiding. They can be used for optimizing catalyst pretreatment in practical circumstances.

12.7 MODELLING

The modelling of TPR patterns has not received much attention. Nevertheless, the results depend on the experimental conditions and modelling will be necessary when the results of different authors have to be compared. Moreover, the kinetic data obtained are useful in optimising pretreatment procedures. We present a convenient method of evaluating kinetic data from TPR patterns.

12.7.1 Theory

The reaction obeys the following stoichiometry:

$$MO + H_2 \rightarrow M + H_2O \tag{12.9}$$

Generally, it is assumed that the rate of a solid gas reaction can be represented by the product of two factors [4], one is only dependent on the temperature and the

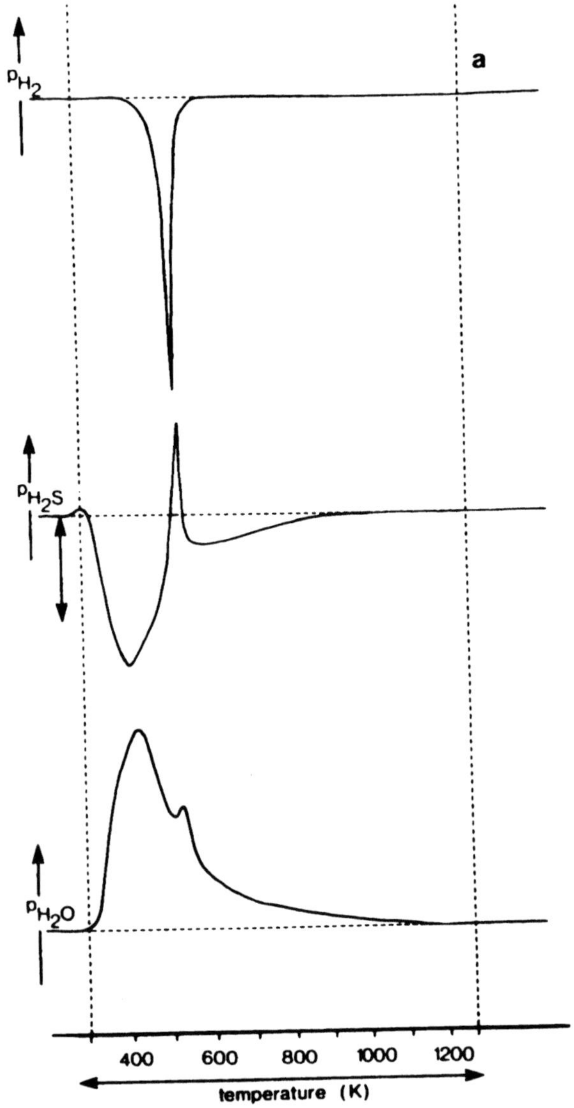

Fig. 12.7. TPS pattern (H_2S, H_2O, H_2) of MoO_3/Al_2O_3, 4.5 atom/nm^2, 100 mg, pretreated in Ar at room temperature.

other is only dependent on the concentrations:

$$\frac{d\alpha}{dt} = k'(T) \cdot f(\alpha) \cdot f'(P_{H_2}, P_{H_2O}) \tag{12.10}$$

α is the degree of conversion of the solid reactant. Table 12.1 gives $f(\alpha)$ functions for several relevant models for reduction kinetics.

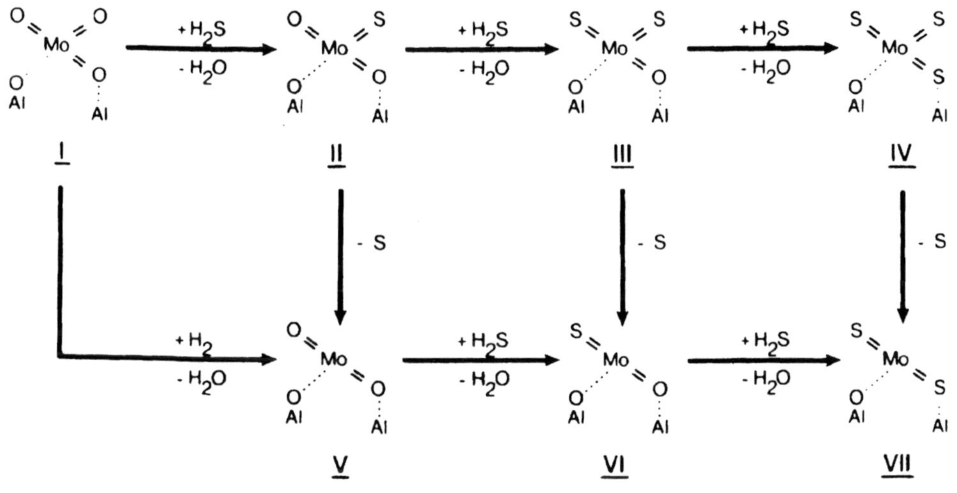

Fig. 12.8. Simplified molecular scheme of sulphiding.

We assume differential conditions, hence the function that depends on gas-phase concentration is constant and is combined with $k'(T)$. Equation (12.10) reduces to:

$$\frac{d\alpha}{dt} = k(T) \cdot f(\alpha) \tag{12.11}$$

For a heating program with a constant heating rate β the temperature is given by:

$$T = \beta t + T_o \tag{12.12}$$

The temperature dependence is assumed to follow the Arrhenius equation:

$$k(T) = A_o \cdot \exp\left(\frac{-E}{RT}\right) \tag{12.13}$$

Combining Eqns. (12.11), (12.12), and (12.13) we obtain

$$\frac{d\alpha}{dT} = \frac{A_o}{\beta} \exp\left(\frac{-E}{RT}\right) \cdot f(\alpha) \tag{12.14}$$

TPR patterns (i.e. $d\alpha/dT$ versus T) can be calculated by solving the differential Eqn. (12.14), giving α explicitly as a function of T and subsequently calculating $d\alpha/dT$. The solution to the differential Eqn. (12.14) can easily be found by separation of variables and integration, which leads to:

$$g(\alpha) = \int_0^{\alpha} \frac{d\alpha}{f(\alpha)} = \int_{T_o}^{T} \frac{A_o}{\beta} \exp\left(\frac{-E}{RT}\right) dT \qquad (12.15)$$

In Table 12.1 $f(\alpha)$ is given, besides functions $g(\alpha)$. The right-hand side of (12.15) can be evaluated numerically by using a 4th order rational approximation.

$$\int_{T_o}^{T} \exp\left(\frac{-E}{RT}\right) dT = \frac{E_R}{R} \frac{\exp(-x)}{x} \frac{x^3 + 18x^2 + 88x + 96}{x^4 + 20x^3 + 240x + 120} \qquad (12.16)$$

This method has been found to be very accurate [5].

12.7.2 Reduction Kinetic Models

TABLE 12.1

Reduction kinetic models: ($f(\alpha)$ differential form, $g(\alpha)$ integral form, Eqn. 12.15)

Model	$f(\alpha)$	$g(\alpha)$
*n*th Order	α^n	$\alpha^{n+1}/(n+1)$
Random nucleation, unimolecular decay law	$(1-\alpha)$	$-\ln(1-\alpha)$
Phase boundary controlled reaction (contracting area)	$(1-\alpha)^{1/2}$	$2(1-(1-\alpha)^{1/2})$
Phase boundary controlled reaction (contracting volume)	$(1-\alpha)^{2/3}$	$3(1-(1-\alpha)^{1/3})$
Two-dimensional growth of nuclei (Avrami–Erofeev)	$2(1-\alpha)(-\ln(1-\alpha))^{1/2}$	$(-\ln(1-\alpha))^{1/2}$
Three-dimensional growth of nuclei (Avrami–Erofeev)	$3(1-\alpha)(-\ln(1-\alpha))^{2/3}$	$(-\ln(1-\alpha))^{1/3}$
One-dimensional diffusion, sharp interface	$1/2\alpha$	α^2
Two-dimensional diffusion, sharp interface	$-1/\ln(1-\alpha)$	$(1-\alpha)\ln(1-\alpha)+\alpha$
Three-dimensional diffusion, sharp interface (Ginstling–Brounshtein)	$3/[2((1-\alpha)^{-1/3}-1)]$	$1-2\alpha/3-(1-\alpha)^{2/3}$
Three-dimensional diffusion (Jander)	$[3(1-\alpha)^{2/3}]/[2(1-(1-\alpha)^{1/3})]$	$(1-(1-\alpha)^{1/3})^2$

12.7.3 Activation energy

The TPR patterns are very sensitive to the value of the activation energy which is needed for the modelling. Fortunately, this can be calculated separately from a series of experiments in which the heating rate is varied. Kissinger introduced this method in DTA measurements. It will be shown that this method can be generally used in TPR modelling. For the maximum of the TPR peak the following equations hold:

$$\frac{d}{dT}\left(\frac{d\alpha}{dT}\right)_{T=T_{max}} = 0 \qquad (12.17)$$

Combination of Eqn. (12.17) with Eqn. (12.14) leads to:

$$\frac{d}{dT}\left(\frac{A_o}{\beta}\exp\left(-\frac{E}{RT}\right)f(\alpha)\right)_{T=T_{max}} = 0 \qquad (12.18)$$

This equation reduces to:

$$\frac{d\alpha}{dT}\left[\frac{E}{RT_{max}^2} + \left(\frac{A_o}{\beta}\exp\left(-\frac{E}{RT_{max}}\right)\right)\cdot\left(\frac{df(\alpha)}{d\alpha}\right)_{T=T_{max}}\right] = 0 \qquad (12.19)$$

Assuming that $f(\alpha)$ and α are independent of the heating rate and that $d\alpha/dT$ is not equal to zero, Eqn. (12.19) can be rewritten as:

$$\ln\left(\frac{\beta}{T_{max}^2}\right) + \ln\left(\frac{E}{RA_o}\right) = -\frac{E}{RT_{max}} + \ln\left(-\frac{df(\alpha)}{d\alpha}\right)_{T=T_{max}} \qquad (12.20)$$

Consequently, plots of $\ln(\beta/T_{max}^2)$ vs $1/T_{max}$ are expected to give straight lines with slope $-E/R$. These plots might be called temperature-programmed `Arrhenius plots'. This equation is used generally. We checked the validity of the assumption for the models of Table 12.1 and investigated the dependence of $df(\alpha)/d\alpha$ on the heating rate at the peak temperature. We found that $df(\alpha)/d\alpha$ was constant for all models. Therefore it can be concluded that the Kissinger method can usually be applied [6].

12.8 EXAMPLE 3: MODELLING OF TPR OF Fe_2O_3

Figures 12.9, 12.10, and 12.11 give TPR patterns of Fe_2O_3 under various experimental conditions [7]. It is interesting to compare the three profiles a–c in Fig. 12.9. The only difference is the sample size.

The larger the sample size the higher the temperature. The explanation is that H_2O inhibits reduction, hence the assumption that the gas phase term is constant is not correct. Figure 12.10 gives results of experiments where the H_2O partial

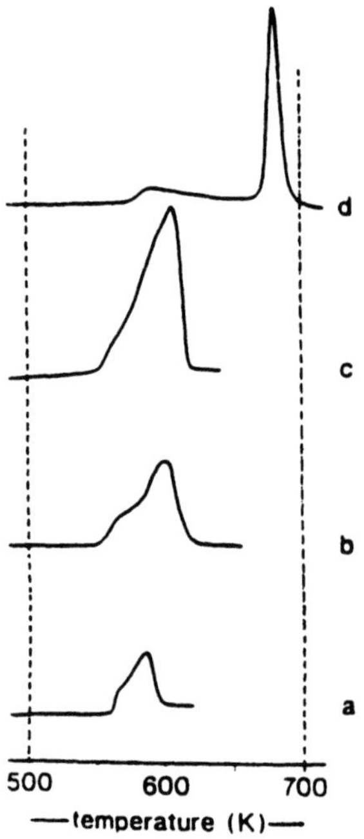

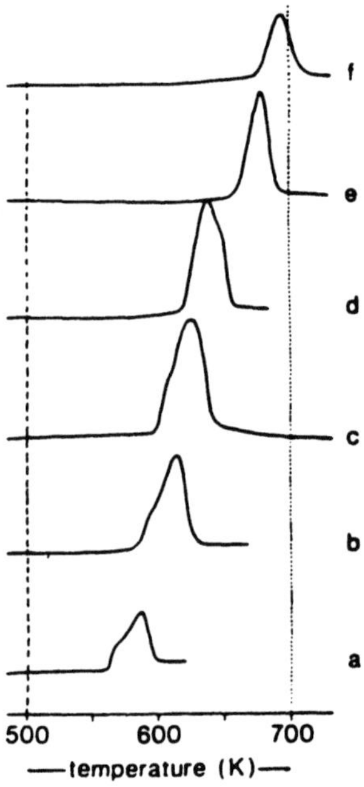

Fig. 12.9 (*left*). TPR patterns (0.2 K/min) of Fe_2O_3 in dry H_2/Ar (a,b,c) and in H_2/Ar saturated with 3% water(d); Fe_2O_3 sample sizes: (a) 3.6 mg; (b) 8.2 mg; (c) 15.9 mg; (d) 7.0 mg.

Fig. 12.10 (*right*). TPR patterns of Fe_2O_3 in dry H_2/Ar as a function of heating rate (dry series): (a) 0.2 K/min (m = 3.6 mg); (b) 0.5 K/min (m = 2.8 mg); (c) 1 K/min (m = 1.8 mg); (d) 2 K/min (m = 0.91 mg); (e) 5 K/min (m = 0.19 mg); (f) 10 K/min (m = 0.08 mg).

pressure was kept as low as possible in order to approach differential conditions with negligible H_2O inhibition ('dry series'). Figure 12.11 gives results of experiments where H_2O was added on purpose. In the `wet' series two peaks are observed, clearly due to two reduction steps, while these peaks are not resolved in the 'dry' series.

On the basis of the results given in Figs. 12.10 and 12.11, three 'Arrhenius plots' can be constructed, as given in Fig 12.12. In view of the wide range of experimental conditions it is concluded that Eqn. (12.20) is followed well, and reliable activation energies can be calculated.

These activation energies can now be used to simulate the TPR profiles. The patterns corresponding to the lowest heating rate were chosen. Figure 12.13 gives simulations for the 'dry' experiment, using the fixed value of the activation energy and T_{max} with Eqn. (12.20) to determine A_0 for the different models given in Table 12.1. Clearly, the different models give distinctly different profiles and a model selection is possible. It appears that the sharpest curve, corresponding to Avrami–Erofeev kinetics, best describes the experimental data.

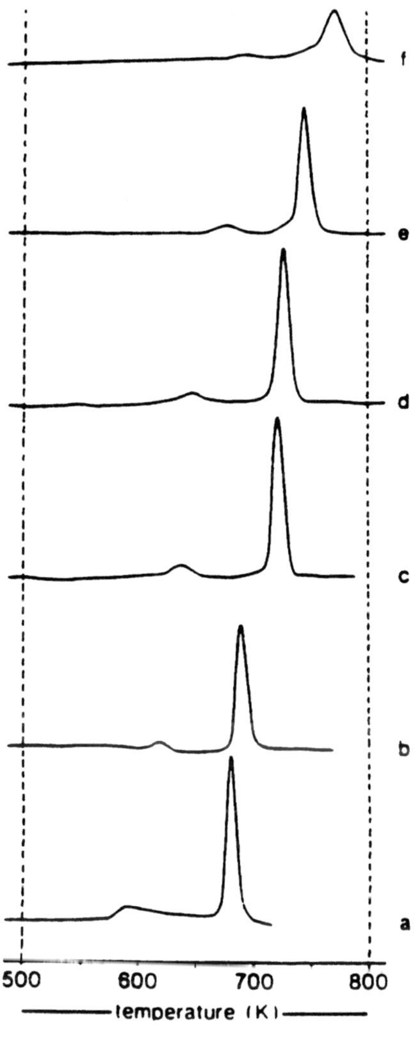

Fig. 12.11. TPR pattterns of Fe_2O_3 in wet H_2/Ar (3% water) as a function of heating rate (wet series): (a) 0.2 K/min (m = 7.0 mg); (b) 0.5 K/min (m = 2.6 mg); (c) 1 K/min (m = 1.5 mg); (d) 2 K/min (m = 0.90 mg); (e) 5 K/min (m = 0.33 mg); (f) 10 K/min (m = 0.17 mg).

Figure 12.14a compares the model based on Avrami–Erofeev kinetics with the experimental curve. The agreement is satisfactory in light of the fact that no adjustable parameter has been used to allow curve-fitting procedures. The difference at the leading edge is attributed to the first reduction step which is not included in the model.

For the wet series, none of the models tried is correct. We attribute this to an inhibition by water at the beginning of the peak (Fig. 12.14b) followed by a reduction process exhibiting a runaway character due to autocatalysis or thermal effect.

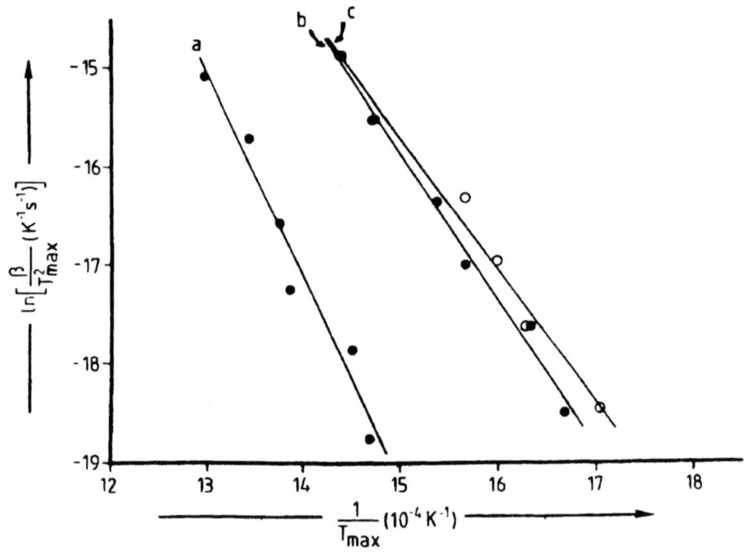

Fig. 12.12. Temperature-programmed Arrhenius plots for the reduction of Fe_2O_3. (a) main peak for wet series; (b) low temperature peak for wet series; (c) main peak for dry series (E = 111 kJ/mol).

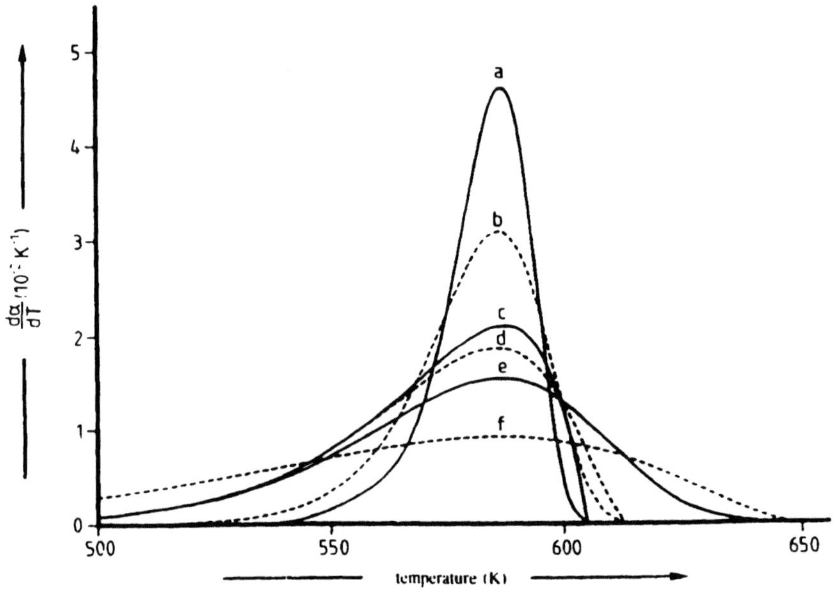

Fig. 12.13. Calculated Fe_3O_4 →Fe metal TPR peaks for six reduction models using E = 111 kJ/mol (TPR on dry H_2/Ar) 0.2 K/min (a) three-dimensional nucleation according to Avrami–Erofeev, (b) two-dimensional nucleation according to Avrami–Erofeev, (c) two-dimensional phase boundary, (d) three-dimensional phase boundary, (e) unimolecular decay, (f) three-dimensional diffusion according to Jander.

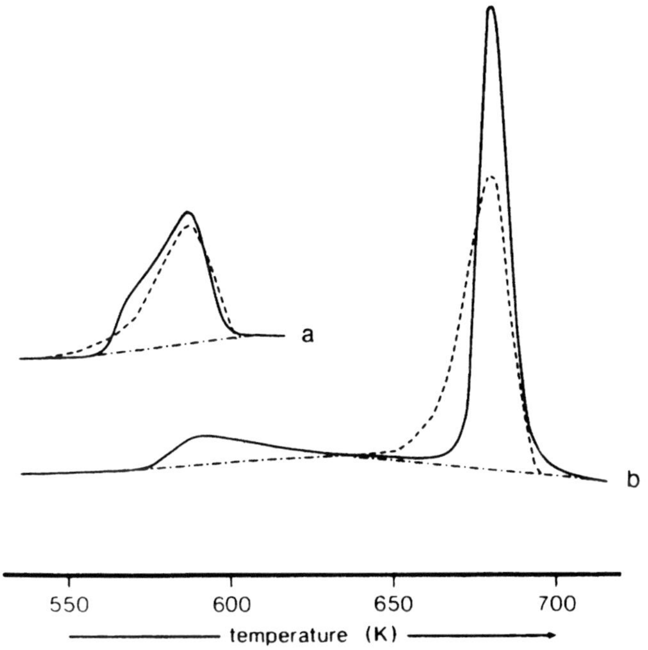

Fig. 12.14. Comparison between measured TPR patterns for Fe_2O_3 and calculated TPR peaks for $Fe_3O_4 \rightarrow$ Fe metal reduction step, using the three-dimensional nucleation model according to Avrami–Erofeev at 0.2 K/min. Calculated dotted line: (a) dry series, (b) wet series, $E = 172$ kJ/mol, $A_0 = 3.5 \times 10^9$ s^{-1}.

In conclusion, this chapter shows the potential of temperature-programmed techniques. Qualitative information on structure is obtained, which allows efficient exploratory research in optimizing pretreatment procedures. Moreover, quantitative kinetic analysis allows optimal process design for catalyst pretreatment, both as a separate step in catalyst manufacture and in the start-up of the reactor.

REFERENCES

1 N.W. Hurst, S.J. Gentry and A. Jones, *Catal. Rev. Sci. Eng.* 24 (2) (1982) 233.
2 P.J. Mangnus, Ph.D thesis, University of Amsterdam, The Netherlands, 1991.
3 P. Arnoldy and J. A. Moulijn, *J. Catal.* 93 (1985) 38–54.
4 P. Arnoldy, J.A.M. van den Heijkant, J.D. de Bok and J.A. Moulijn, *J. Catal.* 92 (1985) 25–55.
5 G.I. Senum and R.T. Yang, *J. Thermal Anal.* 11 (1977) 445.
6 A. Tarfaoui, PhD thesis. Delft University of Technology, 1996.
7 O.J. Wimmers, P. Arnoldy, and J.A. Moulijn, *J. Phys. Chem.* 90 (1986) 1331–1337.

Chapter 13

Adsorption methods for the assessment of the specific surface area, the pore size distribution and the active sites of heterogeneous catalysts

13.1 INTRODUCTION

This chapter aims to give guidelines on how to use adsorption methods for the characterization of the surface area and pore size of heterogeneous catalysts. The information derived from these measurements can range from the total and available specific surface area to the pore sizes and the strength of sorption in micropores. Note that this spans information from a macroscopic description of the pore volume/specific surface area to a detailed microscopic assessment of the environment capable of sorbing molecules. In this chapter we will, however, be confined to the interaction between sorbed molecules and solid sorbents that are based on unspecific attractive and repulsive forces (van der Waals forces, London dispersion forces).

Knowledge of the properties of the texture and pore structure of solid materials is highly important for the development of catalytic materials and catalysts. It should, however, be mentioned that the assessment of the dynamic changes of the material under reaction conditions may be even more important than the assessment under pristine conditions, as the changes in material during use will determine its useful lifetime. Note also that physisorption markedly influences specific properties of porous solids, such as acid strength and catalytic activity which are crucial material parameters for sorption and catalysis.

13.2 PHYSICAL ADSORPTION

Physical adsorption (physisorption) occurs when an adsorptive comes into contact with a solid surface (the sorbent) [1]. These interactions are unspecific and similar to the forces that lead to the non-ideal behavior of a gas (condensation, van der Waals interactions). They include all interactive and repulsive forces (e.g., London dispersion forces and short range intermolecular repulsion) that cannot be ascribed to localized bonding. In analogy to the attractive forces in real gases, physical adsorption may be understood as an increase of concentration at the gas–solid or gas–liquid interface under the influence of integrated van der Waals forces. Various specific interactions (e.g., dipole-induced interactions) exist when either the sorbate or the sorbent are polar, but these interactions are usually also summarized under physisorption unless a directed chemical bond is formed.

The van der Waals forces between molecules and atoms are well known from their introduction in the so-called van der Waals equation empirically describing the behaviour of a real gas in terms of varying pressure and temperature [2]. The term a/V^2 in this equation describes the difference between the ideal and the observed pressure, while b may be interpreted as the volume occupied by the gas molecules

$$(p + a/V^2)(V - b) = RT \tag{13.1}$$

It is important to note that the interaction energy depends on the inverse of the *sixth* power of the distance between the centre of mass of the two molecules:

$$E_{\text{attr}} = -C\,r^{-6} \tag{13.2}$$

C is a coefficient that depends on the nature of the molecule, r the mean distance between the molecules and E_{attr} the total attractive interaction energy.

Repulsive interactions begin to play a role when the molecules closely approach each other, as the nuclear and electronic repulsions and the rising electronic kinetic energy outweigh the attractive forces. The repulsive energy between two molecules can be described by:

$$E_{\text{rep}} = B\,r^{-12} \tag{13.3}$$

where B is an empirical constant.

The total potential energy between two molecules thus becomes:

$$E(r) = -C\,r^{-6} + B\,r^{-12} \tag{13.4}$$

$E(r)$ is often termed the Lennard-Jones potential. A graphic representation of the balance between the two forces and the resulting Lennard-Jones type of potential can be seen in Fig 13.1. It should be emphasized strongly, however, that these

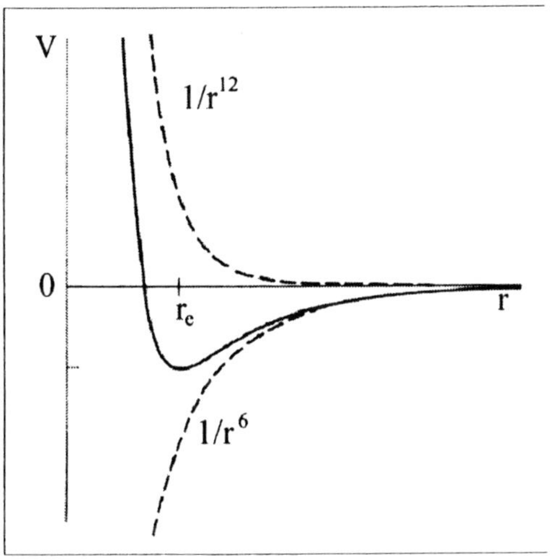

Fig. 13.1. The Lennard-Jones potential and the relation of the parameters to the features of the curve. The dotted lines are the two contributions.

models are oversimplifications and that the exact potential curve may vary significantly depending on the molecules involved in the interaction.
It is important to note that the interaction energies discussed are *weak, short-range forces*. For instance, at 25°C the average interaction energy for pairs of molecules with a dipole moment of 1 Debye unit is about –1.4 kJ mol^{-1} when the separation is 0.3 nm. This should be compared with the average molar kinetic energy of 3/2 RT = 3.7 kJ mol^{-1} at the same temperature.

To apply the approach to the adsorption of a molecule on a surface plane [3], it is necessary to consider the nature of the interaction of the surface atoms of the solid with the sorbed molecules. Intuitively, it is straightforward to see that one needs now to account for the interaction of each surface atom with the sorbed molecule or, more precisely, with each atom in this molecule, in order to describe the interaction correctly. The individual attractive and repulsive contributions need to be summed up. Two consequences can be derived from this approach considering only the above (simplified) formula (13.4):

- The contribution of the directly neighboring atoms to the bonding between the surface and the sorbed molecule will be stronger than that of atoms further apart.
- The curvature of the surface will have an important influence upon the mutual interaction. The larger the curvature (the smaller the pore), the more atoms exist that are close to the sorbed molecule and, hence, the higher the strength of interaction will be (for illustration see Fig.13.2).

It should be especially emphasized that this is not limited to a particular type of interaction and, as a consequence, permanent dipole or quadrupole moments

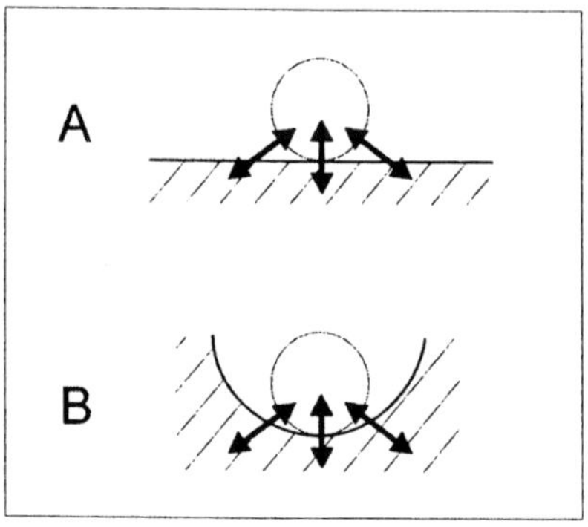

Fig. 13.2. The effect of surface curvature on interaction between surface and molecule.

will tend to increase the overall interaction. Similarly, attractive or repulsive forces between molecules at the surface will influence the stabilization of the sorbed molecules.

13.3 ADSORPTION ISOTHERMS

Adsorption isotherms describe one macroscopic consequence of these interactions, i.e., the relation between the amount of molecules adsorbed on a unit (mass, surface) of the solid and the sorbate equilibrium pressure (or relative pressure) at a given temperature. Below the critical temperature, the pressure is commonly normalized to the saturation pressure p_0 which then leads to a dimensionless expression of the relative pressure p/p_0. The quantity of gas adsorbed is usually expressed as the mass of the sorbate or the volume of gas reduced to STP (standard temperature and pressure) adsorbed per mass of solid sorbent.

The majority of isotherms which result from physical adsorption may conveniently be grouped into five classes, i.e., the five types I to V included in the classification originally proposed by Brunauer, Deming, Deming and Teller [4]. The essential features of these types are indicated in Fig. 13.3.

Isotherm I is typical of adsorption in micropores, e.g., adsorption on molecular sieves and activated carbons. *Isotherm II* represents multilayer physisorption on a flat surface (valid for many nonporous substances). *Isotherms III* and *V* are characteristics of weak gas–solid interactions, e.g., water adsorption on gold. *Isotherm IV* is frequently observed in the study of practical heterogeneous catalysts. Its shape is characteristic of multilayer adsorption accompanied by capillary condensation in mesopores. When the surface of a nonporous adsorbent is energetically uniform the isotherm may have a step-like shape (*Isotherm VI*). A good example of such behaviour is the adsorption isotherm of Kr at 90 K on graphite [5].

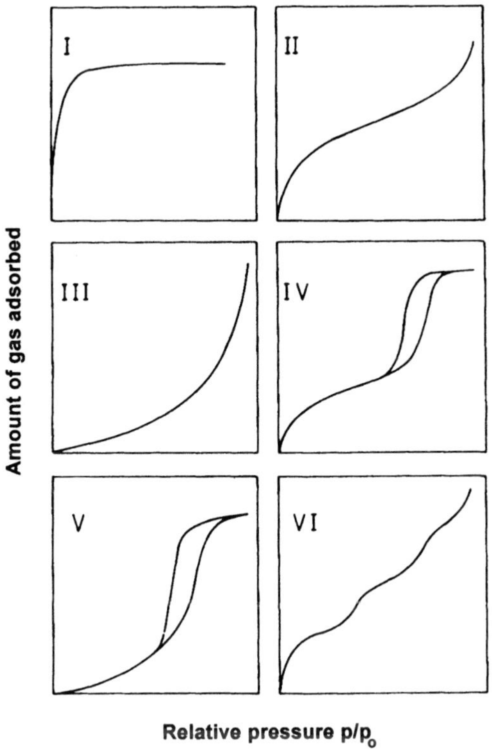

Fig. 13.3. The six types of adsorption isotherms following the Brunauer, Deming, Deming, Teller notation.

In this chapter we will confine ourselves to the application of types II and IV isotherms in the determination of surface areas and pore size distributions. In practice, the range of suitable adsorptives for this purpose is quite narrow, by far the most commonly used one being nitrogen at its boiling point (about 77 K) and, to a lesser extent, Ar.

13.4 CLASSIFICATION OF PORE SIZES

The individual pores in heterogeneous catalysts and in other porous technical substances may vary significantly in size and shape. However, the most basic property remains the average width of the pores, e.g. the diameter of a cylindrical pore, or the distance between the sides of a slit-shaped pore. A classification of pores according to their average width was originally proposed by Dubinin [6] and has now been officially adopted by the International Union of Pure and Applied Chemistry [7]. According to this classification, the widths of micropores range from 0.3 to 2 nm, mesoporous substances have pore sizes from 2 nm up to 50 nm and macropores range from widths of 50 nm up to about 10^5 nm.

Specific surface areas and pore size distributions of mesoporous materials are best probed by nitrogen/argon adsorption and capillary condensation which will be outlined in detail below. It should be emphasized that the concept of specific surface area is not applicable when the size of the sorbed molecules approaches the diameter of the pore. Thus, for microporous substances values for specific surface areas have no physical meaning, but are rather characteristic of the volume of gas adsorbed. Nevertheless, these values are frequently used as practical numbers to compare the quality and porosity of microporous materials. The average pore size of microporous materials has to be probed by size exclusion measurements. For this purpose the uptake of a series of sorbates with increasing minimal kinetic diameter on a solid are explored. The drop in the adsorbed amount with increasing size of the sorbate defines the minimum pore diameter of the tested solid. The method will be described in detail below.

13.5 POROSITY OF POROUS SUBSTANCES

The total porosity of solids can be estimated by simple pycnometric methods using the differences in surface tension between gases and mercury. By definition, the porosity $Pr(\%)$ is given by:

$$Pr = (V_p / V_{tot}) \cdot 100 \tag{13.5}$$

where V_p is the pore volume in $cm^3\ g^{-1}$ and V_{tot} is the total volume of the porous substance, estimated by following the outer boundaries of the particles (usually denoted in $cm^3\ g^{-1}$).

Consequently it follows that the total volume V_{tot} is the sum of volume of the dense material and the pore volume:

$$V_{tot} = V_{dense} + V_p \tag{13.6}$$

V_{dense}, the true specific volume, may be obtained from physicochemical tables, but can also be determined pycnometrically by imbibition of the material in helium. The total specific volume, V_{tot}, is determined pycnometrically by imbibition of the porous material in mercury at 0.1 MPa pressure. Note that most materials are not wetted by mercury and, hence, mercury will not penetrate into the pores at 0.1 MPa pressure.

As $V_{dense} = V_{He}$, $V_{tot} = V_{Hg}$, and $V_p = V_{Hg} - V_{He}$ we may write:

$$Pr = (V_{Hg} - V_{He}) / V_{Hg} \cdot 100\% \tag{13.7}$$

In practice, the porosity varies from about 14%, as found for some zeolites, to about 70% found for activated carbons.

13.6 THE YARDSTICK IN THE DETERMINATION OF SPECIFIC SURFACE AREAS

Intuitively, it is straightforward to see that the amount of sorbed molecules forming a dense monolayer on the surface of a solid can be used to calculate its specific surface area, A. Normalized to the mass of a solid, this quantity is termed monolayer capacity and is defined as the amount of adsorbate that can be accommodated in a completely filled single molecular layer, a monolayer, on the surface of one gram of solid. It is related to the specific surface area A (m^2/g) by the simple equation:

$$A = n_m \, a_m \, L \tag{13.8}$$

where a_m is the average area occupied by a molecule of adsorbate in the completed monolayer and L is the Avogadro constant; n_m, the monolayer capacity ($mol_{sorbate}/g_{solid}$).

While in principle all molecules can be used for this purpose, meaningful values are only obtained if the dimensions of the sorbate are small compared to the diameter of the pores of the solid. Conventionally, most surface area determinations use the area occupied by a nitrogen molecule. The nitrogen adsorption is measured at the boiling point of nitrogen, 77.4 K.

The value of a_m may be estimated following an early suggestion of Emmett and Brunauer [8] from the density of liquid nitrogen at 77.4 K. The estimate uses the assumption that the packing density of adsorbed nitrogen on a surface is the same as in the liquid. This leads to the equation:

$$a_m = 1.091\,(M/\rho_L \, L)^{2/3} \tag{13.9}$$

where $M = 28$, the molecular weight of nitrogen, and 1.091 is a packing factor for 12 nearest neighbours in the bulk liquid and six on the plane surface. The density of liquid nitrogen at 77.4 K, ρ_L, is 0.81 g cm^{-3} and L is equal to 6.022×10^{23} mol^{-1}. Insertion of these quantities into Eqn. (13.9) gives $a_m(N_2) = 16.2$ nm^2 for nitrogen at 77.4 K.

For practical purposes, the average area occupied by other molecules can be determined by comparison of their monolayer capacity with that of nitrogen. Using this method, the areas occupied by argon a_m(Ar) and krypton a_m(Kr) at 77.4 K were then determined to be 0.138 nm^2 and 0.20 nm^2, respectively. For benzene at 273 K, a value 0.40 nm^2 was reported.

13.7 THE LANGMUIR (MONOLAYER ADSORPTION) DESCRIPTION OF ADSORPTION

The simplest isotherm originally dervied by Langmuir [9] is based on the assumptions that:

- all adsorption sites are energetically equivalent;
- the ability of a molecule to bind at a site is independent of whether or not a nearby site is occupied (absence of lateral interactions).

As adsorption takes place on the free part of the surface only, we can state that:

$$r_{ads} = k_a(p/p_o)\,N(1-\theta) \tag{13.10}$$

where r_{ads} is the rate of adsorption on the uncovered part of the surface, p/p_o is the relative pressure, N is the total number of surface sites, and $(1-\theta)$ is the uncovered fraction of the surface; k_a is the rate constant for adsorption.

The rate of desorption is given by:

$$r_{des} = k_d N\theta \tag{13.11}$$

in which k_d is the rate constant for desorption. At equilibrium, $r_{ads} = r_{des}$ and hence:

$$k_a(p/p_o)\,N(1-\theta) = k_d\,N\theta \tag{13.12}$$

The adsorption equilibrium constant K is equal to k_a/k_d and hence:

$$K = \exp\,(-\Delta G^o/RT) = k_a/k_d \tag{13.13}$$

in which ΔG^o is the free standard enthalpy of adsorption. It follows from Eqns. (13.12) and (13.13) that

$$\theta = \frac{Kp/p_o}{1+Kp/p_o} \tag{13.14}$$

which is known as the Langmuir equation.

For practical purposes, it is useful to plot the Langmuir equation in the so-called reciprocal form. By substitution of n/n_m for the coverage θ, in which n is the number of moles adsorbed on 1 g of adsorbent and n_m is the monolayer capacity expressed in the same units, we obtain from Eqn. (13.14):

$$1/n = 1/n_m + 1/(n_m K)\cdot 1/(p/p_o) \tag{13.15}$$

A plot of $1/n$ against $1/(p/p_o)$ should result in a straight-line relationship (see Fig. 13.4). The monolayer capacity can be calculated from the intercept $1/n_m$, and the equilibrium constant is found from tangent $\beta = 1/n_m K$.

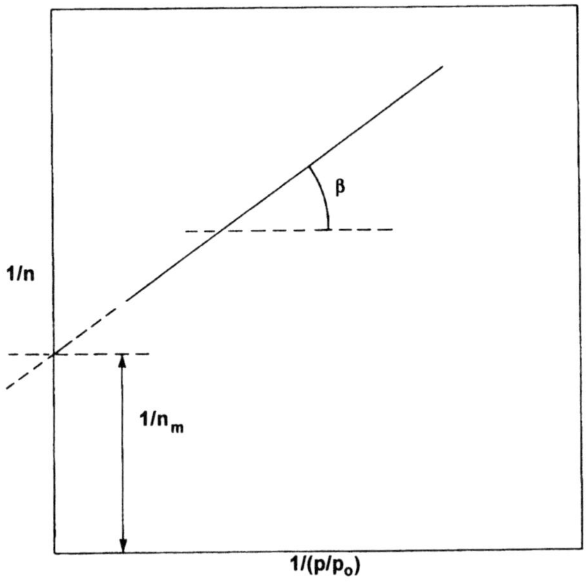

Fig. 13.4. The reciprocal plot of the Langmuir equation.

It should be emphasized that the adsorption on a practical material can seldom be described by the simple Langmuir equation. In most cases, the adsorption sites are not energetically equivalent, neither in a physical, nor in a chemical adsorption. In the physisorption of nitrogen on rutile, for instance, the isosteric heat of adsorption declines from 26 kJ mol^{-1} at zero coverage to only 60% of this value at full coverage [1]. Additionally, considerable mutual interaction may exist between molecules in adsorbed monolayers, especially at coverages above 0.5. In most instances, however, the adsorption can be described by a sum of several Langmuir isotherms and additional factors considering the interactions between molecules. If the heats of adsorption are sufficiently different, the individual isotherms can be assessed by plotting the data in the reciprocal form as described above.

13.8 THE BET (MULTILAYER ADSORPTION) DESCRIPTION OF ADSORPTION

For small inert molecules it can be difficult to assess directly the monolayer, as the heat of adsorption is so small that it approaches the heat of condensation. This leads to parallel adsorption in the first and subsequent layers and therefore does not allow definition of the monolayer capacity. A way to circumvent this difficulty is to determine the adsorption in the presence of multilayers and to relate the sorbed amount to the monolayer capacity. This approach was first formulated by Brunauer, Emmett and Teller [10] and is known as the BET equation. For formulating the relation between the overall concentration of gas adsorbed and the monolayer capacity it is assumed that:

- the heat of adsorption is constant in the first layer;
- the heat of adsorption in the second and all other layers is constant and lower than in the first layer and approximates the heat of condensation; and
- an infinite number of layers can be adsorbed.

These assumptions are justifiable as the heat of adsorption of the small inert sorbate (e.g., N_2 or Ar) is rather low and, hence, differences between sorption sites at the surface will be very small. Similarly, the interaction between the first and the following layers will be close to the heat of condensation, as the effect of polarization by the surface will be small beyond the first layer (screening of the long-range van der Waals forces). From its conception, the BET theory extends the Langmuir model to multilayer adsorption. It postulates that under dynamic equilibrium conditions the rate of adsorption in each layer is equal to the rate of desorption from that layer. Molecules in the first layer are located on sites of constant interaction strength and the molecules in that layer serve as sorption sites for the second layer and so forth. The surface is, therefore, composed of stacks of sorbed molecules. Lateral interactions are assumed to be absent. With these simplifications one arrives at the BET equation

$$\frac{V}{V_{\text{mon}}} = -\frac{c(p/p_o)}{(1-p/p_o)\{1-(1-c)(p/p_o)\}} \tag{13.16}$$

where V is the volume of gas (STP) adsorbed, V_{mon} is the volume of gas (STP) adsorbed in the monolayer and p_o is the vapour pressure above a macroscopically thick layer of the pure liquid on the surface. The quantity c approximates to $\exp -(\Delta H_d - H_{\text{vap}}))/RT$, in which ΔH_d is the enthalpy of adsorption in the first adsorbed layer and H_{vap} is the heat of vaporization, i.e., the negative of the enthalpy of condensation of the gas equalling the heat of adsorption in all following layers. The equation is usually rearranged into the linear form

$$\frac{p}{(p_o - p)V} = \frac{1}{cV_{\text{mon}}} + \frac{(c-1)}{cV_{\text{mon}}}\frac{p}{p_o} \tag{13.16a}$$

For the derivation of the BET equation, we refer to Refs. [2,11].

Notwithstanding shortcomings, the BET equation is still the most generally used method for the determination of specific surface areas. It is applicable in cases where we are dealing with type II or type IV isotherms (see Section 13.3), but the range of partial pressures where theory and practice are in reasonable agreement is limited (from p/p_o 0.05 to 0.3). This is illustrated by Harris and Sing [11], who compared experimental isotherms with the isotherm following from

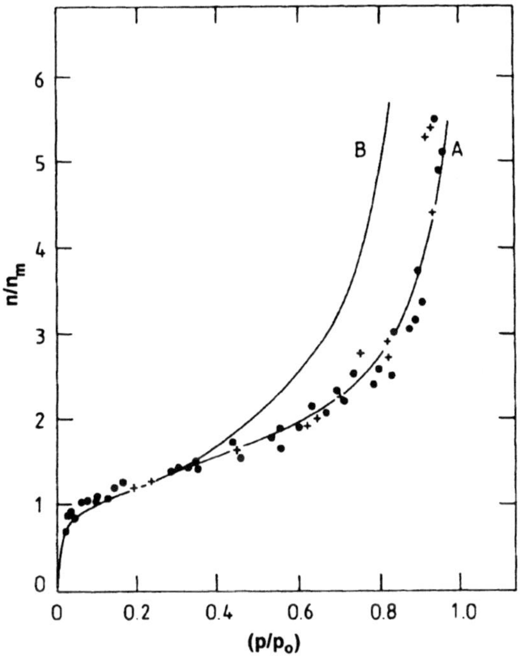

Fig. 13.5. The adsorption of nitrogen on nonporous samples of silica and alumina, the surface areas ranging from 2–11.5 $m^2\,g^{-1}$ for silica and from 5–153 $m^2\,g^{-1}$ for alumina. n/n_m is plotted against p/p_o. n is the number of moles adsorbed per gram of adsorbent and n_m is the monolayer capacity, expressed in the same units. Line A: ● = silica; + = alumina. Line B: the BET isotherm with c values of 100–200.

the BET model (see Fig. 13.5). This figure clearly shows that the BET equation underestimates the adsorption at low pressures and overestimates it at high pressures. When applied in this restricted region, the BET method leads to correct values of the surface area of solids within 5% for the isotherm type II or IV [1]. In the practical application of the BET method, the volume of adsorbed gas is measured at constant temperature as a function of the partial pressure and $p/V(p_o - p)$ is plotted against p/p_o (see Fig. 13.6).

In materials containing micropores (and mesopores), such as molecular sieves and activated carbons, the boundary conditions (multilayer adsorption) are not fulfilled and the BET analysis is not applicable here. When (traditionally) reported in the literature, the BET surface area should be understood as a number proportional to the pore volume rather than the specific surface area. Note also that conceptually the term specific surface area is inapplicable once the solid material surrounds sorbed molecules.

The intercept $1/V_m c$ and the tangent of α, $(c - 1)/V_m c$, readily permit the calculation of V_m and c. n_m, the number of moles of adsorbate per gram of adsorbent in the monolayer, is found from V_m. Finally, A(BET), the BET surface area, is calculated from $A = n_m \cdot a_m \cdot L$ (Eqn. (13.8)).

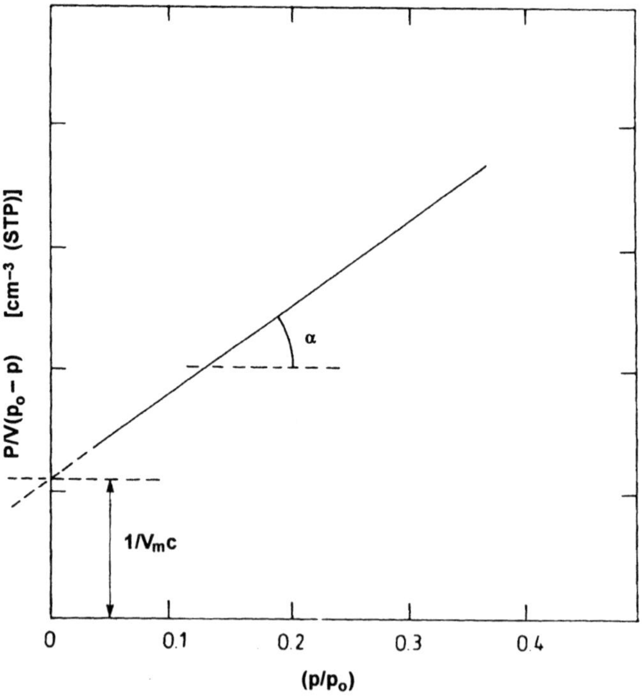

Fig. 13.6. The BET plot.

13.9 THE *t* METHOD, A CONCEPT OF A STANDARD ISOTHERM

The *t* method for the determination of specific surface areas was developed by De Boer et al. [12]. It was shown that the nitrogen adsorption isotherm for a large number of nonporous solids (such as SiO_2, Al_2O_3, ZrO_2, MgO, TiO_2, $BaSO_4$ and many activated carbons) can be represented by a single curve, when the ratio V_a/A (where V_a is the volume of N_2 adsorbed and A is the surface area of the sample) is plotted as a function of the relative pressure. This so-called common *t*-curve, a type II isotherm, is reproduced in Fig. 13.7. On the right-hand side *y* axis the volume (STP) of adsorbed N_2 per surface areas (at 78 K) is indicated. The specific surface areas of the nonporous solids were determined by applying the BET method.

For the partial pressures indicated on the ordinate we have chosen the symbol p'/p_o in order to indicate that we are dealing here with equilibrium pressures above approximately flat surfaces. This will be seen to be important in the discussion of capillary condensation in Section 13.10. The left-hand side abscissa gives the statistical thickness of the adsorbed multilayer, *t*, in nm.

The *t* values on the abscissa are calculated as follows. The number of nitrogen molecules adsorbed in a monolayer on 1 m^2 corresponds to 0.23 cm^3 N_2(STP).

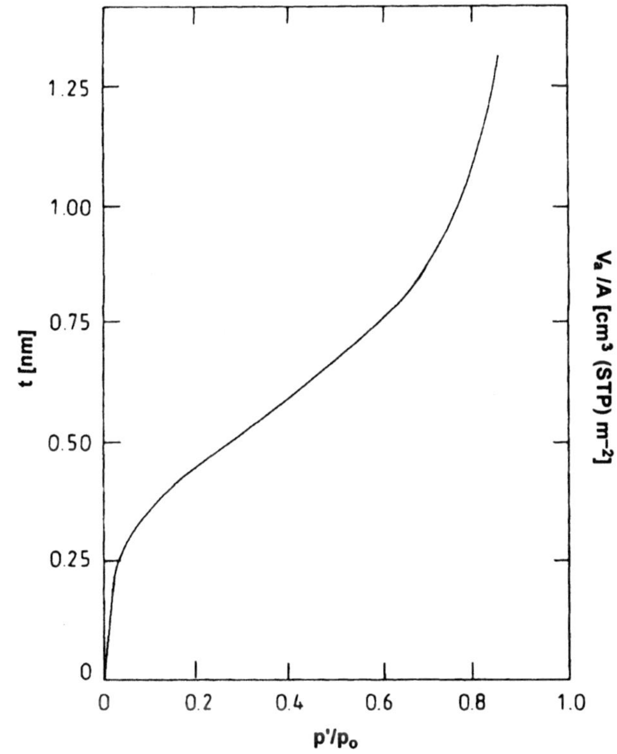

Fig. 13.7. The common t curve. Nitrogen adsorption isotherm at 78 K on nonporous solids.

Hence for a t material (i.e. a material that is ideally described by the t-method) with surface area $A(t)$ it holds that:

$$V_m = 0.23 \cdot A(t) \qquad \text{cm}^3\ \text{N}_2\text{(STP)} \tag{13.17}$$

When a volume V_a of N_2 is adsorbed, the statistical number of adsorbed nitrogen layers in equal to V_a/V_m. As the thickness of one layer of adsorbed nitrogen is 0.354 nm, the statistical layer thickness corresponding to V_a/V_m layers is given by:

$$t = \frac{V_a}{V_m} * 0.354 = \frac{V_a * 0.354}{A(t) * 0.23} \tag{13.18}$$

and hence:

$$A(t) = 1.54 \cdot V_a/t \qquad \text{m}^2 \tag{13.19}$$

It follows from Eqn. (13.19) that, for every nonporous t material, a plot of V_a against t should result in a straight line passing through the origin. Such a plot is

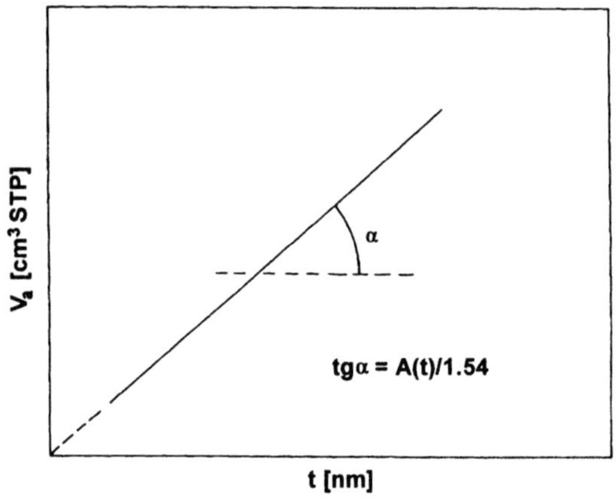

Fig. 13.8. General form of the *t* plot.

called the *t* plot and the slope of the straight line is a measure of $A(t)$ (see Fig. 13.8). Following this approach, the surface area $A(t)$ of a solid is determined by first measuring the N_2 adsorption isotherm at 78 K. Next, this isotherm is converted into a *t* plot by substituting each p/p_o value for the corresponding *t* value. These *t* values are given by the common *t* curve. It is, however, also possible to calculate the *t* values from the equation:

$$t(\mathrm{nm}) = 0.1\left(\frac{13.99}{-\log(p/p_o)+0.034}\right)^{1/2} \tag{13.20}$$

In the range of partial pressures from 0.1 to 0.8, Eqn. (13.20) gives a good description of the course of the common *t* curve.

t Plots often deviate strongly from the ideal situation seen in Fig. 13.8. This deviation allows the nature of pores to be deduced and the micropore volume to be determined (see Fig. 13.9). Curve a in Fig. 13.9 is linear and extrapolates through the origin. In this case we may conclude that we are dealing with a *t* material and $A(t)$ can be calculated from the tangent. Curve b, however, shows a strong upward deviation from linearity at a certain pressure p/p_o, i.e., more N_2 is taken up than might be expected. This indicates that starting from t_1 capillary condensation takes place in addition to adsorption. In contrast, if some narrow pores are filled by multilayer adsorption, further adsorption does not occur in this part, as that surface has become unavailable. Then, the points on the *t* plot begin to deviate downwards from the straight line (see point t_2 in Fig. 13.9). This situation is, for instance, encountered in the presence of slit-shaped pores.

The presence of micropores is indicated by line d in Fig. 13.9. A straight line is found, and the surface area $A(t)$ can be calculated from its tangent of the line. The positive intercept at the *y* axis is caused by a relatively large nitrogen uptake at a

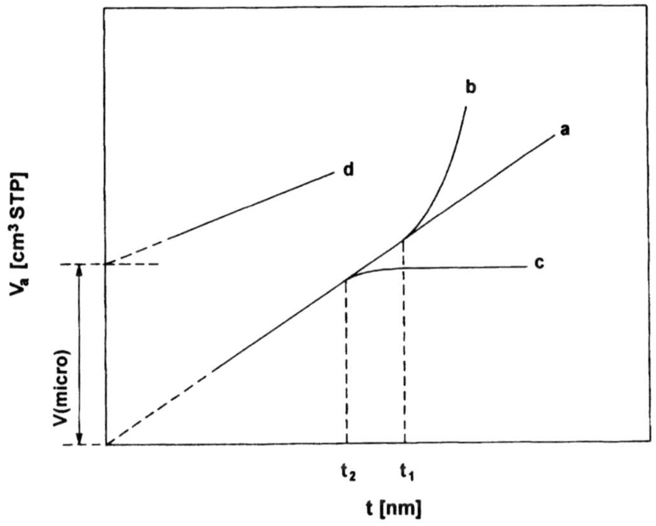

Fig. 13.9. Four different courses of the *t* plot.

very low *t* value (corresponding to a low partial pressure). Such high N_2 uptake at very low pressure is due to the stronger adsorption in micropores and, hence, the micropore volume V_{micro} can be calculated from the intercept.

13.10 ASSESSMENT OF MESOPORE RADII AND VOLUMES VIA THE KELVIN EQUATION

Pore radii and pore volume distributions as a function of the pore radii may be calculated from the relative pressures at which the pores are filled and emptied. For the case of pore filling with liquid nitrogen, the relation between the pore radius, r, and the relative pressure at which filling starts, p_a/p_o, is given by the Kelvin equation for pores with circular cross section:

$$\frac{p_a}{p_b} = e^{-\frac{\gamma V_L}{RT(r-t)}} \tag{13.21}$$

with γ, the surface tension of liquid nitrogen at 77.4 K, equal to 8.72×10^{-7} J cm^{-1} and V_L, the molar volume of liquid nitrogen at 77.4 K, equal to 34.68 cm^3 mol^{-1}, R, the gas constant equal to 8.314 J mol^{-1} K^{-1}, and T = 77.4 K we calculate, starting from Eqn. (13.21):

$$(r-t) = 0.4078/\log(p_o/p_a) \qquad [(r-t) \text{ in nm}] \tag{13.22}$$

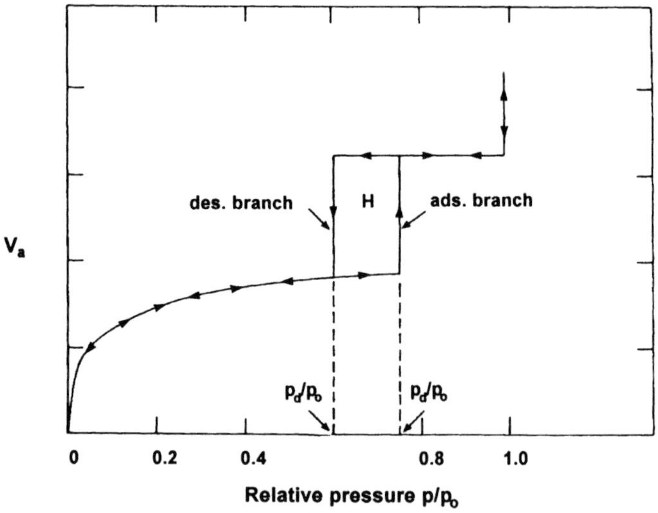

Fig. 13.10. Adsorption and desorption isotherms for a porous material, containing cylindrical pores with radius r.

At each p_a/p_o value, t can be calculated from Eqn. (13.20). Hence from Eqn. (13.22) the radius of a pore can be calculated from the relative pressure p_a/p_o at which capillary condensation starts.

Similarly it can be derived (for details see Ref. [13]) that the pressure at which desorption from the filled pores will start is given by

$$\frac{p_d}{p_o} = e^{-\frac{2\gamma V_L}{RT(r-t)}} \tag{13.23}$$

Kelvin equations describing capillary evaporation from slit-shaped pores or the filling and emptying of "ink bottle" pores can be calculated quite easily by using the appropriate expressions (see Ref. [14]).

From what is said above, it follows that condensation of a gas in the cylindrical pores of a porous material will occur at a different relative pressure p/p_o than desorption, leading to a different dependence of V_a vs. p/p_o during adsorption and desorption (see Fig. 13.10). This leads to a hysteresis loop. From the appropriate Kelvin equation, Eqn. (13.21) and Eqn. (13.23), it follows that, for a t material with identical pores of the diameter r, the point of pore filling and emptying are related by equation (13.24).

$$2\ln(p_a/p_o) = \ln(p_d/p_o) \tag{13.24}$$

Thus when capillary condensation starts, for instance, at $p_a/p_o = 0.8$ (the 'adsorption branch'), emptying of the pores will start at $p_d/p_o = 0.64$ (the 'desorption branch'). In practice, several types of hysteresis loops are found. Their shapes are indicative of the nature of pores and equations equivalent to those for the cylindrical pores can also be derived (in the case of cylindrical pores that are closed at one end the hysteresis does not occur).

13.11 THE CORRECTED KELVIN EQUATION

For deriving the Kelvin equation one uses the changes in the free energy ΔG upon changing the number of moles in the gas phase and the condensed phase. When dN moles of a substance change between the condensed phase and the gas phase at equilibrium conditions we can write

$$dG_{p,T} = \mu_a \, dN - \mu_g \, dN + \gamma \, dA = 0 \tag{13.25}$$

where μ_a is the thermodynamic potential of the adsorbed phase, μ_g the thermodynamic potential of the gas phase, γ the adsorbate/vapour surface tension and dA the change in the adsorbate/vapour interface associated with the transfer of dN moles of vapour to the adsorbed phase. If we treat the vapour like an ideal gas we can write for its thermodynamic potential

$$\mu_g = \mu_{o,g} + RT \ln(p/p_o) \tag{13.26}$$

For an ideal case one assumes that the chemical potential of the liquified gas μ_L is equal to the chemical potential of the sorbed layer μ_a

$$\mu_a = \mu_L = \mu_{o,g} \tag{13.27}$$

For the change of the surface area A during desorption from a concave meniscus we can write

$$dA/dN = -2\, V_L/(r-t) \tag{13.28}$$

for capillary condensation

$$dA/dN = -\, V_L/(r-t) \tag{13.29}$$

which leads to the Eqns. (13.21) and (13.23).

Figure 13.7 shows, however, that this assumption made in Eqn. (13.27) cannot be correct (see also Ref. [14]). At each t value lower than approximately 2 nm, the relative equilibrium pressure is lower than 1, the equilibrium pressure of the liquefied gas.

The correct value of the thermodynamic potential of the adsorbed multilayer is a function of t and is derived from the common t curve

$$\mu_a(t) = \mu_{o,g} + RT \ln p'/p_o(t) = \mu_{o,g} - F(t) \tag{13.30}$$

Equation (13.30) leads to the 'corrected' Kelvin equations for the adsorption and desorption branch:

$$\frac{p_a}{p_o} = e^{-\left(\frac{\gamma V_L}{RT(r-t)} + \frac{F(t)}{RT}\right)} \tag{13.31}$$

$$\frac{p_d}{p_o} = e^{-\left(\frac{2\gamma V_L}{RT(r-t)} + \frac{F(t)}{RT}\right)} \tag{13.32}$$

This equation can be used to explain that stable adsorption on the inner wall of a capillary tube is possible up to a certain critical thickness. Capillary condensation starts from this critical thickness. It appears that capillary condensation in the cylindrical pores with radii between 2 and 7 nm is well described by the corrected Kelvin equation. The deviation from the ideal behavior can amount to up to +20%. Nevertheless, it is difficult to estimate the limits of the applicability of the Kelvin equation. It seems likely that distortions of the meniscus in small pores may occur and that the local pore geometry may have a marked influence [1,15]

For evaluating the mesopore size distribution, the use of N_2 is recommended as the form of the multilayer adsorption isotherm is largely insensitive to differences in the surface structure of the sorbent and the specific surface area and the mesopore size distribution can be calculated from the same isotherm.

Although the corrected Kelvin equation has been shown to give better results, still the most widespread method used is that proposed earlier by Barret, Joyner and Halenda [16]. It is based on a stepwise reducing of p/p_o which leads to emptying of the pores. The derived mesopore size distribution is expressed graphically as relation between $\Delta V_p/\Delta r_p$ versus r_p (= pore radius). The mesopore volume is assumed to be completely filled with N_2 at high p/p_o which corresponds approximately to an r_p of 50 nm. The application of such a method assumes, of course, that the Kelvin equation is valid over the entire range and that the desorption is only controlled by the equation (i.e., distorting effects of the sample are absent).

13.12 MERCURY POROSIMETRY

Mercury porosimetry is used for the determination of the total pore volume and the size distribution of meso- and macropores. The technique uses progressive increase of the external pressure to force mercury into the pores of the sample studied. An inverse relationship exists between the pressure applied P and the pore radius r_p. For cylindrical pores this is the Washburn equation [17]

$$r_p = (2\gamma/P)\cos\theta \quad (13.33)$$

where γ is the surface tension and θ the contact angle. Usually the surface tension of pure mercury and a contact angle of 141° are assumed for calculations. During an experiment the external pressure is gradually increased and the volume change (i.e., the volume of mercury entering the pores of the solid) is measured. The cumulative volume change yields the total volume and the derivative the distribution of pores with a particular diameter. The pressure can be varied from 0.1 to 2000 bar corresponding to radii from 75 μm to 3.5 nm. Note that during expansion, Hg may remain trapped in the pores and that the volume change during Hg intrusion and extrusion needs to be considered to study the nature of the pore distribution in some depth. Technical details of the technique are described in Ref. [18].

13.13 ASSESSING MICROPOROSITY

As outlined above, micropores are characterized by pore diameters smaller than 2 nm. Well-known examples of solids containing micropores are zeolites and activated carbons. Type I isotherms are observed with these materials (see Fig. 13.3) which are characterized by a large uptake of the adsorbate at a very low pressure, i.e., up to a p/p_o value of about 0.15. A sharp bend in the isotherm is found at $p/p_o = 0.0-0.15$ and (if mesopores are absent) subsequently a horizontal plateau up to about $p/p_o = 0.95$. In that case, the uptake at the horizontal plateau may be taken as the micropore volume.

It is generally agreed that the strong interactions in the narrow pores (and the filling of the pores at low p/p_o) are related to the overlap of the adsorption forces. The theoretical treatment of Gurfein et al. [19] and Everett et al. [20] showed conclusively that the strongest interactions were observed when the pore size and the size of the molecules adsorbed matched closely. This concept was also formulated later by Derouane et al. [21,22] who related primarily the radius of curvature of the surface around the sorption site to the strength of interaction. However, the polarizability of the lattice might also markedly influence the strength of interaction with the sorbate. Mortier et al. [23] showed in this respect that the polarizability of the zeolite lattice increases in parallel with the density of the framework atoms, suggesting a trend similar to that observed in the calculations by Derouane et al. [21,22]. New theoretical studies (energy minimization techniques [24], molecular dynamics studies [25] and Monte Carlo simulations [26]) addressing the energetics and siting of alkanes in siliceous zeolites uniformly indicate an increasing strength of interaction with decreasing pore size.

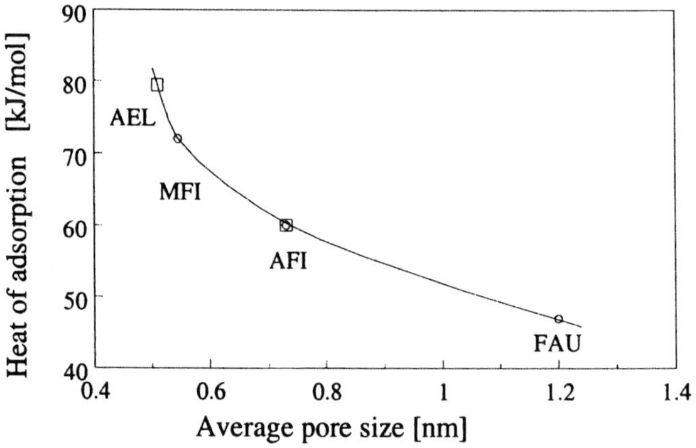

Fig. 13.11. Heat of adsorption of *n*-hexane correlated to the pore size of materials.

This is also experimentally well documented as one sees a continuous decrease of the heat of adsorption of *n*-hexane as the pore size increases (see Fig. 13.11 from Ref. [27]). Note that the curve also suggests that the pore size is more important for the strength of interaction than the chemical composition of the porous material.

In a more formal way, these relations were developed by Dubinin and co-workers [28] who proposed a general equation for micropore filling

$$\frac{n^a}{n^a_p} = e^{-\left(\frac{A}{E}\right)^2} \tag{13.34}$$

where n_a is the amount adsorbed at any relative pressure and n^a_p is the adsorption capacity of the micropore, $A = -RT \ln p_o/p$ and E is the characteristic energy of the adsorption system. After the introduction of a scaling factor β we can formulate

$$n^a = n^a_p * e^{-\left(\frac{(RT \ln p_o/p)^2}{(\beta E)^2}\right)} \tag{13.35}$$

In the logarithmic form this is the Dubinin–Radushkevic (DR) equation

$$\log_{10} n^a = \log_{10} n^a_p - D\log^2_{10}(p_o / p) \tag{13.36}$$

The constant D = 0.434 $B(T/\beta)^2$ and the constant B = $(2.303R/E)^2$. From the intercept with the y axis it is possible to derive n^a_p, the slope of the line should yield the value for D. Later, a more general form was proposed that substitutes the square in the exponential term of Eqn. (13.36) by an empirical parameter m.

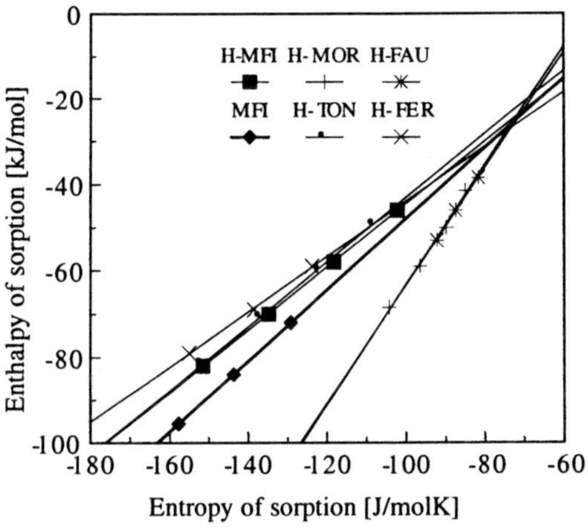

Fig. 13.12. Interdependence of the sorption enthalpy and entropy for a series of n-alkanes on several zeolites.

The DR model has been further refined by Stoeckli et al. [29], who propose to sum over several DR equations each describing a distinct group of micropores. While this is an elegant and realistic approach (there is seldom only one size of micropore present), it should be emphasized that a summation of Langmuir isotherms also suffices to describe complex isotherms.

The effective pore size of molecules in complex microporous materials can also be assessed through the relation between the adsorption enthalpy and the adsorption entropy. This has been shown by Eder et al. [30] using a series of *n*-alkanes in molecular sieves to define the relation between the initial adsorption enthalpy (measured directly with calorimetry) and the adsorption entropy (calculated from the adsorption constant measured by gravimetry). Figure 13.12 shows the lines for several molecular sieves. It is noted that a linear relation exists between the adsorption enthalpy and entropy which indicates a compensation effect, i.e., for a given microporous material the stronger bound molecule also loses the largest amount of entropy compared to the gas phase. The slope for the various molecular sieves is, however, markedly different. The smaller the pore size of the material the higher is the entropy loss (the lower is the entropy of sorption) for a given sorption enthalpy. This clearly emphasizes that the confines of the micropore exert a larger influence on the entropic term of the sorption than on the enthalpic term.

13.14 DISTRIBUTION OF MICROPORES

The most direct way of assessing the distribution of the micropore size is to compare the sorption isotherms of molecules with different minimum kinetic diameters. This approach is well illustrated by the work of Barton et al. [31] and widely used for characterizing molecular sieves. These crystalline microporous silicates and phosphates are named after their ability to prevent molecules larger than their pore diameter from reaching the catalytically active sites inside the pores. Usually, molecules ranging from N_2 to substituted aromatic compounds such as trimethylbenzenes are used as probes. In the presence of distinct functional groups, the use of *in situ* spectroscopy can also serve to evaluate what fraction of sites is located inside and outside the particles of the microporous material. The natural limit of pore sizes to be probed depends on the size and the vapour pressure of the sorbate (approximately 0.8 nm).

Two factors make the use of this direct method problematic. As the interactions in micropores are strong, relaxations of the lattice (and consequently changes in the effective pore diameter of the material) are observed upon adsorption. The accessibility is strongly temperature dependent. For example it is impossible for *m*-xylene to enter the pores of H-ZSM5 at ambient temperatures, while at 573 K the whole pore volume is accessed [32]. This results from the fact that, in the region of tight fit between the molecules and the lattice, the van der Waals radii are important. While this poses a significant constraint for the molecule to enter at 300 K, the barrier can be overcome at elevated temperatures.

Alternatively, the method of Horvath and Kawazoe [33] can be applied which relies on high resolution adsorption measurements at very low pressures. It is certainly a very attractive approach as it can be easily combined with conventional measurements of N_2 and Ar adsorption isotherms. It is strongly suggested, however, that the method needs to be calibrated against well-known and well-defined microporous materials before using it for unknown samples.

The most recent developments involve similar approaches but now based on density functional theory using statistical thermodynamics [34]. Although the majority of results obtained are related to slit-shaped carbon adsorbents [34,35] the approach holds great potential for rapid and widely applicable assessment of the micropore distribution.

13.15 GENERAL CONCLUSIONS AND RECOMMENDATIONS

The specific surface area and the pore size distribution of sorbents and catalysts are of central importance for their properties. For most cases it is sufficient to use the BET theory for the determination, although it will face limitations at very low surface areas and in the presence of micropores. The most successfully used gas is N_2 but, for low surface area materials, the use of a more easily condensable gas, such as Ar, has advantages. It is strongly recommended, however, that these methods be calibrated against a well-known sample, as the packing density of the various gases cannot be extrapolated directly from their molar volumes. The concept of surface area is not applicable for micropores, i.e., when the pore size and the size of the sorbed molecule approach each other.

For determination of the overall porosity, pycnometric methods are recommended that use imbibition of the material in a light inert gas and mercury. Mercury porosimetry is the method of choice for assessing macro- and mesopores. The use of intrusion and extrusion measurements is necessary to understand more complex pore structures.

The assessment of mesopores and their size distribution is only sensible with type IV isotherms. In that case the Kelvin equation can be used and can be well adapted to various pore shapes. The modification of the Kelvin equation by Broekhoff and de Boer allows for a more realistic assessment than the classic method. Using the *t* method data treatment, it is possible to assess the micropore volume and, from its deviations from ideality, the geometry of possible micro and meso pores.

Sorption in micropores can be described by the Dubinin-Radushkevic formalism that has been adapted by Stoeckli et al. This is a largely empirical approach and it should be emphasized that the use of a combination of Langmuir types isotherm leads to similar quantitative results. For evaluation of the distribution of micropores, one can either rely on high-resolution measurements of mostly nitrogen adsorption as suggested by Horvath and Kawazoe or use a combination of probe molecules of different minimum kinetic diameter. More recently, approaches based on density functional theory are put forward.

Especially for the micropores, the use of calorimetry and sorption measurements can lead to better insights about the distribution of the pores and the accommodation of sorbed molecules therein. It seems that the entropic effects and effects of ordering [26] may play an important role in these materials which in turn may be used to characterize unknown samples. The results with these materials suggest that physisorption contributes significantly to the catalytic action (by providing a higher concentration of molecules around active sites) and also influences the binding at the active sites markedly.

ACKNOWLEDGEMENTS

This chapter relies in part on a revised and expanded version of the contribution of J.J.F. Scholten written for the first edition of this book. Help by Dr. C. Paganini, H. Heydenrych and Bert Geerdink is gratefully acknowledged.

REFERENCES

1 K.S.W. Sing and J. Rouquerol, in *Handbook of Heterogeneous Catalysis*, G. Ertl, H. Knözinger and J. Weitkamp, Eds. Vol. 2, p. 427, Wiley-VCH, Weinheim (1997).
2 P.W. Atkins, *Physical Chemistry*, Oxford University Press, Oxford (1997).
3 S.J. Gregg and K.S.W. Sing, in *Adsorption, Surface Area and Porosity* (2nd ed.), Academic Press, London, 1982, p. 6.

4 S. Brunauer, L.S. Deming, W.S. Deming and E. Teller, *J. Am. Chem. Soc.*, 62 (1940) 1723.
5 C.A. Amberg, W.B. Spencer and R.A. Beebe, *Can. J. Chem.*, 33 (1955) 305.
6 M.M. Dubinin, *Chem Rev.*, 60 (1960) 235.
7 IUPAC Manual of Symbols and Terminology, Appendix 2, Part 1, Colloid and Surface Chemistry. *Pure Appl. Chem.*, 31 (1972) 578.
8 P.H. Emmett and S. Brunauer, *J. Am. Chem. Soc.*, 59 (1937) 1553.
9 I. Langmuir, *J. Am. Chem. Soc.*, 38 (1916) 2221.
10 S. Brunauer, P.H. Emmett and E. Teller, *J. Am. Chem. Soc.*, 60 (1938) 309.
11 M.R. Harris and K.S.W. Sing, *Chem. Ind.*, (1959) 487.
12 B.C. Lippens, B.G. Linsen and J.H. de Boer, *J. Catal.*, 3 (1964) 32.
13 J.C.P. Broekhoff and R.H. van Dongen, in *Physical and Chemical Aspects of Adsorbents and Catalysts*, Ch. 1, Academic Press, London, 1970, 103 pp.
14 B.V. Derjaguin, in *Proceedings of the Second International Congress on Surface Activity*, Part II, Butterworths, London, 1957, p. 154.
15 D.H. Everett in *Characterization of Porous Solids*, Eds. S.J. Gregg, K.S.W. Sing, H.F. Stoeckli, SCI, London (1979) pp 229
16 P. Barrett, L.G. Joyner, P.P. Halenda, *J. Am. Chem. Soc.*, 73 (1951) 373.
17 E. W. Loghbrown, *Proc. Nat. Acad. Sci. USA*, 7 (1921) 115.
18 A.A. Liabastre and C.J. Orr, *Colloid Interface Sci.*, 64 (1978) 1.
19 N.S. Gurfein, D.P. Dobychin, L.S. Koplienko, *Russ. J. Phys. Chem.*, 44 (1970) 411.
20 D.H. Everett, D.P. Powl, *J. Chem. Soc. Faraday Trans.* I, 72 (1976) 619.
21 E.G. Derouane, J.B. Nagy, C. Fernandez, Z. Gabelica, E. Laurant, P. Maljean, *Appl. Catal.*, 40 (1988), L1–L10.
22 E.G. Derouane, J.M. Andre, A.A. Lucas, *J. Catal.*, 110 (1988) 58.
23 B.G. Baekelandt, J.L. Lievens, W.J. Mortier, R.A. Schoonheydt, *Proc. 9th Intern. Zeolite Conf.*, 1 (1992) 579.
24 J.O. Titiloye, S.C. Parker, F.S. Stone, C.R.A. Catlow, *J. Phys. Chem.*, 95 (1991) 4038.
25 R.L. June, A.T. Bell, D.N. Theodorou, *J. Phys. Chem.*, 96 (1992) 1051.
26 B. Smit, J.I. Siepmann, *J.Phys.Chem.*, 98 (1994) 8443.
27 F. Eder, J.A. Lercher, *J. Phys. Chem.*, 100 (1996) 16460.
28 M.M. Dubinin, *Chem. Rev.* 60 (1960) 235.
29 H.F. Stoeckli, J. Ph. Houriet, A. Perret, U. Huber, in *Characterization of Porous Solids* 1979, SCI, London, p. 31.
30 F. Eder and J.A. Lercher, *J. Phys. Chem.*, 101 (1997) 1273.
31 S.S. Barton, M.J.B. Evans and B.H. Harrison, *J. Colloid Interface Sci.*, 40 (1974) 462.
32 G. Mirth, J. Cejka and J. A. Lercher, *J. Catal.*, 139 (1993) 24.
33 G. Horvath and K. Kawazoe, *J. Chem. Eng. Jpn.*, 6 (1983) 470.
34 N.A. Seaton, J.P.R.B. Walton and N. Quirke, *Carbon*, 27 (1989) 853.
35 J.P. Olivier, *J. Porous Mat.*, 2 (1995) 9.

Subject Index

STUDIES IN SURFACE SCIENCE AND CATALYSIS

Volume 1	**Preparation of Catalysts I.** Scientific Bases for the Preparation of Heterogeneous Catalysts. Proceedings of the First International Symposium, Brussels, October 14–17, 1975 edited by **B. Delmon, P.A. Jacobs and G. Poncelet**
Volume 2	**The Control of the Reactivity of Solids.** A Critical Survey of the Factors that Influence the Reactivity of Solids, with Special Emphasis on the Control of the Chemical Processes in Relation to Practical Applications by **V.V. Boldyrev, M. Bulens and B. Delmon**
Volume 3	**Preparation of Catalysts II.** Scientific Bases for the Preparation of Heterogeneous Catalysts. Proceedings of the Second International Symposium, Louvain-la-Neuve, September 4–7, 1978 edited by **B. Delmon, P. Grange, P. Jacobs and G. Poncelet**
Volume 4	**Growth and Properties of Metal Clusters.** Applications to Catalysis and the Photographic Process. Proceedings of the 32nd International Meeting of the Société de Chimie Physique, Villeurbanne, September 24–28, 1979 edited by **J. Bourdon**
Volume 5	**Catalysis by Zeolites.** Proceedings of an International Symposium, Ecully (Lyon), September 9–11, 1980 edited by **B. Imelik, C. Naccache, Y. Ben Taarit, J.C. Vedrine, G. Coudurier and H. Praliaud**
Volume 6	**Catalyst Deactivation.** Proceedings of an International Symposium, Antwerp, October 13–15, 1980 edited by **B. Delmon and G.F. Froment**
Volume 7	**New Horizons in Catalysis.** Proceedings of the 7th International Congress on Catalysis, Tokyo, June 30–July 4, 1980. Parts A and B edited by **T. Seiyama and K. Tanabe**
Volume 8	**Catalysis by Supported Complexes** by **Yu.I. Yermakov, B.N. Kuznetsov and V.A. Zakharov**
Volume 9	**Physics of Solid Surfaces.** Proceedings of a Symposium, Bechyňe, September 29–October 3, 1980 edited by **M. Láznička**
Volume 10	**Adsorption at the Gas–Solid and Liquid–Solid Interface.** Proceedings of an International Symposium, Aix-en-Provence, September 21–23, 1981 edited by **J. Rouquerol and K.S.W. Sing**
Volume 11	**Metal-Support and Metal-Additive Effects in Catalysis.** Proceedings of an International Symposium, Ecully (Lyon), September 14–16, 1982 edited by **B. Imelik, C. Naccache, G. Coudurier, H. Praliaud, P. Meriaudeau, P. Gallezot, G.A. Martin and J.C. Vedrine**
Volume 12	**Metal Microstructures in Zeolites.** Preparation - Properties – Applications. Proceedings of a Workshop, Bremen, September 22–24, 1982 edited by **P.A. Jacobs, N.I. Jaeger, P. Jirů and G. Schulz-Ekloff**
Volume 13	**Adsorption on Metal Surfaces.** An Integrated Approach edited by **J. Bénard**
Volume 14	**Vibrations at Surfaces.** Proceedings of the Third International Conference, Asilomar, CA, September 1–4, 1982 edited by **C.R. Brundle and H. Morawitz**
Volume 15	**Heterogeneous Catalytic Reactions Involving Molecular Oxygen** by **G.I. Golodets**

Volume 55 **New Developments in Selective Oxidation.** Proceedings of an International Symposium, Rimini, Italy, September 18–22, 1989
edited by **G. Centi and F. Trifiro**

Volume 56 **Olefin Polymerization Catalysts.** Proceedings of the International Symposium on Recent Developments in Olefin Polymerization Catalysts, Tokyo, October 23–25, 1989
edited by **T. Keii and K. Soga**

Volume 57A **Spectroscopic Analysis of Heterogeneous Catalysts. Part A: Methods of Surface Analysis**
edited by **J.L.G. Fierro**

Volume 57B **Spectroscopic Analysis of Heterogeneous Catalysts. Part B: Chemisorption of Probe Molecules**
edited by **J.L.G. Fierro**

Volume 58 **Introduction to Zeolite Science and Practice**
edited by **H. van Bekkum, E.M. Flanigen and J.C. Jansen**

Volume 59 **Heterogeneous Catalysis and Fine Chemicals II.** Proceedings of the 2nd International Symposium, Poitiers, October 2–6, 1990
edited by **M. Guisnet, J. Barrault, C. Bouchoule, D. Duprez, G. Pérot, R. Maurel and C. Montassier**

Volume 60 **Chemistry of Microporous Crystals.** Proceedings of the International Symposium on Chemistry of Microporous Crystals, Tokyo, June 26–29, 1990
edited by **T. Inui, S. Namba and T. Tatsumi**

Volume 61 **Natural Gas Conversion.** Proceedings of the Symposium on Natural Gas Conversion, Oslo, August 12–17, 1990
edited by **A. Holmen, K.-J. Jens and S. Kolboe**

Volume 62 **Characterization of Porous Solids II.** Proceedings of the IUPAC Symposium (COPS II), Alicante, May 6–9, 1990
edited by **F. Rodríguez-Reinoso, J. Rouquerol, K.S.W. Sing and K.K. Unger**

Volume 63 **Preparation of Catalysts V.** Scientific Bases for the Preparation of Heterogeneous Catalysts. Proceedings of the Fifth International Symposium, Louvain-la-Neuve, September 3–6, 1990
edited by **G. Poncelet, P.A. Jacobs, P. Grange and B. Delmon**

Volume 64 **New Trends in CO Activation**
edited by **L. Guczi**

Volume 65 **Catalysis and Adsorption by Zeolites.** Proceedings of ZEOCAT 90, Leipzig, August 20–23, 1990
edited by G. **Öhlmann, H. Pfeifer and R. Fricke**

Volume 66 **Dioxygen Activation and Homogeneous Catalytic Oxidation.** Proceedings of the Fourth International Symposium on Dioxygen Activation and Homogeneous Catalytic Oxidation, Balatonfüred, September 10–14, 1990
edited by **L.I. Simándi**

Volume 67 **Structure-Activity and Selectivity Relationships in Heterogeneous Catalysis.** Proceedings of the ACS Symposium on Structure-Activity Relationships in Heterogeneous Catalysis, Boston, MA, April 22–27, 1990
edited by **R.K. Grasselli and A.W. Sleight**

Volume 68 **Catalyst Deactivation 1991.** Proceedings of the Fifth International Symposium, Evanston, IL, June 24–26, 1991
edited by **C.H. Bartholomew and J.B. Butt**

Volume 69 **Zeolite Chemistry and Catalysis.** Proceedings of an International Symposium, Prague, Czechoslovakia, September 8–13, 1991
edited by **P.A. Jacobs, N.I. Jaeger, L. Kubelková and B. Wichterlová**

Volume 70 **Poisoning and Promotion in Catalysis based on Surface Science Concepts and Experiments**
by **M. Kiskinova**

Volume 71 **Catalysis and Automotive Pollution Control II.** Proceedings of the 2nd International Symposium (CAPoC 2), Brussels, Belgium, September 10–13, 1990
edited by **A. Crucq**

Volume 72 **New Developments in Selective Oxidation by Heterogeneous Catalysis.** Proceedings of the 3rd European Workshop Meeting on New Developments in Selective Oxidation by Heterogeneous Catalysis, Louvain-la-Neuve, Belgium, April 8–10, 1991
edited by **P. Ruiz and B. Delmon**

Volume 73 **Progress in Catalysis.** Proceedings of the 12th Canadian Symposium on Catalysis, Banff, Alberta, Canada, May 25–28, 1992
edited by **K.J. Smith and E.C. Sanford**

Volume 74 **Angle-Resolved Photoemission. Theory and Current Applications**
edited by **S.D. Kevan**

Volume 75 **New Frontiers in Catalysis, Parts A-C.** Proceedings of the 10th International Congress on Catalysis, Budapest, Hungary, 19–24 July, 1992
edited by **L. Guczi, F. Solymosi and P. Tétényi**

Volume 76 **Fluid Catalytic Cracking: Science and Technology**
edited by **J.S. Magee and M.M. Mitchell, Jr.**

Volume 77 **New Aspects of Spillover Effect in Catalysis. For Development of Highly Active Catalysts.** Proceedings of the Third International Conference on Spillover, Kyoto, Japan, August 17–20, 1993
edited by **T. Inui, K. Fujimoto, T. Uchijima and M. Masai**

Volume 78 **Heterogeneous Catalysis and Fine Chemicals III.** Proceedings of the 3rd International Symposium, Poitiers, April 5 - 8, 1993
edited by **M. Guisnet, J. Barbier, J. Barrault, C. Bouchoule, D. Duprez, G. Pérot and C. Montassier**

Volume 79 **Catalysis: An Integrated Approach to Homogeneous, Heterogeneous and Industrial Catalysis**
edited by **J.A. Moulijn, P.W.N.M. van Leeuwen and R.A. van Santen**

Volume 80 **Fundamentals of Adsorption.** Proceedings of the Fourth International Conference on Fundamentals of Adsorption, Kyoto, Japan, May 17–22, 1992
edited by **M. Suzuki**

Volume 81 **Natural Gas Conversion II.** Proceedings of the Third Natural Gas Conversion Symposium, Sydney, July 4–9, 1993
edited by **H.E. Curry-Hyde and R.F. Howe**

Volume 82 **New Developments in Selective Oxidation II.** Proceedings of the Second World Congress and Fourth European Workshop Meeting, Benalmádena, Spain, September 20–24, 1993
edited by **V. Cortés Corberán and S. Vic Bellón**

Volume 83 **Zeolites and Microporous Crystals.** Proceedings of the International Symposium on Zeolites and Microporous Crystals, Nagoya, Japan, August 22–25, 1993
edited by **T. Hattori and T. Yashima**

Volume 84 **Zeolites and Related Microporous Materials: State of the Art 1994.** Proceedings of the 10th International Zeolite Conference, Garmisch-Partenkirchen, Germany, July 17–22, 1994
edited by **J. Weitkamp, H.G. Karge, H. Pfeifer and W. Hölderich**

Volume 85 **Advanced Zeolite Science and Applications**
edited by **J.C. Jansen, M. Stöcker, H.G. Karge and J.Weitkamp**

Volume 86 **Oscillating Heterogeneous Catalytic Systems**
by **M.M. Slin'ko and N.I. Jaeger**

Volume 87 **Characterization of Porous Solids III.** Proceedings of the IUPAC Symposium (COPS III), Marseille, France, May 9–12, 1993
edited by **J.Rouquerol, F. Rodriguez-Reinoso, K.S.W. Sing and K.K. Unger**